ADVANCES IN
X-RAY ANALYSIS
Volume 29

ADVANCES IN
X-RAY ANALYSIS

Volume 29

Edited by

Charles S. Barrett

University of Denver
Denver, Colorado

Jerome B. Cohen

Northwestern University
Evanston, Illinois

John Faber, Jr.

Argonne National Laboratory
Argonne, Illinois

Ron Jenkins

JCPDS–International Centre for Diffraction Data
Swarthmore, Pennsylvania

Donald E. Leyden

Colorado State University
Fort Collins, Colorado

John C. Russ

North Carolina State University
Raleigh, North Carolina

and

Paul K. Predecki

University of Denver
Denver, Colorado

Sponsored by
University of Denver Research Institute
and
JCPDS–International Centre for Diffraction Data

PLENUM PRESS • NEW YORK AND LONDON

The Library of Congress cataloged the first volume of this title as follows:

Conference on Application of X-ray Analysis.
Proceedings 6th-   1957–   [Denver]

v. illus. 24-28 cm. annual.
No proceedings published for the first 5 conferences.
Vols. for 1958–   called also: Advances in X-ray analysis, v. 2-
Proceedings for 1957 issued by the conference under an earlier name: Conference on Industrial Applications of X-ray Analysis. Other slight variations in name of conference.
Vol. for 1957 published by the University of Denver, Denver Research Institute, Metallurgy Division.
Vols. for 1958–   distributed by Plenum Press, New York.
Conferences sponsored by University of Denver, Denver Research Institute.
1. X-rays — Industrial applications — Congresses. I. Denver University. Denver Research Institute   II. Title: Advances in X-ray analysis.
TA406.5.C6   58-35928

ISBN 0-306-42287-5

Proceedings of the Thirty-fourth Annual Conference on Applications of X-Ray Analysis,
held August 5-9, 1985, in Snowmass, Colorado

© 1986 University of Denver
Plenum Press is a division of Plenum Publishing Corporation
233 Spring Street, New York, N.Y. 10013

FOREWORD

The 34th Annual Denver Conference was the site of a major conference on X-ray measurements of residual stresses, and a workshop to introduce newcomers to this area. Interest in this application of diffraction to quality control has grown considerably in the last few years, primarily due to developments in the U.S. and abroad that now provide us with a much clearer understanding of the meaning of the results, how to sample the entire stress tensor if need be, as well as the ability to estimate errors. Also, automation and special detectors have reduced the measuring time, and permitted the development of instruments suitable for in-field analysis; pulsed-neutron sources supplement our information. With all these advances we are now capable of measuring all six stress components in each phase of a multiphase material and separating them into micro and macro stresses.

Despite this progress, there is no "home" in the U.S. for professional exchange of information in this area. With the site of this Denver meeting now in the mountains of beautiful Snowmass, Colorado, (and the excellent organization of Dr. Paul Predecki and his staff) it seemed appropriate to attempt a start at a biannual gathering of this community. The enclosed papers represent this initial meeting—which was indeed a very good one! Many of us are now looking forward to 1987.

J. B. Cohen
Northwestern University
Evanston, IL 60201

PREFACE

       This volume constitutes the proceedings of the 1985 Denver Conference
on Applications of X-ray Analysis and is the 29th in the series. The
conference was held at Snowmass Resort, Colorado, from August 5 to 9, 1985.
The general chairmen were D. E. Leyden, Colorado State University, and
P. K. Predecki, University of Denver, with C. S. Barrett of the University
of Denver as honorary chairman. The conference advisory committee this
year consisted of

              C. S. Barrett, University of Denver
              R. Jenkins, JCPDS-ICDD
              D. E. Leyden, Colorado State University
              J. C. Russ, North Carolina State University
              C. O. Ruud, The Pennsylvania State University
              P. K. Predecki, University of Denver

In this year's conference, in which the subject of diffraction was emphasized, special sessions on this and other subjects were organized by the
individuals listed below:

1. X-ray Stress Determinations: Techniques, Elastic Constants, chaired
    by J. B. Cohen, Northwestern University, and C. P. Gazzara, AMMRC,
    Watertown, MA
2. Application of XRD and XRF to Thin Films, chaired by C. Goldsmith,
    IBM, Hopewell Junction, NY and J. Willis, Tracor Xray, Inc., Mountain
    View, CA
3. Application of XRD to Geological Problems, chaired by J. J. Fitzpatrick,
    USGS, Reston, VA
4. Search/Match Procedures, Quantitative XRD Analysis, chaired by G.
    Fischer, Corning Glass Works, Corning, NY and G. J. McCarthy, North
    Dakota State University, Fargo, ND
5. Neutron Powder Diffraction Applications, chaired by J. Faber, Jr.,
    Argonne National Laboratory, Argonne, IL and F. K. Ross, University of
    Missouri, Columbia, MO
6. Non-Ambient Applications of XRD, XRD Instrumentation, chaired by
    R. Jenkins, JCPDS-ICDD, Swarthmore, PA
7. XRD and XRF Applications to Polymers, Organics and Forensics, chaired
    by H. D. Bennett, NASA, Kennedy Space Center, FL and T. G. Fawcett,
    The Dow Chemical Co., Midland, MI
8. Quantitative XRF Analysis, chaired by J. Criss, Criss Software, Largo,
    MD and J. Willis, Tracor Xray, Mountain View, CA
9. New XRF Instrumentation and Techniques, chaired by D. E. Leyden, Colo-
    rado State University, Fort Collins, CO and J. V. Gilfrich, Naval
    Research Laboratory, Washington, DC.

This was the first year that a session has consisted entirely of neutron diffraction; 6 papers on the subject are printed in this volume. The program included a total of 79 diffraction papers, and although fluorescence was not a subject of major emphasis this particular year there were, nevertheless, 41 papers in this field on the program.

Tutorial workshops on various XRD and XRF topics were held the first two days of the conference. These are listed below with the names of the workshop organizers and instructors.

"Phase Identification Through Powder Diffraction Methods and the JCPDS-ICDD Databases - I." G. J. McCarthy, North Dakota State University (Chair); C. M. Foris, Dupont Co.; R. Jenkins, JCPDS-ICDD; and C. O. Ruud, The Pennsylvania State University.

"XRF Sample Preparation Methods, Particularly Geological - I and II." V. Buhrke, The Buhrke Co. (chair); R. Bostwick, Spex Industries; M. Garbauskas, General Electric Co.; R. Johnson, USGS; and J. Taggart, Jr., USGS.

"Neutron Powder Diffraction in Materials Science and Chemistry." J. Jorgensen, Argonne National Lab (Chair); F. K. Ross, University of Missouri (chair); F. J. Rotella, Argonne National Lab; P. Rudolf, Texas A & M University; and W. B. Yelon, University of Missouri.

"Phase Identification Through Powder Diffraction Methods and the JCPDS-ICDD Databases - II." R. Jenkins, JCPDS-ICDD (chair); R. G. Garvey, North Dakota State University; G. J.McCarthy, North Dakota State University; and J. Stalick, NBS.

"How to Search the X-Ray Literature, XRD and XRF." R. Jenkins, JCPDS-ICDD (chair)

"Quantitative X-Ray Spectrometry Round Table." G. R. Lachance, Geological Survey of Canada (chair); J. Criss, Criss Software; D. J. Kalnicky, Princeton Gamma-Tech; and M. J. Rokosz, Ford Motor Co.

"Practical Aspects of XRD Stress Measurements." C. O. Ruud, The Pennsylvania State University (chair); R. W. Buenneke, Caterpillar Tractor Co.; C. P. Gazzara, AMMRC, Watertown, MA; M. R. James, Rockwell International; and P. S. Prevey, Lambda Research, Inc., Cincinnati, OH.

"Sources for X-Ray Fluorescence Spectrometry." D. A. Gedcke, EG&G Ortec (chair); J. Croke, Philips Electronic Instruments, Inc.; R. W. Ryon, LLNL; and J. Schindler, ASOMA Instruments.

"Microcomputers in X-Ray Diffraction." R. G. Garvey, North Dakota State University (chair); D. W. Beard, Siemens-Allis, Inc.; R. P. Goehner, Siemens-Allis, Inc.; G. Marquart, Fein-Marquart Associates; G. J. McCarthy, North Dakota State University; and Q. Johnson, Materials Data, Inc.

"XRD Characterization and Identification of Clay Minerals." P. Hauff, The Clayschool (chair); R. Giese, Jr., State University of New York; and R. C. Reynolds, Dartmouth College.

The total registration for the conference was 338, of whom 253 registered for at least one of the workshops. We are indebted to the workshop organizers and instructors who have given generously and extremely effectively of their time and skills in planning and conducting these workshops.

The organizing committee wishes to take this opportunity to thank the planners, organizers and chairmen of the sessions and workshops, mentioned above, and also the other chairmen of contributed paper sessions: J. Parker, J. Renault, M. J. Rokosz, and J. D. Zahrt. The invited speakers put much effort and skill into preparing their material for the conference, and their contributions are deeply appreciated. The list of these follows:

"Evaluation of Macro- and Micro-Residual Stresses on Textured Materials by X-Ray, Neutron Diffraction and Deflection Measurements," V. M. Hauk, Institut für Werkstoffkunde, Aachen, W. Germany

"Double Crystal X-Ray Diffractometry of Inhomogeneous, Multilayer and Very Thin Epitaxial Films," B. K. Tanner and M. J. Hill, Durham University, Durham, England

"Assessment of Multiepilayer III-V Compound Semiconductors by Synchrotron Radiation Diffractometry," D. K. Bowen and S. T. Davies, University of Warwick, Coventry, U.K.; and S. Swaminathan, Roorkee University, India.

"Characterization of Epitaxial Films by X-Ray Diffraction," A. Segmüller, IBM Thomas J. Watson Research Center, Yorktown Heights, NY.

"Rietveld Refinements of Manganese Oxide Minerals," J. E. Post, Smithsonian Institution, Washington, DC; and D. L. Bish, LANL, Los Alamos, NM.

"X-Ray Diffraction Analysis of Mixed-Layer Illite/Smectite: Applications to Geological Problems," S. P. Altaner, USGS, Reston, VA.

"Use of the Rational Strain Definition in Residual Stress Measurements by X-Ray Diffraction," K. Badawi, ENSAM, Centre de Chalons-sur-Marne, France; and L. Castex and J. M. Sprauel, ENSAM, Centre de Paris, France.

"The Identification of Unknown Phases Using Search/Match Procedures," R. P. Goehner, Siemens-Allis, Cherry Hill, NJ; and M. F. Garbauskas, General Electric Co., Schenectady, NY.

"High Resolution Powder Diffraction Studies at a Pulsed Neutron Source," J. Faber, Jr. and R. L. Hitterman, Argonne National Laboratory, Argonne, IL.

"Recognition and Treatment of Background Problems in Neutron Powder Diffraction Refinements," W. H. Baur and R. X. Fischer, University of Illinois, Chicago, IL.

"XRF Geochemistry in an Academic/Service Environment," J. Renault, New Mexico Bureau of Mines and Mineral Resources, Socorro, NM.

Special thanks are due the people of the University of Denver who helped in many ways to make the conference run smoothly and very successfully from start to finish: Dorothy Barrett, Penny Eucker, Jurgen Haas, Jeff Hayden, Alan Lankford, Dorothy Predecki and especially the conference secretary, Mildred Cain, whose typing and polishing of countless manuscripts has helped authors in many of these volumes from Volume 3 on, and who in addition has served as conference secretary since 1975, maintaining all files of organizers, chairmen, speakers, exhibitors, manuscript submissions (or lack thereof) and conference registrants with a level of skill and patience that is truly incredible.

    This happens to be a year in which we can insert a special
acknowledgement of Paul Predecki's efforts, because some new responsibilities in the College of Engineering of the University of Denver have
suddenly left him no time since the conference to write this Preface.
The conference advisory committee members have always recognized and now
assert that the success of every conference in the five years in which
he has been chairman must be attributed directly to his remarkable
insight, enthusiasm, careful attention to details of every kind, and to
his managerial abilities; we look forward to his expected return in a few
months to his planning and managing the next conference.

                                        Charles S. Barrett
                                        For the advisory committee

Unpublished Papers

    The following papers were presented but are not published here for a
variety of reasons.

"Correlation of X-Ray Determined Residual Stresses and Manufacturing Processes of Inconel Alloy 600 U-Bend Tubing," D. P. Ivkovich, C. O. Ruud,
and H. D. Cassel, The Pennsylvania State University, University Park, PA.

"Residual Stresses in Cold Bent Tubes and Bars," B. Pathiraj, W. H. J.
Bruis and B. H. Kolster, Twente University of Technology, Enschede, The
Netherlands.

"A Comparative Study of the Measuring Procedures of Full Width at Half
Maximum in X-Ray Diffraction Line Profile," K. Kawasaki, M. Matsuo and
H. Takechi, Nippon Steel Corp., Kawasaki, Japan; and Y. Shirasuna,
Sophia University.

"The Influence of Deformation-Induced Residual Stresses in the Wear
Behaviour of Aluminum Bronzes," G. Carro and J. J. Wert, Vanderbilt
University, Nashville, TN.

"A Three-Point Bending Device for the Calibration of X-Ray Residual
Stress Elastic Constants," R. W. Hendricks and E. B. S. Pardue, Technology
for Energy Corp., Knoxville, TN.

"Influence of Tensile Deformation on the X-Ray Elastic Constants of Some
Austenitic Steels," B. Pathiraj and B. H. Kolster, Twente University of
Technology, Enschede, The Netherlands.

"Assessment of the Direct X-Ray Response of a Charge Coupled Device Image
Sensor with Minimal Cooling," R. A. Holt, Canada Centre for Mineral and
Energy Technology, Ottawa, Ontario, Canada; and C. D. Anger, R. J. Soffer
and G. A. Klassen, ITRES Research Ltd.

"In-Situ High-Temperature XRD of Polycrystalline Thin Films for Semiconductor Technology: Phase Kinetics, Stress and Microcrystallnity,"
H. E. Goebel, Siemens Research Laboratory, München, W. Germany.

"Examination of Ordered and Disordered Thin Films Using Imaging Proportional Counters," I. Robinson, Bell Labs, Murray Hill, NJ and R. Burns,
Xentronics Co., Inc., Somerville, MA.

"Rietveld Refinements of Manganese Oxide Minerals," J. E. Post, Smithsonian
Institution, Washington, DC; and D. L. Bish, LANL, Los Alamos, NM.

"X-Ray Diffraction Analysis of Mixed-Layer Illite/Smectite: Applications to Geological Problems," S. P. Altaner, USGS, Reston, VA.

"Quantitative XRD Analysis of Clay Minerals: Problems and Methods," P. L. Hauff, The Clayschool, Conifer, CO; and S. Blassingame, Exxon Production Research, Houston, TX.

"Use of Clay Mineral Composition for Stratigraphic Analysis," H. D. Glass and R. E. Hughes, Illinois State Geological Survey, Champaign, IL.

"Multiple Regression Approach to Quantitative XRD Analysis," J. Renault, New Mexico Bureau of Mines and Mineral Resources, Socorro, NM.

"Compton Scattering for Mass Absorption Correction in the Quantitative Determination of Minerals by X-Ray Diffraction," J. S. Crighton, S. S. Iyengar, and G. J. Havrilla, The Standard Oil Co. (Ohio), Cleveland, OH.

"Use of the Rational Strain Definition in Residual Stress Measurements by X-Ray Diffraction," K. Badawi, ENSAM, Centre de Chalons-sur-Marne, France; and L. Castex and J. M. Sprauel, ENSAM, Centre de Paris, France.

"Development of Residual Stresses in Aluminum Single Crystals Subjected to Different Kinds of Deformation Processes," B. Pathiraj and B. H. Kolster, Twente University of Technology, Enschede, The Netherlands.

"The Identification of Unknown Phases Using Search/Match Procedures," R. P. Goehner, Siemens-Allis, Cherry Hill, NJ; and M. F. Garbauskas, General Electric Co., Schenectady, NY.

"Binary Bit Map Search Scheme for Electron Diffraction Analysis," M. J. Carr and W. F. Chambers, Sandia National Labs, Albuquerque, NM.

"The JCPDS PDP-11 Mini-Search Program," H. D. Bennett, NASA Kennedy Space Center, FL and M. A. Holomany, JCPDS-ICDD, Swarthmore, PA.

"Quantification of Graphite in Materials," L. A. Green, R. A. Viator, and C. G. Rayborn, Martin Marietta Energy Systems, Inc., Oak Ridge, TN.

"Rietveld Analysis of Multiple Phase Samples at IPNS," F. J. Rotella, Argonne National Laboratory, Argonne, IL.

"A Fourth Generation Fully Automated High Temperature In Situ X-Ray Diffractometer with Programmable Multiple Gas Flow Control for Studies of Gas/Liquid/Solid Interactions," G. A. Jones and V. Ivansons, E. I. du Pont de Nemours & Co., Inc., Wilmington, DE.

"High Temperature Kinetic Studies Using an Automated Powder Diffractometer," D. A. Carpenter, K. W. Kaiser, J. J. Dunigan, and C.M. Davenport, Martin Marietta Energy Systems, Inc., Oak Ridge, TN.

"High Speed Data Acquisition and Analysis of Laue Patterns Using Imaging Proportional Counters," J. Quigley and R. Burns, Xentronics Co., Somerville, MA.

"Rotating Capillary Stage for Mounting Small Amounts of Powder Samples on an X-Ray Diffractometer," C. G. Rayborn, R. A. Viator, and L. A. Green, Martin Marietta Energy Systems, Inc., Oak Ridge, TN

"Application of Imaging Proportional Counters to Polymer Studies," R. Desper, AMMRC, Watertown, MA; and J. Quigley, Xentronics Co., Somerville, MA.

"Quantitative Analysis of Chloramphenicol Palmitate Polymorphs: Application of a Computing Integrator to X-Ray Powder Diffraction," W. H. De Camp Food and Drug Administration, Washington, DC.

"Application of Energy-Dispersive XRF Analysis to Micro Sampling and Irregular Geometries in Forensic Applications," J. R. Bogert and T. J. Durnick, Kevex Corp., Foster City, CA.

"X-Ray Fluorescence Determination of Molybdenum in Molybdenite Using Compton Scattering," G. J. Havrilla, The Standard Oil Co. (Ohio), Cleveland, OH.

"XRF Analysis of Low Level Sodium Content in Silica Gels," E. M. Sabino and M. Messner-King, The PQ Corp., Lafayette Hill, PA

"Alternative Designs for Wavelength-Dispersive X-Ray Spectrometers," J. V. Gilfrich and D. J. Nagel, Naval Research Laboratory, Washington, DC.

"Design and Performance of a New Twin-Excitation-Energy Total-Reflection X-Ray Fluorescence Spectrometer," H. Schwenke, J. Knoth and H. Schneider, GKSS-Forschungszentrum Geesthacht GmbH, W. Germany.

"Automatic Sample Preparation Using the Perl-X Bead Machine," J. Petin, Laborlux S.A., Luxembourg; and J. Kikkert, Philips I & E Almelo, The Netherlands.

"XRF Geochemistry in an Academic/Service Environment," J. Renault, New Mexico Bureau of Mines and Mineral Resources, Socorro, NM.

"A New Hybrid Method for Multiplet of K X-Ray Fitting," E. Okada and K. Hiromura, Himeji Institute of Technology, Hyogo, Japan.

"XRF Simulation Package," R. W. Green, General Electric Co., Schenectady, NY.

"Determination of Minor and Trace Elements in Petroleum Coke Using Wavelength Dispersive X-Ray Fluorescence," J. S. Crighton and A. W. Varnes, The Standard Oil Co. (Ohio), Cleveland, OH.

"Applications of the Dual Channel Sequential X-Ray Spectrometer," J. A. Nicolosi, F. J. Croke and D. Merlo, Philips Electronic Instruments, Inc., Mahwah, NJ; and R. Jenkins, JCPDS-ICDD, Swarthmore, PA.

"Multi-Element Analysis Using Energy Dispersive X-Ray Fluorescence With a Thin Film Sample Preparation Technique," J. S. Crighton, The Standard Oil Co. (Ohio), Cleveland, OH.

"Routine Analysis with the Dual Anode Tube," J. N. Kikkert, Philips I & E, Almelo, The Netherlands.

CONTENTS

I. XRD STRESS METHODS, EQUIPMENT, AND APPLICATIONS

II. NEUTRON POWDER DIFFRACTION AND ITS APPLICATIONS

III. SEARCH/MATCH PROCEDURES, QUANTITATIVE XRD ANALYSIS

EVALUATION OF MACRO- AND MICRO-RESIDUAL STRESSES ON TEXTURED MATERIALS

BY X-RAY, NEUTRON DIFFRACTION AND DEFLECTION MEASUREMENTS

Viktor M. Hauk

Institut für Werkstoffkunde, Rheinisch Westfälische

Technische Hochschule, D-5100 Aachen, F. R. Germany

ABSTRACT

    To evaluate the state of macro- and micro-residual stresses in cold
rolled materials of 0.2 to 3 mm thickness the following test methods were
used: Measurements of lattice spacing in different directions (φ, ψ) with
X-rays and neutron-rays on many diffraction peaks. The X-ray method was
employed on the surface of the sheet and of the thinned piece after
removing several layers from both sides. The neutron-ray method allows
measurements in the interior of a specimen and the average of the strain
distribution of the total cross section. The etching deflection method
determines the macro-residual stresses. The calculation of the macro- and
micro-residual stresses using X-ray elastic constants (XEC) for polycrystalline material and the compliances of single crystals for groups of
grains are described. Homogeneous and heterogeneous materials like steels,
($\alpha + \beta$)-brass, nickel and copper with 3 % silver were investigated. The
state of residual stresses on the surface and with the depth over the
cross section was evaluated or is in the final process of investigation.

INTENSITIES- AND LATTICE STRAIN-POLEFIGURES

    The measuring parameters are Θ (Bragg's angle), D (interplanar
spacing, usually given as $D_{\{100\}}$ in the figures for comparing the results
of different planes) and ε (strain). Fig.1 shows the definitions of the
angles φ, ψ (φ azimuth, ψ angle between the normals to the specimen and
the reflecting plane, $\psi \gtrless 0$). The figure shows the rolling (RD), transverse (TD) and the normal direction (ND). There may result one of the 4
basic types of D vs. $\sin^2 \psi$ dependencies, also called lattice strain
distributions (Fig.2) or a combination of them[1,2].

    A linear dependency is the predominant result of measurements on
mechanically isotropic homogeneous and heterogeneous materials. A ψ-
splitting is found on materials that have undergone severe surface defor-
mations with tangential forces like grinding, roll peening, honing,
turning. The evaluation of the residual stresses is well covered by the
published method[3]. The third of the basic types Fig.2c belongs to textured
materials; there also may be a combination of types b) and c) in directions inclined to the RD. The last of the basic types the curved D vs.
$\sin^2 \psi$ dependency is due to gradients mainly in the stress free inter-

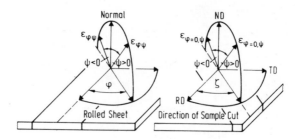

Figure 1. Definitions of angles and directions.

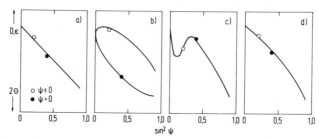

Figure 2. Basic types of lattice strain distributions[1,2].

planar spacing D_o due to alloying of surface layers with the bulk, when they are of different materials. In a minor sense, steep stress gradients may result in a curved D vs. $\sin^2 \psi$ distribution as recently investigated[4,5].

Usually the texture of a material will be shown as the density of the poles orientation in a polefigure. The numbers given are the intensity of the plane {hkl} relative to the intensity for random orientation. In recent years, according to the development of the X-ray measuring techniques[6,7,8] strain polefigures are available to the benefit of this area of research[8,9,10]. In the following a new, successful and hopeful approach to evaluate residual stresses on materials having strongly preferred orientations will be introduced. In discussing texture and stresses one must differentiate between special cases, Fig.3. Experience shows that in the case of a weak texture, defined by intensity I_{max} / I_{min} between 1 to 2 in the I vs. $\sin^2 \psi$ dependency there may exist a linear or a non-linear D vs. $\sin^2 \psi$ distribution. In general if D vs. $\sin^2 \psi$ for {h00} and {hhh} peaks show linearity, the {hkl} non-linearities are caused by elastic anisotropy only[11,12]. In this case the methods of texture free directions, inverse polefigures, anisotropic XEC (X-ray Elasticity Constants) can be used[2,13]. In the case I_{max} / I_{min} > 2, a strong preferred orientation of the grains of the material, there may exist approximately linearity in D vs. $\sin^2 \psi$ only for special peaks, for example {732 + 651} for steels[10,14,15]. In the following the worst case may exist, non-linearity for most of the peaks of the material, showing oscillations due to both elastic and plastic anisotropy.

SEPARATION AND COMPENSATION OF MACRO- AND MICRO-RESIDUAL STRESSES

In textured materials several kinds of stresses macro- (type I) and micro- (type II) residual stresses contribute to the shift of the diffraction peak. They cause linear and non-linear D vs. $\sin^2 \psi$ distributions.

	$\frac{i_{max}}{i_{min}} \leq 2$	$\frac{i_{max}}{i_{min}} > 2$
	D, sin²ψ	
{hkl}	linear, nonlinear	nonlinear
{h00}, {hhh}	linear, nonlinear	nonlinear
{hkl}$_{special}$	linear	linear

Figure 3. Linear and nonlinear D vs. $\sin^2 \psi$ dependencies of different {hkl} reflections for weak and strong textured materials.

Fig.4 shows the stresses contributing to the lattice strain to be measured with X- and neutron-rays on the surface, in the interior or averaged over the total cross section of a specimen and the kind of stresses which are responsible for the shift of the peak. Generally macro- and micro-residual stresses in the long. and in the direction perpendicular to the surface contribute to linear D vs. $\sin^2 \psi$ dependencies. In textured materials there exist micro-residual stresses due to elastic and plastic anisotropy and they show up as oscillations. The relative magnitude of oscillations due to elastic and plastic anisotropy may vary, but all the results for strong textured materials after cold rolling reveal non-linearity of the {h00} and {hhh} peaks, which means that plastic aniso-tropy cannot be neglected. Therefore, all the good ideas for calculating the lattice strain with the aid of inverse polefigures do not result in a practical answer to the problem. Using X-rays the influence of σ_3^{II} can generally be neglected due to small values in the near surface regions and the relatively small penetration depth as many results on thin walled specimen prove. In a very thin specimen the macro-residual stresses are zero and only micro-residual stresses are present[2].

Measuring with neutron-rays on rolled sheets there are two possible procedures, one is to radiate different parts of the bulk material and the second is to broaden the ray bundle and average the macro-residual stres-ses to zero over the whole cross section.

To evaluate the residual stress state of a specimen there are the experimental possibilities shown in Fig.4. In many cases the macro-residual stress distribution of a rolled sheet gained from the etching deflection method helps the evaluation and interpretation of the results got by the diffraction methods. Fig.5 shows the compensation of macro- and micro-residual stresses in a two phase material (q is the volume percentage of the α-phase and 1 - q the volume percentage of the β-phase) with the depth from the surface of the specimen[2]. The formulae enables

Measurement		Principal Formula
General		$(\sigma_1^I + \sigma_1^{II}) - (\sigma_3^I + \sigma_3^{II}) + \sigma_{el}^{II} + \sigma_{pl}^{II}$
X-ray	surface	$\sigma_1^I + \sigma_1^{II} \qquad\qquad + \sigma_{el}^{II} + \sigma_{pl}^{II}$
	thinned specimen	$\sigma_1^{II} \qquad\qquad\qquad + \sigma_{pl}^{II}$
Neutron-ray	interior	$\sigma_1^I + \sigma_1^{II} \qquad - \sigma_3^{II} + \sigma_{el}^{II} + \sigma_{pl}^{II}$
	total cross section	$\sigma_1^{II} \qquad\qquad - \sigma_3^{II} + \sigma_{pl}^{II}$

Figure 4. Macro (I)- and micro (II) -residual stresses contributing to the lattice strain $\varepsilon_{\varphi,\psi} = (D_{\varphi,\psi} - D_o)/D_o$ in the rolling direction (RD).

Normal components:

$$q\sigma_\alpha^I + (1-q)\sigma_\beta^I = \sigma^I$$

$$q\sigma_\alpha^{II} + (1-q)\sigma_\beta^{II} = 0$$

$$q\int_0^P \sigma_\alpha^{I \cdot II} dz + (1-q)\int_0^P \sigma_\beta^{I \cdot II} dz = 0$$

Figure 5. The stress compensation conditions of normal and shear components[2,16,17].

Shear components and σ_{33} in each depth:

$$q\sigma_{i3,\alpha}^{II} + (1-q)\sigma_{i3,\beta}^{II} = 0 \quad (i=1,2,3)$$

also a separation of macro- and micro-residual stresses if peaks of both phases can be analysed[16,17].

EVALUATION OF RESIDUAL STRESSES

The evaluation of the residual stresses from linear D vs. $\sin^2\psi$ dependencies shall use the XEC which can be calculated for homogeneous material from the compliances of the single crystals[18] and which are tabulated in papers[19]. In a recent paper[20] we found good agreement between calculated and experimental values of XEC on a steel in two states of tensile strength measured with 3 wavelengths. There are also XEC known for heterogeneous materials[21]. The same procedure holds for averaging the oscillation to a quasi linear dependency. It must be clearly stated that this method may give false results if the plane used and the range of interplanar spacings are not the correct ones. But in the case of the {211} peak of Cr-radiation and measured in the $\sin^2\psi$ range up to 0.9 ($-70° \leq \psi \leq +70°$), residual stresses on rolled steels show approximately the same value as is obtained with the linear D vs. $\sin^2\psi$ distribution of the {732+651} peak of Mo-radiation.

In the very recent years there have been successful tests to evaluate the residual stresses within destinctive groups of grains which contribute to different peaks. Fig.6 shows the poles of the {211} and {732+651} peak

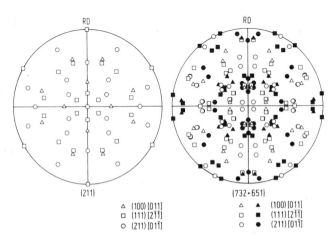

△	(100) [011]
□	(111) [2̄1̄1̄]
○	(211) [011̄]

△	▲	(100) [011]
□	■	(111) [2̄1̄1̄]
○	●	(211) [011̄]

Figure 6. The poles of the {211} and {732+651} reflection planes of the ideal positions of rolled steels.

$$\varepsilon_{hkl} = \sum_{i=1}^{3} \left[s_{12} + s_0(\alpha^2\alpha_i^2 + \beta^2\beta_i^2 + \gamma^2\gamma_i^2) + \tfrac{1}{2}s_{44}(\alpha\alpha_i + \beta\beta_i + \gamma\gamma_i)^2 \right]\sigma_i$$

Figure 7. The strain formula of single crystals [22].

Ideal orientation	$\varepsilon_{\varphi=0,\psi}$	Zone-axis	Phase solid solution	{hkl}
{211}⟨01̄1̄⟩	$(s_{12}+\tfrac{1}{6}s_0)\sigma_1+(s_{12}+\tfrac{1}{3}s_0)\sigma_2+(s_{12}+\tfrac{1}{2}s_{44}+\tfrac{1}{2}s_0)\sigma_3 +\tfrac{1}{2}\left[(s_{44}+\tfrac{2}{3}s_0)\sigma_1-(\tfrac{2}{3}s_0+s_{44})\sigma_3\right]\sin^2\psi$	[1̄11]	Fe	{110} {211} {220} {541} {642}
			Cu Zn	{211}
{011}⟨21̄1̄⟩		[1̄1̄1]	Cu / Ni	{220} {844}
{111}⟨21̄1̄⟩	$(s_{12}+\tfrac{1}{3}s_0)\sigma_1+(s_{12}+\tfrac{1}{3}s_0)\sigma_2+(s_{12}+\tfrac{1}{2}s_{44}+\tfrac{1}{3}s_0)\sigma_3 +\tfrac{1}{6}\left[(3s_{44}+s_0)\sigma_1-s_0\sigma_2-3s_{44}\sigma_3\right]\sin^2\psi +\tfrac{1}{6}\sqrt{2}\,s_0(\sigma_1-\sigma_2)\sin2\psi$	[01̄1]	Fe	{110} {200} {211} {220} {222}
{100}⟨011⟩	$s_{12}\sigma_1+s_{12}\sigma_2+s_{11}\sigma_3 +\tfrac{1}{2}\left[(s_{44}+s_0)\sigma_1+s_0\sigma_2-(2s_0+s_{44})\sigma_3\right]\sin^2\psi$	[01̄1]	Fe	{110} {200} {211} {222}
{211}⟨1̄11⟩	$(s_{12}+\tfrac{1}{3}s_0)\sigma_1+(s_{12}+\tfrac{1}{6}s_0)\sigma_2+(s_{12}+\tfrac{1}{2}s_{44}+\tfrac{1}{2}s_0)\sigma_3 +\tfrac{1}{2}\left[s_{44}\sigma_1+\tfrac{1}{3}s_0\sigma_2-(\tfrac{1}{3}s_0+s_{44})\sigma_3\right]\sin^2\psi +\tfrac{1}{6}\sqrt{2}\,s_0(\sigma_2-\sigma_3)\sin2\psi$	[01̄1]	Ni / Cu	{220} {311} {222} {331}

Ideal orientation	$\varepsilon_{\varphi=90,\psi}$	Zone-axis	Phase solid solution	{hkl}
{211}⟨01̄1̄⟩	$(s_{12}+\tfrac{1}{6}s_0)\sigma_1+(s_{12}+\tfrac{1}{3}s_0)\sigma_2+(s_{12}+\tfrac{1}{2}s_{44}+\tfrac{1}{2}s_0)\sigma_3 +\tfrac{1}{2}\left[\tfrac{1}{3}s_0\sigma_1+s_{44}\sigma_2-(\tfrac{1}{3}s_0+s_{44})\sigma_3\right]\sin^2\psi +\tfrac{1}{6}\sqrt{2}\,s_0(\sigma_1-\sigma_3)\sin2\psi$	[01̄1]	Fe	{110} {200} {211} {220} {222}
			Cu Zn	{211}
{011}⟨21̄1̄⟩	$(s_{12}+\tfrac{1}{6}s_0)\sigma_1+(s_{12}+\tfrac{1}{3}s_0)\sigma_2+(s_{12}+\tfrac{1}{2}s_{44}+\tfrac{1}{2}s_0)\sigma_3 +\tfrac{1}{2}\left[\tfrac{1}{3}s_0\sigma_1+s_{44}\sigma_2-(\tfrac{1}{3}s_0+s_{44})\sigma_3\right]\sin^2\psi$	[21̄1]	Ni / Cu	{220} {311} {420}
{111}⟨21̄1̄⟩	$(s_{12}+\tfrac{1}{3}s_0)\sigma_1+(s_{12}+\tfrac{1}{3}s_0)\sigma_2+(s_{12}+\tfrac{1}{2}s_{44}+\tfrac{1}{3}s_0)\sigma_3 +\tfrac{1}{6}\left[-s_0\sigma_1+(3s_{44}+s_0)\sigma_2-3s_{44}\sigma_3\right]\sin^2\psi$	[21̄1̄]	Fe	{110} {222}
{100}⟨011⟩	$s_{12}\sigma_1+s_{12}\sigma_2+s_{11}\sigma_3 +\tfrac{1}{2}\left[s_0\sigma_1+(s_{44}+s_0)\sigma_2-(2s_0+s_{44})\sigma_3\right]\sin^2\psi$	[011]	Fe	{110} {200} {211} {222}
{211}⟨1̄11⟩	$(s_{12}+\tfrac{1}{3}s_0)\sigma_1+(s_{12}+\tfrac{1}{6}s_0)\sigma_2+(s_{12}+\tfrac{1}{2}s_{44}+\tfrac{1}{2}s_0)\sigma_3 +\tfrac{1}{2}\left[(s_{44}+\tfrac{2}{3}s_0)\sigma_2-(\tfrac{2}{3}s_0+s_{44})\sigma_3\right]\sin^2\psi$	[1̄111]	Ni / Cu	{220} {844}

Figure 8. Formulae for strain stress relation of crystals having defined orientations, RD and TD (to be published with R.W.M. Oudelhoven).

Material	Thickness mm	Main composition-elements Weight %	$R_{p0.2}$ N/mm²	R_m N/mm²	Phase Solid solution	Tests X- N- D-	References
Steel St 52	1.2	0.10 C , 0.34 Si , 1.34 Mn	792	834	Fe	x x	25
Steel R St4	1	0.04 C , 0.02 Si , 0.26 Mn	742	754	Fe	x x x	27,28
(α+β) Brass	3	60 Cu , 40 Zn	422	510	Cu,CuZn	x x x	29
Nickel	0.5	—	830	840	Ni	x x	30
Cu 3 Ag	0.2	3 Ag	497	508	Cu	x x	31

Figure 9. Data of the materials used in this investigation and the tests carried out.

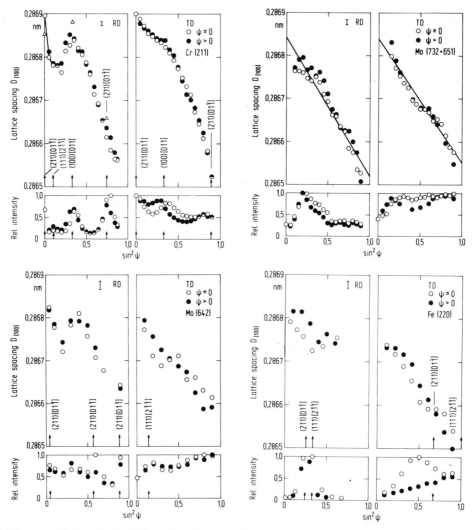

Figures 10 to 13. Lattice strain- and intensity-distributions of a rolled steel in RD and TD for different radiations and peaks[25].

due to different ideal orientations. The basic formula[22] of the strain of
a single crystal is given in Fig.7. The equation will be simplified for
the longitudinal (rolling, drawing) and for the transverse direction, also
for specific ideal orientations of the texture of the material[23,24].

Fig.8 contains the formula for RD ($\varphi = 0$) and TD ($\varphi = 90$) for the
strains on different {hkl} of a crystallite having defined orientation
loaded by stresses σ_1, σ_2 and σ_3 in the specimen system. In formulae with
$\sin 2\psi$ the sign of ψ must be appropiate to the orientation of the {hkl}
permutation involved.

To evaluate the residual stresses the compliances of the single
crystal should be used, $S_0 = S_{11} - S_{12} - 1/2 \, S_{44}$. When measuring with
neutron-rays in the interior of the specimen the crystallites which
contribute to different peaks will have a triaxial stress state. The third
stress component, the micro-residual stress in the direction normal to the
surface, must be zero if summed over all the crystallites of the different
phases within each infinitesimal layer, at each depth.

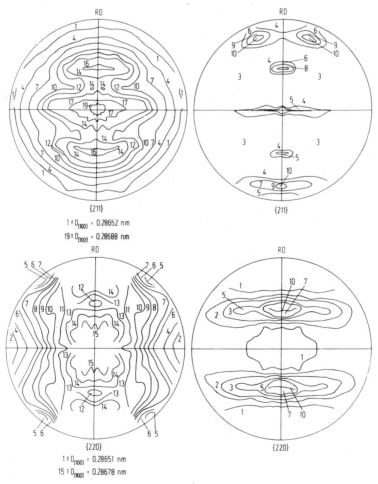

Figures 14 and 15. The strain- and intensity-polefigures of rolled steel
for {211} and {220} peaks[25].

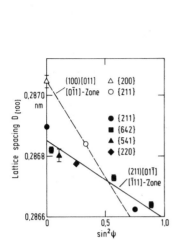

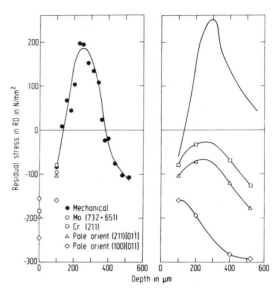

Figure 16. The evaluation of residual stress within crystallites
belonging to two zone axis measured on different {hkl} reflections[25].

Figure 17. The distribution of the
residual stress over the cross
section of the rolled steel sheet
(left end surface, right end center
of the specimen)[25].

EXAMPLES OF EVALUATION OF RESIDUAL STRESSES IN ROLLED SHEETS

In the following the essential issue of many results on different
materials, Fig.9, will be presented. Part of the program has been published,
is in the press or is under way. All materials have been cold rolled and
were available as sheets 0.2 to 3 mm thick. Tests were made by X-ray and
neutron-ray stress measurements and with the etching deflection method as
a reference method to get the macro-residual stress distribution over the
cross section of the sheet.

Steel St52[25]. Figs. 10-13 show the D vs. $\sin^2\psi$ and the I vs. $\sin^2\psi$
distributions in planes which enclose the RD (rolling direction) and the
TD (transverse direction). The ideal orientations are named as well as an
error bar for 0.01° in 2θ as the calibration accuracy. The strain and the
intensity polefigures for the lattice planes {220} and {211} appear in
Figs. 14 and 15. There is some correlation between the high density of the
poles. The D vs. $\sin^2\psi$ dependency for the poles of {211}<011> ideal
orientation and the $[11\bar{1}]$-zone axis show a good linear correlation
(Fig.16). Fig.17 includes the final results; at the left the results of the
measurements and at the right the analysis. The assessment of all these
tests is as follows.

The lattice strain distributions of the planes {hkl} are, in the RD,
nonlinear in most cases with strong oscillations in the region of $\sin^2\psi$
values up to 0.3. In the TD they are linear or consist of two straight
lines. An exception is the {732+651} peak for which the strain distribution is linear in both the RD and TD. The reason for that is the large
number of poles and the mean orientation factor nearly 0.6 which holds
for the mechanical value (the average of all possible orientations). The
same values of residual stresses are obtained by linear regression of the

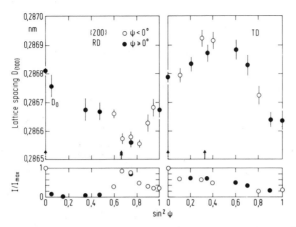

Figure 18. Neutron diffraction results on a rolled steel sheet, {200} peak, RD and TD[27].

{211} strain (in the range $\sin^2\psi$ up to 0.9) and of the {732+651} strain. Approximately the same distribution of residual stresses on surfaces after etching away many layers is obtained if the evaluation of residual stresses has been employed for the poles of crystallites with planes belonging to one zone axis. The differences between macro-residual stresses mechanically obtained and the X-ray results is due to micro-residual stresses due to deformation. They are of the order of -150 N/mm², a reasonable value[26].

Steel RSt4[27,28]. The experimental procedure used was nearly the same as with St52. D vs. $\sin^2\psi$ distributions were measured on the {211} and {732+651} peaks on the surface and after etching several layers and show in principle the well known dependencies. The macro-residual stress distribution over the cross section was verified also by the etching deflection method. In addition extensive neutron diffraction studies have been undertaken. Fig.18 reveals that the oscillations of the {200} peak are also due to plastic anisotropy. From Fig.18 it turns out that measuring in textured materials only up to $\sin^2\psi \leq 0.5$ there may follow false results. Fig.19 shows the special method of the neutron studies in which the strain distribution, the {211} peak in RD as an example, can be measured in the interior a), summed up over the total cross section b) and gives as the difference of both strains c) further details for the stress evaluation. The measurements with neutron rays in the interior of the specimen on different peaks allow one to verify the residual stresses of different groups of crystallites. The result shows a variety of values of triaxial stresses, but when summed up according to the density the component in the thickness direction becomes zero.

The results of all three investigations will be explained for the RD according to Fig.20, shown are the actual stresses vs. the distances from the surface. The difference between macro-residual and the stresses measured with X-ray are the micro-residual stresses in the ferrite phase. The value -100 N/mm² in the centre region of the specimen seems again combateble with previous tests[26].

(α+β) brass. A rolled sheet 3 mm thick of 60/40 weight % Cu/Zn brass with 77 volume % of α-phase and 23 % of β-phase was investigated by X-rays at the surface (o.1 mm etching on each side) and at the surface of a 0.22 mm thin specimen that had been etched on both sides. And the entire

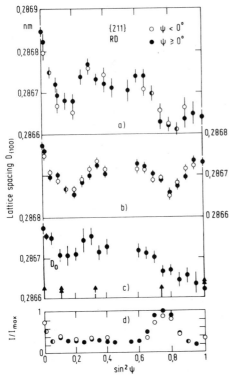

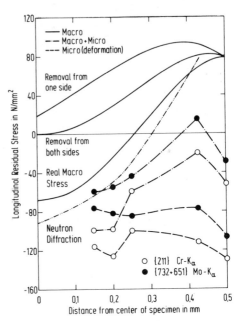

Figure 19. Neutron diffraction results on a rolled steel sheet, {211} peak, RD; a) center area of the specimen b) averaged over the entire cross section c) difference of strain a) and b)[27].

Figure 20. The distribution of the residual stress over the cross section of a rolled steel sheet (left side center, right side surface of specimen)[28].

cross section of a 3 mm thick specimen was etched from one side and the deflection was measured to reveal the macro-residual stresses. The very strong texture enables D-measurements only in certain $\sin^2 \psi$-regions mainly at the poles, see Fig.21 as an example. Therefore and while the oscillations were very high only the crystallite method could be used. Fig.22 shows the result of the analysis of the residual stresses in the rolling direction. At the left are the results of the measurements and at the right the macro-residual stresses and the micro-residual stresses in both phases. The mechanically obtained residual stress distribution reveals the stresses that are present at the corresponding surface after etching. Thus the comparison with the macro-residual stress is possible and shows good agreement.

The results of the neutron diffraction studies, Fig.23 show besides the oscillations in both azimuths practically no residual stress at all, in good agreement with the X-ray result on the α-phase taking into account the fact that the macro-residual stresses are summed to zero. The example proofs the possibility of analysing the residual stress state in heterogeneous, strongly textured materials with X-rays. Whether or not the neutron diffraction measures micro-residual strain or micro-residual stress

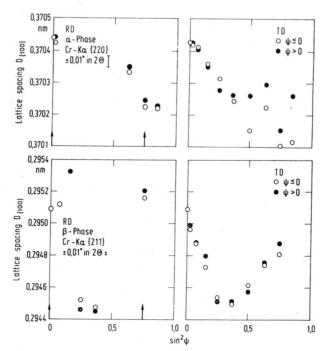

Figure 21. Lattice strain distribution of a very
thinned specimen of rolled (α+β) brass; α- and
β-phase, RD and TD[29].

components in the thickness direction of the specimen could not be determined with these tests.

Nickel[30]. Fig.24 shows the D vs. $\sin^2 \psi$ dependencies of four {hkl}
planes measured with the radiations stated and reveals the relatively
small oscillations. The {420} peak shows bigger oscillations. Polefigures
prove the very sharp texture. Using the crystallite method to evaluate

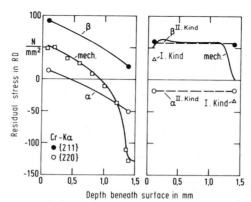

Figure 22. The distribution of the residual stress over the
cross section of a rolled (α+β) brass sheet (left side
surface, right side center of specimen); left figure result
of measurement, right figure analysis of the macro- and
micro-residual stresses in RD[29].

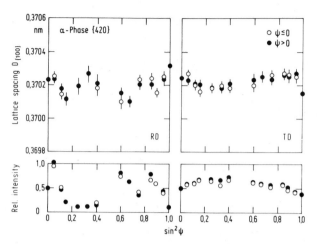

Figure 23. Neutron diffraction results on a rolled (α+β) brass sheet; α-phase, RD and TD, averaged over the entire cross section[29].

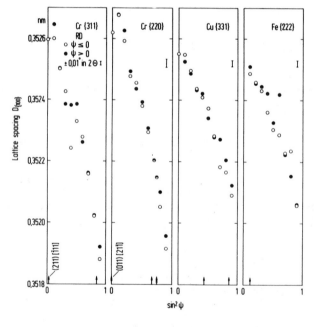

Figure 24. Lattice strain distributions of a rolled nickel sheet in RD for different radiations and peaks[30].

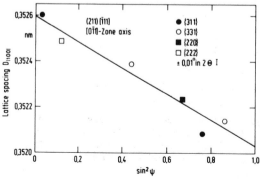

Figure 25. The evaluation of residual stress within crystallites belonging to one zone axis measured on different {hkl} reflections[30].

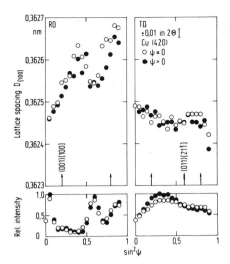

Figure 26. Lattice strain- and intensity-distributions of
a rolled Cu 3 Ag sheet, RD and TD[31].

the residual stress, see Fig.25, the following residual stresses on the
surface of one group of crystallites are present: RD, -320 N/mm²; TD,
-90 N/mm²; and ND, 10 N/mm². The procedure to elaborate the stress state
in detail will be the same as with the (α+β) brass sheet.

Cu 3 Ag[31]. The thin sheet of Cu with 3 % Ag proved to be very anisotropic according to preliminary results[32]. Again the procedure of the residual state evaluation will follow both examples just reported. As an
example Fig.26 shows the strain distribution of the {420} peak with big
oscillations and contrary to all previous results a mean tension residual
stress.

Final remarks. The evaluation of the residual stress state has made
a great step forward, but it is time consuming. Many measurements should
further be made to get the results on different materials with different
textures. We are working on a regression analysis to elaborate the residual stress state from all poles using the entire intensity and the
strain polefigures. We hope to identify general laws which are valid for
all materials and find strain distributions vs. $\sin^2\psi$

a. with a linear relationship as seen with Mo-radiation and the
{732+651} peak for steels,
b. with a quasi-linearity as obtained with Cr-radiation and the
{211} peak for steels and find
c. specific kinds of oscillating curves obtained on different {hkl}
reflections in different other materials.

ACKNOWLEDGEMENT

Thanks are expressed to the organizers of the conference for inviting the
author to present this paper. The Deutsche Forschungsgemeinschaft has
granted the support for this journey.

REFERENCES

1. H. Dölle and V. Hauk, System of possible lattice strain distributions
   on mechanically loaded metallic materials (in German)
   Z. Metallkde. 68: 725 (1977).
2. V. Hauk, Residual stresses. Their importance in science and technology
   (in German), Conference April 1983, in: "Eigenspannungen, Entstehung - Messung - Bewertung", Edited by E. Macherauch and V. Hauk,
   DGM Oberursel, vol. 1: 9 (1983).
   V. Hauk, Stress evaluation on materials having non-linear lattice strain
   distributions, Adv. X-Ray Anal. 27: 101 (1984).
3. P. D. Evenschor and V. Hauk, On non-linear distributions of lattice
   interplanar spacing at X-ray strain measurements (in German),
   Z. Metallkde. 66: 167 (1975).
   H. Dölle and V. Hauk, Evaluation of residual stress systems arbitrarily oriented by X-rays (in German), HTM 31: 165 (1976).
4. H. Dölle and V. Hauk, The theoretical influence of multiaxial depth-
   dependent residual stresses upon the stress measurement by X-rays
   (in German), HTM 34:272 (1979).
5. V. Hauk and W. K. Krug, The theoretical influence of depth-dependent
   States of residual stresses upon the stress evaluation by X-rays II
   (in German), HTM 39: 273 (1984).
6. H. Krause and M. Mathias, An automatic measuring equipment for tex-
   ture- and stress-analysis by X-rays (in German), HTM 38: 129 (1983).
7. C. N. J. Wagner and N. S. Boldrick, The psi-differential and integral
   methods for residual stress measurements by X-ray diffraction,
   presented 32. Denver X-Ray Conference (1983).
8. J. Hoffmann, H. Neff, B. Scholtes and E. Macherauch, Plane polefigures
   and lattice strain polefigures of textured materials (in German),
   HTM 38: 180 (1983).
9. V. Hauk, X-ray methods for measuring residual stress, in "Residual
   stress and stress relaxation", Edited by E. Kula and V. Weiss,
   Plenum Press New York and London: 117 (1982).
10. V. Hauk and G. Vaessen, Evaluation of residual stresses by X-rays on
    textured steels (in German), in: "Eigenspannungen, Entstehung -
    Messung - Bewertung", Edited by E. Macherauch and V. Hauk, DGM
    Oberursel, vol. 2: 9 (1983).
11. P.D. Evenschor and V. Hauk, X-ray elastic constants and distributions
    of interplanar spacings of materials with preferred orientation
    (in German), Z. Metallkde. 66: 164 (1975).
12. V. Hauk and H. Sesemann, Deviations from linear distributions of lat-
    tice interplanar spacings in cubic metals and their relation to
    stress measurement by X-rays (in German),
    Z. Metallkde. 67: 646 (1976).
13. V. Hauk and G. Vaessen, Evaluation of non-linear lattice strain disri-
    bution (in German), in: "Eigenspannungen und Lastspannungen, Moder-
    ne Ermittlung - Ergebnisse - Bewertung", Edited by V. Hauk and
    E. Macherauch, HTM-Beiheft, Carl Hanser Verlag München Wien:
    38 (1982).
14. V. M. Hauk, R. W. M. Oudelhoven and G. J. H. Vaessen, The state of
    residual stress in the near surface region of homogeneous and he-
    terogeneous materials after grinding, Metallurgical Transactions
    13 A: 1239 (1982).
15. V. M. Hauk and G. J. H. Vaessen, Residual stress evaluation with X-
    rays in steels having preferred orientation, Metallurgical Trans-
    actions 15 A: 1407 (1984).
16. I. C. Noyan, Equilibrium conditions for the average stresses measured
    by X-rays, Metallurgical Transactions 14 A: 1907 (1983).

17. V. Hauk and P. J. T. Stuitje, Phase specific investigations of resi-
    dual stresses in heterogeneous materials by X-rays after plastic
    deformations, part I and II (in German),
    Z. Metallkde. 76: 445 and 471 (1985).
18. F. Bollenrath, V. Hauk and E. H. Müller, Calculation of the elasticity
    constants of polycrystalline materials from the compliances of
    single crystals (in German), Z. Metallkde. 58: 76 (1967).
19. V. M. Hauk and E. Macherauch, A useful guide for X-ray stress evalu-
    ation(XSE), Adv. X-ray Anal. 27: 81 (1984).
20. V. Hauk, H. J. Nikolin and H. Weisshaupt, X-ray elasticity constants
    of a low alloyed steel in two states (in German),
    Z. Metallkde. 76: 226 (1985).
21. V. Hauk and H. Kockelmann, Calculation of stress distribution and of
    XEC of two-phase materials (in G.), Z. Metallkde. 68: 719 (1977).
22. H. Möller and G. Martin, Elastic anisotropy and stress measurement by
    X-rays (in German), Mitt. K. W. I. Eisenforsch. Düsseldorf 21:
    261 (1939).
23. P. F. Willemse, B. P. Naughton and C. A. Verbraak, X-ray residual
    stress measurements on cold-drawn steel wire,
    Mat. Science Engg. 56: 25 (1982).
24. V. Hauk and G. Vaessen, Residual stresses in groups of crystallites of
    steels having preferred orientation (in German),
    Z. Metallkde. 76: 102 (1985).
25. V. Hauk, G. Vaessen and B. Weber, The evaluation of residual stresses
    with rolling-texture (in German), HTM 40: 122 (1985).
26. V. Hauk, Residual stresses by deformation (in German), in: "Eigenspan-
    nungen und Lastspannungen, Moderne Ermittlung - Ergebnisse - Bewer-
    tung", Edited by V. Hauk and E. Macherauch, HTM Beiheft, Carl Han-
    ser Verlag München Wien: 92 (1982).
27. L. Pintschovius, V. Hauk and W. K. Krug, to be published.
28. K. Feja, V. Hauk W. K. Krug and L. Pintschovius, to be published.
29. V. Hauk, P. J. T. Stuitje and A. Troost, Analysis of the state of re-
    sidual stresses in a rolled sheet of Cu Zn 40 (in German),
    Materialprüfung 27: (1985).
30. V. Hauk and R. W. M. Oudelhoven, to be published.
31. V. Hauk and B. Weber, to be published.
32. V. Hauk and H. Kockelmann, Lattice strain distributions of plastically
    deformed specimens of pure and silveralloyed copper (in German),
    Z. Metallkde. 71: 303 (1980).

X-RAY ELASTIC CONSTANTS AND THEIR MEANING FOR Aℓ AND Fe

Rui Mei Zhong[*], I. C. Noyan and J. B. Cohen

Department of Materials Science and Engineering
The Technological Institute
Northwestern University
Evanston, IL 60201

INTRODUCTION

In the measurement of residual stresses via diffraction (using x-rays or neutrons) it is strains that are actually determined, by employing the interplanar spacing ($d_{hk\ell}$) of the {hkℓ} planes as an internal strain gauge. The change in this spacing is measured from the shift of diffraction peaks (and Bragg's law) at several orientations of the sample to the incident beam, and the resultant strains are converted to stresses with the "diffraction elastic constants", $S_1(hk\ell)$ and $S_2(hk\ell)/2$.[1] While these take on the values ($-\nu/E$) and $(1+\nu)/E$ respectively for an isotropic solid, in anisotropic materials their values depend on many factors: preferred orientation, shape and orientation of second phases, interaction between grains. In fact there are reports of variation these constants with plastic deformation[2] and theory predicts variations with morphology[3]. While it is possible to calculate approximate values for these constants from theory and the single crystal elastic constants[4], S_1 and $S_2/2$ are really not elastic constants in the strictest sense because of these other factors, and it is best to measure them. One of us (I. C. Noyan) has recently examined this problem in some detail[5], and we summarize his results here.

In a measurement of diffraction elastic constants (on a piece of material as identical as possible to the piece whose stress is sought) a series of loads are applied in the elastic range. At each load, the inter-planar spacing is measured vs. $\sin^2\psi$, where ψ is the tilt of the specimen normal from the bisector of the incident and diffracted beams. The slope (β) is obtained at each applied stress, and plotted vs. this stress. The slope of this second plot is related to one of the desired constants:

$$\beta = \sigma^0_{11} \, \tfrac{S_2}{2} \, [S_{ijk\ell}, \, K_i(\phi)] + C(\varepsilon^r_{ij}) \qquad (1)$$

Here σ^0_{11} is the applied stress, $S_{ijk\ell}$ are elastic constants and the K_i are the stress interaction constants between grains, which depend on their orientation, and hence ϕ. The term C is a complex function of any residual strain (ε^r_{ij}) present in the specimen. It is now well established that

---

[*] Currently at Beijing Research Institute of Mechanical and Electrical Technology, Beijing, China

there can be oscillations in d vs. sin² ψ, either due to fluctuations in residual stress[7] or due to preferred orientation and elastic anistropy[6-8] Even if such fluctuations are present, β obtained by a least-squares fit is linear vs. the applied load[5]. In fact, Eq. 1 can be exploited to establish the cause of these oscillations. If plastic deformation changes only the residual stress distribution and not the texture or second phase morphology, S_2 is unchanged by the process. However, if texture or morphology are altered, S_2 changes. In this paper we discuss two examples of these effects.

EXPERIMENTAL PROCEDURES

Flat tensile specimens were prepared from 1100 Aℓ plate after annealing at 648°K for 2 hrs., followed by reduction in thickness by rolling on a two-high mill, 0.1mm per pass, to 65,82 and 90 pct reduction. Diffraction elastic constants were determined with V filtered CrK_α radiation in the apparatus described in Ref. 9, by applying loads up to 12,000 psi, in 2000 psi increments. The 311 reflection was employed.

RESULTS AND DISCUSSION

Some typical plots of d vs. sin²ψ at various loads are shown in Fig. 1; oscillations are apparent. The slopes vs. applied load are given in

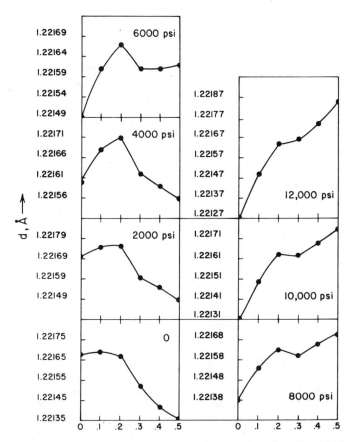

Fig. 1   Interplanar spacing, "d" vs. sin² ψ for 1100 Aℓ reduced in thickness 65pct by cold rolling. Each figure is for the indicated applied load.

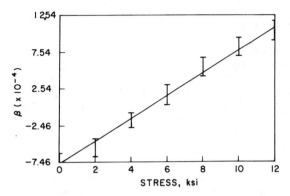

Fig. 2   The constant β in Eq. 1 vs. applied stress.
          1100 Aℓ reduced in thickness 81.7 pct by
          cold rolling.

Fig. 2, and linearly is reasonable. The evaluated elastic constants are summarized in Table I, where they are compared to calculated values. The agreement is quite good and there is no significant variation with deformation and its resultant texture. As Aℓ is nearly isotropic, Eq. 1 predicts that the elastic constants should be those calculated from theory ignoring the K_i terms because these are zero, and this is what is observed in Table I. The oscillations in 2θ vs. $\sin^2\psi$ are due to local fluctuations in residual stress.[7]

On the other hand for Fe it is known that oscillations occur (in d vs. $\sin^2\psi$), that the β vs. applied load is linear, but that the diffraction elastic constant varies with deformation and does not agree with theory that neglects K_i.[2] Again, this is expected; the variation in preferred orientation with deformation causes changes in the K_i in this case.

It has been suggested that when oscillations occur in 2θ or "d" vs. $\sin^2\psi$, that h00 or hhh reflections will not show this, if anisotropic elasticity is the cause.[6,10] This is only the case in the Reuss limit and

TABLE I

DIFFRACTION ELASTIC CONSTANTS FOR 1100 Aℓ

REDUCTION IN THICKNESS	$S_2/2 \times 10^{-8}$	ERROR DUE TO COUNTING STATISTICS $\times 10^{-9}$ [a]
65 pct	12.76	2.8
81.7 pct	12.60	1.3
90 pct	13.08	1.4
Calculated Value[b]	13.24	------

a. See Ref. 9 for equations to calculate errors due to counting statistics.

b. Average of Reuss and Voight limits.

without grain interaction,which is generally not the case in deformed materials. In fact, there are reported cases where the oscillations do not vanish for h00 or hhh peaks.[7] Also Al is nearly isotropic and yet, as we report here, there are oscillations. The source of these oscillations is local fluctuations in residual stress. In such a case (because the depth of penetration varies with ϕ) an average value of stress is of little use,[11] and attempts to eliminate the oscillations by increasing the depth of penetration[12] (changing the wavelength) simply averages over the (important) fluctuations. The tests described here provide a means to decide on the source of the oscillations.

ACKNOWLEDGEMENTS

This research was supported by ONR under Grant No. N00014-80-C116. We thank Paul Rudnik for assistance with the diffraction measurements.

REFERENCES

1. M. E. Hilley, J. A. Larson, C. F. Jactzak and R. E. Rickefs (eds), "Residual Stress Measurement by X-ray Diffraction", SAE Information Report, J. 784a, (2nd edition), SAE, Inc, New York (1971).

2. R. H. Marion and J. B. Cohen, The Need for Experimentally Determined X-ray Elastic Constants, Adv. in X-ray Analysis, 20:355(1977).

3. T. T. Wu, The Effect of Inclusion Shape on the Elastic Moduli of a Two-Phase Material, Int. J. Solids and Structures, 2:1(1966).

4. H. Dölle, Influence of Multiaxial Stress States, Stress Gradients and Elastic Anisotropy on the Evaluation of (Residual) Stresses by X-rays, J. Appl. Cryst., 12:489(1979).

5. I. C. Noyan, Determination of the Elastic Constants of Inhomogeneous Materials with X-ray Diffraction, Mat. Sci & Eng., in press.

6. H. Dölle and V. Hauk, Einfluss der Mechanischen Anisotropie des Vielkristalls (Textur)auf die Rontgenographische Spannungser Mittlung, Z. Metallk., 69:410(1978).

7. I. C. Noyan and J. B. Cohen, Determining Stress in the Presence of Nonlinearities in Interplanar Spacing vs. $\sin^2 \phi$, Adv. in X-ray Analysis, 27:129(1984).

8. R. H. Marion and J. B. Cohen, Anomalies in Measurement of Residual Stress by X-ray Diffraction, Adv. in X-ray Analysis, 18:446(1975).

9. K. Perry, I. C. Noyan, P. J. Rudnik and J. B. Cohen, The Measurement of Elastic Constants for the Determination of Stresses by X-rays, Adv. in X-ray Analysis, 27:159(1984).

10. H. Dölle and J. B. Cohen, Evaluation of (Residual) Stresses in Textured Cubic Metals, Met. Trans., 11A:831(1980).

11. I. C. Noyan and J. B. Cohen, The Use of Neutrons to Measure Stresses, Scripta Metall., 18:627(1984).

12. V. M. Hauk and G. J. H. Vaessen, Residual Stress Evaluation with X-rays, Met. Trans., 15A:1407(1984).

ESTIMATION OF ANISOTROPY OF X-RAY

ELASTIC MODULUS IN STEEL SHEETS

Shin-ichi Nagashima, Masaki Shiratori and Ryuichi Nakagawa

Yokohama National University
Department of Engineering
156 Tokiwadai, Hodogaya-ku, Yokohama, 240 Japan

INTRODUCTION

The oscillation from a linear relation in the 2θ vs. $\sin^2\psi$ diagram has been a most important problem in X-ray stress measurement. There are, therefore, a number of papers concerned with the X-ray elastic constant, lattice strains under stresses and evaluation of stresses of textured materials.[1-8]

The purpose of the present study is to analyze the three-dimensional orientation distribution of steel sheets by means of the Vector method proposed by Ruer and Baro,[11] and to calculate the elastic modulus of textured sheets by means of a finite element method (FEM) using the three-dimensional orientation distribution, and then to calculate the strain/stress ratios vs. the directions defined by the angles between the specimen normal and the normal to the diffracting planes.

EXPERIMENTAL PROCEDURE

Specimen

In the present study, ferritic stainless steel sheets of type AISI 430 have been used. The chemical composition of the steel is 0.05 wt.% carbon and 16.40 wt.% chromium.

Measurement of Young's Modulus

Young's modulus was measured along the directions 0°, 45° and 90° to the rolling direction by means of dynamic and static methods. The dynamic Young's modulus was determined by an oscillation method using annealed sheet specimens of the following size: about 100 mm long, 9 mm wide and 0.8 mm thick. The static Young's modulus was determined by means of a tensile test, from the averaged value of proportional coefficient of the stress vs. strain curve of strains less than 0.1%. The specimen size was 60 mm in length in the parallel region, 50 mm in the gauge length, 25 mm wide and 0.8 mm thick.

Calculation of Young's Modulus[9],[10]

Young's modulus along the directions 0°, 45° and 90° to the rolling direction was calculated by the methods of Reuss and Voigt as well as by means of a finite element method (FEM), using preferred orientation data.

Finite Element Method (FEM)[13],[14]

For the calculation of Young's modulus by means of the FEM, a thin sheet of steel was approximated by an aggregate of triangular finite elements, each pair of which, that composes a square, corresponds to a crystal grain with arbitrary orientation; the dimension of the specimen was taken to be 150 mm x 150 mm x 1 mm. The specimen was divided into 450 elements with 256 nodal points and 225 grains. Since the elastic constants calculated by this method have a statistical nature depending on the orientation of each grain, the results of 30 iterations were averaged.

Determination of Preferred Orientation

The pole figure was measured for 200 and 110 diffraction by the Schulz reflection method, using Co Kα radiation with an Fe filter. Then, the three-dimensional distribution of texture was determined by means of the Vector method,[11] and the result, i.e. the distribution of the texture vector was represented by a method proposed by one of the authors.[12]

X-Ray Strain Measurement

As the purpose of the present study is to clarify the effect of preferred orientation on the elastic strain of individual grains in the deformed bulk specimen, the diffraction angles 2θ for a series of ψ angle values were measured applying tensile stress to the specimen. Two kinds of diffraction planes, (211) and (310), were examined with chromium and cobalt Kα radiation. Tensile specimens were cut out from the sheets, the longitudinal axes of which were 0°, 45°, 90° from the rolling direction.

Estimation of Anisotropy of Specimen

Calculation of strain in a single crystal. The elastic strain of a single crystal along an arbitrary direction can be calculated in the following way. The normal strain ε_ψ to the plane normal to the direction OP, shown in Fig. 1, is given by

$$\varepsilon_\psi = a^2 \varepsilon_{x'} + b^2 \varepsilon_{y'} + c^2 \varepsilon_{z'} + bc\,\gamma_{y'z'} + ca\,\gamma_{z'x'} + ab\,\gamma_{x'y'} \qquad (1)$$

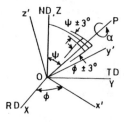

Fig.1. Co-ordinate systems of specimen, O–XYZ,
and crystal, O–x'y'z'. ϕ, ψ, α are shown.

where a, b, c are the direction cosines of OP with the co-ordinate system of the crystal O-x'y'z'.

When a uniaxial stress is applied to the specimen along the direction X which is one of the principal axes of the co-ordinate system of the specimen O-XYZ, the strain ε_ψ is given by

$$\varepsilon_\psi / \sigma_1 = (s_{11} - s_{12} - (s_{44} / 2)) \, M_{11} + s_{12} + (s_{44} /2) \, \sin^2\psi$$

where, $M_{11} = a^2 l_1^2 + b^2 m_1^2 + c^2 n_1^2$, and l_i, m_i, n_i, (i = 1, 2, 3) are the direction cosines between O-XYZ and O-x'y'z', and s_{11}, s_{12}, s_{44} are the elastic compliance for cubic metals. In the present study, the value for α-iron was used.

Calculation of Strain in a Textured Specimen

The Reuss model (the uniform local stress model). As the Voigt model assumes equal strain for grains of any orientation, the local strains concerning the diffraction plane defined by the angles ϕ and ψ (allowing angular deviations ± 3°) were calculated by the use of the Reuss uniform local stress model.

At first, the orientation of all the grains which contain such planes as (211) and (310) satisfying the diffraction condition were determined about the angle α from 0° to 360°. For the determination of grains, the texture vector proposed by Ruer and Baro was helpful. Then, the strain values for the series of angle ϕ, ψ, under the given stress, σ_1, were calculated for all the possible diffraction planes about the angle α, and the result showed remarkable oscillation on the ε_ψ vs. α diagram. (As the Voigt model assumes equal strain for all grains, the strain shows no change with α). This means that the resultant strain of a randomly oriented specimen, ε_ψ, is given by averaging the strain value about the angle α from 0° to 360°, while that of a textured specimen must be averaged for limited grains, and the strain will be a certain value between the max. and min. values of the ε_ψ vs. α curve. In the calculation, the orientations of the grains were determined by the texture vector of the specimen under consideration.

The FEM model. In the textured specimen, the orientations of grains which cause hkl diffraction about the direction defined by ϕ and ψ are limited and determined by the texture vector as described above. And the averaged strain, ε_ψ, caused by the applied uniaxial stress, σ_1, can be calculated taking into account the orientations as well as the volume fraction of grains. The FEM models for calculation are shown in Fig. 2. At first, Young's modulus along the tensile direction and the averaged strain along normal direction to the sheet plane were calculated applying a prescribed uniform displacement to the right end of the Model (a). In the calculation of Young's moduli, major texture components of the steel sheets were distributed randomly onto the 225 grains. The three dimensional orientation distribution of grains was again determined by the texture vector.

Then the averaged local strain of diffraction planes that are defined by the angles ϕ and ψ, was calculated using models shown in Fig. 2(b). In the Model (b), a group of grains, G(d), which contain the diffraction planes under consideration, is surrounded by the textured matrix. The volume fraction of the grains, G(d), of the specimen is shown in Table 1.

Table 1. Volume fraction (%) of G(b) for 211 and 310 Diffraction

hkl	ϕ	$\psi=3°$	9°	15°	21°	27°	33°	39°	45°	51°
211	0°	1.8	1.8	4.0	11.1	7.1	11.1	11.1	7.1	4.0
	45°	1.8	1.8	1.8	11.1	7.1	7.1	7.1	7.1	4.0
	90°	1.8	4.0	7.1	11.1	7.1	4.0	11.1	4.0	1.8
310	0°	1.8	1.8	1.8	4.0	4.0	7.1	7.1	4.0	7.1

EXPERIMENTAL RESULTS

Determination of Preferred Orientation

The three-dimensional orientation of the specimen can be determined using only one incomplete reflection pole figure as well as from the complete pole figure by means of the Vector method.[11] Fig. 3(a) shows the measured (100) pole figure and 3(b) is the recalculated one using the analyzed texture vectors. The major texture components are summarized into 6 groups, A-F, shown in Fig. 4 and Table 2.

Calculation of Young's Modulus

Young's modulus of a steel was calculated assuming that the whole preferred orientation distribution is approximately represented by 6 texture components shown in Fig. 4. Calculation was made by means of FEM and the method of Voigt and Reuss.[9,10] The result is shown in Fig. 5.

The calculated values lie between the E values obtained by the dynamic and static methods, and the oscillation of E values obtained by FEM was larger than that of the Voigt (uniform local strain) model, and the Reuss (uniform local stress) model gives the lower values.

Variation of $\varepsilon_\psi/\sigma_1$ vs. $\sin^2\psi$

As described above, the averaged strain of a randomly oriented specimen and those of the textured specimen used in the present study have been calculated for the series of angle α. The resulting strain/stress ratios, $\varepsilon_\psi/\sigma_1$, vs. $\sin^2\psi$ are shown in Fig. 6 together with the observed values. In the same figures, the upper and lower limits of values in the $\varepsilon_\psi/\sigma_1$ vs. α diagram are plotted by the broken lines. It is clear that the calculated and observed values agree very well and are located within the upper and lower bounds.

DISCUSSION AND CONCLUSION

Young's Modulus

By the use of FEM, Young's modulus of textured metallic materials can be simulated for various types of models, such as an equal strain, equal stress or randomly distributed models. In the present study, Young's modulus was calculated for angles ϕ = 0°, 45°, and 90° to the rolling direction, assuming the whole orientation distribution is represented approximately by 6 major components. The result agreed with the averaged values of dynamic and static measurements.

Preferred Orientation

The three-dimensional orientation distribution, the texture vector, can be analyzed by the Vector method using incomplete reflection pole figures of (110) and (111) if they cover the range of α values from 18°

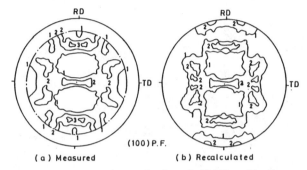

Model number	Number of grains	Volume fraction
b-1	1	0.004
b-2	4	0.018
b-3	9	0.040
b-4	16	0.071
b-5	25	0.111
b-6	36	0.160

Model a   Model b

Fig. 2. FEM models for the calculation of local strain.

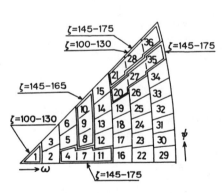

(a) Measured   (b) Recalculated

Fig.3. Measured and recalculated (100) pole figures.

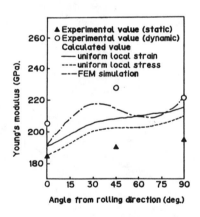

Fig.4. Major components of annealed sheet. 800 C x 1hr.

Fig.5. Variation of Young's modulus in the sheet plane.

Table 2. Major components of the sheet shown in Fig. 4

Group	Box Number	Orientation	Volume Fraction(%)
A	1	$(001)[1\bar{2}0] \sim (001)[1\bar{1}0]$	7
B	7	$(0\bar{1}3)[931]$	13
C	9, 10	$(2\bar{3}9)[3\bar{2}0] \sim (1\bar{1}3)[52\bar{1}]$	7
D	27	$(2\bar{3}4)[52\bar{1}]$	20
E	28, 36	$(3\bar{3}4)[373] \sim (1\bar{1}1)[121]$	20
F	28, 36	$(3\bar{3}4)[733] \sim (1\bar{1}1)[321]$	33

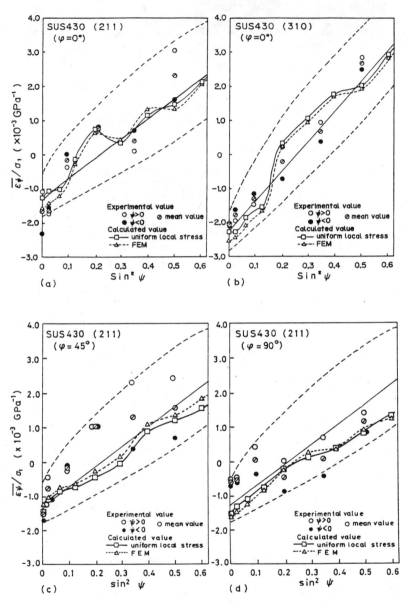

Fig. 6. Variation of local strain/stress ratio vs. $\sin^2\psi$ (AISI430 steel)

to 90°.[11] In the case of (100) pole figure, if a (100) [uvw] orientation component has a deviation larger than 15° about its ideal orientation, three-dimensional analysis is possible, and the recalculated pole figure using the texture vector agreed well with the measured one as shown in Fig. 3. Among the major components listed in Table 2, A and F are the same as those of low carbon steel sheets, but others are little different.

Variation of $\varepsilon_\psi/\sigma_1$ vs. $\sin^2\psi$

Hauk et al.[3,4] have studied on the variation of strain as well as strain/stress ratio vs. $\sin^2\psi$ due to preferred orientation. In the years around 1975, the measurement of pole figures by the use of diffractometers was the most precise method and they determined three major texture components such as (211)[011], (111)[211] and (100)[011]. And they calculated the strain by the use of the Reuss model.

As it is pointed out by Hill,[15] Young's modulus calculated by Voigt and Reuss models give the upper and lower bounds, respectively. Honda et al. have studied the oscillation of the strain/stress ratio vs. $\sin^2\psi$ curve due to texture by the method of FEM, but they only dealt with the fiber texture.[7,8]

In the present paper, the authors have analyzed the three-dimensional texture distribution by the use of the Vector method, and further the effect of texture on the strain/stress ratio was calculated using FEM as well as the Reuss model.

It is clear that the oscillation of the $\varepsilon_\psi/\sigma_1$ curve calculated by the Reuss (equal local stress) model and by the FEM method is nearly the same. As the former model shows the effect of grain orientation on the anisotropy of local strain, while the latter gives the effect of volume fraction as well as orientation of specially oriented grains, it can be concluded that the effect of volume fraction on the anisotropy of local strain is minor.

The calculated values of $\varepsilon_\psi/\sigma_1$ agreed fairly well with the observed ones for 211 diffraction. But for 310 the correlation between calculated and observed values is not clear.

ACKNOWLEDGEMENT

The authors are indebted to Professors D. Ruer and R. Baro for giving us the full computer program of the Vector Method and helpful advice to make use of it. The authors also wish to thank, for assistance, S. Murayama, T. Nakagomi and K. Yamashita, who contributed their experimental skill and sustained efforts.

REFERENCES

1. G. Taira and K. Hayashi, X-Ray Investigation of Polycrystalline Materials (On the Effect of Fiber Texture on the Elastic Constants of α-Iron, Proc. 13th Japan Cong. on Materials Research 20-24 (1970).
2. T. Shiraiwa and Y. Sakamoto, The X-Ray Stress Measurement of the Deformed Steel Having Preferred Orientation, Proc. 13th Japan Cong. on Materials Research 25-32 (1970).
3. V. Hauk, D. Herlach und H. Sesemann, Über nichtlineare Gitterebenenabstandsverteilungen in Stählen, ihre Entstehung, Berechnung und Berücksichtigung bei der Spannungsermittlung, Z. Metallkde 66:734-737 (1975).

4. V. Hauk und H. Sesemann, Abweichungen von linearen Gitterebenenabstands-
    verteilungen in kubischen Metallen und ihr Berücksichtigung beider
    röntgenographischen Spannungsermittlung, Z. Metallkde, 67:646–650
    (1976).

5. R. H. Marion and J. B. Cohen, The Need for Experimentally Determined
    X-Ray Elastic Constants, Advances in X-Ray Analysis 20:355–377
    (1977).

6. H. Dölle, The Influence of Multiaxial Stress States, Stress Gradients
    and Elastic Anisotropy on the Evaluation of (Residual) Stresses by
    X-Rays, Appl. Cryst. 12:489–501 (1979).

7. K. Honda, N. Hosokawa and T. Sarai, Effect of Texture on Stress Mea-
    sured by X-ray Diffraction Method, J. NDT, Japan 26:539–545 (1977).

8. K. Honda and T. Sarai, X-Ray Strain Analysis and Elastic Deformation
    of Polycrystalline Metals, J. Soc. Mat. Sci. Japan 33:367–373 (1984).

9. W. Voigt, Lehrbuch der Kristallphysik (Leipzig, Teubner Verlag), 716
    (1928).

10. A. Reuss, Z. Angew. Math. Mech., 9:49 (1929)

11. D. Ruer and R. Baro, A New Method for the Determination of the Texture
    of Materials of Cubic Structure from Incomplete Reflection Pole
    Figures, Advances in X-Ray Analysis 20:187–200 (1977).

12. S. Nagashima, A New Method for the Three Dimensional Analysis of Pre-
    ferred Orientation of Metallic Materials, J. Korean Inst. Metals
    22:41–52 (1984).

13. S. Nagashima, M. Shiratori and T. Fujiu, The Estimation of Elastic
    Constants for the X-Ray Stress Analysis of Textured Metallic
    Materials, Proc. ICOTOM 6: Vol. II, 1148–1157 (1981).

14. S. Nagashima, M. Shiratori and K. Matsukawa, "The Effect of Grain Dis-
    tribution on the Elastic Moduli in Metallic Materials Which Give
    the Same Pole Figure, Trans. ISIJ, 23:B–26 (1983).

15. R. Hill, Proc. Phys. Soc., A65:349 (1952).

A COMPREHENSIVE APPROACH TO IN SITU STRESS MEASUREMENT

R.A. Holt

Physical Metallurgy Research Laboratories
Canada Centre for Mineral and Energy Technology, Ottawa
Energy, Mines and Resources, Canada

INTRODUCTION

Recent developments in position sensitive detectors (PSD's), solid state power and computer technologies make it possible to design accurate instruments for in-situ stress measurement[1-3]. Such instruments require compromises in the interests of portability, size and speed which may limit accuracy and/or versatility. Furthermore, extraction of a stress tensor from X-ray data is not always straightforward and considerable research is required before an instrument for X-ray stress measurement can be treated as a "black box" to be given to an uneducated operator.

The development of a new instrument for in-situ stress measurement is described in this paper. It was conceived as a field instrument with two position-sensitive proportional counters (PSPC's) for use in the single exposure mode (SET). and incorporates precise angular control of the incident X-ray beam and data analysis to eliminate irregular Bragg peaks owing to coarse grain structure[4]. An experimental instrument was built to test the concept (1).

DESIGN CONSIDERATIONS

Single vs. Dual Detector

The use of two PSPC's, combined with the ω-geometry allows the use of SET which is insensitive to errors from displacement of the specimen surface from the nominal axis of rotation of the instrument[5,6]. Although the SET is desirable for field applications a multiple exposure ($\sin^2\psi$) capability is essential in some circumstances[7].

(1) A commercial prototype of the instrument incorporating these features and the capability of multiple exposure ($\sin^2\psi$) measurements has also been built by Proto Mfg. Ltd. of Oldcastle, Ontario and is currently under test.

Effect of Crystal Orientation Distribution on X-ray Line Profile with PSD

The use of PSD's for stress measurement is dictated by their high rate of data acquisition. With the relationship between the X-ray beam, specimen and detector fixed, the Bragg profile observed in the detector arises from a range of crystal orientations corresponding to half the range of 2θ over which diffracted radiation is observed. A coarse grain structure produces irregularities in the orientation distribution and hence irregular Bragg reflections. The effect is more pronounced if there are grain interaction stresses. Even if the grain size is fine, a gradient in crystallographic texture will produce a systematic error in the location of the Bragg peak because the orientation distribution changes systematically along the arc intercepted by the detector. With broad peaks, apparent shifts of tenths of degrees in 2θ can be produced by texture gradients of 10-15% per degree.

The Bragg peaks can be "smoothed" by rotating the X-ray source relative to the specimen by a small amount in ψ or ω to sample a larger number of grains[7] and summing the counts on a common 2θ scale. However, the systematic effect of a texture gradient remains. When ω-rotation is used, a diffraction profile can be compiled corresponding to the average for a true radial sampling in reciprocal space of a range of crystal orientations by limiting the range of intensities summed to those lying in a virtual window[4] in the detector corresponding to this range of orientations. The compilation of such a "scanned" profile eliminates the systematic error due to texture gradients.

α-Rotation

When the rotation of the source about the ω-axis is accompanied by a similar rotation of the detectors, the 2θ scale remains fixed in the detector. If the detector is fixed to the specimen and only the source rotates (α-rotation, Fig. 1) then the 2θ scale moves relative to the detector in angular increments equal to that of the X-ray source. "Scanned" profiles can be assembled in the computer by shifting the sub-profiles into registry on the 2θ scale.

Provision of α-rotation has several advantages, i.e.:

- it is relatively easy to provide the angular precision required to produce a "scanned" profile over a small angular range (± 5-$8°$);

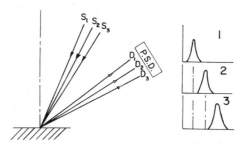

Fig. 1   Illustration of **α**-rotation showing positions of diffracted beams D_1-D_3 for three positions of the source S_1-S_3 and a schematic representation of the Bragg profiles in the P.S.D.

- α-rotation can be used to calibrate the angular sensitivity and linearity of the detector;
- it reduces the requirement for stress-free standard specimens;
- α-rotation can be used to access a limited range of $\sin^2\psi$ when ω is fixed, and hence assess severe non-linearity in $\sin^2\psi$ plots;
- finally, when quartz/carbon filament PSPC's are used, the life of the filaments is prolonged.

ω vs ψ Geometry

The potential advantages of a ψ-diffractometer over an ω-diffractometer have been tabulated by Macherauch and Wolfsteig[8]. For an instrument intended for field use there are considerations not included in their comparison.

For example, the ψ-geometry is incompatible with the dual detector, SET, mode of operation, and the ω-geometry has the advantage of allowing correction for grain size and texture gradient effects as described earlier. The ω-geometry is also consistent with the provision of α-rotation on a common axis.

When ω-rotation is confined to one side of the specimen normal with a single detector $+\ \psi$ can be sampled by rotating the instrument about the specimen normal ($\overline{\phi}$ rotation). In this case, the ω geometry can be less sensitive to errors arising from specimen displacement from the true rotation axis than the ψ geometry.

FEATURES OF THE EXPERIMENTAL INSTRUMENT

The experimental instrument (Fig. 2) is fitted with two TEC 205 sealed PSPC's and incorporates α-rotation driven by Solsyn stepping motors with an accuracy of $+0.005°$ over a range of up to $\pm 7.5°$. The X-ray source is a miniature KEVEX water-cooled tube and the source radius is 200 mm. It is

Fig. 2. The experimental model of the CANMET portable stress diffractometer.

controlled by a DEC MINC 11/23 computer and uses ORTEC analysis
electronics, a Selena A/D converter, Tennelec analog router and TEC
interface. It operates only in the single exposure and limited $\sin^2\psi$ modes
and ω is adjusted manually. The height of the instrument is adjustable
relative to the specimen over a range of +12 mm. The instrument can be
rotated 360° about a vertical axis and traversed +25 mm in two orthogonal
directions in the horizontal plane. All four motions are driven by Slosyn
stepping motors and may be operated manually or under computer control.

INSTRUMENT CONTROL AND DATA REDUCTION

The control and data reduction software is designed to:

- collect and examine profiles for preselected α positions.
  This allows detector set-up and preliminary assessment of unknown
  specimens;
- calibrate the angular sensitivity of the detectors for a range of
  α-positions and a fine grained stress-free specimen;
- record variations in response along the length of both detectors
  when uniformly illuminated by scatter from a glass slide or by a
  radioactive source.
- perform absolute 2θ calibration of detectors;
- measure stress by the SET or limited $\sin^2\psi$ methods and "scanned"
  profiles;
- review raw data, and carry out data reduction using alternative
  data reduction options.

EXPERIENCE WITH EXPERIMENTAL INSTRUMENT

The experimental instrument has been used to assess the accuracy of
measurements made by the SET, to verify the lack of sensitivity of the
setting of the axis of the instrument relative to the specimen surface and
to evaluate the overall performance of the SET by measuring stress in a
number of ferrous components using the "scanning" procedure.

The 2θ positions observed in the PSPC's are affected by drift in the
analog signal processing chain which results in an error in stress of up to
40-60 MPa and must be compensated for to obtain reproducible results. With
compensation, the standard deviation of 2θ positions of the maximum of a
sharp peak for a "scanned" profile compiled from "raw" profiles at 15
α-positions is 0.006-0.009°. This is comparable to the accuracy achieved
without scanning, indicating that the use of the α-drive has a negligible
effect on the standard deviation of peak positions. The reproducibility
achieved in a ferritic steel exhibiting a peak breadth of 0.85° 2θ is shown
as a function of measuring time in Table 1.

Advantage has been taken of the "scanning" procedure in making stress
measurements in 316 stainless steel with a coarse grain structure. When
the orientation of the source was fixed, the Bragg peaks were irregularly
shaped, because of a rapid variation in the volume of diffracting material
with ψ, or absent altogether. Fig. 3 shows a regularly shaped "scanned"
profile and a distorted "raw" profile for (311) using CrK_β radiation.

Table 2 shows examples of ferrous metal components to which the
experimental diffractometer has been applied, the conditions used and the
reproducibility achieved. Examples of results obtained on Inconel 600
nuclear stream generator tubing are shown in Fig. 4. They agree very well
with measurements made with a Philips diffractometer equipped with a bulk
specimen stage, detector arm slide and diffracted beam monochromator[9].

Table 1. Reproducibility as a function of counting time for a ferritic steel with a peak breadth of 0.85° 2θ (after Rachinger correction)

Counting Time S		Counts Peak / BG	Std. Deviation	
Actual	Effective		$\Delta 2\theta°$	Stress −MPa(ksi)
60	32	2600 / 750	.008	7.7 (1.1)
30	16	1300 / 380	.014	13.3 (1.9)
30	.30	2600 / 800	.009	8.8 (1.3)

CONCLUSIONS

A new instrument for in-situ stress measurement has been described. It includes several new features. The instrument operates in the single and limited multiple exposure modes, and compensates for errors arising from coarse grain structure or gradients in crystallographic texture.

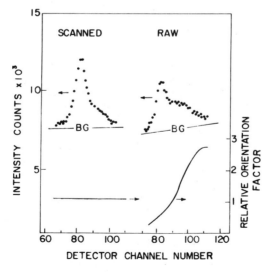

Fig. 3. Irregular "raw" profile resulting from variation in orientation factor in coarse grained 316-55 and a smooth "scanned" profile produced using α-rotation and an orientation "window".

Table 2. Examples of materials and experimental conditions for components examined with experimental model of CANMET Stress Diffractometer

Material	Description	(hkl)	Bragg Angle °	ψ Window °	Reproducibility MPa (ksi)
SA 106 B	Cold formed nuclear feeder pipe	(211)	156.1	3	15(2)[a]
Inconel 600	Ground, cold-formed nuclear steam generator tube	(311)	153.1	4.5	40(6)[a,b,c]
316 SS	Stress relieved plate	(311)	148.1	4	18(3)[b,d]

[a] broadened peak profiles
[b] low peak to background ratio
[c] effective spot size 1.5 x 1.5 mm^2, all others 1.5 x 10 mm^2
[d] coarse grain structure

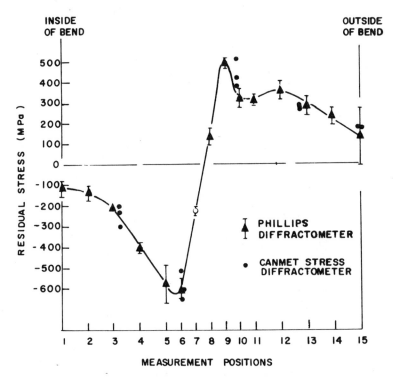

Fig. 4. Longitudinal residual stress as a function of circumferential position in an Inconel-600 nuclear steam generator tube formed into a "U-bend".

ACKNOWLEDGEMENTS

The original concept for the CANMET portable stress diffractometer was developed by C.M. Mitchell[4] before his retirement in January, 1984. The experimental instrument was built under his guidance with CANMET funding by PROTO MFG Ltd. of Oldcastle, Ontario.

I would like to thank D.W.G. White for his valuable support and guidance and E.J.C. Cousineau and P. Fryzuk for experimental assistance.

REFERENCES

1. M. Jones and J.B. Cohen, "Pars - A Portable X-ray Analyzer for Residual Stress", Journal of Testing and Evaluation, Vol. 6 (178), pp. 91-97, (see also U.S. Patent 4095103).

2. C.O. Ruud, "X-ray Analysis and Advances in Field Instrumentation", Journal of Metals, Vol. 13, No. 6 (1979), pp. 10-15.

3. L. Castex, J.M. Spraud and M. Banal, "A New, In-Situ Automatic, Strain Measuring X-ray Diffraction Appartus with PSD", Advances in X-ray Analysis, Vol. 27 (1984), pp. 267-272.

4. C.M. Mitchell, "The CANMET Portable Stress Diffractometer", Proc. International Conference on Pipeline Inspection, Edmonton, Alberta (1983), Government of Canada Publishing Centre, Ottawa, Canada, (see also U.S. patent 4561062).

5. J.T. Norton, "X-ray Stress Measurement by the Single Exposure Technique", Advances in X-ray Analysis, Vol. 11 (1968), pp. 401-410.

6. C.M. Mitchell, "A Dual Detector Diffractometer for Measurement of Residual Stress", Advances in X-ray Analysis, Vo. 20 (1977), pp. 379-391.

7. G. Maeder, J.L. Lebrun and J.M Spraud, "Present Possibilities for the X-ray Diffraction Method of Stress Measurement", NDT International, October (1981), pp. 235-247.

8. E. Macherauch and U. Wolfsteig, "A Modified Diffractometer for Stress Measurements", Advances in X-ray Analysis, Vol. 20 (1977), pp. 369-377.

9. J.E. Winegar, "X-ray Measurement of Residual Stress in Metals at Chalk River Nuclear Laboratories", Atomic Energy of Canada Ltd. Report #AECL 6961, June (1980).

SPACE SHUTTLE MAIN ENGINE HYDROGEN PUMP IMPELLER RESIDUAL STRESS MEASURE-

MENTS USING PARS (PORTABLE APPARATUS FOR MEASUREMENT OF RESIDUAL STRESS)

E.B.S. Pardue, M. V. Mathis,
R. W. Hendricks

Technology for Energy Corp.
Knoxville, TN 37922

E. M. Stangeland

Rockwell International
Rocketdyne Division
Canoga Park, CA 91304

INTRODUCTION

    X-ray residual stress analysis was performed on four space shuttle
hydrogen pump impellers made from Ti-5Al-2.5Sn alloy. Five locations on
the outer rim, near the vanes, were measured on each impeller, identified
as #1, #2, #3, and #4. Impellers #1 and #4 were test-fired impellers,
impeller #2 was a new impeller that had been spin tested, and impeller
#3 was a new impeller that had not been spin tested. The measurement
locations on these impellers corresponded to areas of critical stress
importance. The purpose of the measurements was to compare the stresses
at these locations as a function of impeller processing variables.

    A description of the impellers, the test parameters and procedures,
stress analysis results, and a technical discussion of these results are
presented in this report.

IMPELLER DESCRIPTION

    The Space Shuttle Main Engine (SSME) high pressure fuel turbopump
(HPFTP) is, in principle, a three-stage impeller pump coupled by a shaft
to a two-stage hot gas drive turbine.

    The HPFTP impellers pump liquid hydrogen at a temperature of -423°F
to the main combustion chamber, providing the fuel for the engine. At
the normal operating speed of 36,740 rpm, the impellers produce a volume
flowrate of 16,242 gpm at an outlet pressure of 7040 psia. The HPFTP
turbine input power level to drive the impellers is 77,310 horsepower.

    Due to the extreme demands presented by the HPFTP operating con-
ditions and weight limitations, a titanium alloy with extra-low
interstitial content was selected for impeller fabrication. The titanium
alloy, Ti-5Al-2.5Sn ELI, was selected for the impeller application
because of the low density, high strength, toughness and good fatigue
properties at cryogenic temperatures.

As with most aerospace applications, the materials used and the environments that they are exposed to are unique. Information about material properties and behaviors in these unique conditions must generally be generated by the firm developing the particular application.

The HPFTP impellers are extremely critical components of the SSME and substantial amounts of research have been done to analyze all aspects of their use. However, one aspect of the experimental analysis was lacking concerning the HPFTP impellers – the residual stresses possessed by HPFTP impellers at sequential time stages during fabrication, testing and operation are unknown. Knowledge of residual stress is vital for dynamic component fatigue life calculations. Many processes used during HPFTP impeller fabrication influence the magnitude and sign of the residual stress. Two processes of particular concern were proof spinning and balance grinding.

Impellers with substantial hot-fire test histories were used for determining the effect, of operational fatigue on residual stress. Two separate configurations of high time hot-fired impellers were tested as represented by impellers #1 and #4.

EQUIPMENT AND TEST PROCEDURES

The TEC Model 1620 Portable X-Ray Stress Analysis System (Figure 1) was used in this work. This equipment employs the $\sin^2\psi$, or multiple tilt, technique. The block diagram in Figure 2 shows the steps used in collecting and analyzing data.

For this work, a copper x-ray tube was used to analyze the (213) planes of the titanium alloy. The power level was 45kV and 1.5mA. Acquisition times ranged from 200s to 800s at ψ=0. Four ψ angles, 0°, 15°, 30°, 45°, were used for each measurement. A 2 mm diameter collimator was used for these measurements. The x-ray elastic constant, $\frac{1 + \upsilon^*}{E}$ used for this work was 7.6 x 10^{-8} psi^{-1}.

$^*\upsilon$=Poisson's ratio
E=X-ray elastic modulus

Figure 1.   TEC Model 1620 Portable X-Ray Analysis System.

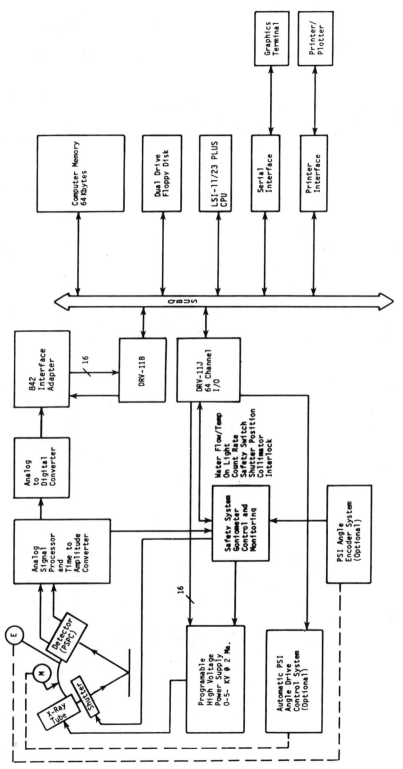

Figure 2.   PARS System Schematic.

MEASUREMENT LOCATIONS

The critical stress locations were denoted as position #1, #1Z, #2, #3, and #4 and typical locations are shown in Figure 3. Position #1 is at the right side of a full vane, position #1Z is near the right side of the adjacent vane, position #2 is beneath the center of the vane, position #3 is between vanes, and position #4 is inside the vane near the external surface. Similar locations were measured on each impeller. Measurements were made in both the circumferential (tangential) and longitudinal (axial) directions wherever physically possible. Geometrical limitations restricted measurements inside the vane (position #4) to the circumferential direction only. Positive ψ angles (0°, 15°, 30°, 45°) were used in all cases. All operating parameters were held constant so that valid comparisons could be made between the different locations and between impellers.

RESULTS

A summary of the stress analysis results is listed in Table 1. An example of the computer-generated stress analysis $\sin^2\psi$ versus d graph is shown in Figure 4.

The stresses were all compressive and ranged from -17 to -106 ksi. Impeller #4 had the smallest range, from about -40 to -55 ksi, while the average range for the remaining impellers was from approximately -20 or -35 ksi to -70 or -80 ksi.

DISCUSSION

All measurements on these impellers showed compressive stresses at the surface. The large scatter in the data made it difficult to make simple comparisons between measurement locations and between impellers. Averaged values do suggest that the stresses cover a smaller range on

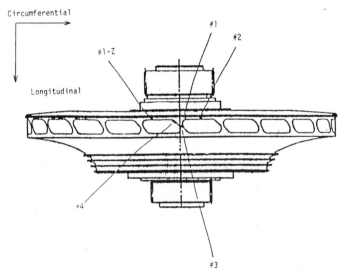

Figure 3. Typical Measurement Locations.

Table 1. Summary of Stress Analysis Results for Impellers #1, #2, #3, and #4

Locations	Direction	Average Stress, ksi			
		Impeller #1 (Test-Fired)	Impeller #2 (Spin Tested)	Impeller #3 (New)	Impeller #4 (Test-Fired)
#1, #1Z, #2	Longitudinal	-37.7 ± 12.1	-48.0 ± 20.2	-78.3 ± 24.6	-52.0 ± 21.4
	Circumferential	-50.9 ± 8.6	-24.4 ± 6.7	-37.1 ± 9.6	-54.3 ± 15.0
#3	Longitudinal*	-22.9	-67.1	-48.2	-40.8
	Circumferential*	-46.6	-19.0	-51.7	-55.8
#4	Circumferential*	-72.5	-77.6	-73.1	-39.4

*One measurement only

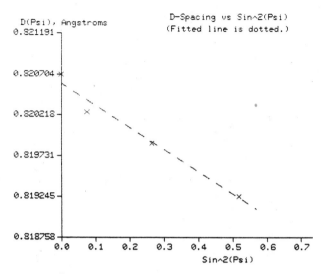

Figure 4. $\text{Sin}^2\psi$ versus d-spacing for Impeller #1, Position 2,
Tangential, $\sigma = -42.7 \pm 2.4$ ksi.

the impeller #4 compared with the others. There was no obvious correlation between stress value and measurement direction. Nor is there an
obvious indicator that residual stresses are different for the different
conditions tested. If the scatter in the data is assumed to be small
for position #4 (inside the vane) then there may be a difference in
stress between the impeller #4 and the others. It should be noted,
though, that only one measurement per impeller was made at position #4
and there is considerable scatter in the data at other positions.

Many of the d-spacing versus $\text{sin}^2\psi$ plots were linear. The non-
linear plots were attributed to texture in the impellers. There does
not appear to be a correlation between measurement location or direction
and plot linearity. This implies that there is no correlation between
physical location on the impeller and preferred orientation or grain
size.

CONCLUSIONS

Operational fatigue, as induced by significant amounts of operational time, appears to have had no affect on the residual stresses of
either of the two impeller configurations. In fact, the residual
stresses of impellers #1 and #4 were virtually no different than the
residual stresses of the new impeller before proof spinning and balance
grinding.

The impeller residual stresses were not significantly changed
during either of the fabrication and testing processes checked. Proof
spinning and balance grinding had no affect on the residual stresses when
compared to a new impeller before proof spinning and balance grinding.

COMPARISON OF RESIDUAL STRESS MEASUREMENT

AND BREAKING STRENGTH ON ALUMINA RODS

D.J. Snoha, C.O. Ruud, and H.D. Cassel

Materials Research Laboratory
The Pennsylvania State University
University Park, PA 16802

ABSTRACT

X-ray diffraction (XRD) utilizing copper K-alpha radiation and the single exposure technique (SET) was used for residual stress measurement on two different thermal treatments of Al-300 alumina rods. The measurements were performed in the axial and tangential directions on forty halves of twenty broken rods. The twenty rods were intially annealed for 8 hours at 1000°C; then ten were slow cooled and ten were equilibrated at 1450°C then force-air quenched. A fine mesh sapphire (alumina) powder was used as a zero stress standard for the determination of calibration validity of the position sensitive scintillation detector based (PSSD) stress measuring instrument used for this study. This instrument was developed at the Materials Research Laboratory of The Pennsylvania State University.

OBJECTIVE

The objective of this investigation was to demonstrate the viability of XRD stress analysis as a means for quantitatively confirming the improved residual stress condition due to thermal treatment of ceramic components, and to predict the breaking strength in alpha alumina rods from measured residual stresses.

EXPERIMENTAL PROCEDURES AND RESULTS

Two heat treatments of alpha-alumina rod specimens were selected for XRD residual stress measurement. These were forty halves of twenty broken rods, approximately 0.2 inches (5 mm) in diameter by 2.0 inches (51 mm) long [2]. Copper K-alpha radiation diffracted from the (1 0 16) crystallographic planes was chosen for this investigation because it provided the most rapid collection times and highest precision of the resultant data. Table 1 lists these as well as other parameters utilized for stress measurement on the annealed and quenched rods in the tangential and axial directions.

Stress readings over a specimen-to-detector distance (R_0 in Fig. 1) of 1.57 ± 0.08 inches (40 ± 2 mm) were performed on a −400 mesh sapphire powder to confirm proper instrument calibration. Precise control of R_0 is not necessary with the Penn State PSSD-based stress measuring instrument, thereby

43

Table 1.   Analysis Parameters for the Al-300 Rods

	Annealed		Quenched	
	Tangential	Axial	Tangential	Axial
Calibration Confirmation on Zero Stress Alpha-Alumina Powder in KSI (MPa)	-3 ± 3 (-21 ± 21)	1 ± 6 (7 ± 41)	-3 ± 3 (-21 ± 21)	2 ± 6 (14 ± 62)
Irradiated Area in inches (mm)	0.5 x 0.04 (13.0 x 1.0)	0.12 x 0.10 (3.0 x 2.5)	0.5 x 0.04 (13.0 x 1.0)	0.12 x 0.10 (3.0 x 2.5)
Full-Wave Rectified Power Supply Settings	40 KV 13 ma	40KV 16 ma	40KV 13 ma	40KV 16 ma
Data Collection Time in seconds	30	15	30	15
Average Three-Point-Bend Breaking Strength in lbF (N)	113 (504)		243 (1083)	
Characteristic Radiation, Miller Indices, Bragg Angle	Copper K-Alpha, (1 0 16), 75.3°			
Bulk Elastic Constants	$E = 46.0 \times 10^6$ psi (317×10^3 MPa), $\nu = 0.21$			

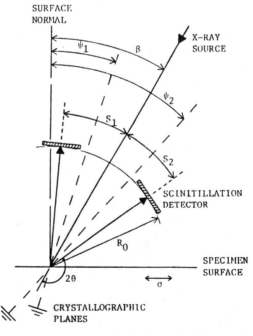

Fig. 1. Single Exposure Technique (SET) Arrangement.

Table 2. XRD Residual Stress and Three-Point-Bend
Breaking Strength Data from the Al-300 Rods

Rod Identification	Measured Tangential Stress in KSI	Measured Axial Stress in KSI	Breaking Strength in lb$_F$
Annealed, A1	18	-12	110
2	17	-25	113
3	20	-10	112
4	17	-27	102
5	23	-24	110
6	13	-20	114
7	18	-19	121
8	10	-21	131
9	17	-12	121
10	16	-16	100
Quenched, Q1	9	-32	274
2	5	-29	233
3	2	-31	154**
4	-2	-30	272
5	7	-40	232
6	8	-34	234
7*	13	-27	204
8	6	-26	216
9	2	-50	268
10	2	-59	258

*Only one half provided.
**Pre-cracked (data not included in plots)

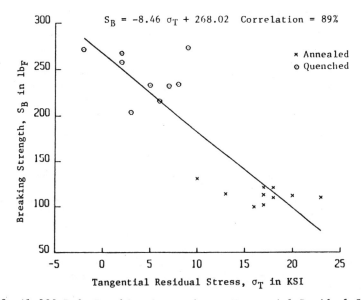

Fig. 2. Al-300 Rods Breaking Strength vs. Tangential Residual Stress.

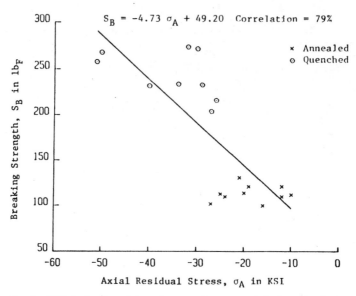

Fig. 3. Al-300 Rods Breaking Strength vs. Axial Residual Stress.

allowing for ease of application [3]. Table 1 shows the mean and standard
deviation of the readings obtained from the zero stress powder prior to and
following the residual stress measurements on the rods.

Tangential residual stress measurements were performed upon the broken
halves of the twenty specimens by locating a 0.5 x 0.04 inch (13.0 x 1.0 mm)
line-shaped irradiated area on the specimen such that the long dimension was
parallel to the rod axis and near the center of the broken half of the rod.
Measurements were made with the rod halves in a horizontial position and
rotating at 0.15 rps so as to obtain the average stress around the circumference. Table 2 lists the measured residual stress data obtained from the
Al-300 rods in the tangential and axial directions along with the three-
point-bend breaking strength data provided with the specimens. This breaking
strength data is plotted versus the tangential XRD data from both the
annealed and quenched rods in Fig. 2. A linear regression fit to the data
yields a correlation of 89%. The XRD data indicate the annealed and quenched
rods possessed an average tangential stress of about 17 KSI (117 MPa) and 5
KSI (34 MPa) respectively, indicating that the residual stresses on the
surface of the quenched rods were 12 KSI (83 MPa) more compressive than the
annealed.

Axial residual stress measurements were also performed on the
twenty broken halves. An almost square 0.12 x 0.10 inch (3.0 x 2.5 mm)
irradiated area was used with the slightly longer dimension located along
the rod axis, i.e., in the direction of the measured stress. The rods were
once again measured in the horizontal position and rotated at 0.15 rps. The
residual stress data obtained from the annealed and quenched rods in the
axial direction are listed in Table 2. Figure 3 shows a plot of the same
breaking strength data previously mentioned versus this measured axial data.
A linear regression fit to the data yields a correlation of 79%. The average
axial residual stress in the annealed and quenched rods was -18 KSI (-124

MPa) and -35 KSI (-241 MPa) respectively. Again, the residual stresses on the surface of the quenched rods were more compressive by 17 KSI (117 MPa) than on the surface of the annealed rods.

Analysis of the linear regression fit to the residual stress data show that the breaking strength of the rods can be predicted, within an approximately 80% confidence level, to within 32 and 43 lb_F respectively for the tangential and axial measurements. It is reasonable that the axial stress measurements would have provided a closer estimate of breaking strength than the tangential measurements since the rods were broken across the axial direction. This was not the case, and may be accounted for by the fact that the XRD measurements were made several centimeters remote from that portion of the rod that provided the breaking strength data. Further, there was considerable variance in the data from each end of the broken rod, an indication that the residual stress conditions of the rods were not particularly homogeneous.

The fair predictions of breaking strength from the tangential and axial residual stress data is an indication that XRD measurements, if performed in the region where subsequent breaking might occur, could be an excellent non-destructive testing method for ceramic components. Also, this study has provided a quantitative measurement (few of which have been undertaken, as noted by Cockbain [4]) of surface stress improvement due to thermal treatment; with unprecedented precision, speed, and ease of measurement. The proficiency of the measurements on ceramics reported herein may be compared to a sparse number of others performed with conventional XRD instrumentation [5, 6, 7].

CONCLUSIONS

1. X-ray diffraction residual stress analysis can be used to quantify the surface stress condition in alpha-alumina ceramics.

2. The diffraction peak from the (1 0 16) crystallographic planes provides for good precision and repeatability of XRD stress measurements.

3. Tangential and axial residual stress measurements predicted breaking strength within 32 and 43 lb_F respectively.

ACKNOWLEDGMENTS

The authors wish to thank Professor D.P.H. Hasselman of the Virginia Polytechnic Institute and State University for providing the alpha-alumina rods. Also, Dr. J.W. McCauley and C.P. Gazzara of the Army Materials and Mechanics Research Center (AMMRC) are thanked for their advice and encouragement. Portions of this investigation were performed under contract DAAG46-83-K-0036 from AMMRC, Watertown, MA.

REFERENCES

1. C.O. Ruud, "Position Sensitive Detector Improves X-Ray Powder Diffraction," Ind. Res. and Dev., pp. 84-87, Jan. (1983).
2. Y. Tree, A. Venkateswaran, and D.P.H. Hasselman, "Observations on the Fracture and Deformation Behavior During Annealing of Residually Stressed Polycrystalline Aluminum Oxides," J. Mat. Sci., 18:2135-2148 (1983).

3. C.O. Ruud and D.J. Snoha, "Displacement Errors in the Application of
       Portable X-Ray Diffraction Stress Measurement Instrumentation,"
       J. of Met., 36(2):32-38 (1984).
4. A.G. Cockbain, "Strain in Hot Pressed Dense Silicon Nitride," Proc.
       Brit. Cer. Soc. 25, 253-259 (1975).
5. D.K. Smith and S. Weissmann, "Residual Stress and Grain Deformation in
       Extruded Polycrystalline BeO Ceramics," J. Am. Cer. Soc. 51(6),
       330336 (1968).
6. M.R. James, D.J. Green, and F.F. Lange, "Determination of Residual
       Stresses in Transformation-Toughened Ceramics," Adv. in X-Ray
       Anal. 27, 221-228 (1984).
7. C.W. Semple, "Residual Stress Determination in Alumina Bodies," AMMRC
       TR 70-74, D/A Proj. IT062105A330, AMCMS Code 502E.11.296, AMMRC,
       June 1970.

X-RAY STRESS ANALYSIS OF NICKEL-PLATED COMPONENTS

USING DIFFERENT RADIATION WAVELENGTHS

R. H. McSwain

Naval Air Rework Facility
Naval Air Station
Pensacola, Florida   32508

E.B.S. Pardue, R. W. Hendricks, and M. V. Mathis

Technology for Energy Corporation
Knoxville, Tennessee   37922

INTRODUCTION

X-ray stress analyses were performed on seven (7) nickel-plated camshafts from helicopter transmissions. The purpose of this investigation was to determine if this residual stress technique was useful in determining the state of stress in load-bearing plated surfaces of critical reworked aircraft components. Specifically, it was desired to characterize six different stages of nickel plating processing. These were: as-plated, plated and baked, machined, machined and baked, ground, and ground and baked. Three different radiation wavelengths with different penetrations were used in this investigation. CrK_α, CoK_α, and CuK_α radiations were each used on all of the camshafts. These wavelengths were selected to determine if there were stress gradients in the plated layers. This paper presents and discusses the results of these studies.

PROCESSING HISTORY

The base material of the camshafts was 9310 steel. Any previous plating material was removed, the surface was rough ground and then stress relieved at $275°F \pm 25°F$ for 5 hours. The areas of the camshaft that were not to be plated were masked, and the camshafts were submerged in the nickel sulfamate electrolytic solution. Two plating conditions

49

were investigated--a normal condition requiring 50 A/ft^2 and a test
condition requiring 75 A/ft^2. In all cases the applied plating was
0.010 inches thick.

After plating, several of the camshafts were given post-processing
treatments. These treatments included: (1) baking; (2) baking followed
by machining to 0.005 inches thick; (3) baking, machining, and baking;
(4) baking and grinding to 0.005 inches thick; and (5) baking, grinding,
and baking. All baking operations were conducted at 275°F ± 25°F for
5 hours for the purpose of stress relieving the plating. The processing
conditions for each of the camshafts used in this study are summarized
in Table 1.

TECHNIQUE

The x-ray residual stress measurements were performed with the
Technology for Energy Corporation Models 1610 and 1620 Portable
Apparatus for Residual Stress (PARS) systems using the sin$^2\psi$, or
multiple tilt, technique. This instrumentation incorporates a low power
(100 W maximum) x-ray tube, a position-sensitive proportional counter
(PSPC), and a centerless diffractometer specially designed for multiple
ψ-angle measurements on large components.

For the first set of measurements CrK$_\alpha$ radiation was used to
analyze the (220) planes from nickel. A 4.0 mm diameter collimator was

Table 1. Processing Conditions for Hard Nickel-Plated Camshafts

Sample Number	Plating Condition (A/ft^2)	Further Processing
1	75	As plated and baked
2	50	As plated
3	50	As plated and baked
4	50	Ground
5	50	Ground and baked
6	50	Machined
7	50	Machined and baked

used, and the power level was 35 kV at 1.7 mA. Data were acquired at eight ψ-angles with an acquisition time of 150 s each.

CoK$_\alpha$ radiation was employed to obtain data from the (222) planes. A 2.0 mm diameter collimator was used, and the power level was 40 kV at 1.5 mA. Data were acquired at six ψ-angles with an acquisition time of 90 s each.

The (420) planes were investigated with CuK$_\alpha$ radiation. Six ψ-angles were used with acquisition times ranging from 60 s to 90 s each. A 2.0 mm diameter collimator was used in each measurement with power levels of 45 kV at 0.4 mA and 45 kV at 1.7 mA.

All stress measurements were made in the axial direction of the camshafts in the center of the plated region, and all were measured at the same location for each radiation.

RESULTS

Typical x-ray residual stress measurements of d versus $\sin^2\psi$ are shown for each x-ray wavelength in Figures 1-4. The residual stresses

Table 2. Longitudinal X-ray Residual Stress Measurements on Camshafts with Various X-ray Wavelengths

	Stress Measured with		
	CrK$_\alpha$	CoK$_\alpha$	CuK$_\alpha$
	Plating Penetration Depth, %		
Camshaft ID	10%	15%	30%
1. Plated and Baked (75 A/ft^2)	-112.3 ± 15.4	-110.6 ± 22.5	-119.3 ± 34.3
2. Plated	-74.3 ± 15.2	$+6.3 \pm 20.3$	-44.3 ± 23.9
3. Plated and Baked	-30.5 ± 18.8	$+17.2 \pm 27.0$	-26.6 ± 8.0
4. Ground	-12.4 ± 11.5	-5.6 ± 21.1	-6.2 ± 10.3
5. Ground and Baked	$+2.3 \pm 13.0$	-3.2 ± 26.4	-8.6 ± 10.2
6. Machined	-99.7 ± 14.7	-40.8 ± 16.1	-68.7 ± 18.6
7. Machined and Baked	-18.1 ± 21.0	-25.8 ± 32.7	-32.6 ± 17.9

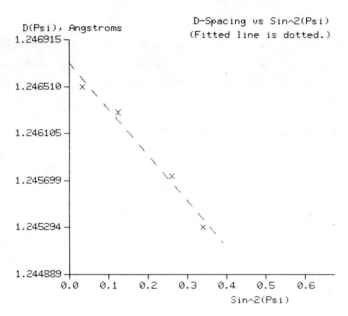

Figure 1. Machined plating, CrK$_\alpha$ radiation (σ = -99.7 ± 14.7 ksi).

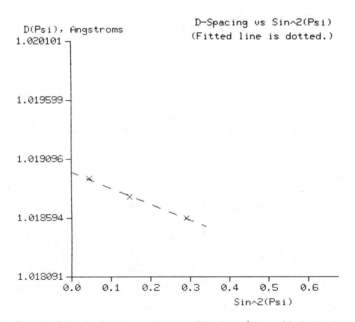

Figure 2. Machined plating, CoK$_\alpha$ radiation (σ = -40.8 ± 16.1 ksi).

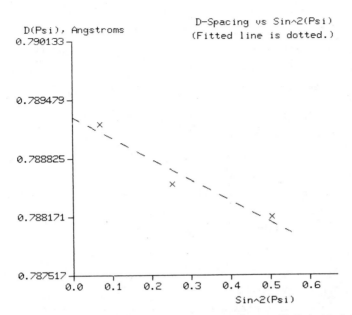

Figure 3. Machined plating, CuK$_\alpha$ radiation (σ = -68.7 $\pm$ 18.6 ksi).

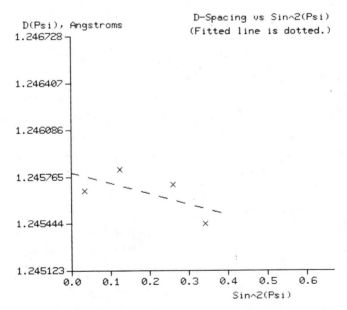

Figure 4. Machined plating followed by baking, CrK$_\alpha$ radiation
(σ = -18.1 $\pm$ 21.0 ksi).

and their standard deviations based on counting statistics and goodness-of-fit to the straight line are shown in Table 2. It is seen that these plots are relatively linear, as expected from the biaxial stress theory for data acquired with CoK_α and CuK_α; while, for the majority of the measurements with CrK_α radiation, there are oscillations in the curve (Figure 4). These are discussed in the next section.

All of the data for the CuK_α radiation tests indicate compressive stresses, whereas data for the CrK_α and CoK_α radiation tests show both compressive and tensile stresses. The only sample with a uniform stress for all three radiations is the one that was plated at 75 A/ft^2 and baked. In this case the stress was approximately -114 ksi (compressive stress). This was also the greatest magnitude of stress found.

DISCUSSION

The x-ray stress data obtained from nickel-plated camshafts are complex. The first and most detailed set of measurements was performed with CrK_α radiation which has a total depth of penetration of only 10% of the nickel deposition layer of 0.005 to 0.010 inches. Stresses deeper in the nickel deposit were sampled by using the harder (shorter wavelength) radiation of CoK_α and CuK_α. Note that the total penetration depth for CoK_α is about 15%, while for CuK_α it is about 30% of the plating thickness.

An examination of Table 2 suggests several interesting features. First, the stress measured on the 75 A/ft^2 sample is essentially the same for all three x-ray wavelengths. This suggests that this plating/baking condition produces a relatively homogeneous deposition layer which appears to have a high surface stress. What is puzzling is the fact that bulk nickel cannot support stresses anywhere near the magnitude indicated in Table 2 (-120 ksi). This implies that either* the x-ray elastic constants used in this study, which were those for bulk nickel, are substantially larger than those representative of plated nickel, or that there are d-spacing variations which result from effects other than stress (e.g., solid solubilities of impurities and/or interstitials). At this time, it is not possible to distinguish between these various effects. However, an independent measure of the

---

*Since the d-spacing is the direct observable in x-ray residual stress measurements and because we have calibrated our instruments against known standards, errors in d-spacing changes have been eliminated.

x-ray elastic constant in as-plated nickel (at 50 A/ft^2) indicates that
the elastic constant may be about 60% of that for bulk nickel. This is
insufficient to explain the stress values obtained at 75 A/ft^2.

For the remaining samples, all plated at 50 A/ft^2, the data
obtained with CrK$_\alpha$ radiation (least penetration) and CuK$_\alpha$ radiation
(deepest penetration) appear to be logical. The as-plated sample shows
compressive residual stress, which is reduced by about a factor of two
on baking. Machining produces a large compressive residual stress which
is reduced substantially by baking. Grinding appears to reduce the
compressive stresses remaining after the first baking operation. The
stresses (as determined at all wavelengths) in the ground samples and
the ground and baked samples are more tensile than the plated and baked
condition. The above trends are similar for data obtained with CoK$_\alpha$
radiation with the exception of the as-plated and as-plated and baked
samples. In these cases, the stresses were slightly tensile. The large
differences between the results obtained with three x-ray wavelengths
suggest that there may be significant stress gradients in the materials
plated at 50 A/ft^2, that the various plating conditions may not be
reproducible, and/or that there are impurity gradients in the plated
layers superimposed upon the d-spacing variations which result from
stress. Further compounding these problems was the observation of a
significant variation in diffracted intensity of CrK$_\alpha$ radiation as a
function of ψ-angle. This implies a significant preferred orientation
near the surface. That the effect was not observed with CoK$_\alpha$ radiation
is consistent with theory which predicts that (hhh) planes should be
independent of preferred orientation.

CONCLUSIONS

X-ray residual stress measurements in nickel-plated camshafts
produced complex results. It is suggested that unusually high "stress"
values obtained in materials prepared at 75 A/ft^2 may be the result of
d-spacing changes caused by impurity gradients in the plated layer.
Measurements for materials prepared with normal plating currents
(50 A/ft^2) yield stress results which are consistent with expectation.
Baking operations had a tendency to make the stresses less compressive
for each condition. Preferred orientation of grains was indicated in
most cases where CrK$_\alpha$ radiation was used. It is unknown at present if
these data are contaminated by an impurity effect as is hypothesized for
the 75 A/ft^2 material.

X-RAY STRESS DETERMINATION IN A SINGLE LAP JOINT

CONTAINING A DEBONDED REGION*

A. Lankford, C. S. Barrett and Paul Predecki

University of Denver

Denver, CO   80208

INTRODUCTION

XRD has been found to be a useful technique for investigating both surface[1] and interior[2,3] stresses in adhesive bonded joints. For the interior stresses, to gain access to the joint interface, adherends were chosen such that one adherend was relatively transparent to the X-radiation used and the other was not. Incident X-rays then penetrated the first adherend and the adhesive, and were diffracted from just below the surface of the second adherend.

In prior work[3] it was shown that the measured stresses due to an applied load agreed quite well with stresses calculated for the same joint using the TEXGAP-2D finite element program, except at one extremity of the bond. One explanation proposed for the discrepancy was that a small debond was present at this extremity. In the present study, therefore, an investigation was made of a joint containing an intentional debond at this extremity.

EXPERIMENTAL PROCEDURE

The method used for preparing the single lap-joint sample has been described elsewhere.[3] Be was used for the transparent adherend, FM-73M** for the adhesive and 5052 aluminum in the H32 condition (non-heat-treatable, cold-work-relieved, nominally at 345°C, with H32 temper as for sheet metalwork) for the opaque adherend. The 5052-H32 Al was further annealed 30 min. at 220°C. In this condition it was finer grained than the 6061-T6 used previously[3] and the annealing improved the $K\alpha_1\alpha_2$ peak resolution, Fig. 1.

A single lap joint sample was made from commercial 3.18 mm (1/8 in.)

* Work supported by the U.S. Army Research Office on Grant # DAAG29-81-0150.

**Rubber-modified epoxy with polyester mat, American Cyanamide Co., Havre de Grace, MD.

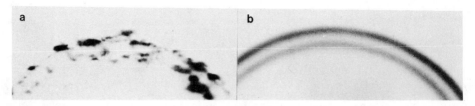

Fig. 1. Debye rings due to 333 + 511 CuK$\alpha_{1,2}$ from a 1 mm diameter irradiated spot. (a) From a specimen of Al alloy 6061 in the aged condition T6
as used in earlier studies[1,2] in which the large grains required specimen
oscillation. (b) From Al alloy 5052 in the solution-hardened (non-aged)
condition H32 followed by a stress relief anneal to sharpen the rings, as
used in this study. No graininess is seen after a 15 min. anneal at 310°C
or after a 30 min. anneal at 240°C or at 220°C (with rings identical in
appearance to Fig. 1(b)). Specimen stationary; oscillation unnecessary.

sheet milled along edges to dimensions shown in Fig. 2. A non-bonded
region was created by inserting a .025 mm thick teflon strip to a depth
of 1.59 mm from the end of the Be. After curing, the depth of the
debond was found to be only about 0.5 mm.

  X-ray diffraction measurements were made on this sample using the
same procedure developed earlier for non-debonded samples,[2] i.e. an omega
type diffractometer was used (psi tilts being about the 2θ axis) with a
diffracted beam monochromator and scintillation counter. The position of
the 511 + 333 peak from Al at $\sim$ 160.5°2θ with CuKα_1 radiation was determined by computerized step scanning and least squares fitting a parabola
to 5 steps, each 0.2° 2θ wide, over the top 50% of the α_1 peak.

  Attempts were made to increase the speed of peak determination by (1)
replacing the monochromator and scintillation detector with a position
sensitive detector (Braun wire type) or (2) using a monochromated incident
beam and the PSD to detect the diffracted beam. Both methods gave reduced

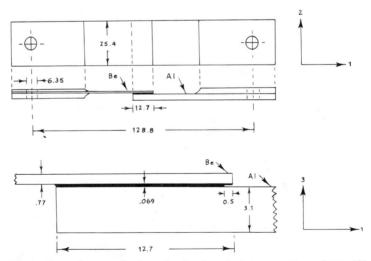

Fig. 2. Dimensions in mm of the single lap joint specimen (ARO-68) used
for stress determination. Tab thicknesses at the ends were such that the
central plane of the section at each clevis pin was coplanar with the
central plane of the 0.069 mm thick FM-73M adhesive. Debond (.5 mm) is
at end of Be.

peak to background ratios (Fig. 3) due to the strong Compton scattering
from the Be and adhesive, and were therefore abandoned.

Strain measurements were made along the 3' direction which was
related to the 1,2,3 specimen axes (Fig. 2) by the angles ϕ and ψ in the
usual manner.[4] ψ was the tilt of 3' from the 3 direction and ϕ the angle
between 1 and the projection of 3' onto the 1-2 plane. Measurements were
taken along the centerline of the bond at ϕ = 0 and ψ = 0, ± 45, and near
the bond extremities at ψ = 0, 30, 45. Beam width was 1.6 mm (.56 mm near
the bond extremities). In the ϕ = 90° orientation, psi values used were
ψ = ± 45 and the beam width was 1.6 mm. Measurements were made before and
after application of a 2669N (600 lb) load in the 1 direction.

Measurements were also made on the Al adherend prior to bonding to
obtain "stress free" values of the peak position for all the ϕ and ψ values

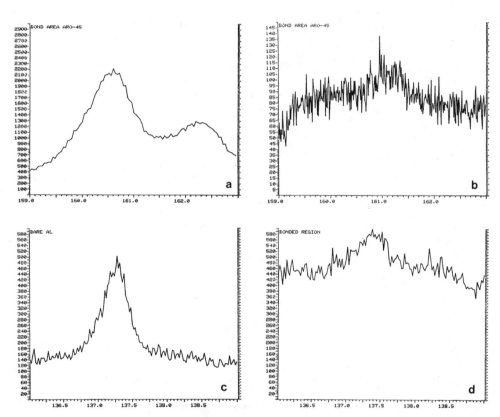

Fig. 3. Profiles of 5052-H32, stress relieved 30 min. at 220°C with ¼°
divergence slit, 0.024 in. receiving slit at 173 mm from specimen. (a)
With diffracted beam graphite monochromator on Siemens Crystalloflex IV
with 0.04° 2θ step scanning, 2 sec counts, scintillation counter. Profile
after penetrating Be and adhesive. (b) With Braun position sensitive
detector at 173 mm from specimen, same divergence slit, after penetrating
Be and adhesive. (c) Without Be or adhesive, 422 CuKα₁ reflection ∿ 137.5°
2θ with Huber Guinier system, Braun PSD tangent to Guinier circle of
radius ∿ 100 mm. (d) With beam penetrating Be and adhesive, Huber Guinier
system as in (c).

listed above. All peak positions were corrected for Lorentz polarization and absorption and were normalized to 25°C. Sample temperature was held constant to ± .1°C using a thermostated hot air blower and a specimen enclosure.

RESULTS AND DISCUSSION

The peak positions were converted to residual and net applied stresses using the same equations, assumptions and elastic constants used for the non-debonded sample.[2,3]

The curing stresses obtained are shown in Figs. 4 and 5. These stress distributions are generally similar to those obtained without the

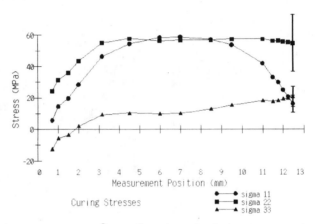

Fig. 4. Curing stresses $\sigma_{11}{}^c$, $\sigma_{22}{}^c$ and $\sigma_{33}{}^c$ along the bonded area of specimen ARO-68. The standard deviation decreases with decreasing stress.

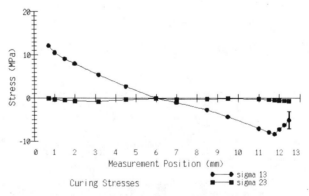

Fig. 5. Curing stress $\sigma_{13}{}^c$ and $\sigma_{23}{}^c$ along the bonded area of ARO-68. The standard deviation decreases with decreasing stress.

debond,[2] except for curing stresses $\sigma_{11}{}^c$ and $\sigma_{33}{}^c$. Stress $\sigma_{11}{}^c$ is about
30% higher and $\sigma_{33}{}^c$ about a factor of 2 higher at the debond, near the end
of the Be adherend, than were obtained in this region without the debond.
The difference, particularly in $\sigma_{33}{}^c$, probably results from the stress
concentration produced by the debond. There is not much evidence of slope
changes at the debond in Figs. 4 and 5 due to the small debond depth
(0.5 mm) comparable to the beam width (.56 mm).

The net stresses due to the 2669N load applied are shown in Figs. 6
and 7. Again the stress distributions are similar to those obtained without a debond present[3] except that an abrupt slope change is now evident in
all distributions at the edge of the debond. Also the maximum values of
the stresses σ_{11}, σ_{22} and σ_{33} at this edge are higher by about 38, 23 and
10% respectively than the corresponding stresses a distance of ~ 0.5 mm
from the end of the Be in the non-debonded sample.[3]

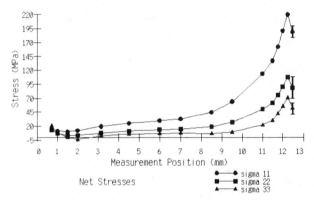

Fig. 6. Net stresses σ_{11}, σ_{22} and σ_{33} due to applied load on ARO-68 of
2669N (600 lbs force), i.e. stresses under load minus stresses in cured,
unloaded specimen at each position. The standard deviation decreases
with decreasing stress.

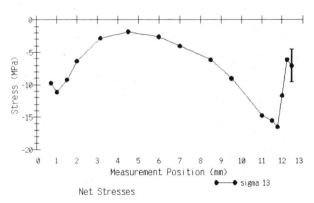

Fig. 7. Net shear stresses σ_{13} in ARO-68 due to applied load of 2669N.
The standard deviation decreases with decreasing stress.

Since the elastic stresses near a crack tip and in the crack plane increase as $(a/2r)^{\frac{1}{2}}$, where a is crack depth (debond depth) and r the distance from the crack tip,[5] the observed increases due to the debond are reasonable assuming a = 0.5 mm and the effective value of r where the stress measurements are made is estimated to be in the range .1 to .3 mm. (A more accurate comparison is not possible because of the large beam width). However, for the non-debonded sample the ratio of the measured to the calculated applied stresses is about a factor of 2 for σ_{22} and a factor of 5 for σ_{33}.[3] This would require an "a" value > .8 mm if it were due to a debond. Such a large "a" value would have been detected as an abrupt slope change in the stress distributions, as seen for example in Fig. 6.

CONCLUSIONS

We conclude that:

(1) Debonds are detectable as abrupt slope changes in the applied stress distributions provided the debond depth is roughly equal to or greater than the beam width. Debonds are also detectable in the residual stress distributions, but less readily so.

(2) Debonds produce measureable increases in applied stresses in the vicinity of the debond edge.

(3) The discrepancy between the measured and calculated stresses in a non-debonded sample[3] are probably too large to be accounted for by an accidental debond.

(4) Position-sensitive detectors, either with or without a monochromator in the primary beam, provide no advantage for the types of measurements made in this study.

REFERENCES

1. C. S. Barrett and Paul Predecki, X-ray Diffraction Evaluation of Adhesive Bonds and Stress Measurements with Diffracting Paint, Adv. in X-ray Anal. 24:231-238 (1981).
2. Paul Predecki and C. S. Barrett, Stress Determination in an Adhesive Bonded Joint by X-ray Diffraction, Adv. in X-ray Anal. 27:251-260 (1984).
3. Paul Predecki, C. S. Barrett, A. B. Lankford and D. Gutierrez-Lemini, Stresses in an Adhesive Bond at an Adhesive/Adhesive Interface Under Load, J. Adhesion (1986) in press (12 pages).
4. See, for example, J. B. Cohen, H. Dolle and M. R. James in Natl. Bureau of Standards Special Pub. 567: "Accuracy in Powder Diffraction." S. Block and C. R. Hubbard, eds. pp. 453-477 (1980).
5. See, for example, D. Broek, "Elementary Engineering Fracture Mechanics." Martinus Nijhoff, p. 9 (1982).

X-RAY FRACTOGRAPHY ON FATIGUE FRACTURE SURFACES OF

AISI 4340 STEEL[*]

Yukio Hirose[**], Zenjiro Yajima[***] and Toshio Mura

Department of Civil Engineering and Material Research Center,
Northwestern University, Evanston, ILL., 60201, U.S.A.

** On leave from Kanazawa University, Japan
*** On leave from Kanazawa Institute of Technology, Japan

INTRODUCTION

The X-ray fractographic technique was applied to fatigue fracture surfaces of tempered AISI 4340 steel. Residual stresses and half-value breadths were measured by the X-ray diffraction.

In the present paper, the residual stresses and plastic strains on fatigue fracture surfaces and some parameters in the fracture mechanics were investigated. A simple model of mechanics was proposed to explain these experimental results.

A SIMPLE MECHANICAL MODEL FOR RESIDUAL STRESSES ON FRACTURE SURFACES

Residual stresses beneath a fatigue fracture surface are caussed by plastic strain ε^p which is a residual (plastic) compressive strain ($\varepsilon^p < 0$). For mathematical simplicity, the plastic strain is assumed to be uniformly distributed in the hatched domain in Fig. 1. The residual stress σ_z is assumed to be constant σ^p in h and constant σ in H as shown in Fig. 1(d). Then

$$\sigma^p h + \sigma H = 0 \qquad\qquad (1)$$

where h is the width of the plastic domain and H is the remaining part. Assuming the uniform strain in the z direction to be ε, we can write

$$\sigma^p = (\varepsilon - \varepsilon^p)E, \qquad \text{in } y \leq h \qquad\qquad (2)$$

$$\sigma = \varepsilon E, \qquad \text{in } h \leq y \leq H \qquad\qquad (3)$$

where E is Young's modulus.

---

* This work was supported by NSF-MRL Grant DMR-8216972 and partly by U.S. Army Research Grant DAAG-29-85-K-0134.

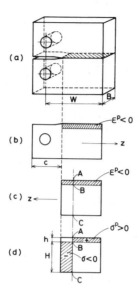

Fig. 1.   Test specimen and simple mechanical model.

When σ and ε are eliminated from the above three equations, we have

$$\sigma^p = -\varepsilon^p E / \{\ 1 + (\ h/c\ )(\ c/H\ )\ \} \tag{4}$$

where c is an arbitrary constant. When c is taken as the crack length, the BCS model[1] gives

$$h/c = (\pi^2/8\ )(\ \Delta K/\sigma_Y\ )^2 \tag{5}$$

where ΔK is the range of stress intensity factor and σ_Y is the yield strength.

Substituting (5) into (4), we have

$$\sigma^p = -\varepsilon^p E / \{\ 1 + (\ \pi^2/8\ )(\ \Delta K/\sigma_Y\ )^2\ \} \tag{6}$$

where c/H has been taken as a unit for simplicity. When the stress ratio $R = \sigma_{min}/\sigma_{max}$ is introduced, (6) becomes

$$\sigma^p = -\varepsilon^p E / \{\ 1 + (\ \pi^2/8\ )(\ 1 - R\ )^2 (\ K_{max}/\sigma_Y\ )^2\ \} \tag{7}$$

where K_{max} is the maximum stress intensity factor.

SPECIMEN AND EXPERIMENTAL PROCEDURES

The material used in the experiments was AISI 4340 steel (wt% : 0.39C, 0.28Si, 0.74Mn, 1.38Ni, 0.78Cr, 0.23Mo and 0.14Cu) which was normalized at 880°C for 1 hour, then austenized at 850°C for 1 hour and oil quenched. The tempering processes were conducted at 200°C and 400°C for 2 hours respectively. The mechanical properties are shown in Table 1.

The test pieces were C.T. specimens in Fig. 1 with W=50.8 mm and B=15 mm. An electro-hydraulic closed loop testing machine was used and the crack length was measured by a traveling microscope with sensitivity of 0.01 mm.

Table 1. Mechanical Properties of Test Material

Tempering temperature	Yield strength	Tensile strength
T.T.(°C)	σ_Y(MPa)	σ_B(MPa)
200	1530	1880
600	951	1050

The fatigue tests were conducted under constant range of stress intensity factors (constant ΔK). For the constant ΔK fatigue tests the load was decreased stepwise to maintain the ΔK value within 3% limits. The crack closure was measured by the unloading compliance technique to calculate the effective stress intensity factor ΔK_{eff}. Fatigue crack propagation tests were conducted under stress ratios R of 0.05 and 0.5.

The residual stresses and half-value breadths of X-ray diffraction profiles were measured at and beneath the fracture surfaces by a parallel beam X-ray stress measurement system. The method of measurement applied was the standard $\sin^2\psi$ method[2] by using parallel beams of $Cr-K_\alpha$ X-ray as described in previous papers.[3,4] The area irradiated by X-ray was of 1 mm width and 8 mm length at the middle of the fracture surfaces. The distribution of the residual stresses and the half-value breadths in the depth direction y were measured by removing the surface layer successively by electro-polishing.

In parallel with the above tests, tensile tests of the same material were carried out to obtain the correlation between the X-ray line broadening and plastic strains.[5]

RESULTS

Figure 2 shows the relation between the crack growth rates and the range of stress intensity factor ΔK. The relation between the crack growth rates and the effective range of stress intensity factor ΔK_{eff} is also indicated in Fig. 2 by solid symbols. The effect of stress ratio R was observed near the threshold stress intensity factor for the 600°C tempered specimen. It can be said that a higher R gives a faster crack growth rate. For the 200°C tempered specimens the effect of R was not observed. It is interesting to note that the effect of R on the crack growth disappeared when ΔK_{eff} was used in stead of ΔK. ΔK_{eff} was calculated from the compliance technique.

Figure 3 shows the relation between the maximum stress intensity factor K_{max} and the residual stresses on fracture surfaces. The residual stresses was tension on fatigue fracture surfaces. The residual stress increased with K_{max} for the 200°C tempered specimens, while it has a maximum value at about K_{max}=30 MPa√m for the 600°C tempered specimens.

It is known that the residual stress on the fracture surfaces of mild steel in fatigue tests has the tendency to decrease with increasing K_{max} after K_{max}=30 MPa√m.[6] The effect of stress ratio R is observed clearly in Fig. 3. At a given K_{max} value, the residual stresses were higher for the higher R.

Figures 4 and 5 show the distribution of residual stresses and half-value breadths near the fracture surfaces. The residual stress in the

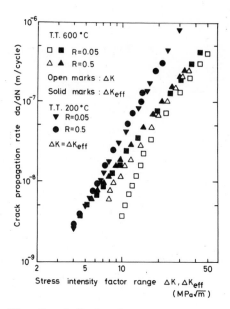

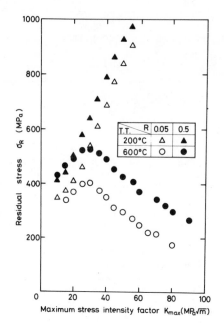

Fig. 2. Relation between crack
propagation rate and
stress intensity
factor range.

Fig. 3. Relation between residual
stress and maximum stress
intensity factor.

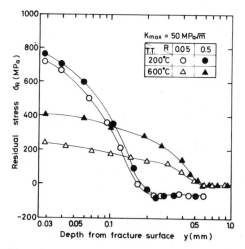

Fig. 4. Residual stress distribution
near fracture surface.

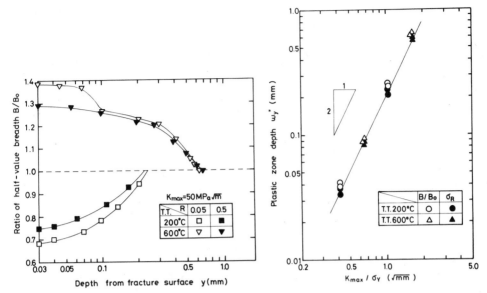

Fig. 5.   Half-value breadth distribution   Fig. 6.   Relation between plastic
          near fracture surface.   zone depth and stress
                                                         intensity factor divided
                                                         by yield strength.

vicinity of the fracture surface was tension. The tensile residual stresses
increased just inside the fracture surfaces and then gradually diminished.
The tensile residual stress is caused by the local plastic deformation near
the crack tip.[7] The drop of value of the residual stress at $y=0$ may be
reduced to the stress relief by the surface roughness.[7] The half-value
breadths measured near the fracture surfaces were larger than the initial
value (B_o=1.9 deg) measured before the fatigue test in the 600°C tempered
specimens. On the other hand, the half-value breadths decreased near the
fracture surfaces in the 200°C tempered specimens.

The plastic zone ω_y^* was obtained as the value of y in Fig. 5 where
B/B_o=1 and also the value of y in Fig. 4 where the stresses change the sign.
In Fig. 6, the depth was plotted against the maximum stress intensity factor
K_{max} divided by the yield strength σ_Y.

The value of ω_y^* is proportional to the square of K_{max}/σ_Y, i.e.

$$\omega_y^* = \alpha \ (K_{max}/\sigma_Y)^2 \tag{8}$$

where α is 0.19. The proportional constant α obtained by the fracture
test of the same material was 0.13[8] which agreed with the finite element
analysis for perfect plastic material by Levy et al.[9]

DISCUSSION

Figure 7 shows the relation between the half-value breadth ratio B/B_o
and the von Mises equivalent plastic strain ε_f obtained from tensile tests,
where B_o is the initial half-value breadth measured before the tests
(B_o=5.7×10^{-2}rad, 1.9×10^{-2}rad for 200°C and 600°C tempered specimens,
respectively) and B the value after deformation. In the 200°C tempered
specimens, B/B_o had the minimum value near ε_f=1%. On the other hand, B/B_o

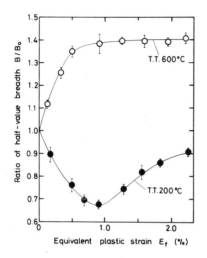

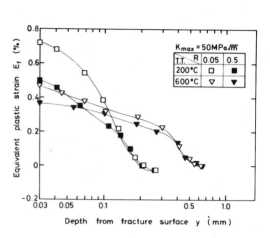

Fig. 7.   Relation between half-
          value breadth and
          plastic strain.

Fig. 8.   Plastic strain distribution
          near fracture surface.

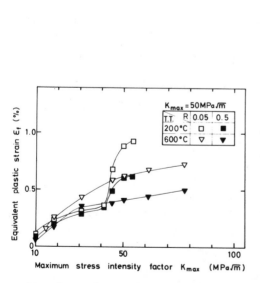

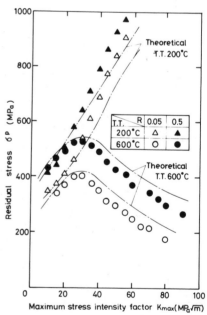

Fig. 9.   Relation between plastic
          strain and maximum stress
          intensity factor.

Fig. 10. Relation between residual
          stress and maximum stress
          intensity factor.

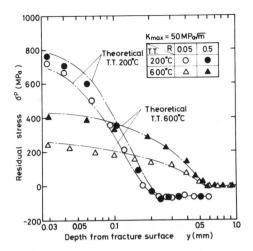

Fig. 11. Residual stress distribution near fracture surface.

increased up to about ε_f=0.5% at first and then saturated in the 600°C
tempered specimens.

    The plastic strain ε_f on the fatigue fracture surface can be estimated
from the measurement of the half-value breadth by using Figs. 7 and 5.
The value of ε_f is expected to correspond with the maximum strain build up
in the material at the crack tip before fatigue fracture.

    Figures 8 and 9 show the relation between ε_f and y and K_{max}. For the
200°C tempered specimen ε_f decreased with y in Fig. 8 and increased gradually
with K_{max} and then increased suddenly over K_{max}=45 MPa$\sqrt{m}$ in Fig. 9. On the
other hand, for the 600°C tempered specimen ε_f gradually decreased with y in
Fig. 8 and increased with K_{max} in Fig. 9.

    When ε^p in Eq (7) is regarded as ε_f, we can calculate σ^p from Eq (7).
Figures 10 and 11 show the theoretical value of residual stress.
Here E=2×10^6 MPa was taken. The theoretical value agreed well with the
experimental data. The effect of R could be well explained by Eq (7).

CONCLUSION

    The main results obtained in the present study are summarized as
follows:

    (1) The residual stresses on fatigue fracture surfaces were tension.
The residual stresses increased with K_{max} for the 200°C tempered specimens,
while it has a maximum value at about K_{max}=30 MPa$\sqrt{m}$ for the 600°C tempered
specimens. A simple model of mechanics was proposed to explain these
experimental results and the model agreed with well the experimental data.

    (2) The maximum depth of the plastic zone ω_y^* determined from residual
stresses and half-value breadths distribution was correlated to K_{max} and the
yield strength σ_Y as follows

$$\omega_y^* = 0.19 \ (\ K_{max}/\sigma_Y \ )^2$$

The width of X-ray diffraction profiles was suggested to be the appropriate
parameter for plastic zone size determination as well as for the residual
stress.

REFERENCES

1. B. A. Bilby, A. H. Cottrel and K. H. Swinden, "The Spread of Plastic
    Yield from a Notch," Proc. Roy. Soc., A272:304 (1963).
2. Committee on Mechanical Behavior of Materials, "Standard Method of
    X-Ray Stress Measurement," J. Soci. Mat. Sci. Jap., 13 (1973).
3. Y. Hirose, Z. Yajima, and K. Tanaka, "X-Ray Fractography on Stress
    Corrosion Cracking of High Strength Steel,"
    Advances in X-Ray Analysis, 27:213 (1984).
4. Y. Hirose, Z. Yajima, and K. Tanaka, "X-Ray Fractographic Approach to
    Fracture Toughness of AISI 4340 Steel,"
    Advances in X-Ray Analysis, 28:289 (1985).
5. Z. Yajima, Y. Hirose, and K. Tanaka, "X-Ray Diffraction Observation
    of Fracture Surfaces of Ductile Cast Iron,"
    Advances in X-Ray Analysis, 26:291 (1983).
6. Y. Sekita, S. Kodama, and H. Misawa, "X-Ray Fractography on Fatigue
    Fractured Surface," J. Soci. Mat. Sci. Jap., 32:258 (1983).
7. K. Tanaka, and N. Hatanaka, "Residual Stress Near Fatigue Fracture
    Surfaces of High Strength and Mild Steels Measured by X-Ray Method,"
    J. Soci. Mat. Sci. Jap., 31:215 (1982).
8. Z. Yajima, Y. Hirose, and K. Tanaka, "X-Ray Diffraction Observation
    of Fractured Surface of Fracture Toughness Specimen of High Strength
    Steel," J. Jap. Soc. Strength and Fracture of Materials, 16:59 (1981).
9. N. Levy, P. V. Marcal, W. J. Ostergren, and J. R. Rice, "Small Scal
    Yielding Near a Crack in Plane Strain: A Finite Element Analysis,"
    Int. J. Frac. Mech., 7:143 (1971).

INTERNAL STRAIN (STRESS) IN AN SiC/Al PARTICLE-REINFORCED COMPOSITE

H. M. Ledbetter and M. W. Austin

Fracture and Deformation Division
Institute for Materials Science and Engineering
National Bureau of Standards, Boulder, Colorado 80303

ABSTRACT

Silicon carbide and 6061 aluminum alloy possess very different thermal-expansion coefficients: 3.3 and $22.5 \cdot 10^{-6} K^{-1}$, respectively. Thus, one expects large internal strains and stresses in these composites because the two constituents form interfacial bonds at high temperatures and are cooled to ambient temperatures. From a simple elastic model, one expects a hydrostatic tensile stress in the aluminum matrix and a hydrostatic compressive stress in the silicon-carbide particles. Using conventional diffraction geometry, using Cu $K\alpha$ radiation, we studied three surfaces of a plate specimen. For both phases, we determined the unit-cell dimensions for two situations: unmixed and mixed in the final composite. The silicon-carbide particles showed a compressive stress and the aluminum matrix a tensile stress, seventy-five percent of the yield strength. Measurements show that both stress tensors are approximately hydrostatic.

INTRODUCTION

This study considers the internal strain and accompanying stress in a two-phase material: an aluminum-alloy matrix reinforced with silicon-carbide particles. The stress arises thermally from three conditions: (1) a coherent particle-matrix interface, (2) composite consolidation at a high temperature, and (3) dissimilar particle-matrix thermal expansivities, α. A small temperature change generates large stress. For example, the temperature decrease change, ΔT, required to cause a hydrostatic tensile stress equal to the aluminum-alloy yield strength, σ_y^m, is only

$$\Delta T = \sigma_y^m / 3 B^m (\alpha^m - \alpha^p) = 64 \text{ K} \qquad (1)$$

where B denotes bulk modulus, superscript m matrix, and superscript p particle. Here, we use the ambient-temperature physical properties: $B^m = 74.9$ GPa, $\alpha^p = 3.3 \cdot 10^{-6} K^{-1}$, $\alpha^m = 22.5 \cdot 10^{-6} K^{-1}$, and $\sigma_y^m = 274$ MPa.

Previous studies on internal stress in composites include those by Barrett and Predecki [1], Tsai et al. [2], and Arsenault and Taya [3].

Barrett and Predecki studied stress in noncrystalline polymers and in poly-
mer-matrix composites by inserting (during material manufacture) fine powders
of diffracting materials such as aluminum. Tsai et al. studied a graphite-
fiber-reinforced aluminum. In the aluminum, they detected a tensile stress
ranging from 33 to 228 MPa. Arsenault and Taya studied a silicon-carbide-
fiber-reinforced aluminum and reported compressive stress in the aluminum.

The present study differs from previous ones principally in studying a
particle-reinforced composite and obtaining internal strains for both the
occluded phase and the matrix phase.

X-RAY DIFFRACTION ANALYSIS

In 1976, Evenschor and Hauk [4,5] gave a general expression for the
lattice strain:

$$\varepsilon'_{33} = \frac{d_{\phi\psi} - d_o}{d_o} \tag{2}$$

$$= S_1[\sigma_{11} + \sigma_{22} + \sigma_{33}] + (S_2/2)[\sigma_{11}\cos^2\phi + \sigma_{12}\sin2\phi + \sigma_{22}\sin^2\phi]\sin^2\psi$$

$$+ (S_2/2)\sigma_{33}\cos^2\psi + (S_2/2)[\sigma_{13}\cos\phi + \sigma_{23}\sin\phi]\sin2\psi.$$

Here $d_{\phi\psi}$ denotes interplanar spacing for the direction $\phi\psi$, d_o denotes spacing
for the nonstressed case, and S_1 and $S_2/2$ denote the x-ray elastic constants.
Note that equation (2) contains nine terms. Older models [9] contain only
six terms because they omit all the σ_{i3} stress components.

For the usual Bragg-Brentano parafocusing x-ray diffraction geometry,
$\phi = \psi = 0$ and equation (2) becomes

$$\varepsilon_{33} = S_1(\sigma_{11} + \sigma_{22} + \sigma_{33}) + (S_2/2)\sigma_{33}. \tag{3}$$

Assuming elastic isotropy,

$$S_1 = S_{12} = -\nu/E \tag{4}$$

and

$$S_2/2 = S_{11} - S_{12} = 2S_{44} = (1 + \nu)/E \tag{5}$$

where S_{ij} denote the Voigt-notation elastic compliances, E denotes Young
modulus, and ν Poisson ratio. Thus,

$$\varepsilon_{33} = (1/E)[\sigma_{33} - \nu(\sigma_{11} + \sigma_{22})]. \tag{6}$$

If the stress is hydrostatic ($\sigma_{11} = \sigma_{22} = \sigma_{33}$), then

$$\varepsilon_{33} = [(1 - 2\nu)/E]\sigma_{33} = (B/3)\sigma_{33} \tag{7}$$

where B denotes the bulk modulus.

For a $\sin^2\psi$-type analysis [8], choosing scanning in the x_1-x_3 plane so that $\phi = 0$, equation (2) becomes

$$\epsilon'_{33} = S_1(\sigma_{11} + \sigma_{22} + \sigma_{33}) + (S_2/2)\sigma_{33}$$
$$+ (S_2/2)(\sigma_{11} - \sigma_{33})\sin^2\psi + (S_2/2)\sigma_{13}\sin 2\psi. \qquad (8)$$

Then, the slope of equation (8) equals

$$\frac{\partial\epsilon'_{33}}{\partial\sin^2\psi} = \epsilon_{11} - \epsilon_{33} + 2\epsilon_{13}\cot 2\psi. \qquad (9)$$

If no shear strains exist, then a graph of ϵ'_{33} versus $\sin^2\psi$ appears linear with slope

$$\frac{\partial\epsilon'_{33}}{\partial\sin^2\psi} = \epsilon_{11} - \epsilon_{33} = (S_2/2)(\sigma_{11} - \sigma_{33}). \qquad (10)$$

One can obtain the shear strains, ϵ_{ij}, by considering equation (8) for both plus and minus ψ at $\psi = 45°$. One obtains

$$\Delta\epsilon'_{33} = \epsilon'_{33}(+\psi) - \epsilon'_{33}(-\psi) = 2\epsilon_{13}. \qquad (11)$$

From whence:

$$\sigma_{13} = [E/(1 + \nu)]\epsilon_{13}. \qquad (12)$$

Similar relations follow accordingly for σ_{12} and σ_{23}.

EXPERIMENT

A.   Material

We obtained materials from a commercial supplier in the form of 1-cm plates. The supplier started with commercially available Aℓ-6061 and SiC particles. These powders were blended, compacted, and sintered to produce billets measuring 25 x 30 x 4 cm. Billets were hot rolled at 700-783 K with a reduction per pass ranging from 10 to 50 percent. Rolled plates were subjected to a standard T6 heat treatment: solution treat at 800 K (980°F) for 2 h, water quench, age at 436 K (325°F) for 18 h. The plate contained nominally 30 volume percent SiC. We chose the coordinate system x_3 = plate normal, x_1 = rolling direction, x_2 = in-plate, perpendicular to rolling direction. Figure 1 shows a photomicrograph of the composite.

B.   Measurements

The measuring apparatus consisted of a commercial horizontal Bragg-Brentano-focusing computer-automated diffractometer with a 22-cm diffracto-meter radius. The diffractometer contained a scintillation counter and a monochromator, a curved-graphite (Johann) type, located between the specimen and the counter. We used Cu Kα radiation excited at 45 kV and 40 mA.

Fig. 1.   Photomicrograph of SiC-particle-reinforced aluminum-alloy-matrix
           composite. Particle volume fraction equals thirty percent. Parti-
           cles range in size up to 5 μm.

     We prepared flat specimen surfaces by chemically polishing the com-
posite with a mixture of phosphoric (6:9), sulphuric (2.25:9), and nitric
(1:9) acids for 15 minutes at the boiling point. Typically 0.1 mm of surface
was removed from the studied surface. Three orthogonal faces were studied
corresponding to principal directions in the sample.

     Aluminum lines 111 through 422 and discernible silicon-carbide lines
[10i(i = 1, 2, 4, 7, 9), 208, 209, 213, 10.15, 219, 20.15] were scanned in
0.01-degree steps for 60 seconds. We obtained peak positions by fitting to
a Pearson type-VII function:

$$y = I_{max}\left[1 + \frac{(\theta - \theta_{max})^2}{2W^2}\right]^{-m} \qquad (13)$$

where I_{max} denotes maximum intensity; θ, diffraction angle; and W, full-peak
width at half-maximum intensity. We chose m = 2, making $y(\theta)$ a Lorentzian.

     Both Cu $K\alpha_1$ and $K\alpha_2$ lines were measured using a deconvolution proce-
dure, and the measured lattice spacings were least-square fitted versus the
$\cos\theta\cot\theta$ extrapolation function.

     In both the incident beam and the diffracted beam, we used 1-mm hori-
zontal Soller-slit packs with 1.5-degree vertical divergence slits and
0.2-mm receiving slits.

     The uncontrolled temperature in the diffractometer chamber varied
from 26 to 29°C. For aluminum, a variation of ±1.5°C produces an error of
Δa = ±0.00014 Å, too large for careful study, but not fatal to the present
one and its conclusions. The present study found Δa's ranging up to 0.0044,
thirty times larger than the temperature-induced error.

RESULTS

Table 1 contains the study's principal results. For both the face-centered-cubic aluminum matrix and the hexagonal inclusion, Table 1 shows unit-cell dimensions, a and c; the associated strains $\varepsilon = \Delta a/a$; the volume strain; and the stress estimated from

$$\sigma = B\Delta V/V \tag{14}$$

where $B^P = 223.4$ GPa and $B^m = 74.9$ GPa.

Table 1. For both matrix and particles: unit-cell dimensions, strains, and calculated stresses

	Matrix				Particle					
Face	a	ε	$\Delta V/V$	σ_{ii}	a	c	ε_a	ε_c	$\Delta V/V$	σ_{ii}
	(A)	(10^{-3})	(10^{-3})	(MPa)	(A)	(A)	(10^{-3})	(10^{-3})	(10^{-3})	(MPa)
x	4.0532	0.896	2.692	202	3.0805	15.1171	-0.357	-0.479	-1.193	-266
	4.0536	0.990	2.974	223	3.0808	15.1180	-0.282	-0.418	-0.982	-219
y	4.0532	0.894	2.684	201	3.0804	15.1195	-0.409	-0.321	-1.139	-254
	4.0540	1.077	3.233	242	3.0806	15.1217	-0.308	-0.176	-0.792	-177
z	4.0498	0.047	0.141	11	3.0809	15.1198	-0.247	-0.298	-0.791	-177
	4.0495	-0.035	-0.104	- 8	3.0807	15.1217	-0.308	-0.176	-0.792	-177
	4.0510	0.343	1.030	77	3.0806	15.1208	-0.334	-0.237	-0.905	-202
Ref.	4.0496				3.0816	15.1243				

Figure 2 shows a $\Delta d/d_o$-versus-$\sin^2\psi$ diagram for one of the six principal diffraction geometries. Table 2 contains numerical results for $\sigma_{ii}-\sigma_{jj}$, differences in principal stress components, and for shear stresses $\sigma_{ij}(i \neq j)$. We obtained these from results in Fig. 4 and its companion figures.

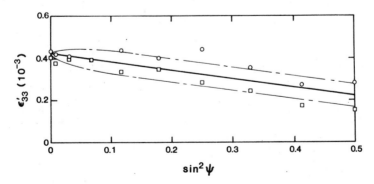

Fig. 2.   Lattice strain ε'_{33} versus $\sin^2\psi$ for the 420 reflection from the aluminum matrix. This diagram represents diffraction from the x_1 surface scanned in the x_1-x_2 plane. Thus, it corresponds to the stress $\sigma_{22} - \sigma_{11}$.

Table 2. For the matrix phase, results of $\sin^2\psi$ analysis.

ij (10^{-3})	$\varepsilon_{ii} - \varepsilon_{jj}$ (MPa)	$\sigma_{ii} - \sigma_{jj}$ (10^{-3})	ε_{ij} (MPa)	σ_{ij}
31	-0.827	-44	-0.060	-3.2
21	-0.403	-21	-0.063	-3.4
23	-0.459	25	-0.051	-2.7

From results in Tables 1 and 2, we construct the following stress tensors, represented as three-by-three matrices, for the matrix and the particle:

$$\overline{\sigma}^m_{ij} = \begin{bmatrix} 228 & -3 & -3 \\ -3 & 207 & -3 \\ -3 & -3 & 182 \end{bmatrix}, \rightarrow \begin{bmatrix} 206 & 0 & 0 \\ 0 & 206 & 0 \\ 0 & 0 & 206 \end{bmatrix} \quad (15)$$

and

$$\overline{\sigma}^p_{ij} = \begin{bmatrix} -243 & 0 & 0 \\ 0 & -248 & 0 \\ 0 & 0 & -185 \end{bmatrix} \rightarrow \begin{bmatrix} -219 & 0 & 0 \\ 0 & -219 & 0 \\ 0 & 0 & -219 \end{bmatrix} \quad (16)$$

Matrices on the left represent measurements. Matrices on the right represent
the imposed constraint of hydrostatic stress. In the particle case, we did
not measure the shear strains or stresses, ε_{ij} or σ_{ij} ($i \neq j$). In the matrix
case, we neglected the z-face results in Table 1; these results disagree
dramatically with our other measurements; through further study, we hope to
resolve this dilemma. We emphasize that equations (15) and (16) refer to a
cartesian (measurement) coordinate system and represent the average over a
volume containing very many particles.

DISCUSSION

First, we note that $\overline{\sigma}^m_{ij}$ is approximately a hydrostatic tensile stress.
Approximately, $\overline{\sigma}^m_{ij} = \sigma\delta_{ij}$ where σ denotes a scalar stress and δ_{ij} the
Kronecker delta. An elastic sphere-in-hole model predicts such a stress.
Although such stress states occur infrequently in composites, they do occur.
Garmong [7] remarked that "sintered powder alloys may be the best examples of
such structures."

Second, we caution that $\overline{\sigma}^m_{ij}$ represents a macroscopic average of a
volume large relative to particle size. Furthermore, $\overline{\sigma}^m_{ij}$ represents an
average over all directions. Locally, one should describe the stress in
spherical coordinates. Taking $r = x_1$, $\theta = x_2$, and $\phi = x_3$, then

$$\bar{\sigma}^m_{r\theta\phi} = \begin{bmatrix} \bar{\sigma}^m_{rr} & 0 & 0 \\ 0 & \bar{\sigma}^m_{\theta\theta} & 0 \\ 0 & 0 & \bar{\sigma}^m_{\phi\phi} \end{bmatrix} . \tag{17}$$

(Here the bar over sigma means an average value between r_0 and R, where r_0 is the radius of the nonconstrained particle, and R represents the radius of the total-volume sphere.) One can use several approaches to obtain numerical values for $\bar{\sigma}^m_{rr}$ and $\bar{\sigma}^m_{\theta\theta} = \bar{\sigma}^m_{\phi\phi}$. Here, we consider two. First, one can calculate the average stress

$$\bar{\sigma}^m_{rr} = \int_{r_0}^{R} \sigma_{rr}dr / \int_{r_0}^{R} dr$$

$$= \frac{4G^m C}{R^3} \left[1 - \frac{R}{2r_0} - \frac{R^2}{2r_0^2} \right] . \tag{18}$$

Similarly,

$$\bar{\sigma}^m_{\theta\theta} = \bar{\sigma}^m_{\phi\phi} = \frac{4G^m C}{R^3} \left[1 + \frac{R}{r_0} + \frac{R^2}{4r_0^2} \right]. \tag{19}$$

Thus, if one knows R/r_0, one can calculate $\bar{\sigma}^m_{rr}$ and $\bar{\sigma}^m_{\theta\theta}$. We estimate the upper bound on r_0/R (and probably a good estimate of r_0/R) by assuming a close-packed arrangement of spheres. Then,

$$(r_0/R)^3 = (6/\sqrt{2}\pi)f = 0.405, \tag{20}$$

where f denotes the particle volume fraction, 0.3. This gives $r_0/R = 0.74$ and (in MPa)

$$\bar{\sigma}^m_{r\theta\phi} = \begin{bmatrix} -121 & 0 & 0 \\ 0 & 369 & 0 \\ 0 & 0 & 369 \end{bmatrix} . \tag{21}$$

In a second approach, we require mechanical equilibrium at the particle-matrix interface for the radial stress component:

$$\bar{\sigma}^m_{rr} = \bar{\sigma}^p_{rr} = -219 \text{ MPa}. \tag{22}$$

At the particle-matrix interface this gives (in MPa)

$$\bar{\sigma}^m_{r\theta\phi} = \begin{bmatrix} -219 & 0 & 0 \\ 0 & 419 & 0 \\ 0 & 0 & 419 \end{bmatrix} , \tag{23}$$

where we require that trace σ = trace $\bar{\sigma}$, which one can prove for the elastic case. Equations (21) and (23), derived by different physical assumptions, show surprising compatibility.

Third, we remark that stress components in equations (21) and (23) exceed the matrix yield strength, 274 MPa. This suggests severe local work hardening in the matrix, especially near the particle-matrix interface.

CONCLUSIONS

From this study, four conclusions emerged:

1.   Conventional Bragg-Brentano-focusing x-ray diffraction shows that the matrix exists in tension, approximately hydrostatic, on average 206 MPa, 75 percent of the matrix yield strength.

2.   Similar diffraction measurements show that the particles exist in compression, on average 219 MPa. These two results fail to satisfy mechanical equilibrium.

3.   Measurements by the $\sin^2\psi$ method provide two results that support an approximately hydrostatic matrix stress: small differences between the diagonal stress components; negligibly small shear strains.

4.   Transforming the stress tensor into spherical-polar coordinates reveals stress components ($\bar{\sigma}^m_{\theta\theta} = \bar{\sigma}^m_{\phi\phi}$) that exceed the matrix yield strength.

ACKNOWLEDGMENT

Ming Lei, guest scientist from the Institute for Metals Research, Shenyang, People's Republic of China, contributed several calculations and a critical reading.

REFERENCES

1.   C. S. Barrett and P. Predecki, Adv. X-Ray Anal. 23, 331-332 (1980). See references therein.

2.   S.-D. Tsai, D. Mahulikar, H. L. Marcus, I. C. Noyan, and J. B. Cohen, Mater. Sci. Eng. 47, 145-149 (1981).

3.   R. J. Arsenault and M. Taya, Proceedings ICCM-V (San Diego, July 1985), forthcoming.

4.   P. D. Evenshor and V. Hauk, Z. Metallk. 66, 164-168 (1975).

5.   H. Dölle, J. Appl. Cryst. 12, 489-501 (1979).

6.   C. S. Barrett and T. B. Massalski, Structure of Metals (McGraw-Hill, New York, 1966), pp. 466-485.

ERRORS DUE TO COUNTING STATISTICS IN THE TRIAXIAL STRAIN (STRESS) TENSOR DETERMINED BY DIFFRACTION

P. Rudnik and J. B. Cohen

Department of Materials Science and Engineering
The Technological Institute
Northwestern University
Evanston, IL 60201

INTRODUCTION

Knowledge of the errors in a diffraction measurement of residual strains and stresses is useful information, not only in its own right, but also because it permits automation of a measurement to an operator-specified precision.[1] There are three sources of these errors:

1) Instrumental effects, primarily due to sample displacement, separation of the θ and 2θ axes of the diffractometer, and beam divergence. All three can be estimated[2]; or minimized by employing parallel beam geometry.[3]

2) Uncertainties in x-ray elastic constants, which can now be evaluated.[4]

3) Errors in the diffraction peak position related to counting statistics. Equations to evaluate this source have been developed in Ref. 1 for the case of a stress state for which all σ_{i3} (i = 1,2,3) = 0, with the direction "3" normal to the sample surface, see Fig. 1. This means that the stresses lie only in the surface, e.g., a biaxial stress state $\begin{vmatrix} \sigma_{11} & \sigma_{12} \\ \sigma_{12} & \sigma_{22} \end{vmatrix}$. There are, however,

numerous situations when the normal components are appreciable in an x-ray measurement[5,6] and this is generally the case for neutron diffraction because with neutrons the beam can sample a sizeable volume, at a significant depth below the surface[7]. It

79

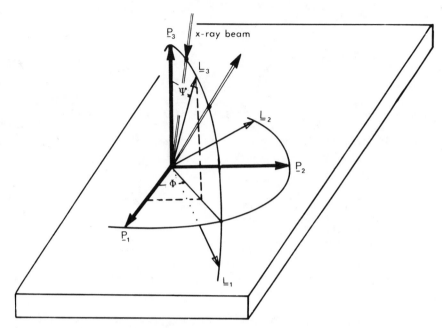

FIG. 1:   The axial system. Strains are measured with diffraction by
measuring the change in spacing of planes normal to the L_3
direction. (The axes P_i define the specimen surface.)

TABLE I: STRESS TENSORS (AND STANDARD DEVIATIONS)
FOR SPECIMEN C3,* REF. 5

DATA SET 1

539.74	(161.94)	-24.03	(78.81)	-39.15	(4.58)
		552.04	(161.66)	2.30	(3.56)
				78.29	(130.57)

DATA SET 2

520.60	(137.25)	-4.03	(66.60)	-34.17	(3.21)
		555.19	(137.03)	0.11	(2.69)
				82.20	(110.67)

DATA SET 3

535.03	(158.28)	-20.13	(77.06)	-40.19	(5.72)
		555.98	(157.99)	-0.98	(4.56)
				86.66	(127.69)

DATA SET 4

538.53	(146.23)	-30.63	(70.95)	-38.03	(3.83)
		565.37	(146.14)	0.76	(3.89)
				88.18	(117.92)

AVERAGE

532.95	-19.69	-37.89
	556.65	0.55
		83.83

REFERENCE 5

541	-20	-38
	565	1
		86

*   values given in MPa; $V(d_0)^{\frac{1}{2}} = 0.00016$ Å

is the purpose of this paper to derive equations to evaluate the counting
statistical error for the entire three dimensional strain (or stress) tensor,

$$\begin{vmatrix} \sigma_{11} & \sigma_{12} & \sigma_{13} \\ \sigma_{12} & \sigma_{22} & \sigma_{23} \\ \sigma_{13} & \sigma_{23} & \sigma_{33} \end{vmatrix}.$$

BASIC EQUATIONS

We begin with the general equation for the strains (ε_{ij}) and how these
affect the interplanar spacing "d". (Refer to Fig. 1 for the axial system.)
The measurement is made in the Φ direction, with a sample tilted ψ from the
normal position (which is with the surface normal bisecting incident and
scattered beams). Primed quantities refer to strains in the L_i co-ordinate
system, unprimed terms are in the P_i system.

$$<\varepsilon'_{33}>_{\Phi\psi} = (d_{\Phi\psi} - d_0)/d_0 = [<\varepsilon_{11}> \cos^2\Phi + <\varepsilon_{22}> \sin^2\Phi + <\varepsilon_{12}> \sin2\Phi$$

$$-<\varepsilon_{33}>] \sin^2\psi + <\varepsilon_{33}> + [<\varepsilon_{13}> \cos\Phi + <\varepsilon_{23}> \sin\Phi]\sin2\psi \qquad (1)$$

Note that the stress-free spacing, d_0, is involved. While this term
can be eliminated for a biaxial stress state, this is not possible for a
general strain or stress tensor, and the reader may consult Ref. 8 for a
discussion of problems associated with the measurement of this quantity.
When ε_{23} or $\varepsilon_{13} \neq 0$, ε'_{33} is not linear with $\sin^2\psi$ and has different curvature
for $+\psi$ and $-\psi$. The carats imply that the strain values are averaged over the
depth of penetration of the incident x-ray (neutron) beam and this is to be
understood in what follows, as this additional notation is eliminated below.

Next, we define terms which involve measurements of $d_{\Phi,\psi}$ at plus and
minus ψ tilts of the surface normal.[5]

$$a_1 \equiv 1/2[\varepsilon'_{\Phi\psi+} + \varepsilon'_{\Phi\psi-}] = \{d_{\Phi\psi+} + d_{\Phi\psi-})/2d_0 \}-1$$

$$= \varepsilon_{33} + [\varepsilon_{11} \cos^2\Phi + \varepsilon_{22} \sin^2\Phi + \varepsilon_{12} \sin2\Phi - \varepsilon_{33}] \sin^2\psi \qquad (2a)$$

Clearly, a_1 should be linear with $\sin^2\psi$, and ε_{33} is the intercept, regardless
of Φ.

$$a_2 \equiv 1/2[\varepsilon_{\Phi\psi+} - \varepsilon_{\Phi\psi-}] = (d_{\Phi\psi+} - d_{\Phi,\psi-})/2d_0$$

$$= [\varepsilon_{13} \cos\Phi + \varepsilon_{23} \sin\Phi]\sin|2\psi|. \qquad (2b)$$

Therefore, a_2 is linear vs. $\sin|2\phi|$.

Let:

$$\ell_1 = da_1/d\sin^2\psi,\tag{3a}$$

$$\ell_2 = da_2/d\sin|2\phi|\tag{3b}$$

Then, at: $\Phi = 0°$, $_0\ell_1 = \varepsilon_{11} - \varepsilon_{33}$,

$\Phi = 90°$, $_{90}\ell_1 = \varepsilon_{22} - \varepsilon_{33}$,

$\Phi = 45°$, $_{45}\ell_1 = 1/2(\varepsilon_{11} + \varepsilon_{22}) + \varepsilon_{12} - \varepsilon_{33}$,

$$= \varepsilon_{12} + 1/2(_0\ell_1 + _{90}\ell_1).\tag{3c}$$

and similarly:

at $\Phi = 0°$: $_0\ell_2 = \varepsilon_{13}$,

at $\Phi = 90°$: $_{90}\ell_2 = \varepsilon_{23}$. $\qquad\qquad$ (3d)

Knowledge of the strain tensor permits the calculation of the stress components (σ_{ij}) from:

$$\sigma_{ij} = [1/2S_2(hk\ell)]^{-1}\{\varepsilon_{ij} - \delta_{ij}\{S_1(hk\ell)/[3S_1(hk\ell)$$

$$+ 1/2S_2(hk\ell)]\}\cdot[\varepsilon_{11} + \varepsilon_{22} + \varepsilon_{33}]\}.\tag{4}$$

Here, δ_{ij} is the Kronecker delta function and S_1 and $1/2S_2$ are the x-ray elastic constants which depend on the indices of the diffraction peak, $hk\ell$. (For an isotropic solid these values are $-\nu/E$ and $(1 + \nu)/E$ respectively.)

VARIANCES DUE TO COUNTING STATISTICS

For a function $X = f(x_1, x_2, x_3 \ldots)$, assuming the x_n are independent, the variance (V) is[9]:

$$V(X) = (\frac{dX}{dx_1})^2 V(x_1) + (\frac{dX}{dx_2})^2 V(x_2) + (\frac{dX}{dx_3})^2 V(x_3) + \ldots\tag{5}$$

For the straight line, $y = mx + b$, the slope and intercept is given by:

$$m = \frac{\sum_i (x_i - \bar{x})(y_i - \bar{y})}{\sum_i (x_i - \bar{x})^2}\tag{6a}$$

$$b = (\Sigma y_i - m\Sigma x_i)/N\tag{6b}$$

where N is the number of data points.

Employing Eq. (5):

$$V(m) = \left[\frac{\sum\limits_i (y_i - \bar{y})}{\sum\limits_i (x_i - \bar{x})^2}\right]^2 V(x) + \left[\frac{\sum\limits_i (x_i - \bar{x})}{\sum\limits_i (x_i - \bar{x})^2}\right]^2 V(y) \qquad (6c)$$

Therefore:

$$V(b) = \frac{\sum(x_i - \bar{x})^2}{N} \cdot V(m) = \frac{\sum(x_i - \bar{x})^2}{N} \cdot \left[\frac{\sum(y_i - \bar{y})}{\sum(x_i - \bar{x})^2}\right]^2 V(x)$$

$$+ \left[\frac{\sum\limits_i (x_i - \bar{x})}{\sum(x_i - \bar{x})^2}\right]^2 V(y)\} \qquad (6d)$$

Therefore, in terms of a_1 vs. $\sin^2 \psi$:

$$V(\ell_1) = \left[\frac{\sum\limits_i (a_{1\,i} - \bar{a}_1)}{\sum\limits_i (\sin^2 \psi_i - \overline{\sin^2 \psi})^2}\right]^2 V(\sin^2 \psi)$$

$$+ \left[\frac{\sum\limits_i (\sin^2 \psi_i - \overline{\sin^2 \psi})}{\sum\limits_i (\sin^2 \psi_i - \overline{\sin^2 \psi})^2}\right]^2 V(a_1) \qquad (7)$$

The variance in ψ is negligible, so the first term can be ignored.

Also, from Eq. (2a):

$$V(a_1) = \left[\frac{da_1}{d(d_{\Phi\psi+})}\right]^2 V(d_{\Phi\psi+}) + \left[\frac{da_1}{d(d_{\Phi\psi-})}\right]^2 V(d_{\Phi\psi-}) + \left[\frac{da_1}{d(d_0)}\right]^2 V(d_0) \qquad (8)$$

Writing Bragg's law in the form $d = \frac{\lambda}{2\sin\theta}$, adopting the convention that θ^+, θ^- are the θ values (in degrees) for the peaks at $+\psi$, $-\psi$ respectively, and employing Eq. (5):

$$V(d_{\Phi,\psi+}) = (\pi/180)^2 (\lambda\cos\theta^+/2\sin^2\theta^+)^2 \, V(2\theta^+)/2 \qquad (9)$$

and similarly for $V(d_{\Phi,\psi-})$. Recalling Eq. (2a):

$$\left[\frac{da_1}{d(d_{\Phi\psi+})}\right]^2 = \left[\frac{da_1}{d(d_{\Phi\psi-})}\right]^2 = \frac{1}{4d_0^2}, \qquad (10a)$$

$$\frac{da_1}{d(d_0)} = \frac{[d_{\Phi\psi+} + d_{\Phi\psi-}]^2}{4d_0{}^4} = d_+{}^2/4d_0{}^4 \ . \tag{10b}$$

Thus, we may rewrite Eq. (7):

$$V(\ell_1) = \left[\frac{\sum_i(\sin^2\psi_i - \overline{\sin^2\psi})}{\sum_i(\sin^2\psi_i - \overline{\sin^2\psi})^2} \right]^2 \frac{1}{4d_0{}^2}\{(\pi/180)^2\ \frac{\lambda^2}{8}\ [(\frac{\cos\theta^2}{\sin^2\theta^+})^2 V_i(2\theta^+)$$

$$+ (\frac{\cos\theta^-}{\sin^2\theta^-})^2\ V_i(2\theta^-)] + (d_+^2/d_0^2)V(d_0)\} \tag{11}$$

In a similar manner for a_2 vs. $\sin|2\psi|$, where $a_2 \equiv (d_{\Phi\psi-} - d_{\Phi\psi-})/2d_0 = d^-/2d_0$:

$$V(\ell_2) = \left[\frac{\sum_i \sin|2\psi_i| - \overline{\sin|2\psi|})}{\sum_i(\sin|2\psi_i| - \sin|2\psi|)^2} \right]^2 \frac{1}{4d_0{}^2}\{(\frac{\pi}{180})^2\ (\frac{\lambda^2}{8})[(\frac{\cos\theta^-}{\sin^2\theta^-})^2 V_i(2\theta^+)$$

$$+ (\frac{\cos\theta^-}{\sin^2\theta^-})^2\ V_i(2\theta^-)] + (d_-^2/d_0^2)\ V(d_0)\} \tag{12}$$

We now propagate these values into the strain and stress tensors.

THE STRAIN TENSOR

Abbreviating the intercept of a_1 vs. $\sin^2\psi$ as I, then at any Φ:

$$\varepsilon_{33} = I \ (\text{of } a_1 \ \text{vs.} \ \sin^2\psi), \tag{13a}$$

$$V(\varepsilon_{33}) = V(\ell_1) + V(I) \tag{13b}$$

$$V(I) = \frac{\sum(\sin^2\psi_i - \overline{\sin^2\psi})^2}{N} \cdot \left[\frac{\sum(\sin^2\psi_i - \overline{\sin^2\psi})}{\sum(\sin^2\psi_i - \overline{\sin^2\psi})^2} \right]^2 V(a_{1\ i})$$

$$= \frac{1}{N} \frac{[\sum(\sin^2\psi_i - \overline{\sin^2\psi})]^2}{\sum(\sin^2\psi_i - \sin^2\psi)^2} V(a_{1\ i}) \tag{13c}$$

Now, from Eqs. (3c), at $\Phi = 0°$:

$$_0\ell_1 = \varepsilon_{11} - \varepsilon_{33} = \varepsilon_{11} - I, \tag{14}$$

$$V(\varepsilon_{11}) = 2V(_0\ell_1) + V(I).$$

Similarly, for $\Phi = 90°$:

$$\varepsilon_{22} = {}_{90}\ell_1 + I,$$

$$V(\varepsilon_{22}) = V({}_{90}\ell_1) + V({}_0\ell_1) + V(I),$$ (15)

and for $\Phi = 45°$:

$$\varepsilon_{12} = {}_{45}\ell_1 + \varepsilon_{33} - 0.5 (\varepsilon_{11} + \varepsilon_{22})$$

$$= {}_{45}\ell_1 - 0.5 ({}_0\ell_1 + {}_{90}\ell_1),$$

$$V(\varepsilon_{12}) = V({}_{45}\ell_1) + 0.25 [V({}_0\ell_1) + V({}_{90}\ell_1)].$$ (16)

From Eqs. (3d):

$$V(\varepsilon_{13}) = V({}_0\ell_2),$$ (17)

$$V(\varepsilon_{23}) = V({}_{90}\ell_2).$$ (18)

THE STRESS TENSOR

We define $Q = S_1/(3S_1 + 1/2S_2)$ (which is $[{}^{-\nu}/1-2\nu]$ for an isotropic solid). Then Eq. (4) may be written, for the diagonal stress components, as:

$$\sigma_{ij} = (1/2S_2)^{-1}[(1-Q)\varepsilon_{ii} - Q\varepsilon_{kk} - Q\varepsilon_{jj}].$$ (19)

Here $i = 1,2,3$; $j = 2,3,1$; $k = 3,1,2$.

From Eq. (19):

$$V(\sigma_{ii})^{\frac{1}{2}} = (1/2S_2)^{-1} \{(1-Q)^2 V(\varepsilon_{ii}) + Q^2[V(\varepsilon_{kk}) + V(\varepsilon_{jj})]\}^{\frac{1}{2}}.$$ (20)

Therefore, with Eqs. (13-15):

$$V(\sigma_{11})^{\frac{1}{2}} = (1/2S_2)^{-1} \{(2-4Q + 4Q^2)V({}_0\ell_1) + Q^2 V({}_{90}\ell_1)$$
$$+ (1-2Q + 3Q^2) V(I)\}^{\frac{1}{2}},$$ (21)

$$V(\sigma_{22})^{\frac{1}{2}} = (1/2S_2)^{-1} \{(1-2Q + 4Q^2)V({}_0\ell_1) + (1-2Q + Q^2)V({}_{90}\ell_1)$$
$$+ (1-2Q + 3Q^2)V(I)\}^{\frac{1}{2}},$$ (22)

$$V(\sigma_{33})^{\frac{1}{2}} = (1/2S_2)^{-1}\{(1-2Q + 4Q^2)V({}_0\ell_1) + Q^2 V({}_{90}\ell_1)$$
$$+ (1-2Q + 3Q^2)V(I)\}^{\frac{1}{2}}$$ (23)

Similarly:

$$V(\sigma_{12})^{\frac{1}{2}} = (1/2S_2)^{-1}[V(_{45}\ell_1) + 0.25 [V(_0\ell_1) + V(_{90}\ell_1)]^{\frac{1}{2}}, \tag{24}$$

$$V(\sigma_{13})^{\frac{1}{2}} = (1/2S_2)^{-1}V(_0\ell_2)^{\frac{1}{2}}, \tag{25}$$

$$V(\sigma_{23})^{\frac{1}{2}} = (1/2S_2)^{-1}V(_{90}\ell_2)^{\frac{1}{2}}. \tag{26}$$

EXAMPLES

To illustrate the typical magnitudes of the errors due to counting statistics, we employed data from Ref. 5, for a ground steel specimen, that is we used the peak positions and the variances in these positions with Eq. (9). [Formulae to calculate this variance for the parabolic fit employed in Ref. 5 are given in Ref. 1; for other types of fits the appropriate equation may be substituted.] The resultant errors are given in Tables I-III. For the first two tables it was assumed that the error in $d_{\Phi,\psi}$ was the actual measured value. If there is no preferred orientation, the intensity of the peak changes little with the ψ-tilt. In this case, Tables I and II show the effect of the uncertainty in d_0; reducing this error all the stress components by the same proportion, except σ_{13}, σ_{23}, which remain relatively unaffected, because the role of the error in d_0 is damped by $(d-)^2$ in Eq. (12).

If there is preferred orientation, the peak intensity can vary greatly with ψ and there will be large variances contributing to $V(\ell_1)$ from weak peaks. This was minimized in the following way. The average variance, σ_i, in the 2θ peak position for $+\psi$ and $-\psi$ was obtained and the weighting factor c_i was formed:

$$c_i = (1/\sigma_i^2) / \sum_i (1/\sigma_i^2) \tag{27}$$

The Eqs. 11 and 12 were then altered to multiply $V_i(2\theta^+)$, $V_i(2\theta^-)$ terms by this weighting for Table III. There is only a small difference (between Tables II and III) because of the lack of texture in the specimen; the peak intensity changed only by about 8 pct with ψ. With more severe preferred orientation the effect will be larger.

TABLE II: STRESS TENSOR AND STANDARD DEVIATIONS WHEN
$V(d_0)^{\frac{1}{2}} = 0.00004$ Å *

		DATA SET 1			
539.74	(48.24)	-24.03	(24.50)	-39.15	(4.58)
		552.16	(47.26)	2.30	(3.56)
				80.41	(38.96)
		DATA SET 2			
520.60	(40.52)	-4.03	(19.95)	-34.17	(3.21)
		555.19	(39.73)	0.11	(2.69)
				82.20	(32.75)
		DATA SET 3			
535.03	(47.81)	-20.14	(24.35)	-40.19	(5.72)
		555.98	(46.81)	-0.98	(4.56)
				86.66	(38.84)
		DATA SET 4			
538.53	(42.84)	-30.63	(21.10)	-38.03	(3.83)
		565.37	(42.51)	0.76	(3.89)
				88.18	(34.67)

*values given in MPa

TABLE III: WEIGHTED STRESS TENSOR AND STANDARD DEVIATIONS*

		DATA SET 1			
536.24	(48.56)	-24.62	(24.56)	-38.33	(4.59)
		554.65	(47.60)	2.90	(3.65)
				80.68	(39.22)
		DATA SET 2			
520.29	(41.84)	-6.04	(20.11)	-34.82	(3.55)
		560.22	(40.60)	1.56	(2.73)
				85.29	(33.76)
		DATA SET 3			
532.56	(48.15)	-7.90	(24.93)	-39.66	(5.81)
		549.83	(47.16)	-2.81	(4.57)
				82.21	(39.16)
		DATA SET 4			
539.23	(42.88)	-31.22	(21.23)	-38.28	(3.85)
		565.17	(42.65)	-0.59	(3.94)
				88.98	(34.68)

* $V(d_0)^{\frac{1}{2}} = 0.00004$ Å; values given in MPa

CONCLUDING REMARKS

There are now adequate equations for calculating errors in stress measurements due to instrumental effects, counting statistics and in the x-ray elastic constants. We would like to conclude this paper with a plea to the community making stress measurements via diffraction to regularly report these errors with their data. It is all too common for investigators to repeat a measurement (of stress or an elastic constant) once and to use the difference as an error estimate. Another practice is to dust a stress-free powder on the specimen surface and to use a (single) measurement of the stresses measured with this powder as an error estimate. Finally, some report an error in a slope vs. $\sin^2 \psi$ obtained by least-squares, but ignore the uncertainty in each point in this plot in estimating errors. None of these procedures is particularly satisfying in a statistical sense. Of course, if time permits, the average of, say, ten repetitions of a measurement is the best of all error estimates. If this cannot be done, error estimates from the available equations are far more satisfactory than the currently all - too common procedures.

ACKNOWLEDGEMENTS

This research was supported by ONR under contract No. N00014-80-C116. We thank Dr. I. C. Noyan for his advice.

REFERENCES

1) M. R. James and J. B. Cohen, Study of the Precision of X-ray Stress Analysis, Adv. X-ray Analysis, 20:291(1977).

2) R. H. Marion, "X-ray Stress Analysis in Plastically Deformed Metals", Ph.D. Thesis, Northwestern University (1973).

3) S. Taira, T. Abe and T. Ehiro, X-ray Study of Surface Residual Stress Produced in Fatigue Process of Annealed Metals, Bull J.S.M.E., 12:53, 947(1969).

4) K. Perry, I. C. Noyan, P. J. Rudnik and J. B. Cohen, "The Measurement of Elastic Constants for the Determination of Stresses by X-rays", Adv. X-ray Analysis, 27: 159(1984).

5) H. Dolle and J. B. Cohen, Residual Stresses in Ground Steels, Met. Trans., 11A:831(1980).

6) I. C. Noyan, Equilibrium Conditions for the Average Stress Measured by X-rays, Met. Trans., 14A:1907(1983).

7) A. D. Krawitz, J. E. Brune and M. J. Schmank, Measurement of Stress in the Interior or of Solids with Neutrons in: "Residual Stresses and Stress Relaxation", E. Kula and V. Weiss eds., Plenum Press New York (1982).

8) I. C. Noyan, Determination of the Unstressed Lattice Parameter, "a_0" for (Triaxial) Residual Stress Determined by X-rays, Adv. X-ray Analysis, 28:281-288 (1985).

9) O. Davies and P. Goldsmith, Statistical Methods in Research and Production, Hafner Publ. Co., New York (1952).

PROFILE FITTING IN RESIDUAL STRESS DETERMINATION

T. J. Devine[*] and J. B. Cohen

Department of Materials Science and Engineering
The Technological Institute
Northwestern University
Evanston, IL 60201

INTRODUCTION

Of major importance in the determination of residual stress via diffraction is the accuracy of the measurement of the scattering angle $(2\theta_p)$ of a Bragg peak. This determines the accuracy of the interplanar (d) spacing and hence the strain and stress. In the U.S., the most commonly accepted method of determining peak position is a parabolic fit near the top of a peak. (While a diffraction peak is not parabolic, this is a satisfactory function near the maximum[1].) The error in this procedure has been derived[2] and tested, and it has been shown that a multipoint fit with a least 7 points is rapid and as precise or more precise than the centroid, the bisector of the half width, or cross correlation[2,3], except for sharp peaks in which case the centroid or cross correlation are slightly better. Thus a parabolic fit is generally useful and, since a least-squares fit to this function is readily carried out on modern micro-processors, automation of a stress measurement is possible, including evaluation of errors. In this procedure, with intensities across a peak at i $2\theta_i$ values, the variance in peak position, $\sigma^2(2\theta_p)$, is:

$$\sigma^2(2\theta_p) = \sum_i (\frac{\partial 2\theta}{\partial I_i})^2 \; \sigma^2(I_i). \tag{1}$$

The first term on the right-hand side is then evaluated for a parabola,[2] and the second has been shown by Wilson[4] to be:

$$\sigma^2(I_i) = I_i/t \quad \text{for a fixed time, t, at each position,} \tag{2a}$$

$$= I_i^2/c \quad \text{for fixed count, c, at each position.} \tag{2b}$$

That these equations are correct is indicated in Fig. 1. Typical examples of the effect of error in $2\theta_p$ on stress determinations are illustrated in Figs. 2 and 3.

These figures illustrate one particular goal of this paper. There is

---

[*] Currently Research Assistant, Department of Metallurgical and Mining Engineering, University of Wisconsin, Madison, WI 53706

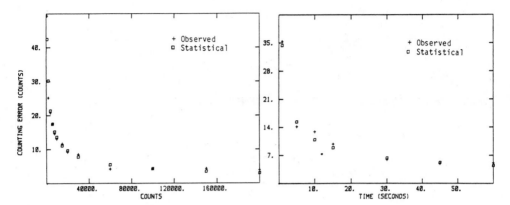

FIG. 1:   Comparison of statistical error in intensity measurements with
          observed error. Left: Fixed counts, Eq. 2a, Right: Fixed time
          measurements, Eq. 2b. Twenty-five replications of a point on a
          211 ferrite reflection from a 1008 steel were employed with
          filtered CuK_α radiation.

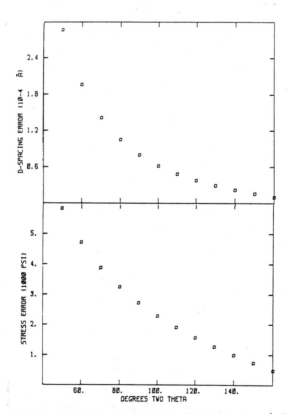

FIG. 2:   Estimated statistical error in: a) "d"-
          spacing and b) stress, as a function of
          diffraction angle. A seven point para-
          bola was assumed.

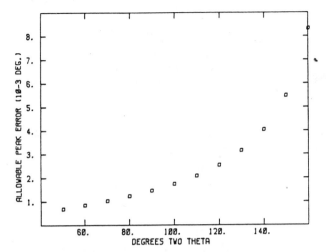

FIG. 3.   Allowable error in peak 2θ for a maximum
          1000 psi error in stress, as a function
          of diffraction angle. Seven point para-
          bolic fit.

increasing interest in measuring the stresses in second phases as well as
the matrix in multiphase materials, or in the strengthening phases in composites. Because of their low volume fraction, and their structure and its
perfection, peaks with reasonable intensities are available only at intermediate angles, where the errors may be too large to obtain reliable stress
values. Perhaps greater precision can be obtained by curve fitting the
entire diffraction peak to some suitable function other than a parabola.
Furthermore, the speed of stress measurements has been greatly reduced by
the use of position sensitive detectors[5,6], by means of which an entire
peak profile is recorded at the same time and no detector motion is required – a kind of digitized return to film! Why not use a fit to the
entire peak? The data is already available in the same time it takes to
obtain the information for a parabolic fit.

It is with these two points in mind that we have examined the use of
profile fits in the measurement of a peak position.

FUNCTIONS

Two functions were chosen to compare to the parabolic fit. The first
of these is a Modified Lorentzian, proposed by Mignot and Rondot[7]:

$$I^{K_{\alpha_1}}(2\theta_i) = I_0 \left[\cos \pi \frac{(2\theta_i) - 2\theta_p - \delta}{a} \right]^n \frac{K^2}{K^2 + (2\theta_i - 2\theta_p)^2} \cdot \qquad (3)$$

Measurements are made at many $2\theta_i$ and a solution is sought for a, $2\theta_p$, I_0
the maximum intensity, K which is related to the peak width, δ to account
for small peak asymmetry and n. The cosine term forces the function to
fall more rapidly with 2θ in the tail of the peak than a pure Lorentzian
function. This equation can be modified to include a $K_{\alpha_1} - K_{\alpha_2}$ doublet
(with separation Δ):

$$I_i^{total} = I_i^{K_{\alpha_1}}(2\theta_i) + \frac{1}{2} I_i^{K_{\alpha_2}}(2\theta_i - \Delta) . \tag{4}$$

The term δ was chosen as zero, because preliminary tests (multiple scans of peaks) indicated that it led to large variations in $2\theta_p$ without much improvement in the fit to the entire shape. Both singlet and doublet forms were tried.

The second function was a Pearson Type VII distribution[8].

$$I_i(2\theta_i) = I_0[1 + \frac{(2\theta_i - 2\theta_p)^2}{ma^2}]^{-m} . \tag{5}$$

The terms I_0 and a are sought in the profile fitting procedure. The parameter m varies with peak shape For narrow peaks m = 1 is appropriate, in which case Eq. 5 is a Lorentzian. When m approaches infinity it can be shown that Eq. 5 approaches a Gaussian function. Values of m = 1-3 and infinity were tried. Again, two terms can be added to form a doublet, and this was attempted for m = 3.

Background was subtracted. For the case of the parabolic fit, this was measured at $3°$ 2θ (before) the peak and this constant value was sub-tracted from all data points. This was adequate because only points in the vicinity of the peak (the top 15 pct) were employed in the fit. For the other two functions a linear variation was assumed and the slope and inter-cept were sought in the solution, or the values were fixed from the data. If these were included in the solution, there were 7 variables for the Modified Lorentzian (these two, plus I_0, K, a, n and $2\theta_p$). With the Pearson Type VII there were four, (two for background plus I_0, and a).

While a least-squares solution is possible for a parabolic fit, this is not as simple for the other two functions (except for the simpler Pearson Type VII forms). A modified simplex method was adapted in this study[9,10]. Convergence was tested against the significance ratio between the best and worst points of the simplex, whose points were defined in terms of a goodness of fit, or reliability index, R:

$$R = \frac{\sum_{i=1}^{n}[I_i^{obs} - I_i^{calc}]^2}{\sum_{i=1}^{n}[I_i^{obs}]^2} \times 100 . \tag{6}$$

In the simplex procedure, a multidimensional R space is formed (n + 1 dimensions, where n is the number of unknowns), one point from initial guesses at parameters and the others from fractional changes in each of the "guesses". Changes are then made in these values in a systematic way to reduce the range of these R values until by some test it is found that all values are essentially identical. This was judged by forming r, the ratio of the reliability index of the worst point in the simplex, to that of the best. This ratio was subjected to an F test[11], that is r was calculated as:

$$r_{p,n-p,\alpha} = [\frac{p}{n-p} F_{p,n-p,\alpha} + 1]^{\frac{1}{2}} \tag{7}$$

Here p is the number of parameters, and α is the probability of incorrectly concluding that the best and worst points in the simplex are different. Iterations were continued until $r < r_{p,n-p,\alpha}$. The test value was 1.00018, which corresponds to an α value less than 0.005. Values even closer to unity were attempted, but beyond this value the error in $2\theta_p$ was not substantially improved. It is important in using the simplex procedure to make appropriate first guesses at the change in variables, so that the R values are far apart, and to accelerate convergence only changes in the parameters which decreased R were accepted; if an improvement occured, the change was increased in the same direction. In effect, the point with the worst R is moved through the centroid of the (n + 1) sided polygon of R values to lower R. Finally, the entire simplex was contracted by moving all points in any iteration half the length of a side of the polygon toward the best point. Even with all these precautions, typically nearly 5 minutes were required on PDP 11/34 to accomplish the ~150 iterations in the case of a Modified Lorentzian with 7 parameters. This was reduced to ~1.5 minutes for 5 parameters, and was ~45 seconds for the Pearson Type VII (with 5 parameters).

STATISTICAL ANALYSIS

Our procedure was to remeasure a peak many times (typically 10) and to examine the variation in fitting parameters. Accordingly, we employed various statistical tests to judge the results. The mean of $2\theta_p$ and R for any set (n) of peaks fit by a single method was obtained, as well as the observed variance, S^2.

$$S^2 (<2\theta_p>) = \frac{1}{n-1} \sum_i (2\theta_i - <2\theta_p>)^2 , \tag{8}$$

and as well the true variance $\sigma^2 = S^2/n$. The range around the mean with 95% confidence limits was formed as $<2\theta_p> \pm 1.96\sigma (2\theta_p)$. The confidence limits for the variance was established with a chi-squared test (C) at the $1-\alpha$ confidence level as follows:

$$\frac{\sigma^2}{C_\alpha^2} <\sigma^2 <\frac{\sigma^2}{C_{1-\alpha}^2} . \tag{9}$$

For a given peak, the use of different profile fitting methods could lead to different $2\theta_p$. A test was made to ascertain if such differences were real or due to counting statistics. Let ℓ be the number of different profile methods for any one peak. Then the "profile method" variance was defined as:

$$S_\ell^2 (<2\theta_p>) = \frac{1}{\ell-1} \sum_\ell (2\theta_{\ell p} - <2\theta_p>)^2 , \tag{10}$$

where $<2\theta_p> = \frac{1}{\ell} \sum_\ell 2\theta_{\ell p}$; ℓ is the number of peak positions for the different methods for the same peak.

The pooled variance for all methods is defined as:

$$S_p^2 = \frac{1}{\ell} \sum_\ell S^2 \tag{11}$$

With n peaks for each method, there are then (n-1) degrees of freedom for each method, and ℓ (n-1) for S_p (and [ℓ -1] for S_ℓ). The ratio:

$$F = \frac{n S^2_\ell}{S^2_p} ,$$

(12)

was then tested. If Eq. 12 is less than the tabulated F value for some confidence level, the difference in mean position is due to random counting statistics (with that confidence).

EXPERIMENTAL METHODS

Samples were chosen to represent a typical range of peak intensities and peak widths, and to include second-phase peaks. Their preparation and the peak characteristics are described in Table I. The diffraction conditions are in Table II. The data were obtained on a microprocessor controlled diffractometer. The parabolic fit was performed on line, and data were obtained in a three stage point-counting procedure described in Ref. 2, to locate the peak and the range of 2θ covering the top 15 pct. Seven points in this range were employed. For the other functions, data was obtained with the same system (without removing the specimen) at $.02^\circ 2\theta$ intervals (each counted for 15 seconds) across the entire profile, and transferred to a PDP 11/34 minicomputer for data processing. The data were not corrected for the Lorentz-polarization factor or scattering factor variation, as would be needed in actual stress analysis; only the fit to the shape and the value (and error) of $2\theta_p$ were of interest here.

RESULTS AND DISCUSSION

As with the parabolic fit, the Pearson Type VII distribution fits only the main part of the peak, but falls more rapidly in the tails than the observations, the difference being larger as "m" increases. (The region fit was actually similar to that for a parabola.) On the other hand, the Modified Lorentzian fits the entire shape quite well. A comparison of the various fits with ten recordings of three different peaks is given in Table III, and samples of the fits are shown in Fig. 4. A fixed linear background was employed for the results in this table; solutions with a variable background had worse errors in peak position, although with the significance tests at the 95 pct level these apparent differences were not necessarily real. More importantly, the errors in the fitting parameters showed drastic decreases, and the values of K and I_o approached the measured peak width and intensity when the background was fixed prior to the solution for the other parameters.

It is evident from the table, that the Modified Lorentzian and parabolic fit provides similar error values for all peak shapes. This is true of the Pearson Type VII function as well but the value of m is different for each peak type. Thus, some knowledge of the peak shape would be required prior to fitting, which would make automation more difficult than for the other functions. Also, the Goodness-of-Fit values are quite high with this function.

Comparing the method and pooled variances indicated that the slight differences in peak positions in Table III for the various functions are significant in all cases except for the broad peaks. In this latter case the low intensity leads to larger scatter so that any difference is masked.

Note especially that the parabolic fit gives the lowest errors of any method for weak or broad peaks.

TABLE I

TYPICAL TEST CURVE FEATURES

MATERIAL	PREPARATION[1]	PEAK[2] INTENSITIES (CPS)	BACKGROUND[3] INTENSITIES (CPS)	PK/BKD RATIO	FWHM (°2θ)	PEAK TYPE
STEEL 1008, 110$_\alpha$	As rolled and grit blasted	1221	427	2.86	.65	Broad
1008, 110$_\alpha$	Same + 2 hrs. 723°K	1546	250	6.18	.42	Sharp
1074, 021/112 }Fe$_3$C	Cast, grit blasted + 723°K, 2hrs.	151	122	1.23	.40	Weak
BRASS 60/40, 211$_\beta$ }	Grit blasted + 1 hr. 673°K	398	167	2.38	.48	Doublet ψ =0°
60/40, 211$_\beta$ }		256	114	2.24	1.02	Doublet ψ =45°

1. All samples were polished with 600 grit paper to remove oxidation.
2. Peak intensity is that for K$_{\alpha_1}$ radiation.
3. Background intensity is the cps value at the peak, obtained from a straight line regression using intensity values beyond each tail.

TABLE II

OPERATING CONDITIONS

FEATURE	SETTING	SAMPLES
Divergent Slit	1^O	All
Receiving Slit	$.15^O$	Annealed 1008, 1074, Brass
Beam Size on Sample at $\psi = 0^O$ (approximate)	2.5mm x 2.5mm	All
Tube Target	Cr	All
Tube Voltage-Current	40kv – 10ma 35kv – 10ma 40kv – 15ma 35kv – 10ma	Annealed 1008 Deformed 1008 1074 Brass
Filter	Vanadium Oxide	All
Soller Slits	None used	All

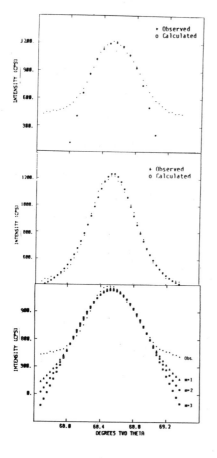

4. Various fits to the 110 ferrite peak, 1008 steel, 15 seconds counting per point. Top: 7 point parabolic fit, top 15 pct. Middle: Modified Lorentzian function. Bottom: Pearson Type VII distributions.

TABLE III

COMPARISON OF FIT METHODS[1]

METHOD	MEAN $2\theta_p$	OBSERVED ERROR $(\times 10^{-2}$ degrees)	95% LIMITS ON ERROR[4] $(\times 10^{-2}$ degrees)	MEAN GOODNESS-OF-FIT
A) Curve Type: Sharp (Annealed) Ferrite 110, 1008 Steel				
Parabolic[2]	68.82193	.287	.197, .514	29.91
Pearson Type VII[3]				
m = 3	68.81368	.473	.326, .863	30.24
m = 2	68.82818	.172	.118, .314	32.43
m = 1	68.82902	.282	.194, .515	41.77
m =	68.80705	.172	.118, .314	33.43
Modified Lorentzian[2]	68.83298	.135	.093, .246	1.75
B) Curve Type: Broad (deformed) Ferrite 110, 1008 Steel				
Parabolic	68.53719	.173	.199, .316	1.16
Pearson Type VII				
m = 3	68.53853	.511	.352, .870	20.53
m = 2	68.54362	.334	.229, .609	9.95
m = 1	68.53595	1.40	.963, 2.56	17.07
m =	68.54447	.277	.191, .506	11.21
Modified Lorentzian	68.54165	.351	.242, .641	1.88
C) Curve Type: Weak Fe_3C 112 + 021, 1074 Steel				
Parabolic	57.42794	1.05	.723, 1.92	2.79
Pearson Type VII				
m = 3	57.35688	.263	.181, .480	26.5
m = 2	57.36416	1.78	1.22, 3.25	17.2
m = 1	57.36556	.785	.540, 1.43	6.98
m =	57.40937	5.15	3.54, 9.40	4.49
Modified Lorentzian	57.40813	1.28	.881, 2.34	2.11

1. 15 Seconds per point, 10 peaks per mean, Fixed background - 5 points either side of peak used to fix line (for parabola, see ref. 2).

2. Parabolic fit is 7 pt. fit to top 15%. A single-valued background is subtracted before fit.

3. All Pearson Type VII and Modified Lorentzian utilized 90 pts. taken at .02° 2 increments, except for Fe_3C, for which 58 points were measured.

4. Values listed are low and high limits respectively.

The 60/40 brass sample exhibited a doublet whose resolution varied with tilt of the sample to the x-ray beam (as would be done in a stress measurement). Examples of fits to this peak are given in Fig. 5, and a summary of results in Table IV. The singlet forms of the Modified Lorentzian and Pearson Type VII functions gave values of $\Delta 2\theta[(\psi = 0^O) - (\psi = 45^O)]$ closer to the values of the parabola, despite the fact that the <R> values were lower with the doublet form. From the analysis of variance at both angles, the differences in peak positions with each fit technique in this table were significant.

Possible limitations in the data that could occur in practice were also explored. Firstly, it may not be possible to record the entire peak; another one nearby may overlap on one side, or the equipment itself could preclude recording the entire peak. Some results are shown in Table V. With a Modified Lorentzian and a sharp peak the "correct" (parabolic) peak position, a low error in this value, and a low R value are all obtained with data that only just reaches the peak's maximum. Similar tests with a broad profile showed that a wider range was necessary, but only a few tenths of a degree 2θ beyond the maximum is adequate. Analysis of variance tests confirmed that beyond 68.82^O, any change in peak position is solely due to random fluctuations, and the peak values obtained with data up to and beyond this angle are not distinguishable from the value for the entire peak, at 95 pct confidence.

Next, with the Modified Lorentzian, the effect of time per data point and the number of data points were explored, Table VI. The error is not statistically different at the shorter counting time, or smaller number of points.

[For further details the reader is referred to Ref. 12.]

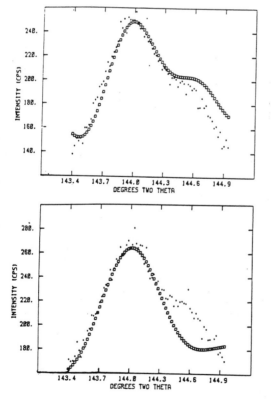

FIG. 5. Fit of the Modified Lorentzian function to the 211 β reflection from 60/40 brass, $\psi = 45°$. a) doublet equation, b) single function. Dots are observations, open figures are fit.

TABLE IV

DOUBLET ANALYSIS[1]

FITTING METHOD	PREDICTED SHIFT WITH TILT	φ = 0°			φ = 45°		
		$\langle 2\theta_p \rangle$ (°2θ)	ERROR (x10^{-2})	$\langle R \rangle$	$\langle 2\theta_p \rangle$	ERROR (x10^{-2})	$\langle R \rangle$
Parabolic	.082	144.07920	.555	.291	143.99692	.802	3.74
Pearson Type VII (m = 3)							
W/Doublet	.078	144.04743	6.206	11.34	143.96992	.659	4.45
W.O./Doublet	.133	144.19838	4.015	12.34	144.06610	8.56	6.81
Modified Lorentzian							
W/Doublet	.088	144.05093	2.665	4.45	143.96318	5.49	3.95
W.O./Doublet	.020	144.07462	1.576	7.86	144.05425	1.19	4.62

1. 60/40 Brass, 211 peak, CrK$_\alpha$ Radiation, 10 scan replications, 78 points per peak.

TABLE V

PARTIAL FIT OF SHARP PEAK (1008 STEEL) TO MODIFIED LORENTZIAN

TWO THETA VALUE OF LAST POINT	$\langle 2\theta_p \rangle$ $(^\circ 2\theta)$	OBSERVED ERROR	$\langle R \rangle$
68.70	68.77712	.296E-1	1.48
68.74	68.79541	.584E-2	1.47
68.78	68.82561	.341E-1	1.70
68.82	68.83322	.101E-2	1.74
68.86	68.83303	.836E-3	1.74
68.90	68.83298	.135E-2	1.75
68.94	68.83226	.113E-2	1.75
68.98	68.83322	.110E-2	1.74
69.02	68.93300	.937E-3	1.72

10 second count per point, 0.02° 2θ intervals.

Lowest two theta value in all cases was 67.50°.

'Value of Last Point' indicates the highest two theta value used in analysis.

Peak value, as determined from the full angular range fit, is 68.83299 ± .00135 $^\circ$2θ.

A fixed background, obtained from 10 data points on the low angle side of the curve, was used.

TABLE VI

EFFECT OF NUMBER OF POINTS—MODIFIED LORENTZIAN, BROAD CURVE (1008 STEEL)

TIME PER POINT (SEC.)	NUMBER OF POINTS[1]	MEAN PEAK POSITION $(^\circ 2\theta)$	OBSERVED ERROR $(\times 10^{-2})$	$\langle R \rangle$
1	22	68.54718	.572	3.09
	90	68.54763	.434	3.59
15	22	68.54224	.360	2.07
	90	68.54165	.351	1.88

1. The 22 points were selected as every fourth point of the full 90 point fit. Background line determined from 5 points on either side of the peak.

CONCLUSIONS

1. The parabolic fit has the best overall ability to determine peak positions over a wide range of shapes.

2. A Modified Lorentzian can be quite helpful if only a part of a peak can be explored, especially for sharp peaks.

3. The errors in a Modified Lorentzian fit are not very sensitive to counting time. With a position sensitive detector precision comparable to that for a parabolic fit can be obtained in about one tenth the time. As all of the data are recorded for either fit with this type of detector, a considerable saving in measurement time is possible with a Modified Lorentzian function. Unfortunately, the fit itself takes the order of 1 minute on a miniprocessor. If this time could be reduced with a specially designed microprocessor for this purpose, this function offers a way of drastically reducing the time for stress measurements in the field, beyond that already achieved by the use of a PSD.

4. Goodness-of-Fit is not necessarily an accurate gage of error in peak position. (For example, in Table III, part (c), similar values of R are associated with widely different peak locations.)

ACKNOWLEDGEMENTS

This research was sponsored by ONR under Grant No. N00014-80-C116. Prof. P. Georgopoulos suggested the use of the simplex procedure. Assistance with the experiments by G. Raykhtsaum, P. Rudnik, P. Zschack and Dr. I. C. Noyan are gratefully acknowledged. This paper represents a portion of a thesis submitted (by Timothy J. Devine) in August, 1984 as partial fulfillment of the requirements for the M.S. degree at Northwestern University.

REFERENCES

1. J. Thomsen and F. Y. Yap, Effect of Statistical Counting Error on Wavelength Criteria for X-ray Spectra, J. Res. NBS, 72A: 198(1968).
2. M. R. James and J. B. Cohen, Study of the Precision of X-ray Stress Analysis, Adv. X-ray Analysis, 20:291(1976).
3. M. Knuutila, "Computer Controlled Residual Stress Analysis and Its Application to Carburized Steel", Dissertation No. 81, Department of Mechanical Eng., Linkoping Univ. Sweden (1982).
4. A. J. C. Wilson, Statistical Variance of Line Profile Parameters, Measures of Intensity, Location and Dispersion, Acta Crystallogr.,23:888(1967).
5. M. R. James and J. B. Cohen, The Application of a Position-Sensitive X-ray Detector to the Measurement of Residual Stresses, Adv. X-ray Analysis, 19:695(1976).
6. M. James and J. B. Cohen, "PARS" - A Portable X-ray Analyzer for Residual Stresses, J. Testing and Evaluation, 6:91(1978).
7. J. Mignot and D. Rondot, Application du Lissage des Raies de Diffraction des Rayons-X a la Separation du Doublet K_{α_1}, K_{α_1}, J. Appl. Crystallogr, 9:460(1976).
8. M. M. Hall, V. G. Veeraghavan, H. Rubin and P. G. Winchell, The Approximation of Symmetric X-ray Peaks by Pearson Type VII Distributions, J. Appl. Crystallogr., 10:66(1977)
9. W. Spendley, G. R. Hext and F. B. Himsworth, Sequential Application of Simplex Designs in Optimization and Evolutionary Operation, Technometrics, 4:441(1962).
10. J. A. Nelder and R. Mead, A Simplex Method for Function Minimization, The Computer Jrnl., 7:308(1965).
11. W. C. Hamilton, "Statistics in Physical Science", Ronald Press Co., New York (1964).
12. T. J. Devine, Comparison of Full Profile Peak Finding Methods For Use in X-ray Residual Stress Analysis, M.S. Thesis, Northwestern University (1985).

THE USE OF PEARSON VII DISTRIBUTION FUNCTIONS

IN X-RAY DIFFRACTION RESIDUAL STRESS MEASUREMENT

Paul S. Prevey

Lambda Research, Inc.
1111 Harrison Avenue
Cincinnati, OH 45214

ABSTRACT

The fitting of a parabola by least squares regression to the upper portion of diffraction peaks is commonly used for determining lattice spacing in residual stress measurement. When $K\alpha$ techniques are employed, the presence of the $K\alpha$ doublet is shown to lead to significant potential error and non-linearities in lattice spacing as a function of $Sin^2\psi$ caused by variation in the degree of blending of the doublet. An algorithm is described for fitting Pearson VII distribution functions to determine the position of the $K\alpha_1$ component, eliminating errors caused by defocusing of diffraction peaks of intermediate breadth. The method is applied to determine the subsurface residual stress distribution in ground Ti-6Al-4V, comparing directly the use of parabolic and Pearson VII peak profiles, and is shown to provide precision better than $\pm 1\%$ in elastic constant determination.

INTRODUCTION

Anomalous non-linearities in lattice spacing as a function of $Sin^2\psi$ have been reported in the literature and observed by the author [11] employing parabolic peak profile fitting $K\alpha$ radiation techniques for residual stress measurement on machined or ground surfaces. The oscillations frequently are seen not in the highly deformed surface layers, but in the undeformed layers exposed by electropolishing, and are not explained by the residual stress models proposed by Marion and Cohen [12] or Dölle and Cohen. [13] The nature and degree of non-linearity can be altered by changing the x-ray optics to affect resolution of the doublet. The phenomenon now appears to be a result of experimental error associated with the least squares fit of parabolic profiles to the top of diffraction peaks of intermediate broadening in the presence of variable blending of the $K\alpha$ doublet caused by defocusing as ψ is changed during stress measurement.

Many practical techniques for x-ray diffraction residual stress measurement employ $K\alpha$ radiations to provide intense diffraction peaks at the high back reflection region. $K\alpha$ techniques have the disadvantage of producing a diffraction peak doublet composed of the $K\alpha_1$ and $K\alpha_2$

components. The presence of the two closely spaced diffraction peaks, which generally cannot be separated instrumentally, can lead to significant uncertainty in the determination of the lattice spacing to the precision required for the calculation of lattice strain.

If the $K\alpha$ diffraction peaks are extremely broad, as in hardened steels, the x-ray wavelength can be considered to be a weighted average for the doublet, and the fully blended doublet located precisely by calculating the vertex of a parabolic profile fitted to the top 15 percent of the combined doublet. [1] When the doublet can be resolved, a parabolic profile can be fitted to the top of the $K\alpha_1$ peak alone. However, for a broad class of specimens encountered in practice, the $K\alpha$ doublet can be neither completely blended nor resolved for all ψ angles required for residual stress measurement. The position of the diffraction peak calculated from the vertex of the fitted parabola will then depend upon both the fraction of the doublet included in the regression analysis and the degree of blending of the doublet caused by defocusing as the sample is rotated in the incident x-ray beam.

The use of centroids or cross-correlation for peak location has the advantage of being independent of the shape of the diffraction peak, but requires integration of the diffracted intensity over the entire peak profile. [10] The accuracy of these integration methods is dependent upon the precision with which the intensity in the tails of the diffraction peak can be determined and upon the range of integration. The diffracted intensity must be measured at small angular increments to provide a precise definition of the entire diffraction peak profile. The Rachinger [2] correction can be applied to separate the $K\alpha$ doublet without assuming a form for the diffraction peak profile. However, accurate correction again requires a large number of data points spanning the entire diffraction peak. Further, the method requires that the angular separation of the doublet be assumed prior to correction, which, in effect, assumes the diffraction angle being measured. The accuracy of the Rachinger correction diminishes rapidly on the $K\alpha_2$ side of the doublet, and may lead to inaccuracies in the determination of the position of the $K\alpha_1$ peak, too large to be tolerated in residual stress measurement. [3]

If the profile of the combined doublet can be adequately approximated by a suitable function for any degree of resolution, the peak position could be determined by least squares fitting of the function to intensity data collected at only a few angles, as in the case of the parabola method. To this end, the fitting of a combined $K\alpha$ doublet peak profile derived from Pearson VII distribution functions was undertaken as proposed by Gupta and Cullity. [3]

APPROXIMATION OF THE $K\alpha$ DOUBLET USING PEARSON VII DISTRIBUTION FUNCTIONS

Pearson VII distribution functions have been used for years to approximate the form of diffraction peaks for a variety of purposes, [4,5,6] but are not commonly applied for the determination of peak position in residual stress measurement. The $K\alpha$ doublet profile may be approximated as a summation of Pearson VII distribution functions, composed of a $K\alpha_1$ peak displaced by a fixed increment (dependent upon the difference between the $K\alpha$ wavelengths) from a $K\alpha_2$ peak of identical shape, but having an intensity which is a fixed fraction of the $K\alpha_1$ intensity. [3]

If $y(x)$ is the intensity as a function of angular position x, the general form of such a function describing the combined $K\alpha$ doublet is

$$y(x) = A f (x - x_o) + C A f (x - x_o - \delta)$$ (1)

where f(x) is the Pearson VII distribution function, A is the intensity of the $K\alpha_1$ peak, x_o is the angular position of the $K\alpha_1$ peak, δ is the angular separation of the doublet, and C is some fixed fraction (typically 0.5) relating the intensity of the $K\alpha_1$ and $K\alpha_2$ peaks.

The general form of the Pearson VII distribution function, ranging from the extremes of Cauchy to Gaussian profiles, is

$$f(x) = [\, 1 + K^2 \, (x - x_o)^2 \, / M \,]^{-M} \tag{2}$$

where K governs the width of the profile, and M the rate of decay of the tails. For M = 1, the profile is purely Cauchy; for M = 2, a Lorentzian; and for M = infinity, the profile is purely Gaussian. A continuous range of profiles may be generated as a weighted summation of Cauchy and Gaussian components.

In order to locate the position of the $K\alpha_1$ line contributing to the unresolved doublet, the combined profile described in Equation (1) must be fitted to data points collected across the diffraction peak by non-linear least squares regression. The task is complicated if M is allowed to be a variable during the regression analysis. Fortunately, the diffraction peak profiles in the high back-reflection region, as are used for residual stress determination, may be closely approximated by a purely Cauchy profile. [7] This observation was verified by fitting both Cauchy and Gaussian profiles to diffraction peaks representing a variety of peak breadths and degrees of doublet resolution. Profiles fitted to data points with N = 3 x 10^5 for a hardened steel and powdered iron are shown in Figures 1 and 2. Significant deviation from the Cauchy peak profile occurs primarily in the tails on the broadest peaks. Purely Gaussian profiles were found to drop away too rapidly in the tails of the diffraction peak for adequate approximation. The sum of the squares of the residuals was approximately twice as large for the Gaussian than for the Cauchy profiles for all of the samples examined. Therefore, a summation of purely Cauchy profiles with M = 1 was adopted as adequate for an approximation of the $K\alpha$ doublet.

An algorithm was developed for fitting the combined Cauchy doublet profile by non-linear least squares regression using the method of linearization and successive approximation. The program returns the values

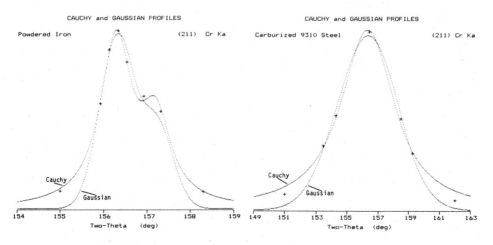

Figure 1   Figure 2

of K, A, x_o and δ for a Cauchy profile, using five or more inverse
intensities measured to high accuracy, spanning the doublet. Between each
iteration of the regression process, the spacing, δ, between the $K\alpha_1$
and $K\alpha_2$ peaks, is refined based upon the position of the $K\alpha_1$ peak
determined in the previous iteration. The intensity of the $K\alpha_1$ and $K\alpha_2$
peaks, given by C in Equation (1), is held constant. The value of C was
taken to be 0.5 for the results presented here, but could be adjusted to
reflect actual measured intensities of the $K\alpha$ components of the doublet.
The program, running on a Z80 microprocessor system in compiled BASIC, will
close to a solution with a variation of less than one part in 10^5 between
iterations in less than ten iterations, requiring approximately 10 seconds.
The method has been used for over 5,000 individual stress measurements on a
broad range of alloys without failure to converge using data collected with
both parafocusing diffractometers and a PSD.

COMPARISON OF PARABOLIC AND CAUCHY PROFILES FOR THE COMBINED DOUBLET

Assuming that a summation of Cauchy profiles provides an accurate
representation of the (211) diffraction peak doublet produced using
chromium $K\alpha$ radiation for a simulated stress measurement in a steel
specimen, a family of profiles representing a range of peak breadths is
shown in Figure 3. The $K\alpha_1$ peak is assumed to be located at precisely
156.0 degrees, and the $K\alpha_1$ and $K\alpha_2$ wavelengths are assumed to be
2.28962 and 2.29351 Å, respectively. Combined profiles are shown for values
of K ranging from 0.5 to 4.0, representing complete blending and separation
of the $K\alpha$ doublet. As seen in Figure 3, the vertex of the diffraction
peak, which would be determined by fitting a parabola, shifts from 156.0
degrees for the resolved $K\alpha_1$ peak to approximately 156.3 degrees as the
doublet is blended.

For diffraction peaks of intermediate width (0.5 < K < 2.5),
substantial error in determining the peak position may result from the use
of a parabolic profile and a fixed weighted average wavelength from two
sources. First, the peak breadth will vary as a result of defocusing as ψ
is changed during the course of stress measurement even if parafocusing is
used. The degree of defocusing and the resulting variation in K,
encountered during a single stress measurement, will be a complex function
of the incident beam divergence, slit and focal spot geometry, the range
and sign of ψ tilts employed, and the diffraction angle. [8] In
general, the degree of defocusing can be expected to increase rapidly as
the diffraction angle decreases, and to be greater for a fixed slit than
for a parafocusing technique. Second, the fraction of the upper portion
of the diffraction peak included in the parabolic regression procedure, in
practice, may be difficult to control. For diffraction from the surface of
highly deformed specimens, where the diffracted intensity is low and the
background intensity high, uncertainties in the precise background
intensity may lead to significant variation in the percentage of the
diffraction peak included in the analysis for different ψ angles.

Figure 4 shows the angular error in the determination of the $K\alpha_1$
peak position for a parabolic profile fitted to the combined Cauchy doublet
using from 10 to 35 percent of the diffraction peak in increments of 5
percent as a function of K. Figure 4 was derived by fitting a parabola by
linear least squares regression to the data points shown in Figure 3 at
increments for K of 0.1. The simulated intensity points are uniformly
spaced at 0.0625 degree increments. No random error was assumed in the
simulation. The number of data points included in the parabolic regression
analysis ranged from 82 points for the 30 percent, K = 0.3 case, to a
minimum of 4 points for the 10 percent, K = 4.0 case. The "+" symbols in
Figure 4 represent the calculated positions for the diffraction peak, using

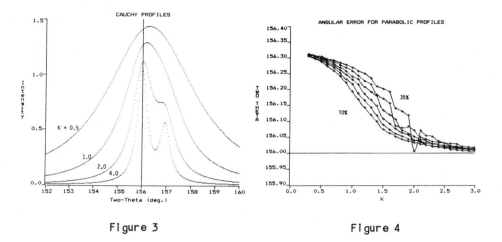

Figure 3   Figure 4

the parabolic fit. The points have been simply connected by straight lines for presentation.

For fully blended or separated doublets (K < 0.5 or K > 2.5), the position of the diffraction peak determined using parabolic regression becomes virtually independent of both peak width and the fraction of the profile included in the analysis. No significant error would result, provided the appropriate wavelengths were used ($K\alpha_1$ or weighted average) to calculate lattice spacing. For intermediate resolution (0.5 < K < 2.5), the position of the peak determined from the parabolic regression procedure is highly dependent upon both the peak breadth and the fraction of the peak included with maximum sensitivity at approximately K = 1.5. Significant error could result from variation in either parameter. As increased fractions of the peak are included in the parabolic analysis of partially resolved doublets (1.5 < K < 2.5), and the "shoulder" formed by the $K\alpha_2$ peak is included, the solution by parabolic regression becomes unstable for fractions in excess of 25 percent.

If the parabolic method of peak location is used to determine absolute rather than relative lattice spacings (as required to determine the full residual stress tensor), quite different results would be expected for d_o determined on the surface of a specimen deformed by grinding or machining, where K is small due to line broadening, and beneath the surface where the material may be stress free and undeformed. The practice of determining d_o beneath the deformed surface using parabolas could then lead to significant error.

An indirect method of correcting for blending of the $K\alpha$ doublet using Pearson VII functions has been described by Dölle and Cohen.[13] A parabola is fitted to the upper portion of the combined doublet and the position of the $K\alpha_1$ peak is calculated from an error function similar to those shown in Figure 4 derived from combined Gaussian profiles representing the $K\alpha_1$ and $K\alpha_2$ components of the doublet. There are several difficulties inherent in this method which are eliminated by fitting the Pearson VII profiles directly. First, the form of the peak profile used to generate the correction function must be assumed (Gaussian for Dölle and Cohen). The correction function is highly dependent upon the form assumed. Although only Cauchy profiles are presented here, the algorithm has been refined to fit generalized Pearson VII functions where M is a variable in the regression analysis. Second, the use of a fixed correction function presupposes both the separation of the doublet and the portion of the diffraction peak included in the parabolic regression analysis, which as seen from Figure 4, is itself a source of error. Third, the width of the diffraction peak, which is also a sensitive function of the fraction of the peak included in the parabolic regression analysis, must be determined separately in order to apply the correction, introducing an additional source of experimental error.

The use of parallel beam optics would reduce the variation in blending of the doublet as a function of ψ tilt during measurement. Whether the variation in the degree of blending would be less than observed with parafocusing or eliminated entirely with the low incident beam divergence achievable is beyond the author's experience. Errors due to variation in the portion of the diffraction peak included in the parabolic regression analysis would still remain.

The method of diffraction peak location by fitting Pearson VII functions proposed here would still be subject to error in the event of defocusing so severe as to cause asymmetry in the individual doublet component profiles, as would peak profile approximation using any symmetrical function. The method also requires accurate determination of the background intensity, which is not required for peak location using simple parabolic regression.

APPLICATION TO RESIDUAL STRESS MEASUREMENT IN GROUND TI-6AI-4V

A direct comparison was made using both parabolic and Cauchy profiles on the identical five data points collected in the top 20 percent of the (21.3) diffraction peak produced with Cu Kα radiation for a positive ψ tilt $\sin^2\psi$ technique on a sample of ground TI-6AI-4V, assuming the plane stress model. Data were collected at the surface and as a function of depth to a maximum of 0.175 mm, removing material for subsurface measurement by electropolishing. Measurements were made on a horizontal GE goniometer, with a 1.0 degree incident beam divergence and a 0.2 degree receiving slit, using a parafocusing technique, Si(Li) detector, and six positive ψ tilts, to a maximum of 45 degrees at even increments of $\sin^2\psi$. The elastic constant, $E/(1+v) = 84.1 +- 0.5$ GPa, was determined empirically using the four-point bending technique previously described. [9]

The inverse intensity was determined at five points in the top 20 percent of the diffraction peak, for $N = 3 \times 10^4$, and corrected for a linearly sloping background intensity, Lorentz polarization and absorption, prior to fitting the parabolic and Cauchy profiles. The lattice spacings for the parabolic profiles were calculated using the weighted average of the Kα doublet, while the Cauchy profiles were calculated for the Kα_1 wavelength alone. The results shown have been corrected for penetration of the radiation into the subsurface stress gradient.

Plots of d (21.3) and the value of K, derived from the fitted Cauchy profile describing the peak breadth, are plotted as functions of $\sin^2\psi$ for selected depths in Figures 5 through 8. The stress derived by fitting a straight line by least squares regression through d as a function of $\sin^2\psi$ is shown for both the parabolic and the Cauchy profiles with the error indicated by one standard deviation, based upon the uncertainty in the slope of the fitted line.

No significant differences are observed in the residual stresses derived from the Cauchy and parabolic profiles at the surface or at 0.01 mm. At these depths, shown in Figures 5 and 6, line broadening is dominated by the plastic deformation of the alloy, and the diffraction peak width, indicated by K, does not vary significantly with ψ. At 0.025 mm, shown in Figure 7, a significant difference is observed in the stress derived from the parabolic and Cauchy profiles, although no pronounced anomalies are evident. The discrepancy of nominally 40 MPa between the results accompanies a linearly varying peak breadth due to defocusing. At the remaining three depths, from 0.048 to 0.175 mm, pronounced anomalies in the d-$\sin^2\psi$ plots derived from the parabolic profile are evident. Typical results are shown for the maximum depth in Figure 6. The Cauchy results yield a constant stress of nominally -34 MPa at the three depths, while the parabolic results give values ranging from + 86 to + 61 MPa, with the standard deviation approaching the magnitude of the stress measured.

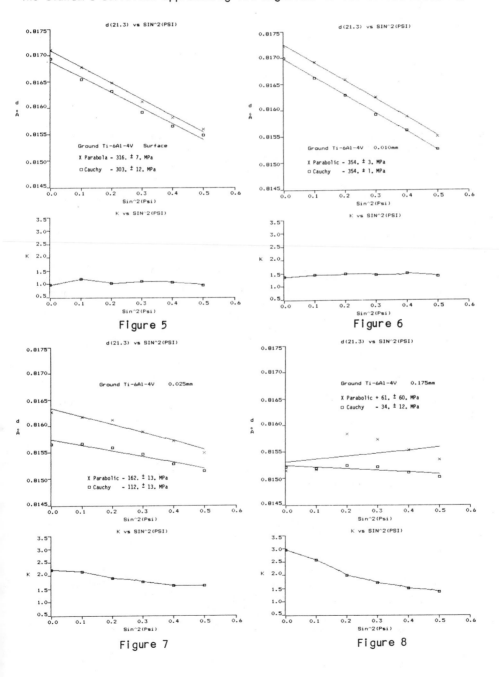

Figure 5

Figure 6

Figure 7

Figure 8

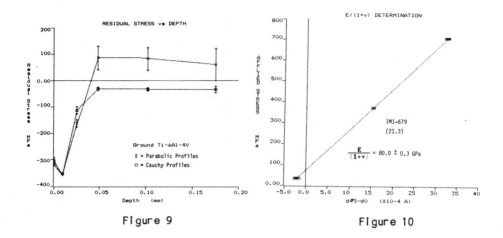

Figure 9   Figure 10

The residual stress distributions, showing the mean value and standard
deviations, are presented for both parabolic and Cauchy profiles as
functions of depth in Figure 9. The sample was in the form of a thick
walled tube approximately 10 cm in diameter with a 3.8 cm wall. The
stresses measured were in the circumferential direction parallel to the
grinding direction. Failure to achieve equilibrium in the range of depths
examined is attributed to hoop stresses of low magnitude in the tube.

X-RAY ELASTIC CONSTANT DETERMINATION

The Cauchy profile procedure for locating the $K\alpha_1$ peak position has
been applied to the determination of x-ray elastic constants by the four-
point bending technique.[9] The results indicate a significant reduction
in the random error in the determination of the lattice spacing and,
therefore, improvement in the precision with which the elastic constant,
$E/(1 + \nu)$ can be determined, compared to the parabolic regression method
previously used. Typical results for titanium alloy IMI-679, showing the
change in lattice spacing for the (21.3) planes as a function of applied
stress, are shown in Figure 10. A total of 13 data points are plotted,
five at the maximum and minimum loads, and three at the intermediate load,
to test linearity. The uncertainty shown is one standard deviation based
upon the line fitted by least squares regression. The x-ray elastic

TABLE I

X-RAY ELASTIC CONSTANTS FOR SELECTED STEELS
Determined Using the Cauchy Peak Profile Method
(211) Cr $K\alpha_1$ 2θ = 156. deg.

Steel	Hardness (Rc)	$E/(1 + \nu)$ ($\times 10^6$ psi)	GPa	S.D.(%)
SAE 1050	56.	26.72 +-0.27	184.2 +-1.9	1.0
SAE 5150	34.	26.98 +-0.08	186.0 +-0.5	0.3
AISI 15B48	44.	25.86 +-0.10	178.2 +-0.7	0.4
AISI 15B48	15.	28.12 +-0.05	193.4 +-0.3	0.2
MIL-S-46850	50.	23.24 +-0.07	160.2 +-0.5	0.3
High C Tool (1.3 C)	27.	28.64 +-0.08	197.5 +-0.5	0.3
AISI 402 SS	22.	26.32 +-0.10	181.5 +-0.7	0.4

constants for the (211) direction for seven steels, determined using the Cauchy profile method, are presented in Table I. The random error, due primarily to uncertainties in peak location, is 1.0% or less as indicated by the standard deviations and represents a reduction in random error to approximately half that routinely achieved with parabolic regression.

CONCLUSIONS

A method of locating the diffraction peak in the high back-reflection region for x-ray diffraction residual stress measurement, using the Cauchy sub-class of Pearson VII distribution functions to approximate the combined $K\alpha$ doublet, has been developed. The position, intensity, and breadth of the $K\alpha_1$ peak can be determined from the intensity measured for as few as five points spanning the $K\alpha$ doublet. The method has been demonstrated to provide a more reliable method of determining lattice spacing in residual stress measurement than the method of fitting parabolic profiles by least squares regression for diffraction peaks of intermediate breadth. Anomalies in d vs. $\sin^2\psi$, which result from variable blending of the doublet caused by defocusing and variation in the portion of the peak included in the analysis, appear to be essentially eliminated by the method.

If the diffraction peak breadth happens to vary nearly linearly with $\sin^2\psi$, significant error may result from the use of parabolic regression, even though anomalies are not evident in the $\sin^2\psi$ plots. Determination of the unstressed lattice spacing, d_o, using parabolic regression in subsurface material for studies of triaxial stresses in deformed surfaces, may result in significant error.

REFERENCES

1. D. P. Koistinen and R. E. Marburger, "Simplified Procedure for Calculating Peak Position in X-Ray Residual Stress Measurements on Hardened Steel," ASM Transactions, Vol. 51, 537 (1959).
2. W. A. Rachinger, J. Sci. Instruments, Vol. 25, 254 (1948).
3. S. K. Gupta and B. D. Cullity, "Problems Associated with K-alpha Doublet in Residual Stress Measurement," Adv. in X-Ray Analysis, Vol. 23, 333 (1980).
4. A. Brown and J. W. Edmonds, "The Fitting of Powder Diffraction Profiles to an Analytical Expression and the Technique of Line Broadening Factors," Adv. in X-Ray Analysis, Vol. 23, 361 (1980).
5. M. M. Hall, "The Approximation of Symmetric X-Ray Peaks by Pearson Type VII Distributions," J. Apl. Cryst., Vol. 10, 66 (1977).
6. S. Enzo and W. Parrish, "A Method of Background Subtraction for the Analysis of Broadened Profiles," Adv. in X-Ray Anal., Vol. 27, 37 (1983).
7. H. P. Klug and L. E. Alexander, X-Ray Diffraction Procedures, 2nd Ed., Wiley, NY, p. 642, (1974).
8. A. J. C. Wilson, Mathematical Theory of X-Ray Powder Diffractometry, Centrex, Eindhoven, (1963).
9. P. S. Prevey, "A Method of Determining Elastic Constants in Selected Crystallographic Directions for X-Ray Diffraction Residual Stress Measurement," Adv. in X-Ray Analysis, Vol. 20, pp. 345-354, (1977).
10. V. M. Hauk and E. Macherauch, "A Useful Guide for X-Ray Stress Evaluation (XSE)," Adv. in X-Ray Anal., Vol. 27, p. 82 (1983).
11. P. S. Prevey, "Comparison of X-Ray Diffraction Residual Stress Measurement Methods on Machined Surfaces," Adv. in X-Ray Anal., Vol. 19, pp. 709-724, (1976).
12. A. H. Marion and J. B. Cohen, "Anomalies in the Measurement of Residual Stress by X-Ray Diffraction," Adv. in X-Ray Anal., Vol. 18, (1975).
13. H. Dölle and J. B. Cohen, "Residual Stresses in Ground Steels," Met. Trans. A., Vol. 11A, pp. 159-164, (1980).

THE CHOICE OF LATTICE PLANES IN X-RAY

STRAIN MEASUREMENTS OF SINGLE CRYSTALS

Balder Ortner

Erich-Schmid-Institut f. Festkörperphysik
d. Österreichischen Akademie d. Wissenschaften
Leoben, Austria

INTRODUCTION

It is well known that all of the six independent components of the strain tensor can be calculated if the linear strains in six appropriate directions are known (e.g.[1-4]). That calculation is to solve a system of linear equations, whose coefficients are defined by the orientations of the measured planes. The strains are determined by lattice plane distance measurements using X-rays.

The linear equation system can only be solved if the matrix of coefficients has rank 6. Whether this condition is met or not can be decided without calculating a determinant, just from geometric relationships among the planes to be measured[1,5,6]. A demand beyond that necessary condition is to make the matrix of coefficients so that the accuracy of the calculated strain tensor is best. From error calculation we know that there exist distinct ratios between the inevitable measurement errors and the errors of the calculated strain components. These ratios depend strongly on the geometric relationship among the lattice planes. It is the purpose of this paper to show how lattice planes should be chosen in order to get these ratios as small as possible i.e. to get a maximum of accuracy at a given number of measurements, or a minimum of experimental effort if a distinct limit of error is to be reached.

ERROR CALCULUS

What is really measured are Bragg's angles of different planes (hkl). From these we get the lattice plane distances (d) and by comparing them with the unstrained distances (d_o) we can calculate the linear strains ($\varepsilon_{(hkl)}$) in the directions normal to the planes (hkl).

$$d_{(hkl)} = \lambda/(2\sin\theta); \quad d_{o(hkl)} = a_o\sqrt{h^2+k^2+l^2} \quad (*) \qquad (1a,b)$$

$$\varepsilon_{(hkl)} = (d_{(hkl)} - d_{o(hkl)})/d_{o(hkl)} \qquad (2)$$

---

*)For convenience equ.1 and the whole deduction is done for cubic crystal systems. But it can easily be shown[6] that the results are valid also for other crystal systems without any restriction. The only condition is that the strain tensor is defined in a cartesian coordinate system, as usual.

The strain must be measured in different directions - the normals to the lattice planes ($h_m\ k_m\ l_m$). We write $\varepsilon_{(m)}$ for $\varepsilon_{(h_m k_m l_m)}$.

If ε_{ij} are the components of the strain tensor then the relationship between ε_{ij} and $\varepsilon_{(hkl)} = \varepsilon_{(m)}$ is given by:

$$\varepsilon_{(m)} = \sum_{i=i}^{6} a_{mi}\varepsilon_i \qquad (3a)$$

ε_i are the strain tensor components in matrix notation according to Voigt's order (equ.3b); a_{mi} is defined by equ.3c.

$$\varepsilon_i = \varepsilon_{11}\ ,\ \varepsilon_{22}\ ,\ \varepsilon_{33}\ ,\ 2\varepsilon_{23}\ ,\ 2\varepsilon_{13}\ ,\ 2\varepsilon_{12} \qquad (3b)$$

$$a_{mi}=h_m/N_m,k_m/N_m,l_m/N_m,k_m l_m/N_m,h_m l_m/N_m,h_m k_m/N_m;\ N_m=\sqrt{h_m^2+k_m^2+l_m^2} \qquad (3c)$$

Equ.3a actually is a system of linear equations which can be resolved for ε_i, provided the matrix a_{mi} has rank 6. If the number of measurements (N) is greater than 6 we use a least square fit[1,7,8]. When doing this we should not forget to take into account the fact that the $\varepsilon_{(m)}$-values usually are not equally accurate, because of their different Bragg's angles. If the weighting factors (p_m) of the measurements are known then the solution of equ.3a reads:

$$\varepsilon_i= \sum_{j=1}^{6} Q_{ij}E_j\ ;\quad Q_{ij}=(\sum_{m=1}^{N} p_m a_{mi}a_{mj})^{-1}\ ;\quad E_j= \sum_{m=1}^{N} p_m a_{mj}\varepsilon_{(m)} \qquad (4a,b,c)$$

If all measurements are liable to statistical errors with Gauss's distribution, the mean statistical errors (standard deviations) of the calculated strain components are :

$$\Delta\varepsilon_i^2=Q_{ii}\,\Delta\varepsilon_M^2 \qquad (5)$$

where $\Delta\varepsilon_M$ can be estimated using equ.6:

$$\Delta\varepsilon_M^2=(\sum_{m=1}^{N} p_m\varepsilon_{(m)}^2 - \sum_{j=1}^{6} E_j\varepsilon_j)/(N-6) \qquad (6)$$

The simplest assumption to get the weighting factors would be that only the reading of Bragg's angles (θ_m) is subject to statistical errors, which are independent of θ_m. Then we have:

$$\Delta\varepsilon_{(m)}= ctg\ \theta_m\ \Delta\theta_m\ ,\ \text{therefore:}\quad p_m=tg^2\theta_m \quad (*) \qquad (7a,b)$$

As already mentioned the idea dealt with in this paper is: The reliability of the calculated strain tensor should not only be estimated at the end of the whole procedure but the lattice planes for the measurement should be chosen so that it will be as high as possible. So the task is to find a sample of lattice planes so that the factors Q_{11} to Q_{44} have values as small as possible.

In most cases we could of course not decide which of two samples is the best one if we simply tried to compare the two sets of Q_{11} to Q_{66}. For example, if Q_{11} of the first sample is smaller than Q_{11} of the second sample, but the contrary relation holds for Q_{22}, then we could not say which sample is the better one. This problem is solved

---

*From the viewpoint of error calculus it would be more reasonable to start with an equation system where the correlation between the measured Bragg's angle and the strain tensor is used directly instead of at first calculating $\varepsilon_{(m)}$ and then using equ.3a. When doing so we would introduce the factor $tg\theta_m$ automatically. Instead of equ.3a we would have: $\theta_m - \theta_{om} = \sum_i a'_{mi}\varepsilon_i\ ;\ a'_{mi}= a_{mi}tg^2\theta_m$

when introducing a condensed value of Q_{11} to Q_{44} by defining a mean value of $\overline{\Delta\varepsilon_i^2}$:

$$\overline{\Delta\varepsilon_i^2} = \Delta\varepsilon_{ij}^2 = (\Delta\varepsilon_1^2 + \Delta\varepsilon_2^2 + \Delta\varepsilon_3^2 + (\Delta\varepsilon_4^2 + \Delta\varepsilon_5^2 + \Delta\varepsilon_6^2)/2)/9 =$$
$$= \Delta\varepsilon_M^2(Q_{11} + Q_{22} + Q_{33} + (Q_{44} + Q_{55} + Q_{66})/2)/9 = Q\,\Delta\varepsilon_M^2 \qquad (8)$$

The meaning of Q is this: Q is the ratio between the mean variance (the probable error) of the calculated strain tensor components and the variance of the measurements (the probable measurement errors). With Q we have a means to evaluate the quality of a configuration of lattice planes, since the smaller Q is, the more accurate is the calculated strain tensor.

The factor 1/2 in equ.8 seems to be somewhat arbitrary, but it can be shown[6] that only by equ.8 a mean value of $\Delta\varepsilon_i^2$ (and therefore of Q_{ii}) is defined which is invariant at a rotation of the coordinate system. It is clear that an expression (here:Q) which is to be used to evaluate the "quality" of a sample of measurement directions, must depend only on this sample, but not on the orientation of the coordinate system.

It would be hard work to find out the best configuration for a given crystal and given measurement conditions. But we can make the problem somewhat easier by the following simplification: Because of equs.7a,b one will always try to take Bragg's angles near to the largest values which are allowed by the experimental setup. Therefore we can take the approximation that all the weighting factors in equs.4,6 are equal, in other words, the accuracy of all $\varepsilon_{(m)}$-measurements is the same. Then Q is only dependent on the orientation relationship of the lattice planes. For that case it is possible to decide generally which of different configurations of lattice planes is the best one.

GOOD CONFIGURATIONS OF DIRECTIONS

To find out the best configurations, a computer-program was designed to generate different configurations and to calculate and compare the values Q. The result of a very great number (many 100,000s) of such tests is:

If the smallest number of measurements (6) is taken then the very best configuration is that shown in Fig.1a - or a configuration which is reached by rotating this configuration.

It is a remarkable configuration, since the directions (the lattice plane normals) 1 to 6 are identical with the body diagonals of the regular icosahedron. In other words, this is the most symmetrical distribution of 6 straight lines in the space, since all the angles enclosed by two of these lines are the same.

It can be shown by analytic calculus, that Q=.75 is the smallest possible value if N = 6. Furthermore we can show the existence of a lower boundary of Q for any number of measurements (N). (The proof of this assertion[6] is, like that of equ.8, too complicated to be given here.)

$$Q = \Delta\varepsilon_{ij}^2/\Delta\varepsilon_M^2 \geq 4.5/N \qquad (9)$$

If N = 12, 18, 24,... it is easy to find a configuration with Q=4.5/N, because we simply have to take 2-,3-,4-,...-times the icosahedral configuration. For N = 7, 8, 9...13, 14.. using the computer

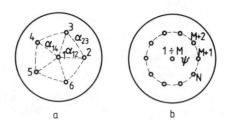

Fig. 1. Good configurations of measurement directions.
(a) "Icosahedron", best sample of 6 directions,
$\Delta\varepsilon_i^2/\Delta\varepsilon_M^2 = .75$; $\cos^2(\alpha_{12}) = \cos^2(\alpha_{23}) = \cos^2(\alpha_{14}) = \ldots = .2$
(b) (N-M)-configuration. $\cos^2(\psi) = .2$, $\psi \cong 63°$.

program mentioned above it was always possible to find different
configurations with Q very close to the lower boundary, yet for
practical purposes only the configurations shown in Fig.1 are needed.
It proved that a (N-M)-fold symmetry configuration as shown in Fig.1b
has $Q \cong 4.5/N$ if $\psi = 63°$, and if M differs not too much from the
condition:

M = int(N/6+.5)   (10)

It is clear that measuring two times a (N-M)-configuration,
would give the same result as a (2N-2M)-configuration, provided that
N-M ≥ 5. For example, if we measured 2 times at the planes of Fig.2a,
we would get the same Q-value as for a (14-2)-configuration with the
same ψ. But the calculation would not work if we took two times a
(5-1)-configuration instead of a (10-2)-configuration. Since in the
2x(5-1)-configuration we have only 5 independent measurements, so the
matrix a_{mi} has rank 5, or Q becomes infinite.

Another simple method to get a good configuration is by simply
taking the icosahedral (5-fold symmetry) configuration for the first
six directions and for the next ones again directions of the same or
a rotated icosahedral configuration. For instance: If we measured in
the directions 1,2,3 of Fig.1a and two times in the directions 4,5,6,
then Q would become 0.56, which is only 12 percent larger than the
lower boundary (4.5/9).

We can, of course, also consider the case where the angle (ψ)
between the plane to be measured and the surface is restricted for
experimental reasons. If, e.g. ψ should be not greater than 45° then
the good configurations are quite similar as in the case of unre-
stricted ψ. The value of Q is of course larger, namely $Q \cong 8.4/N$ for
the (N-M)-configuration. Also, the rule for the distribution of the
points between the center and the small circle is not that of equ.10
but of equ.11.

M = int(N/5+.4)   (if $\psi \leq 45°$)   (11)

APPLICATION TO PRACTICAL WORK

From the foregoing considerations we can conclude: When planning
the experiment one should always try to find lattice planes for the
measurement so that the configuration of the planes is similar to one
shown in Fig.1. If doing this, we have a very good chance that the
inevitable measurement errors do not result in exaggerated errors of
the calculated strain tensor components. It is clear that usually we
can not get one of the symmetric configurations with exactly
$\psi = 63.3°$, because there is only a finite number of planes which are

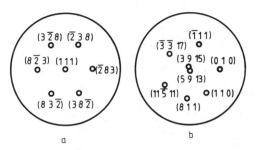

Fig. 2. Two examples of lattice plane configurations with the ratios $\overline{\Delta \varepsilon_i^2}/\Delta \varepsilon_M^2$ fairly close to the least possible value for that number of measurements.
(a) Exact 6 fold symmetry - (7-1)-configuration; $\overline{\Delta \varepsilon_i^2}/\Delta \varepsilon_M^2$ = .78 (= 4.5/7 + 20%).
(b) Sample of planes similar to (8-2)-configuration; $\overline{\Delta \varepsilon_i^2}/\Delta \varepsilon_M^2$ = .70 (=4.5/8 + 25%)

eligible for X-ray measurement. Yet it is not so important to take the very smallest value of Q. One must also keep in mind that a configuration with Q not so close to the best possible value but with large diffraction angles can be much better than one with a smaller value of Q but with smaller diffraction angles. This is because of the dependence of $\Delta \varepsilon_M$ on the diffraction angle.

EXAMPLES

In Fig.2 two examples are shown with quite fair values of Q. In the example of Fig.2a where a (111) orientation was assumed, the three-fold symmetry of the specimen's surface was used to find a sample with a (7-1)-configuration. The other example shows a crystal in (123) orientation. Here the planes have been chosen to get a configuration similar to the (8-2)-configuration. Fig.2b is an example where no exact symmetry is found, where the inclination angles of the oblique planes are not equal to each other and also equ.10 is not met, yet the ratio Q is quite close to its lower boundary. The angles ψ are much smaller than the optimum value 63°, since the planes are chosen to find a compromise between small Q and high diffraction angles.

CONCLUSION

The strain tensor of a single crystal can be calculated from strain measurements in different directions (lattice plane normals). The accuracy of these tensor components does not only depend on the accuracy of the measurements and on their number, but also on the orientation relationship among those directions. Therefore the planes should be chosen so that there will be maximum accuracy at a given number of measurements. Configurations of directions have been found, which would give very small errors. These configurations are the basis for a procedure to find good configurations of lattice planes in practical work.

ACKNOWLEDGEMENTS

The author is indebted to Prof. H.P. Stüwe for many helpful discussions and to Gesellschaft zur Förderung der Metallforschung and Steiermärkische Landesregierung for financial support.

LITERATURE

1. T. Imura, S. Weissmann and J.J. Slade, A study of Age Hardening
    of Al-3.85% Cu by the Divergent X-ray Beam Method,
    Acta Cryst. 15:786 (1962)
2. F. Bollenrath, V. Hauk, E.H. Müller, Röntgenographische
    Verformungsmessung an Einzelkristalliten verschiedener
    Korngröße, Metall 22:442 (1968)
3. S. Taira, "X-ray Studies on Mechanical Behaviour of Materials",
    The Society of Mat.Sci., Japan, Kyoto, (1974) (cited after$_4$)
4. J. Godijk, S. Nannenberg, and P.F. Willemse, A Goniometer for
    the Measurement of Stresses in Single Crystal and Coarse-
    Grained Specimens, J.Appl.Cryst. 13:128 (1980)
5. B. Ortner, Röntgenographische Spannungsmessung an einkristal-
    linen Proben, in: "Eigenspannungen , E. Macherauch, V. Hauk,
    Edts., Deutsche Ges.Metallkunde, Oberursel (1983)
6. B. Ortner, to be published
7. E. Hardtwig, "Fehler und Ausgleichsrechnung", Bibliographisches
    Institut, Mannheim (1968)
8. J.F. Nye, "Physical Properties of Crystals", Clarendon Press,
    Oxford (1969)

NEUTRON POWDER DIFFRACTION AT A PULSED NEUTRON SOURCE:

A STUDY OF RESOLUTION EFFECTS

J. Faber, Jr. and R. L. Hitterman

Argonne National Laboratory
Materials Science and Technology Division
Argonne, IL, 60439

ABSTRACT

The General Purpose Powder Diffractometer (GPPD), a high resolution ($\Delta d/d=0.002$) time-of-flight instrument, exhibits a resolution function that is almost independent of d-spacing. Some of the special properties of time-of-flight scattering data obtained at a pulsed neutron source will be discussed. A method is described that transforms wavelength dependent data, obtained at a pulsed neutron source, so that standard structural least-squares analyses can be applied. Several criteria are given to show when these techniques are useful in time-of-flight data analysis.

I. INTRODUCTION

The GPPD is a high resolution time-of-flight (TOF) neutron scattering instrument at the Argonne Intense Pulsed Neutron Source (IPNS). The IPNS[1,2] is a major user-dedicated national facility that produces pulsed beams of polychromatic neutrons in the thermal regime ($0.15<\lambda<15\text{Å}$). Neutrons are produced by spallation using a uranium target and 450MeV protons. By pulsing the proton accelerator at 30Hz, the neutrons are produced in short bursts. For scattering studies in the thermal regime[3], the fast neutrons produced by spallation are slowed-down by hydrogenous moderator materials (typically polyethylene or liquid methane) to provide white beam thermalized neutron fluxes. The time for a neutron to be emitted from the moderator varies approximately as $1/v$, where v is the neutron velocity. The width of the pulse thus varies from about 5 to 50μs

119

and depends upon the moderator effective temperature and geometry, and the neutron energy, or corresponding wavelength. As we shall see (Section II), steady state methods that require scattering angle scans are replaced with scans in TOF, at constant angle.

In this paper, we shall briefly describe the configuration of the GPPD at IPNS (Section III). In particular, we shall show how high data acquisition rates and high resolution can be realized in the TOF case. To emphasize the high resolution capability of the GPPD, a transformation of the TOF data will be described, that takes into account the special wavelength dependent properties of TOF data obtained from a pulsed neutron source (see Section IV). This particular transformation however, does not obviate the need to recover precise integrated intensity information from the transformed data. We have written a computer code, TOFMANY, that is designed to recover integrated intensity information from TOF data at IPNS. This method is of particular importance when standard Rietveld[4] techniques cannot adequately fit the Bragg profiles. For materials with very complicated crystal structures, the choice of techniques for structural refinement become severely limited. Analysis methods that rely on single peak or multiplet peak measurement techniques must be abandoned, but the Rietveld structure refinement technique[5,6] is available.

II. SEVERAL USEFUL TOF RELATIONS

For crystalline samples, the information about interplanar spacings can be obtained by rewriting Bragg's law in terms of t, the time of flight required for a neutron to travel the total flight path L, from source to detector:

$$\lambda = h/mv = ht/mL = 2d\sin\theta, \tag{1}$$

where d is the interplanar spacing, λ is the neutron wavelength, h is Planck's constant and m is the neutron mass. Equation (1) simplifies to

$$3956t/L = 2d\sin\theta, \tag{2}$$

where t is the time in seconds and L is the total flight path in meters. In the TOF technique, d-space dependent information is obtained by fixing the scattering angle and using standard timing techniques to "scan" TOF. To maximize the efficiency of data acquisition, a very wide wavelength range is desired so that all d-spaces of interest are scanned for each pulse. This is indeed the case for the GPPD at IPNS, where with a room

temperature polyethylene moderator, we are able to obtain 0.2<λ<5Å. In
the case of steady state techniques, the data acquisition systematics
generally involve a scan from low 2θ to high 2θ, or from large to small
d-spacing. Notice (see Equations (1) and (2)) that the data are inverted
in the TOF case, i.e., a logical organization of the data is from short
to long time after each neutron source burst. In this case, the scan is
from small to large d-space.

III. THE GPPD AT IPNS

A schematic diagram of the GPPD at the IPNS facility is illustrated
in Figure 1. The distance between the source and the sample is 19.96m.
Each of the 144 detectors in this instrument is a 10atm ^{3}He gas-filled
cylindrically-shaped proportional counter with 1.27cm diameter and 38.1cm
long. Detectors are located at various scattering angles but the
scattered flight path for all detectors is fixed at 1.5m from the sample
position. Standard timing electronics are sufficient to provide a high
degree of wavelength resolution in the TOF case. For example, from
Equation (2), the velocity for a neutron with λ=1Å is approximately
4000m/s. The standard data acquisition[7,8] timing electronics for the GPPD
at IPNS uses an 8MHz clock, which implies roughly 125ns timing
resolution. For neutron path lengths illustrated here, the electronic
timing resolution is more than adequate. The sample chamber is 2ft in
diameter, thus providing sufficient space for ancillary equipment[9].

When compared to x-ray techniques, neutron experiments are almost
always considered to be flux-limited. This potential disadvantage in the
case of neutrons can be minimized by designing detector banks with large

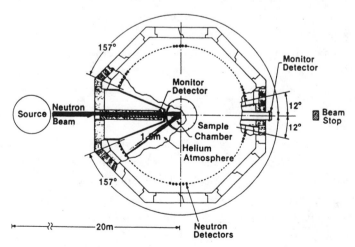

Figure 1. Schematic diagram of the General Purpose Powder
Diffractometer (GPPD) at IPNS. Currently there are 144 detectors
arranged into eight different scattering angle banks.

solid angles to optimize data rates. In the case of the GPPD, banks of detectors are located symmetrically on both sides of the incident beam at 2θ=150, 90, and 60°. Separate single detector banks are located at 30 and 20°. Twice the data rate is realized by simply adding the data together from symmetrical detectors on both sides of the instrument. Within a particular bank of detectors, the TOF values for a particular d-spacing or Bragg peak with Miller indices hkl, appear at slightly different TOF values (Equation 2). For the GPPD, this problem is solved with electronic time-focussing of the data[10].

The time-focussing condition, which is essentially a constant $Q=4\pi\sin\theta/\lambda$ requirement can be expressed as:

$$t'=t_i\sin\theta_o/\sin\theta_i,$$

where t' is the time-focussed time-of-flight value for a neutron event at a detector with Bragg angle θ_i. The reference angle for the entire group of detectors in the bank is defined by θ_o. This means that after the time-focussing algorithm has been applied to each neutron event, a single histogram (neutron intensity vs time-of-flight) representing the collective bank of detectors is obtained. The data acquisition rate is then proportional to the number of detectors in a particular detector bank. One practical limit on solid angle for each detector group and hence count rate for the time-focussed group of detectors is set by desired requirements of acceptable resolution. The scattering angle dependent component of the resolution function of the instrument is proportional to $\cot\theta d\theta$. For example at 2θ=150°, as the solid angle of the bank is extended to smaller scattering angles, significant reductions in resolution must be obtained. The present detector complement at 2θ=150° is 31 on each side of the instrument, which spans a total solid angle of 0.13 steradians. There are 144 detectors in the diffractometer. Cost considerations are still the limiting constraint on detector complement, not resolution considerations. For the configuration of detectors at 2θ=150°, the time-focussed data rate is roughly 60 times that for a single detector.

The beam size for the GPPD at the sample position has rectangular cross section, 1.27cm wide and 5cm high (samples with cylindrical geometry that match these dimensions are typically used). The neutron absorption cross section for most elements is much smaller than the appropriate quantity for x-rays. If the quantity of available sample is not restrictive, the full beam cross section can be employed, thus providing for substantial increases in neutron data rate. A typical sample might require 4-24 hours of data collect time to produce one

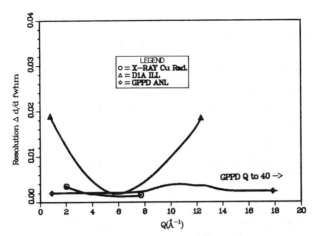

Figure 2. Resolution, $\Delta d/d$ at full width at half maximum, as a function of $Q=4\pi\sin\theta/\lambda$. The first curve denoted in the legend describes the resolution for a standard x-ray powder diffractometer. The experimental data used to generate a smooth curve provide the low-Q limit shown. The second curve in the legend is for D1A, the high resolution powder diffractometer at the ILL. The last curve described is that for the GPPD at IPNS. Notice the enormous Q range available on this instrument.

histogram (intensity vs TOF) for each of the time-focussed banks of detectors.

In Figure 2, the resolution $\Delta d/d$ at fwhm is shown as a function of $Q=4\pi\sin\theta/\lambda$. The first curve marked in the legend of the figure is the experimentally determined resolution obtained for Cu Kα radiation from a sealed x-ray tube using a standard GE x-ray diffractometer with a 1° beam slit, a 0.1° scattered beam slit and a 4° take-off angle. Notice that with Cu radiation, the maximum $Q_{max}=8\text{Å}^{-1}$. For Mo radiation, the limitations on large Q values are less severe and $Q_{max}\sim18\text{Å}^{-1}$. In either of these cases, the K-edge involves a doublet of characteristic lines. For the two other resolution examples in Figure 2, this is not the case.

The second curve marked in the legend of Figure 2 is that for the high resolution D1A powder diffractometer[11,12,13] at the Institute Laue-Langevin (ILL), with $\lambda\sim1.1\text{Å}$. This instrument is often considered the world standard for comparison of constant wavelength neutron powder diffractometers. The minimum value of the instrumental resolution function is determined by the take-off angle of the incident beam monochromator. The Q dependence of the resolution function is controlled mainly by the angular divergences in various parts of the beam path. The D1A instrument has an additional degree of versatility in that the take-off angle on the monochromator can be adjusted at will. The minimum of the resolution function can then be tuned to a particular experiment. The last curve listed in the legend is that for the 150° scattering angle bank of the GPPD at IPNS. Notice the enormous Q range spanned by the instrument. The resolution of this instrument is roughly independent of TOF. This can be understood (see Section I) by considering that the source pulse shape widens as $1/v$, so that resolution, $\Delta t/t = \Delta d/d =$ constant. In practice there are other moderator-dependent contributions to the instrumental resolution function, but these and other particular details of the instrument will be discussed elsewhere. It should be emphasized that neutron scattering lengths as opposed to x-ray atomic scattering factors, $f(Q)$, are scalars. This means that for high resolution neutron powder diffractometers, only thermal vibration effects attenuate the Bragg signal at high Q.

One unique feature of the GPPD is that excellent resolution is obtained at large d-spacings. Standard x-ray methods to measure small strain values emphasize back-scattering or large scattering angle measurements. For a steady state instrument that uses an incident beam monochromator, the emphasis is on Bragg reflections at or near the take-off angle minimum in the resolution function. The corresponding d-space values are intermediate ones. With reasonable freedom to choose the incident beam wavelength, this restriction becomes less severe. In the TOF case, the best strain resolution is obtained at large d-space values. As discussed above, the resolution is approximately independent of d-space and the usual systematics for accumulating data is with constant bin widths in the histogram. At large d values, the density of data points that describe a Bragg peak are largest, thus providing the best strain resolution. As we shall see in the next section, one disadvantage in the TOF case is that the flux-on-sample for largest d values is low. Thus the time required to achieve acceptable statistics is increased.

IV. RECONSTRUCTION OF TOF DATA FOR STANDARD LEAST-SQUARES ANALYSIS

We begin by noting the expression for the intensity of a powder diffraction Bragg peak using the TOF method:

$$P(hkl) = \phi(\lambda)\,\eta(\lambda).(\lambda^4/8\pi r).1_s V.(\rho'/\rho).jN^2|F|^2 A(hkl)E(hkl)LP(\theta). \qquad (3)$$

The incident beam neutron flux is described by $\phi(\lambda)$, $\eta(\lambda)$ is the wavelength dependence of the detector efficiency, r is the distance between sample and detector, 1_s is the length of the detector, V is the volume of the sample in the neutron beam, ρ is the density, j the powder multiplicity for the hkl of interest, N is the number density of unit cells illuminated by the beam, F is the structure factor, A(hkl) is the absorption correction, E(hkl) is the extinction correction,, and $LP(\theta)$ is the Lorentz factor. This expression clearly reflects the λ or TOF

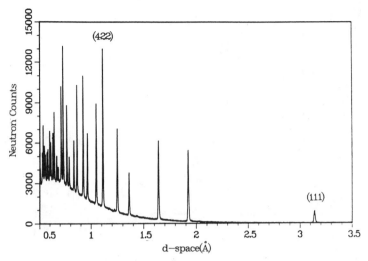

Figure 3. Raw data plot of neutron counts as a function of d-space for a silicon powder sample (2θ=90° detector bank).

relationship between Bragg integrated intensity and the structure factors, F. Note that unlike the usual steady state relation, we must now consider the flux on sample quantity, $\phi(\lambda)$, and the wavelength dependence of the detector efficiency, $\eta(\lambda)$. Figure 3 illustrates the wavelength dependence of the TOF data from silicon powder using a pulsed neutron source. The data are from the 90° scattering angle banks of the GPPD. The nominal resolution, $\Delta d/d=0.0045$. If we can measure $\phi(\lambda)$ and $\eta(\lambda)$, and correct the data for the λ^4 dependence, then we can extract a quantity that is simply proportional to the structure factors, F. This is accomplished by a separate experiment that measures the scattering from a calibrated vanadium sample (to obtain the product of $\phi(\lambda)$ and $\eta(\lambda)$). The scattering from vanadium is dominated by incoherent scattering[14] (4.8 barns) and therefore produces a constant cross section (neglecting multiple scattering and absorption effects) as a function of TOF. The vanadium Bragg peaks make only a miniscule contribution to the total scattering (the scattering length for vanadium[14], $b_v=-0.04 \times 10^{-8}$ cm while typical scattering lengths are 10-30 times greater than this value and intensity is proportional to b^2). We can construct a simple transformation that renormalizes the silicon data in Figure 3 to remove these wavelength dependent features. Figure 4 shows the results of this transformation. Apart from the λ^4 dependence given in Equation (3), the data in Figure 4 now have an appearance similar to that associated with standard steady state techniques. A standard crystallographic least-squares program can then be used to refine structural parameters. However, in the case of overlapped reflections, it is useful to develop a

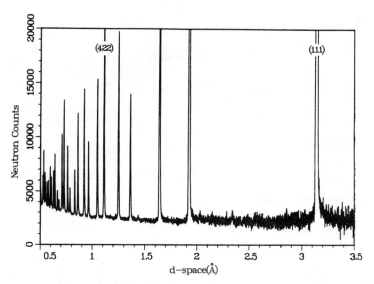

Figure 4. Renormalized data plot of neutron counts as a function
of d-space. Vanadium data (not shown) were used to perform a
point-by-point renormalization of the silicon data shown in
Figure 3. The apparent scatter in the data at large d-space is
due to the multiplicative character of the transformation.

technique to recover the integrated intensities of the individual
profiles.

   A computer code was written to fit multiplets of Bragg profiles using
the transformed silicon data illustrated in Figure 4 as input. This code
has been successfully used to study the effects of grain interaction
stresses[15] in metallurgical samples. The assumption is that the
background is reasonably well-determined on both sides of the multiplets.
A linear background function is assumed in the TOF region of the
multiplet. Figure 4 shows that these assumptions appear reasonable. A
moderate degree of structural complexity is accommodated by using this
code to analyze partially overlapped Bragg reflections. The integrated
intensities of the silicon Bragg reflections are thus obtained. The
results are illustrated in Figure 5, where we plot the integrated
intensities as a function of $h^2+k^2+l^2$, which is proportional to Q^2, where
$Q=4\pi\sin\theta/\lambda$. The very weak h+k+l=4n+2 reflections were not analyzed. A
correction for absorption has been applied to the data. Note that 48
independent Bragg integrated intensities were obtained from the data
illustrated in Figure 4 using TOFMANY to help resolve the partially
overlapped reflections at high Q. Rietveld[4] profile refinement analysis
was also carried out on the data illustrated in Figure 3. The Rietveld
results[5] gave a weighted profile residual R_{wtp}=0.031; the silicon
temperature factor B(Si)=0.43(1)Å^2 and the lattice parameter,

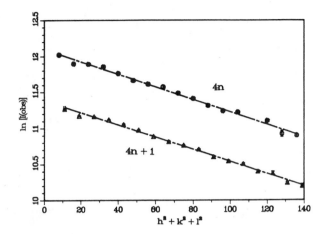

Figure 5. The ln[I(obs)] vs $h^2+k^2+l^2$. Note that $h^2+k^2+l^2 \alpha Q^2$. The silicon integrated intensities were obtained from the data illustrated in Figure 4, using the TOFMANY computer code. The very weak h+k+l=4n+2 reflections were not analyzed.

a_o=5.4309(1). Using the data in Figure 5, $R(F^2)$=0.023 and $B(Si)$=0.45(1)Å^2. The results are equivalent.

Another example of multiplet peak fitting program performance is illustrated in Figure 6. The data are from Al_2O_3. A selected TOF range from the 2θ=150° detector banks was chosen. For the TOF range illustrated, a total of six Bragg reflections were detected. The d–

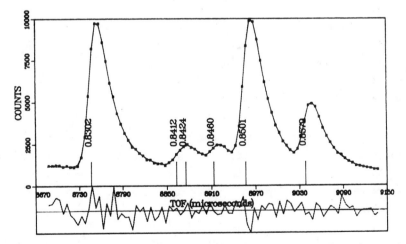

Figure 6. Point by point observed and calculated neutron intensities for a multiplet of six Bragg reflections from Al_2O_3 using the 150° detector bank of the GPPD. The raw data are marked by symbols and the solid line represents the calculated value. The vertical tick marks locate the calculated positions of the peaks in d–space coordinates. The differences between theory and experiment are illustrated in the bottom of the figure. The scale is chosen so that the largest difference fills the allotted vertical space (±300 neutron counts).

Table 1. Integrated intensity comparison for a multiplet of six
Bragg reflections from Al_2O_3. The intensities are normalized to
the strongest reflection in this multiplet (see Figure 6). The
Miller indices of the reflections correspond to a hexagonal unit
cell representation. Estimated standard deviations for the
Rietveld-determined integrated intensities were not given in the
analysis.

hkl	d-space(Å)	I(Rietveld)	I(TOFMANY)
(4,1, 6)	0.8302(2)	955	984(10)
(2,1,13)	0.8412(2)	81	101(21)
(3,2, 7)	0.8424(4)	107	67(20)
(2,0,14)	0.8460(2)	145	136(3)
(3,0,12)	0.8501(2)	1000	1000(4)
(1,3,10)	0.8579(2)	428	430(4)

spacings for the Bragg reflections are marked by ticks in Figure 6. For
the multiplet fitting results, a measure of goodness of fit can be
defined by:

$$\chi^2 = \Sigma(\Delta I^2)/N \qquad\qquad\qquad (4)$$

where ΔI is the weighted differences between theory and experiment,
summed over all data points, and N is the number of degrees of freedom of
the system (the difference between the the number of observations and the
number of adjustable parameters). The results illustrated in Figure 6,
gave $\chi^2 = 1.3$. A question might be raised here concerning the significance
of the fit for a 6-peak fit in the TOF range illustrated in Figure 6. To
answer this question, least-squares analysis was carried out using five
peaks in the TOF region illustrated in Figure 6. In this case, $\chi^2 = 1.8$ was
obtained. The fit was clearly not of the quality illustrated in Figure 6.
Alternatively, when a multiplet of seven reflections was used in the
analysis, convergence was not achieved. From a Rietveld analysis of the
entire TOF range, integrated intensities of these reflections were
obtained. A comparison of the Rietveld and TOFMANY results are shown in
Table 1. The intensity values indicate excellent agreement between the
methods. Values for the d-spaces for the reflections were also determined
in the analysis.

V. DISCUSSION

The method to recover the integrated intensities of Bragg peaks
described above has several other important applications in powder
diffraction studies. Suppose the structure of the material is unknown.

Our multiplet fitting program can still be used as an analytic tool to determine the d-spacings of the Bragg lines present in the pattern. Indexing programs that attempt to assign unit cell size and shape can then be tried. The results of a preliminary study to solve the structure of FeOCl intercalated with TTF are available[1,16]. Consider for example particle size, anisotropic strain[15], compositional inhomogeneity, complex polycrystalline texture, and multi-phase metallurgical effects[17]. All of these may require specialized peak shape descriptions that are not presently implemented in Rietveld structure refinement computer codes. The simplicity of the TOFMANY code allows for straightforward modifications of the peak shape function used to fit the Bragg profiles. The minimization method does not require derivatives of the peak shape function. This means that functions which are not closed analytical expressions in a mathematical sense can be easily implemented in the program. For materials with very complicated crystal structures, the choice of techniques for structural refinement become severely limited. Analysis methods that rely on single peak or multiplet peak descriptions must be abandoned, but the Rietveld structure refinement technique is available. The success of this technique is well-known[1].

ACKNOWLEDGEMENTS

This work was supported by the U. S. Department of Energy, BES-Material Sciences under Contract W-31-109-Eng-38.

REFERENCES

[1] IPNS Progress Report 1983-1985, Argonne National Laboratory, IPNS Division, Argonne, IL 60439 and references contained within. Proposal forms for experiments are also available.

[2] G. H. Lander and D. L. Price, "Neutron Scattering with Spallation Sources", Phys. Today 38, 38 (1985).

[3] J. M. Carpenter, G. H. Lander and C. G. Windsor, "Instrumentation at Pulsed Neutron Sources", Rev. Sci. Instrum. 55, 1019 (1984).

[4] H. M. Rietveld, "A Profile Refinement Method for Nuclear and Magnetic Structures", J. Appl. Crystallogr. 2, 65 (1969).

[5] R. B. Von Dreele, J. D. Jorgensen and C. G. Windsor, "Rietveld Refinement with Spallation Neutron Powder Diffraction Data", J. Appl. Crystallogr. 15, 581 (1982).

[6] J. D. Jorgensen and F. J. Rotella, "High-Resolution Time-of-Flight Powder Diffractometer at the ZING-P' Pulsed Neutron Source", J. Appl. Crystallogr. 15, 27 (1982).

[7] R. K. Crawford, R. T. Daly, J. R. Haumann, R. L. Hitterman, C. B. Morgan, G. E. Ostrowski and T. G. Worlton, "The Data Acquisition System for the Neutron Scattering Instruments at the Intense Pulsed Neutron Source", IEEE Trans. Nucl. Sci. **NS-28**, 3692 (1981).

[8] J. R. Haumann, R. T. Daly, T. G. Worlton, and R. K. Crawford, "IPNS Distributed Processing Data Acquisition System", IEEE Trans. Nucl. Sci. **NS28**, 62 (1982).

[9] J. Faber, Jr., "Sample Environments at IPNS: Present and Future Capabilities", Revue Phys. Appl. **19**, 643 (1984).

[10] J. D. Jorgensen and J. Faber, Jr., "Electronically Focussed Powder Diffractometers at IPNS-I", ICANS-VI Meeting, Argonne National Laboratory, June 27, 1982, ANL Report ANL-82-80, 105 (1983).

[11] A. W. Hewat, "Design for a Conventional High Resolution Powder Diffractometer", J. Nucl. Instrum. Methods **127**, 361 (1975).

[12] A. W. Hewat and I. Bailey, "D1A: A High Resolution Neutron Powder Diffractometer with a Bank of Mylar Collimators", J. Nucl. Instrum. Methods **137**, 463 (1976).

[13] Annual Report 84, Institut Laue Langevin, Grenoble, France, pp 96-97.

[14] L. Koester and A. Steyerl, "Neutron Physics" in Springer Tracts in Modern Physics, **80**, Springer-Verlag, NY 1977, pg 37.

[15] S. R. MacEwen, J. Faber, Jr., and A. P. L. Turner, "The Use of Time-of-Flight Neutron Diffraction to Study Grain Interaction Stresses", Acta Metall. **31**, 657 (1983).

[16] B. A. Averill, S. M. Kauzlarich, B. K. Teo and J. Faber, Jr., "Structural and Physical Studies on a New Class of Low-Dimensional Conducting Materials: FeOCl Intercalated with TTF and Related Molecules", to be published in Molecular Crystals and Liquid Crystals, 1985.

[17] A. D. Krawitz, R. Roberts and J. Faber, Jr., "Residual Stress Relaxation in Cemented Carbide Composites Studied Using the Argonne Intense Pulsed Neutron Source", Advances in X-Ray Analysis **27**, University of Denver-Plenum Press, New York, 1984, pp 239-249.

RECOGNITION AND TREATMENT OF BACKGROUND PROBLEMS

IN NEUTRON POWDER DIFFRACTION REFINEMENTS

W. H. Baur and R. X. Fischer

Department of Geological Sciences, Box 4348
University of Illinois at Chicago
Chicago, Illinois 60680

ABSTRACT

Irregular backgrounds in time-of-flight neutron powder diffraction data and nonlinear backgrounds in angle dependent neutron powder diffraction data. were corrected by fitting fifth degree polynomials to those portions of the data most affected. The anomalous background intensities were in both cases due to non Bragg scattering. The polynomial fitting was carried out over a sufficiently wide range of the profile to avoid interfering with the Bragg peaks. The corrected data gave internally consistent results for the crystallographically nonequivalent (Si,Al)-O(1) and (Si,Al)-O(2) bond lengths in zeolite rho and compare favorably with previous refinements of this zeolite in space group Im$\bar{3}$m. Internal and external inconsistencies of bond lengths are just as diagnostic of refinement difficulties as negative isotropic temperature factors or non positive definite anisotropic temperature factors are.

INTRODUCTION

The analysis of powder diffraction data for the refinement and solution of crystal structures has made great advances in the years since Rietveld (1969) introduced a crystal structure refinement method for crystalline powders using profile intensities instead of integrated intensities. A recent comprehensive summary of the state of the art has been given by Shirley (1984). We wish to address a specific problem, namely the analysis of powder diffraction data contaminated by broad, diffuse and irregular background peaks. This is an important topic when trying to refine the crystal structures of zeolites, which are notoriously difficult to prepare in pure form.

The success of a crystal structure refinement depends on the quality of the data set and on the correct choice of the initial least-squares parameters. Especially in powder diffraction the starting model for the crystal structure must be close to the real structure, otherwise convergence will not be achieved. When the sample is well crystallized peaks can be observed and resolved from the background down to d $\simeq$ 0.9Å and sometimes even further. Unfortunately mainly these regions are affected by irregular background scattering in most of our zeolite samples. These peaks contain valuable information and should not be neglected. Usually the local software based on the Rietveld code and modified for the special purposes of a specific powder diffractometer provides well developed routines for fitting profile shape parameters and for modelling the background due to instrumental effects, incoherent scattering and diffuse scattering from random disorders in the crystal structure. These effects are more or less regular and can be described by simple functions.

Neutron powder diffraction profiles may contain sharp or diffuse peaks which are not due to the sample in which we are interested, but instead to admixed crystalline or amorphous phases which remain after sample preparation. In addition there may also occur peaks, again sharp or diffuse, caused by sample containers, furnaces etc. As long as these additional unwanted contributions are Bragg peaks, they can be either removed from the pattern, or if they come from identifiable phases they can be included in the refinement by using a multiphase Rietveld type refinement code. In the case of broad diffuse peaks this is not possible because we might not have a proper model for them, or if we choose to remove them we have to sacrifice a large portion of our valuable information. If we leave these broad peaks in the pattern, they contribute to an irregular background, which will dramatically affect the results of the refinement.

In this paper we illustrate how the background was treated in a neutron powder diffraction data set taken of zeolite rho (Fischer et al., 1986b) on the Special Environment Powder Diffractometer at the Intense Pulsed Neutron Source (IPNS) at Argonne National Laboratory (Jorgensen & Faber, 1983). The diffraction pattern of this zeolite rho showed broad peaks apparently caused by scattering from an additional amorphous phase present in the sample. We also explain how the background was smoothed by using the data sets collected for zeolite rho and ZK-5 on the high resolution five counter neutron powder diffractometer of the National Bureau of Standards Reactor (NBSR), Washington, DC (Fischer et al. 1986c). In none of these cases was it possible to get a successful refinement without additional background corrections. Richardson et al. (1985) are using Fourier-filtering techniques to remove irregular background in analogous circumstances.

BACKGROUND RECOGNITION AND STANDARD CORRECTION METHODS

The currently available background correction methods fall principally into two categories, in one the background is visually estimated and corrected prior to the least-squares procedure, in the other the fitting of the background is part of the least-squares refinement:
1. The simplest way to correct for the background is to subtract the intensities below a straight line between the left and right tails of the peak. Difficulties in finding the linear background for a single peak are discussed by Enzo and Parrish (1984) who also developed a method of extrapolating the tails of partly overlapping peaks. However, when there are regions in the profile where so many peaks overlap (Fig. 1) that it is impossible to resolve the contribution of the Bragg peaks from the background scattering, the background cannot be estimated by any simple graphical method. The visual assignment of background becomes the more difficult the broader the Bragg peaks are (Fig. 2) and impossible when there is also a contribution of scattering from amorphous matter. This is shown in Fig. 1, where the sample represents an intermediate crystalline state (with a component of scattering from an amorphous phase) as compared to the calculated profiles of purely crystalline phases (Fig. 2).
2. Many programs provide background fitting routines using functions for the determination of the background which model the background giving a reasonably good fit for well defined and pure samples. The code written by Wiles and Young (1981) fits the background with fifth degree polynomials which can compensate for some irregularities in the profile shape. Baerlocher (1984) introduced a new version of his X-ray Rietveld system at the Thirteenth

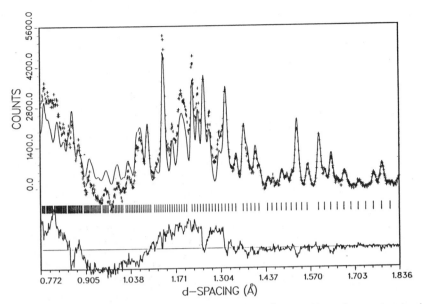

Fig. 1: Observed (crosses) and calculated (line) profile after *standard* correction with Chebychev polynomials, with difference plot underneath (on the same scale as the line profile) showing the deviations between observed and calculated intensities. The calculated profile represents the *final* refinement of the crystal structure of a zeolite rho (Fischer et al., 1986b). Tickmarks indicate positions of Bragg peaks.

International Congress of Crystallography in which he uses orthogonal polynomials for various regions of the profile. Immirzi (1980) fits bell-shaped bumps, caused by amorphous materials in the sample, with a Pearson VII function. The parameters are adjusted in the least-squares refinement.

In the codes used by us the background is described by simple functions. The program in use at NBS (Prince, 1980) is a modification of Rietveld's (1969) original code, adapted to the five detector powder diffractometer at NBS (Prince and Santoro, 1980). It approximates the background by a simple exponential function for data points from the first counter up to $2\theta=30°$. For the remaining four counters the background is assumed to be linear.

The program used for Rietveld type refinement of time-of-flight powder diffraction data is also derived from Rietveld's (1969) program. It was adapted to the time-of-flight method by Von Dreele et al. (1982). It allows the choice between four different functions, the most flexible of which, in terms of adjusting to the most diverse backgrounds, is a shifted Chebychev polynomial (Rotella, 1982; Von Dreele, personal communication).

As said above, a simple way to treat irregular background is to subtract it by linear interpolation between estimated values of the background at a large number of positions. The programming system at IPNS (Rotella, 1982) provides such an option in its data preparation step (written by Von Dreele). The problem we see with this, is that sometimes the peaks overlap to such a degree that the underlying background cannot be defined with any confidence. This happened in the data sets of zeolite rho and Zk-5, which were collected on high resolution instruments. Fig. 1 shows the observed intensities after fitting the background with the standard Chebychev polynomial in the data set of a shallow bed zeolite rho (Fischer et al., 1986b). The calculated profile is based on the final refinements of the crystal structure after removing the

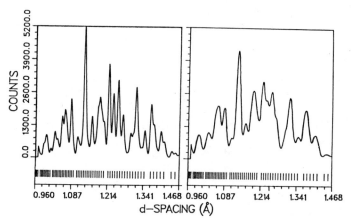

Fig. 2: Comparison of calculated backgrounds for pure phases (without any
amorphous component) assuming sharp well resolved peaks in the left figure and broader peaks in the right figure. The second case has broader
peaks than our zeolite rho samples. Visual estimation would place the
backgrounds at different levels in these two cases.

background by the polynomial fitting method. While calculated and observed
patterns show a reasonably good fit for the region above d ≃ 1.35 Å, the profile for the lower d-values cannot be fitted. The refinement yields positional coordinates which give a highly distorted framework and shows negative
isotropic temperature factors for most atoms. Obviously we see two different
effects here: big humps in the profile are caused by overlapping peaks and
by irregularly increased background intensities. The difference intensity
fluctuates between positive and negative values. This is due to the misfit
of the standard Chebychev polynomial which tends to compensate for the humps
by cutting off parts of the profile (see Figs. 1 and 4 between d = 1.4 and
1.8 Å). A visual estimation of the background in these regions is impossible
and certainly would result in an improperly corrected intensity set.

RECOGNITION OF POOR REFINEMENTS DUE TO BACKGROUND PROBLEMS

It is common to question the validity of a refinement of a crystal
structure when it results in negative isotropic temperature factors or non
positive definite anisotropic temperature factors. The same cannot be said
about checks for reasonableness in the individual and mean bond distances
resulting from the refinement. However, these can be just as diagnostic for
refinement problems as the temperature factors. Refinements in which we find
physically unreasonable values of temperature factors and/or bond lengths and
angles can be due to poor diffraction data or to the use of a wrong model.
The least-squares refinement can converge then into false minima. A model is
wrong when we try to refine a structure in a wrong space group, or when atoms
have been placed into wrong positions, or atoms belonging into the structure
have been left out.

Grid Searches. To guard against the latter possibility we search, after
we have obtained a partial model, the asymmetric unit on a grid with a step
size of 0.3Å. For each of the positions on this grid the occupancy of a
dummy atom is refined. The occupancy factors and the residuals are then
plotted in XY layers of the unit cell. In this fashion we are able to locate
atomic positions which remain stable in subsequent refinements and that are

difficult to locate from Fourier or difference maps (Fig. 3). Such maps are
calculated by using Fourier coefficients obtained from the overlapped peaks
by separation weighted according to calculated structure factors. This means
that they are always biased towards the current model. This is similar to
the situation one encounters when calculating successive Fourier syntheses in
noncentrosymmetric structures in the single crystal case.

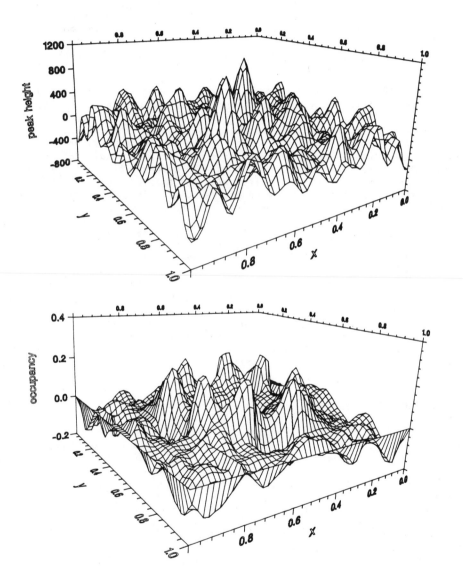

Fig. 3: Comparison of an xy-layer (z=0.34) of a difference Fourier map (top)
with the corresponding layer of a grid search map (bottom) of zeolite
rho. The difference plot has the 'egg carton' appearance typical of a
Fourier synthesis dominated by just a few low order terms. The peaks in
the two maps do not always coincide. Peaks found in a grid search have
a high probability of remaining stable in subsequent Rietveld refinements. The peaks in the two symmetrically independent quadruplets in
the grid search map are at positions which were assigned to deuterium.
Their coordinates remained stable in the refinement.

Once we are reasonably sure to have a complete model, we try to ascertain that the results do not have problems related to the following:

Thermal parameters. Unusual temperature factors, very large, or negative when isotropic, or non positive definite when anisotropic are indications for unsuccessful refinements which are widely accepted as being diagnostic of refinement difficulties. If such unusual thermal parameters are clearly significant one should try and find out the reasons for these values. Incompletely removed background intensities can be partly compensated in the refinement by very small or negative temperature factors. We found repeatedly that negative temperature factors adjusted to normal values after the background correction. Even the refinement of anisotropic temperature factors gave reasonable results (Fischer et al., 1986a).

External consistency of bond lengths. The values of the bond lengths and angles resulting from the refinement should compare favorably with known empirical knowledge. In the case of largely heteropolar compounds, such as most oxides and fluorides, the mean bond lengths from the cations to the anions within a given coordination polyhedron should have values agreeing well with the sums of the corresponding effective ionic radii (Shannon, 1976). Deviations of more than 0.03Å between observation and the sum of the radii, might indicate either problems with the assumed chemical composition or with the refinement. In some cases, especially for small cations with high formal charge it is even possible to calculate on the basis of the observed bond strength distributions the expected individual bond lengths and compare them with the observations (Baur, 1970). The deviations between observed and calculated individual distances should not exceed several hundredths of an Ångstrom. Specifically for the silicates with which we dealt mostly, there are many possible checks on the quality of the refinement.

Internal consistency of bond lengths. Even without a comparison with other structural information it is possible to check for the internal consistency of a refinement by comparing within a given structure the lengths of crystallographically independent but chemically equivalent bonds with each other. We call bonds chemically equivalent when the topology of their neighbors is equivalent (Baur et al., 1983).

BACKGROUND TREATMENT

The neutron powder diffraction data of a zeolite rho (Fischer et al., 1986b) showed a strong, prominent and very broad peak at a d value of about 1.2Å (Fig. 4). The calculated profile in this figure is based on the refinement of scale factor, lattice constant, positional and thermal parameters of the framework atoms, peak shape and background parameters. The curve immediately below the diffraction profile is an approximation of a shifted Chebychev polynomial to the observed experimental background. These polynomials are one of the background functions included in the Rietveld code written by Von Dreele, Jorgensen & Windsor (1982). It can be seen that the region between d ~0.95Å and ~1.4Å cannot be fitted properly (Fig. 4).

Before performing additional background corrections, the crystal structure was refined from intensity data with d>1.43Å, thus omitting the problematic background region. The refined atomic parameters were then held constant in further refinements of the full data range with d>0.95Å. In a separate least-squares program parameters for a polynomial of fifth degree were calculated which fit the $I_{obs}-I_{calc}$ data points in the region between d = 0.95Å and 1.35Å (Fig. 5). The polynomials are of the form

$$B(d) = a_0 + a_1 d + a_2 d^2 + a_3 d^3 + a_4 d^4 + a_5 d^5$$

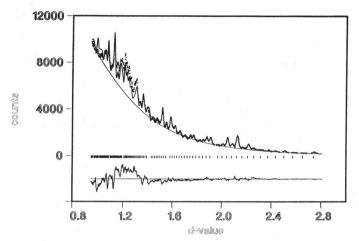

Fig. 4: Observed (dots) and calculated (line) profile before background
treatment. The curved, monotonically decreasing line under the diffraction profile is the refined background, reflecting mostly the distribution of neutron wavelengths in the incident spectrum.

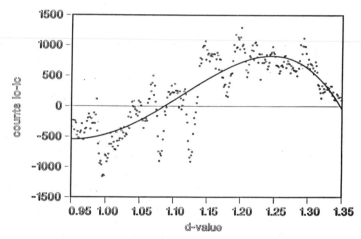

Fig. 5: Difference plot for the range 0.95Å < d < 1.55Å with the curve representing the polynomial of fifth degree.

The values B(d) were subtracted directly from the raw data, and the resulting
difference values ($I_{obs}-I_{calc}$) were again fitted after refining each time the
profile parameters. After three iterations of this procedure the profile was
smoothed, giving a reasonable fit between observed and calculated intensities
(Fig. 6). The final results are given in Table 1 along with the corresponding values before the background subtraction. The same procedure has been
applied to six other data sets collected on zeolite rho at IPNS with similar
results.

    The treatment of the background and the refinement of the crystal struc-
ture of another zeolite rho with step scan data from the five detector neutron powder diffractometer (Prince & Santoro, 1980) has been described by

Table 1. Significant results of the structure refinements of a) the framework of zeolite rho in various stages of the background correction
procedure and b) the complete structure of zeolite rho compared to
previous refinements [T = (Si,Al)]. The definitions
of the residuals are: $R_I = \Sigma_i |I_{oi} - 1/C \cdot I_{ci}|/\Sigma_i I_{oi}$ and $R_{wp} = [\Sigma_i w_i (y_{oi} - 1/C \cdot y_{ci})^2/\Sigma_i w_i y_{oi}^2]^{\frac{1}{2}}$.

a) procedure	No. of data points	No. of Bragg refl.	d-value [A] min	max	R_I [%]	R_{wp} [%]	T-0 (1)	T-0 (2)
framework refined	1842	259	0.955	4.9	24.9	7.1	1.639	1.628
framework refined	1456	82	1.431	4.9	14.4	5.8	1.606	1.651
framework fixed, 1.iteration	1842	259	0.955	4.9	18.8	4.8	-	-
framework fixed, 2.iteration	1842	259	0.955	4.9	17.1	4.5	-	-
framework refined	1842	259	0.955	4.9	15.1	4.1	1.628	1.624
framework refined, 3. iteration	1842	259	0.955	4.9	13.7	3.8	1.627	1.624
framework refined, anisotr. T-fact., peak shape param. changed in data prep. program	3032	259	0.955	4.9	12.6	4.2	1.626	1.623
b) reference								
Parise et al. (1984)	-	48	1.695	19.5	7.5	14.3	1.719 (14)	1.585 (9)
McCusker (1984)	2865	83	1.344	12.1	5.8	10.8	1.624 (5)	1.596 (4)
McCusker & Baerlocher (1984)	2737	127	1.199	11.3	13.4	18.0	1.684 (8)	1.635 (5)
Fischer et al. (1986a)	2094	227	0.941	12.6	13.7	8.7	1.623 (11)	1.596 (8)
Fischer et al. (1986b)	2330	245	0.967	4.9	7.5	3.6	1.625 (5)	1.625 (4)
"	2270	259	0.948	4.9	6.5	2.7	1.628 (4)	1.621 (4)
"	2876	259	0.948	4.9	9.9	3.8	1.632 (6)	1.616 (5)
"	2389	259	0.948	4.9	6.5	3.0	1.633 (5)	1.619 (5)

Fischer et al. (1986a). The background, which usually can be fitted by a
linear function for well crystallized samples in the 2θ-range above 15°,
shows broad peaks due to diffuse scattering from the sample holder. The same
effect is observed in the neutron powder pattern of ZK-5 (Fischer et al.
1986c) as shown in Fig. 7. Visually selected data points, representing the
contribution of the background to the profile, are plotted for the two samples of zeolite ZK-5 (calcined at 500°C and 650°C, respectively) in the range
of the first four counters of the diffractometer. The curves represent polynomials of the form

$$B(\theta) = a_0 + a_1\theta + a_2\theta^2 + a_3\theta^3 + a_4\theta^4 + a_5\theta^5$$

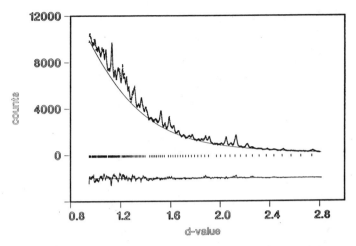

Fig. 6: As Fig. 4, but after background treatment.

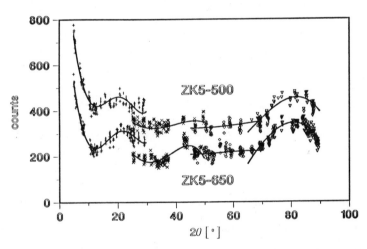

Fig. 7: The contribution of the background to the profile in zeolite ZK-5
calcined at 500°C and 650°C respectively. Data points represent observed
intensities which contribute to the background; the curves represent the
background functions.

The parameters a_i of this function were calculated in a separate least-
squares program for each of the four counters. The parameters a_0 and a_1 were
then refined together with the other profile and structural parameters in a
modified Rietveld program (Prince, 1980). The necessity of the introduction
of nonlinear functions for fitting the background as shown in Fig. 7 is
emphasized and confirmed by the improvement of the results. The residuals
are lower and the refined interatomic distances agree better with accepted
values after the background correction with the polynomial functions has been
performed. The R_I dropped on average by 5% after smoothing the background,
and the maximum differences between crystallographically inequivalent, but
chemically similar (Si,Al)-O distances were reduced from a maximum value of
0.34Å to a maximum value of 0.12Å.

COMPARISONS OF (Si,Al)-O DISTANCES IN ZEOLITES

The zeolite rho samples for which diffraction data were collected at
IPNS have the approximate chemical composition $D_{10}Al_{10}Si_{38}O_{96}$ and crystallize
in space group $Im\bar{3}m$. In part b. of Table 1 we compare their bond lengths
with those observed for all other zeolites rho found to crystallize in $Im\bar{3}m$.
Obviously the differences between T-O(1) and T-O(2) (Table 1) are after our
background treatment the smallest observed so far. We take this as an indication of the high internal consistency of our results. Conversely it is
unlikely that the difference of 0.13Å observed for one of the other samples
is real. Probably this is an artifact of the refinement of not completely
corrected diffraction data or due to the paucity of data after the regions of
irregular background were completely removed.

The geometry of the aluminosilicate framework in our zeolite rho is also
consistent with external evidence. The mean bond length (Si,Al)-O is
1.610(10)Å in a rho calcined at 650°C (Fischer et al., 1986a). In this sample 6.7% of the silicon atoms are substituted by aluminum atoms according to
an NMR analysis. We can use empirically derived bond lengths for comparison.
The mean Al-O bond length for a tetrahedron in which all oxygen atoms are
shared with neighboring Al or Si tetrahedra is 1.746Å (Baur & Ohta, 1982).
For the mean Si-O bond length we use an empirical equation (#8, Baur, 1978):

$$(Si-O)_{mean} = (1.554+0.034NSECM+0.0045CNM)Å$$

where NSECM is the mean value per tetrahedron of the negative secant of the
angle Si-O-Si or Si-O-Al, and CNM is the mean coordination number of all oxygen atoms in the tetrahedron. Taking 2 for CNM and using the
(Si,Al)-O-(Si,Al) angles as observed by us, the estimated mean for this zeolite is 1.613Å. The agreement with the observed mean (Si,Al)-O distances is
also extremely good in all the other zeolites studied by us. Thus the geometry observed for the framework is both internally and externally consistent
with our previous empirical knowledge.

CONCLUSION

In cases where diffuse scattering or an amorphous phase contribute to
the powder diffraction data, the irregular contribution of the background to
the profile requires a background correction which cannot be performed by
simple functions. The fifth degree polynomials used by us for this purpose
are applied to a sufficiently wide region of the profile to avoid interfering
with the Bragg peaks. The treatment of the background as described in this
work was our first attempt to handle such irregularities in powder patterns.
The smoothing of the background with polynomials of fifth degree significantly improved the refinement. After the background corrections it was possible to locate deuterium atoms bonded to the framework oxygen atoms and to
find indications for the non framework positions of the aluminum atoms after
dealumination of the framework.

The case for the usefulness of this background correction is bolstered
by the comparison of the (Si,Al)-O(1) and (Si,Al)-O(2) distances in our
refinements of zeolite rho with those performed previously without taking
into account irregular contributions to the background. We were able to
refine these zeolite crystal structures well, despite the presence of large
background peaks due to amorphous matter. These refinements extend to a much
wider range of reciprocal space than was previously possible. It was possible to utilize reflexions down to d-values of 0.94Å and to include twice as
many I_{hkl} as in any previous refinement of a zeolite rho (Table 1b). In
addition the new refinements give results that are internally and externally
more consistent with acceptable empirical information than previous work.

We are convinced that bond lengths obtained from powder refinements should agree well with the best empirical bond lengths from high quality single crystal studies. If they do not agree, this disagreement is as diagnostic of refinement difficulties as negative thermal parameters are. The results are then unacceptable because large unexplained deviations of mean and individual interatomic distances from accepted values would, at face value, indicate that the interatomic forces which establish the interatomic distances are different, depending on whether we perform a structure determination on single crystals or on powders. This is unlikely.

Though the refinement may be statistically biased by the subtraction of fitted background intensities, as is the case whenever background corrections are made without refining background parameters together with the structural parameters, the results justify the procedure. However, in the future we plan to include the background correction in a Rietveld type least-squares refinement program.

ACKNOWLEDGEMENTS

We thank Drs. J. D. Jorgensen and E. Prince for valuable discussions. This work was supported by grants from E. I. du Pont et Nemours & Co. and from NATO (Research Grant No. 149/84 to WHB, E. Tillmanns, R. D. Shannon and G. D. Stucky). We also acknowledge the award of instrument time at IPNS (Argonne National Laboratory) and at NBSR, and of computer time at the Computer Center of the University of Illinois at Chicago.

REFERENCES

Baerlocher, C., 1984, The X-ray Rietveld System XRS-84, Acta Cryst., A40, C368.

Baur, W. H., 1970, Bond length variation and distorted coordination polyhedra in inorganic crystals, Trans. Amer. Cryst. Assoc., 6, 129.

Baur, W. H., 1978, Variation of mean Si-O bond lengths in silicon-oxygen tetrahedra, Acta Cryst., B34, 1751.

Baur, W. H., and Ohta, T., 1982, The Si_5O_{16} pentamer in zunyite refined and empirical relations for individual silicon-oxygen bonds, Acta Cryst., B38, 390.

Baur, W. H., Tillmanns, E., and Hofmeister, W., 1983, Topological analysis of crystal structures, Acta Cryst., B39, 669.

Enzo, S., Parrish, W., 1984, A method of background subtraction for the analysis of broadened profiles, in "Adv. in X-Ray Analysis, Vol. 27", J.B. Cohen, J.C. Russ, D.E. Leyden, C.S. Barrett, P.K. Predecki, eds., Plenum Press, New York.

Fischer, R. X., Baur, W. H., Shannon, R. D., Staley, R. H., Vega, A. J., Abrams, L., and Prince, E., 1985, Neutron powder diffraction study and physical characterization of zeolite D-rho deep-bed calcined at 500°C and 650°C, in preparation.

Fischer, R. X., Baur, W. H., Shannon, R. D., Staley, R. H., Vega, A. J., Abrams, L., D. R. Corbin, and Jorgensen, J. D. , 1986b, Neutron powder diffraction study and physical characterization of zeolite D-rho shallow-bed calcined at 500°C and 600°C, in preparation.

Fischer, R. X., Baur, W. H., Shannon, R. D., Staley, R. H., Vega, A. J., Abrams, L., and Prince, E., 1986c, Neutron powder diffraction study and physical characterization of zeolite ZK-5 calcined at 500°C and 650°C, in preparation.

Immirzi, A., 1980, Constrained powder-profile refinement based on generalized coordinates, Acta Cryst., B36, 2378.

Jorgensen, J. D., and Faber, J., 1983, Electronically focused powder diffrac-
    tometers at IPNS-I, *in* "Proc. 6th Meeting International Collab. Advanced
    Neutron Sources", Argonne National Laboratory, ANL-82-80.
McCusker, L. B., 1984, Crystal structures of the ammonium and hydrogen forms
    of zeolite rho, Zeolites, 4, 51.
McCusker, L. B., and Baerlocher, C., 1984, The effect of dehydration upon the
    crystal structure of zeolite rho, *In* "Proceedings of the 6th Interna-
    tional Conference on Zeolites, Reno, 1983", D. Olson and A. Bisio, eds.,
    Butterworth: Guildford.
Parise, J. B., Gier, T. E., Corbin, D. R., and Cox, D. E., 1984, Structural
    changes occuring upon dehydration of zeolite rho, J. Phys. Chem. *88*,
    1635.
Prince, E., 1980, Modification to the Rietveld powder refinement program,
    National Bureau of Standards, Technical Note 1117, 8.
Prince, E., and Santoro, A., 1980, The five detector neutron powder diffrac-
    tometer, National Bureau of Standards, Technical Note 1117, 11.
Richardson, J. W., Pluth, J. J., Smith, J. V., and Faber, J., 1985, Fourier-
    filtering techniques for the analysis of high-resolution pulsed neutron
    diffraction data, Abstracts of 34th Annual Denver Conference on Applica-
    tions of X-ray Analysis, Univ. Denver Research Institute, Denver.
Rietveld, H. M., 1969, A profile refinement method for nuclear and magnetic
    structures, J. Appl. Cryst. *2*, 65.
Rotella, F. J., 1982, "Users manual for Rietveld analysis of time-of-flight
    neutron powder data at IPNS", revised 1984, Intense Pulsed Neutron
    Source, Argonne National Laboratory.
Shannon, R. D., 1976, Revised effective ionic radii and systematic studies of
    interatomic distances in halides and chalcogenides, Acta Cryst., A*32*,
    751.
Shirley, R., 1984, Measurement and analysis of powder data from single solid
    phases, *in* "Methods and applications in crystallographic computing", S.
    R. Hall and T. Ashida, eds., Oxford Univ. Press: Oxford.
Von Dreele, R. B., Jorgensen, J. D., and Windsor, C. G., 1982, Rietveld
    refinement with spallation neutron powder diffraction data, J. Appl.
    Cryst. *15*, 581.
Wiles, D.B., and Young, R.A., 1981, A new computer program for Rietveld anal-
    ysis of x-ray powder diffraction patterns, J. Appl. Cryst., *14*, 149.

FOURIER-FILTERING TECHNIQUES FOR THE ANALYSIS OF

HIGH-RESOLUTION PULSED NEUTRON POWDER DIFFRACTION DATA

James W. Richardson, Jr.[+] and John Faber, Jr.[*]

[+]IPNS and [*]Materials Science and Technology Divisions
Argonne National Laboratory
Argonne, Illinois   60439

[+]Department of Geophysical Sciences
University of Chicago
Chicago, Illinois   60637

ABSTRACT

Rietveld profile refinements using high-resolution pulsed neutron powder diffraction data, collected at IPNS, often reveal broad intensity contributions from sources other than the crystalline materials being studied. Such non-crystalline intensity hampers standard Rietveld refinement, and its removal and/or identification is imperative for successful refinement of the crystalline structure. A Fourier-filtering technique allows removal of the non-crystalline scattering contributions to the overall scattering pattern and yields information about the non-crystalline material. In particular, Fourier transformation of residual intensities not accounted for by the Rietveld procedure results in a real-space correlation function similar to a radial distribution function (RDF). From the inverse Fourier transform of the correlation function a Fourier-filtered fit to the diffuse scattering is obtained. This mathematical technique was applied to data for crystalline quartz, amorphous silica, and to a simulated diffraction pattern for a mixture of the two phases.

INTRODUCTION

The development of Rietveld profile analysis[1,2,3] has revolutionized structure determination using neutron powder diffraction. High intensity, high resolution neutron diffraction data, as collected at the Intense Pulsed Neutron Source (IPNS) at Argonne National Laboratory[4], for example, are of sufficient quality to allow complete, detailed refinement of moderately complex structures, often in the presence of impurity crystalline phases. In addition, we have observed for a number of materials, particularly molecular-sieves[5], diffuse non-crystalline scattering components which actually begin to dominate the crystalline pattern at intermediate d-spacings. To facilitate complete and accurate refinement of the crystalline component(s) in such cases, it is necessary to devise a scheme for removing the diffuse components.

Possible sources of non-crystalline scattering include: (1) incompletely crystallized gels used in preparation of synthetic silicates and molecular sieves, (2) sample containers (glass or ceramic, for instance), (3) short-range ordering due to non-stoichiometry, etc., (4) thermal diffuse scattering (TDS) and (5) separate amorphous phase(s). Each of these scattering phenomena can be characterized as interference functions developed from short-range interaction of atoms in the sample. This scattering produces broad oscillations (relative to the Bragg scattering) superimposed on the crystalline pattern. The slowly oscillating nature of this scattering can be expressed in a physically meaningful way as the Fourier transform of a real-space correlation function, which in the case of scattering from an amorphous material, would be related to the atomic radial distribution function (RDF) for that amorphous component.

We report here a Fourier-filtering procedure whereby the anticipated form of the diffuse scattering is used to subtract previously unidentified non-crystalline contributions from the overall diffraction pattern, thus allowing completion of the crystalline refinement, while providing useful real-space information about the non-crystalline component(s), in the form of the calculated correlation function. The development of this procedure was motivated not only by the recognized need with molecular-sieve materials, but also by the recent success of other workers[6] in removing these non-crystalline contributions by fitting them with empirical functions.

Theoretical Considerations

Experimentally observed Rietveld intensities can be expressed, in reciprocal space, in the form:

$$Y_{obs}(Q) = [Y_C^X(Q) + Y_B(Q)] + Y_C^A(Q) \qquad (1)$$

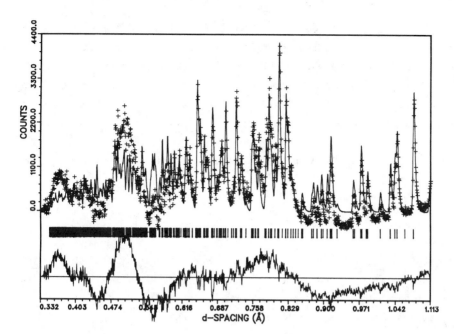

Figure 1. Rietveld profile fit for simulated quartz-silica mixture, from the backscattering
(±150°) data banks on the SEPD time-of-flight powder diffractometer, before Fourier-filtering.
The plus signs (+) are the observed, background subtracted intensities, $Y_{obs}(Q)-Y_B(Q)$. The
solid line represents the calculated crystalline intensities, $Y_C^X(Q)$. The residual intensities, to be fit, are shown at the bottom of the figure. Tick marks below the profile indicate
the positions of the Bragg reflections. Note the poor fit due to the considerable non-crystalline scattering component.

where $Y_C^X(Q)$ are the calculated crystalline (Bragg scattering) intensities, $Y_B(Q)$ are calculated background intensities, $Y_C^A(Q)$ are the non-
crystalline (distinct scattering) intensity contributions to be calculated and $Q = 4\pi\sin\theta/\lambda$ is the scattering vector. In a time-of-flight
experiment data are collected in data banks centered about a number of
fixed angles $2\theta=\pm150°$, $\pm90°$, $\pm60°$, $+30°$ and $-20°$, and the neutron
wavelength is variable. Each data bank has its characteristic range of
reciprocal space as determined by Bragg's equation. A typical Rietveld
profile fit for a sample containing considerable non-crystalline
scattering (a simulated quartz-silica mixture to be described below),
before Fourier-filtering, is shown in Figure 1.

The objective of our Fourier-filtering procedure is to obtain values
for $Y_C^A(Q)$ which are independent of the crystalline refinement, i.e., to

"filter" out sharp features (see Figure 1) in the Rietveld difference function, and remove these contributions from the Rietveld analysis. This necessitates representing the Rietveld difference intensities as the function $Qi(Q)$ (shown in Figure 2 for the quartz-silica mixture) which in this case is related to $(S(Q)-1)$, the distinct scattering part of the amorphous structure factor commonly used in glass diffraction analyses.

The correlation function $D(r)$ is calculated as the 1-dimensional Fourier transform of $Qi(Q)$:

$$D(r) = \frac{2}{\pi} \sum_{Q_{min}}^{Q_{max}} Q\, M(Q)\, i(Q)\, \sin(Qr)\, \Delta Q, \qquad (2)$$

where $i(Q)=[Y_{obs}(Q)-Y_C^X(Q)-Y_B(Q)]/Y_S(Q)$. $M(Q)$ is a modification function which has the form $M(Q)=\sin(\pi Q/Q_{max})/(\pi Q/Q_{max})$. This function serves to ensure proper behavior at high Q values by converging to zero at $Q=Q_{max}$. $Y_S(Q)$ represents the incident white beam spectrum. $D(r)$ contains maxima

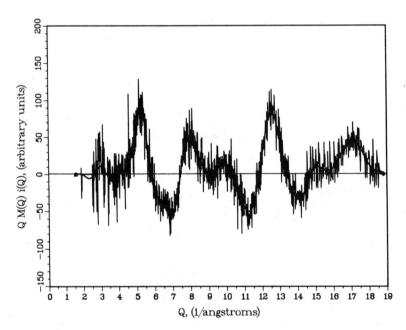

Figure 2. Residual intensities, $QM(Q)i(Q)$, for simulated quartz-silica mixture, from the $\pm150°$ data banks, with an effective Q range of ~1.6-19.0 Å^{-1}. The jagged curve represents the raw data, while the smooth curve is the Fourier-filtered fit. A correlation limit (r_{max}) of 6.5 Å was used.

at interatomic spacings characteristic of the short-range interations
giving rise to the non-crystalline scattering.   D(r), derived from the
quartz-silica mixture to be described below, for amorphous SiO_2 from the
$\pm 150°$ data banks is plotted in Figure 3. Comparable results are found
from calculations using $\pm 90°$ and $\pm 60°$ scattering angle data.

From the inverse Fourier transformation of D(r), a smooth function
$Y_C^A(Q)$ is calculated as:

$$Y_C^A(Q) = Y_S(Q) \sum_{r_{min}}^{r_{max}} D(r) \frac{\sin(Qr)}{(Qr)} \Delta r \tag{3}$$

Filtering is accomplished by setting r_{max} at some appropriate value,
normally around 5-15 Å, depending on the range of significant correlation
in the non-crystalline component.   The smooth Fourier-filtered function
is plotted along with the original Rietveld residual intensities in

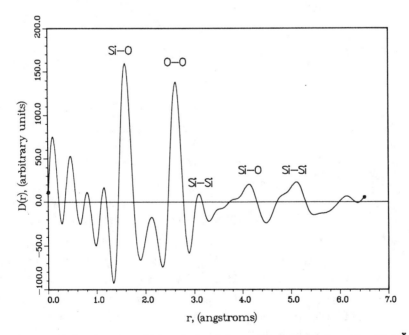

Figure 3. Correlation function, D(r), calculated from residual intensities, $[Y_{obs}(Q)-Y_C^X(Q)-Y_B(Q)]$, for simulated quartz-silica mixture, from $\pm 150°$ data banks. Identifiable interatomic spacings are as follows: Si-O = 1.60, O-O = 2.63, Si-Si = 3.12, Si-O = 4.13, Si-Si = 5.11 Å. Additional unidentified features probably result from multiple scattering contributions, etc., which are currently ignored in this analysis.

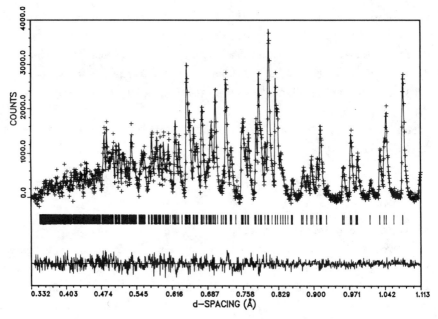

Figure 4. Final Rietveld profile fit for simulated quartz-silica mixture, after two iterations of Fourier-filtering. Note that the residual intensities have been reduced to zero, within the experimental error of the measurements, throughout the range of this plot. Symbol definitions are as described in Figure 1.

Figure 3. Figure 4 shows the final Rietveld profile fit for the simulated quartz-silica mixture after Fourier-filtering using a value of r_{max}=6.5 Å.

ITERATIVE REFINEMENT PROCEDURE

Unfortunately, the crystalline and non-crystalline scattering are not completely decoupled. The overall reliability of the real-space non-crystalline information is strongly related to how well the crystalline model fits the Bragg scattering. Conversely, if there is a large contribution from non-crystalline scattering, refinement of the crystalline model will be effected. In particular, the Rietveld refinement will attempt to compensate for non-crystalline scattering by anomalously adjusting thermal parameters, etc. For these reasons it is important to devise an appropriate iterative procedure which will allow accurate refinement of the crystalline model, while simultaneously retaining information about the non-crystalline scattering.

The procedure used involves initial refinement of those structural parameters least effected by the non-crystalline scattering, using the original uncorrected data. This provides a reasonable starting point for the Fourier-filtering. As the analysis proceeds, crystalline refinements are done on corrected data only, and subsequent Fourier-filtering corrections are made based on intensity differences between the original uncorrected data and the improved crystalline model.

To illustrate the procedure and the accuracy of results, we have applied the Fourier-filtering procedure to data from a "purely crystalline" sample and from a mixed crystalline-amorphous sample, each with known composition. Neutron powder diffraction data were collected on the special environment time-of-flight powder diffractometer (SEPD) at IPNS for samples of crystalline α-quartz (SiO_2) and amorphous silica (SiO_2). A simulated pattern for a mixed sample was produced by summing together the data from these two samples. Conventional Rietveld refinement using the two data sets (α-quartz alone and α-quartz mixed with amorphous silica) revealed that both contained observable non-crystalline scattering components. These amorphous contributions were extracted by Fourier-filtering and the refinements were continued.

The iterative procedure followed in both cases was as follows: (1) refinement of structural parameters, including isotropic thermal parameters using the uncorrected data, (2) Fourier-filtering to produce the first corrected data sets, (3) continued refinement including anisotropic thermal parameters, using corrected data, (4) calculation of second corrected data sets by removal of first corrections and Fourier-filtering based on current crystalline structural models, (5) final refinements using second corrections.

RESULTS

In this and other similar experiments it is important to determine how extensively the crystalline refinement is effected by such manipulation of the diffraction data and how meaningful the derived non-crystalline information is. Final crystalline structural parameters from the refinements of quartz and the quartz-silica mixture (after two iterations of Fourier-filtering) are tabulated along with results from a previous refinement of α-quartz[7] (also using IPNS data) in Table 1. The present refinements are in very good agreement with each other and, in fact, with the previous refinement.

Figure 3 shows that the interatomic spacings derived from the Fourier-filtering are as expected for an amorphous SiO_2 material. The precise structure of amorphous silica is obviously not known, but the first and second neighbor distances 1.60 and 2.63 Å are representative of approximately tetrahedral coordination of oxygen atoms around silicon. Similarly the third neighbor distance of 3.12 Å is consistent with inter-tetrahedral (Si-O-Si) angles of slightly less than 180°. Additional identifications at 4.13 and 5.11 Å are made based on packing of tetrahedra defined by the first, second and third neighbor coordination. From Fourier transform theory the reliability of these data will decrease with increasing correlation distance, so data beyond ~3.0 Å are only approximate. Unlabelled maxima in the correlation function are probably due to multiple scattering and absorption in the amorphous component which are partially compensated by the Rietveld background calculations but otherwise ignored at this point.

Table 1. Comparative refinements of quartz with and without amorphous silica component added (150° data bank, space group $P3_121$, with two independent atoms, Si and O). Two iterations of Fourier-filtering.

Parameter	Quartz	Quartz-Silica Mixture	Reference[6]
Si, x	.4700(3)	.4696(4)	.4700(2)
y	.0000	.0000	.0000
z	.3333	.3333	.3333
β_{11}	.0096(5)	.0095(8)	.0068(3)
β_{22}	.0039(6)	.0053(10)	.0054(6)
β_{33}	.0027(3)	.0025(5)	.0044(6)
β_{13}	-.0018(7)	-.0029(12)	.0006(2)
O, x	.4131(2)	.4133(3)	.4131(2)
y	.2675(2)	.2672(3)	.2677(2)
z	.2151(1)	.2146(2)	.2144(1)
β_{11}	.0148(4)	.0166(7)	.0195(4)
β_{22}	.0130(4)	.0124(6)	.0120(3)
β_{33}	.0070(2)	.0072(3)	.0072(1)
β_{12}	.0097(4)	.0097(6)	.0104(3)
β_{13}	.0016(2)	.0012(4)	.0023(2)
β_{23}	.0026(2)	.0025(3)	.0045(2)
a_0, b_0	4.9137(1)	4.9136(1)	4.9141(1)
c_0	5.4053(1)	5.4054(1)	5.4060(1)
R_{wp}, %	4.41	2.26	3.11
R_{exp}	3.43	1.88	

For this relatively simple crystalline material, two iterations of Fourier-filtering were needed. More than two iterations might be expected for more complex structures, depending on the correctness of the crystalline model in the initial stages of the analysis.

This Fourier-filtering technique is currently being applied in the analysis of $AlPO_4$ and $AlSiO_4$ molecular-sieves. Preliminary results[5] show that Fourier-filtering does facilitate completion of the crystalline refinement of these materials. Is is not known whether the non-crystalline scattering observed in these data is due to short range ordering in the sample or from a separate amorphous phase. The calculated correlation functions are, however, consistent with those expected for amorphous materials having chemical compositions $AlPO_4$ and $AlSiO_4$, respectively.

CONCLUSION

Fundamental to the Fourier-filtering technique is the use of only low order harmonics in the Fourier summation. With this we fit only slowly oscillating features in the residual intensity function which are not due to errors in the crystalline model, thus minimizing correlations between the crystalline and non-crystalline scattering components and allowing us to treat them simultaneously as separate phases. The Fourier-filtering is done on the entire diffraction pattern, providing us with useful structural information about both the crystalline and non-crystalline components.

Results presented here – using data prepared from the summing of separately measured crystalline and amorphous SiO_2 diffraction data – demonstrate the effectiveness of the Fourier-filtering technique. Additional experiments with truly mixed crystalline and amorphous samples will be performed to more thoroughly determine the role of multiple scattering and absorption in these multi-phase samples and to further assess the accuracy of derived non-crystalline correlation data.

ACKNOWLEDGEMENTS

JWR acknowledges financial support from NSF grant CHE 84-05167 and a grant-in-aid from Union Carbide Corporation, both administered by J. V. Smith. Helpful discussions with J. J. Pluth and J. V. Smith are appreciated. This work was supported by the U. S. Department of Energy, BES-Material Sciences under contract W-31-109-Eng-38.

REFERENCES

[1] H. M. Rietveld, "A Profile Refinement Method for Nuclear and Magnetic Structures", J. Appl. Crystallogr. 2:65 (1969).

[2] R B. Von Dreele, J. D. Jorgensen and C. G. Windsor, "Rietveld Refinement with Spallation Neutron Powder Diffraction Data", J. Appl. Crystallogr. 15:581 (1982).

[3] J. D. Jorgensen and F. J. Rotella, "High-Resolution Time-of-Flight Powder Diffractometer at the ZING-P' Pulsed Neutron Source", J. Appl. Crystallogr. 15:27 (1982).

[4] IPNS Progress Report 1983-1985, Argonne National Laboratory, Argonne, Illinois. Available from the IPNS Divisional Office, Argonne National Laboratory, Argonne, Illinois 60439.

[5] J. M. Bennett, J. W. Richardson, Jr., J. J. Pluth and J. V. Smith, "Aluminophosphate Molecular Sieve $AlPO_4$-11: Structure Determination from Powder Using a Pulsed Neutron Source", J. Chem. Soc., Chem. Comm., in press (1985).

[6] W. H. Baur and R. X. Fischer, "Recognition and Treatment of Background Problems in Neutron Powder Diffraction Refinements", this volume (1985).

[7] G. A. Lager, J. D. Jorgensen and F. J. Rotella, "Crystal Structure and Thermal Expansion of α-Quartz SiO_2 at Low Temperatures", J. Appl. Phys. 53:10 (1982).

NEUTRON POWDER DIFFRACTION APPLICATIONS AT

THE UNIVERSITY OF MISSOURI RESEARCH REACTOR

W. B. Yelon, F. K. Ross and R. Berliner

University of Missouri
Research Reactor
Columbia, MO   65211

ABSTRACT

A neutron powder diffractometer at the University of Missouri Research Reactor (MURR) uses a linear position sensitive detector (PSD) which has increased both resolution and data acquisition rates. Rietveld analysis works as well with this system as with more conventional single and multi-Soller slit detector systems. This analysis has been successfully applied to problems involving more than 75 parameters and 1200 reflections and a future instrument upgrade should allow analyses which involve 100–150 parameters. A special advantage of the PSD instrument is that it needs only small (1–2 gm) samples to achieve high statistical accuracy.

INTRODUCTION

The development of the Rietveld method[1] has led to a reinvigoration of powder methods for the analysis of crystal structure, especially for the refinement of complex structures of materials which are not readily prepared as single crystals. Neutron scattering has particularly profited from this renewed interest since experience has shown that the method works more reliably with neutron data than with X-ray data. Furthermore, neutron scattering can provide some unique information which cannot be obtained from X-ray experiments, either single crystal or powder. In response to this new interest, a variety of facilities have been developed at both steady state and pulsed neutron sources, providing improved resolution and higher data acquisition rates.

At the University of Missouri Research Reactor, we have constructed a powder diffractometer which has been in almost continuous use for a wide variety of experiments. This instrument[2] uses a newly developed linear position sensitive detector instead of more conventional Soller slit collimator–detector systems used in single detector or multi-detector systems. More than two years operating experience has been accumulated since the first reports of this instrument and it seems appropriate to review its performance and some of the results obtained with this facility.

A Position Sensitive Detector for Powders

The powder diffractometer has three linear position sensitive detectors constructed by Reuter-Stokes, Inc. and uses a charge division method for

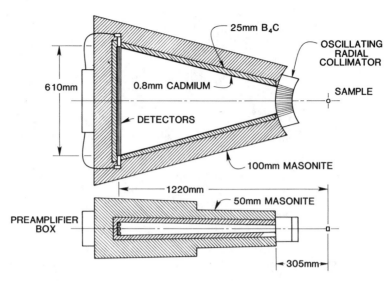

Fig. 1. Schematic drawing of the MURR-PSD powder diffractometer.

position encoding. The detectors are 1" diameter x 24" active length. In
the current configuration, with the three elements stacked one above the
other (Fig. 1), they span approximately 3° vertical x 27° horizontal divergence. With 15' collimation in front of the monochromator and samples of
1/4" diameter or less, the resolution is approximately 0.5% at the minimum
at $2\theta \approx 2\theta_{monochr.} = 60°$. Although the samples used with the PSD are
usually smaller (< 1/4") than those used with the old single detector
system (~ 1/2"), data acquisition rates are more than 50 times higher than
with the old system and good statistical accuracy (> 10^6 cts) is achieved
with small samples (1-3 gms) in as little as 12 hours.

In order to use a PSD, the path between sample and all points on the
detector must be open. This would be expected to give high backgrounds,
but we have installed an oscillating radial collimator between the sample
and the entrance to the detector housing (Fig. 1). The collimator allows
neutrons emerging from the sample to reach the detector, while neutrons
emerging from surfaces more than ~ 40 mm from the instrument center, such
as from cryostats, furnaces, etc., are absorbed in the blades of the collimator. The oscillation ensures that the detector is not shadowed by the
collimator blades. The background recorded with the PSD with beam open and
no sample is ~ 2/3 ct/min/0.1°. This is lower than the background observed
with the previous Soller collimator system.

The linear detector obviously does not lie on an arc, and corrections
are needed in the conversion of detector bin to angle as well as for
changes in the efficiency due to changing sample to detector distance, and
to changes of the angle of incidence of the neutrons on the detector, which
changes the effective gas length. These corrections are verified through a
thorough calibration procedure which limits the systematic errors in position to < 0.01°. Efficiency corrections which are applied when the data is
rebinned into equal angular intervals lead to variations of less than
0.25%. These are as good, or better than the relative efficiencies of
multi-detector systems. One effect of the linear detectors cannot be compensated by simple correction, the angular width of a peak recorded near
the end of the detector is slightly greater than that observed at the detector center. Fortunately, a simple modification to the Rietveld code can

correct for this effect, although with the present resolution, satisfactory refinements are performed without any corrections.

Data is typically collected in 25° angular intervals from 5° to 105° (2θ) for complete refinement, although often one single interval will be followed as a function of time or temperature. A more complete description of the instrument is given elsewhere.[2]

Applications

Much recent attention has focused on the attributes of the higher re-solution neutron powder diffraction instruments at the pulsed neutron sources, as these provide good intensities and resolutions to large q (sin θ/λ). However, for a wide variety of problems, a conventional fixed wave-length instrument may provide comparable or better data. These are prob-lems for which large q data is not usable, for example in poorly diffract-ing material or materials for which small q data is especially important (e.g., in magnetic neutron scattering). These areas have been among the special focuses of research with the MURR PSD diffractomer.

Catalysts. One of the most active areas in neutron powder diffraction is in studies of catalysts. These include oxides, mixed phases, and zeo-lites. For example, a long standing question concerned the $FeSbO_x$ system. At issue was the question of whether the catalytic activity was associated with a two phase system of a particular morphology or whether an interme-diate phase was formed. Our studies[3] show that the diffraction pattern (Fig. 2) can be fitted by a two phase model, with the well known constitu-ents $FeSbO_4$ and Sb_2O_4. Interestingly, the peak widths for Sb_2O_4 are signi-ficantly greater than those of the $FeSbO_4$ suggesting that the Sb_2O_4 forms small particles epitaxially on the $FeSbO_4$ and that the physical relation-ship of the two phases is crucial to the particular catalytic action. Some of the results of the refinement are shown in Table 1. Since this was a

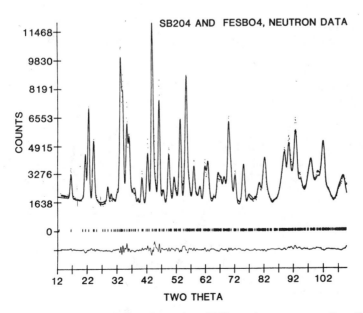

Fig. 2. Observed and calculated powder diffraction pattern for the mixed phase catalyst $FeSbO_x$.

Table 1.   Final Parameters for the Multiphase
Refinements of $FeSbO_4$ and $\alpha-Sb_2O_4$

Constant Wavelength Data:

	$\alpha-Sb_2O_4$	$FeSbO_4$
Space Group	$Pna2_1$, Z=4	$P4_2/mm$, Z=2
Cell Parameters (in Angstroms)		
a	5.441(20	4.6365(6)
b	4.800(1)	4.6365(6)
c	11.766(3)	3.0742(6)
V	307.72	66.09
Halfwidth Parameters		
U	2.22	0.95
V	−1.13	−.34
W	0.61	0.49
Final Agreement Factors		
R_{wp}	5.42%	
R_p	7.05%	
R_{exp}	1.77%	
R_{Bragg}	2.33%(349 refl)	2.49%(49 refl)

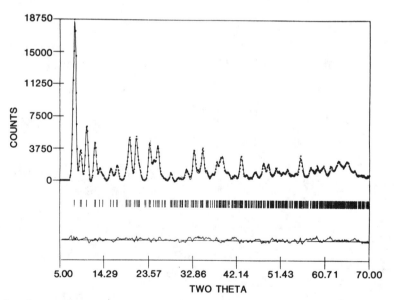

Fig. 3. Observed and calculated powder pattern for a synthetic zeolite;
gallium mazzite.

Table 2. Final Parameters with Estimated Standard
Deviations in Parentheses

Atom	Site	x	y	z	B	Occn.
Si1(T6)	12j	0.4898(10)	0.1594(8)	0.25	1.3(1)	14.2
Si2(T4)	24l	0.0919(6)	0.3584(6)	0.0468(12)	"	27.8
O1	6h	0.7422(6)	2x	0.25	3.36(6)	6.0
O2	6h	0.5774(5)	2x	0.25	"	6.0
O3	12j	0.0975(7)	0.3859(7)	0.25	"	12.0
O4	24l	0.1047(6)	0.4376(5)	0.9251(9)	"	24.0
O5	12k	0.1658(4)	2x	0.9986(14)	"	12.0
O6	12i	0.2744(7)	0.0	0.0	- "	12.0
Na1	6g	0.5	0.0	0.0	4.0	2.2(2)
Na2	12j	0.926(7)	0.164(6)	0.25	4.0	2.6(4)
Na3	6h	0.454(3)	2x	0.25	4.0	2.2(1)

a 18.043(9) Å   c 7.662(2) Å   Vol 2160(4) Å^3
U^a 2.8(3)x10^4 V -1.5(1)x10^4 W 5.7(1)x10^3
        zθ -6.9(4)   asymmetry parameter 9x10^{-6}
R_i (reflection intensity) 5.90 R_p (profile) 8.71
R_{wp} (weighted profile) 8.58 R_{exp} (expected) 4.63

aThe half-width parameters, U, V and W (30) and the counter zero-point
error, zθ, are expressed in units of 1/100 degrees.

multiple phase specimen, it is clear that no single crystal could be obtained to perform this study and furthermore, that the constant wavelength experiment gave more than sufficient data.

A second example in the area of catalysis concerns zeolites, which are notoriously difficult to prepare as single crystals. They are, in addition, rather poor scatterers and signals tend to become indistinguishable from background at rather small q (~ 0.5 Å^{-1} sin θ/λ). They are thus well matched to the characteristics of a steady state neutron powder diffraction instrument. Figure 3 shows observed and calculated powder diagrams for a synthetic gallium mazzite.[4] It is seen that the diagram becomes nearly featureless above 65° (2θ) and that inclusion of more data will not add significant extra information. This structure refines quite well and information can be extracted about the symmetry of the framework atoms, about the partitioning of the Si/Ga atoms and about the occupancy of the cation sites (Table 2). This refinement is the first successful powder neutron study of a non-cubic zeolite and suggests that the field has the potential for significant growth.

Magnetism. Neutron diffraction is a well known probe of magnetic ordering, since the magnetic moment of the neutron responds to the magnetic fields present within a material. Experiments with magnetic materials are complicated by the fact that each magnetic atom has, in addition to the coordinates x,y,z, three components to the magnetic moment vector. Moreover, the symmetry of the magnetic lattice may be different from that of the crystal lattice. Thus the size of the magnetic problem may be much larger than the X-ray crystallographic problem. Furthermore, because of the polycrystalline averaging much of the information is lost for crystals with symmetry higher than monoclinic. Nevertheless, no other method provides the

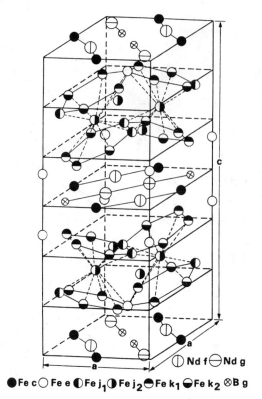

Fig. 4. The structure of $Nd_2Fe_{14}B$, as determined from powder diffraction data.

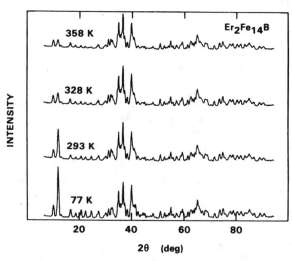

Fig. 5. Temperature dependence of the powder diffraction pattern for $Er_2Fe_{14}B$.

information given by neutrons and despite the difficulties most information about complex magnetic structures is deduced from neutron scattering.

One of the most important recent developments in magnetism is the discovery of a magnetic phase: $Nd_2Fe_{14}B$.[5] The structure of this phase, and its exact stoichiometry were unknown when powder samples were provided to MURR by General Motors. After the space group was determined from the neutron and X-ray diagrams, trial and error methods were combined with the Rietveld method to model the structure. The stoichiometry and structure (Fig. 4) were ultimately determined.[6] This structure with 68 atoms/unit cell (Z=4) is believed to be the largest unknown structure solved by powder methods, and is clearly an indicator that as the quality of powder data improves, more sophisticated methods of analysis can be applied to solve structures from powders.

In $Nd_2Fe_{14}B$, data obtained above the magnetic ordering temperature were used to obtain crystal structure. Below T_c the magnetic structure is found to be ferromagnetic with all moments aligned parallel to the crystallographic c-axis. In $Er_2Fe_{14}B$, on the other hand, a spin reorientation occurs close to room temperature. This is clearly visible in the neutron powder data (Fig. 5), where it is seen that the 002 reflection becomes the strongest peak in the diagram at lower temperatures.[7] This is attributed to the moments remaining ferrimagnetic (Er antiparallel to Fe) but moving to the a-b plane from the c-axis as the temperature is lowered. The details of the refinement appear to confirm this, although slight deviations from ferrimagnetic alignments cannot be ruled out. The combined nuclear and magnetic refinements in these compounds required 30-40 parameters with ~ 400 peaks contributing to the diffraction pattern.

Hydrides. Neutron scattering is particularly useful in the study of materials containing hydrogen or deuterium since both H and D have appreciable coherent scattering amplitudes and contribute significantly to the intensity distribution. Powder diffraction has frequently been applied to hydrides, such as those of the rare-earth transition metal compounds since single crystals of these materials have not been synthesized. For example, $Ho_6Mn_{23}D_{22}$ has a large change in cell parameter on hydriding but remains cubic with the D atoms occupying four unique sites at room temperature.[8] However, at low temperature, the symmetry is lowered from cubic to tetragonal (P4/mmm) and the deuterium atoms are found to occupy 12 sites. In addition, since the magnetic structure is found to be non-colinear, more than 50 parameters must refined in this problem (Table 3). Although the goodness-of-fit measures are not extremely good, it is remarkable that so many parameters can be refined and that physically and chemically meaningful results can be extracted from the powder diffraction experiment.

Phase Transitions and Other Phenomena. The ability to measure a number of reflections simultaneously with a single setting of the position sensitive detector confers special advantages on the MURR instrument. For example, keeping the detector at a fixed setting and varying a parameter such as temperature and following a phase transition becomes quite straightforward. In many cases, adequate intensities can be obtained in only a few minutes. Thus, dynamic as well as static phenomena can be studied.

One example of such use of the PSD is an investigation of the low temperature phase of Li metal. Both the dynamics of the transition and the structure were studied. The resolution of the detector was more than sufficient to detect systematic shifts in peak position for the low temperature 9R phase, a rhombohedral structure with a 9-layer hexagonal stacking. Careful analysis of the peak positions and widths (Fig. 6) has led to a new model for stacking faults which completely describes the experimental data and which can be generalized to a wide class of problems.[9]

Table 3.   Atomic Parameters of $Ho_6Mn_{23}D_{23}$ at 9K in the P4/mmm Structure

fcc Sites	Tetragonal Sites	N	Refined Coordinates			$(\mu_B/atom)$	θ [a] (degrees)
			x	y	z		
Ho e	Ho g	2	0.022	0.222	0.201	3.4	51.4
	h	2	0.5	0.5	0.304	3.4	128.6
	j	4	0.222	0.222	0.0	3.5	100.1
	k	4	0.299	0.299	0.0	3.5	79.9
Mn b	Mn b	1	0.0	0.0	0.5	3.5	29.5
	c	1	0.0	0.5	0.0	3.5	150.5
Mn d	Mn e	2	0.0	0.5	0.5	2.3	67.7
	f	2	0.5	0.0	0.0	2.3	112.3
	r_1	8	0.246	0.246	0.762	0.0	---
Mn f_1	Mn s_1	8	0.329	0.0	0.161	2.8	117.5
	t_1	8	0.103	0.5	0.312	2.8	62.5
Mn f_2	Mn s_2	8	0.256	0.0	0.399	1.5	30.3
	t_2	8	0.266	0.5	0.150	1.5	149.7
D a	D a	1	0.0	0.0	0.0		
	d	1	0.5	0.5	0.5		
D f_3	D s_3	3.2	0.152	0.0	0.059		
	t_3	8	0.270	0.5	0.387		
D j	D p	3.2	0.191	0.449	0.0		
	q	2.6	0.199	0.995	0.5		
	r_2	8	0.144	0.144	0.326		
D j	D r_3	0	---	---	---		
	r_4	4	0.335	0.335	0.059		
	r_5	0	---	---	---		
D k	D s_4	1.4	0.238	0.0	0.117		
	t_4	1.4	0.192	0.5	0.458		
	u_1	0.4	0.188	0.134	0.162		
	u_2	8	0.277	0.380	0.322		

Cell Parameters,   a = 8.981 Å,   c = 12.638 Å
R_{wp} (weighted profile) 5.71% R_{exp} (expected) 3.02%
R_n (nuclear) 5.73% R_m (magnetic) 13.3%
χ = 1.9

[a]θ is the angle between the magnetic moment and the c-axis.

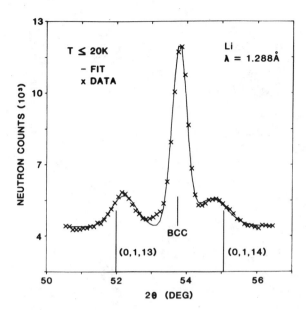

Fig. 6. Observed and calculated pattern for Li at 20 K showing both BCC
and 9R peaks.

The ability to observe the total peak profile with a fixed detector
setting has also been used to particular advantage in studies of applied
and residual stresses. This will be discussed in detail in the next paper,
and will not be further mentioned here.

Future Prospects

The present instrument has only modest resolution and sufficient intensity for most problems. Some immediate improvements will lead to improved resolution with only a small loss in intensity. In particular, a
7.5' Soller collimator is available to install between source and monochromator. When combined with a small reduction in electronic noise, it
will lead to a system resolution of ~ 0.25% at $2\theta = 60°$. Although integrated peak areas will decrease by ~ 0.5, peak heights should be reduced by
only a small amount in the focusing region. The parallax effects discussed
in sec. 2 (PSD description) will however become more important.

The next generation machine at MURR will most likely have two banks of
detectors, each spanning ~ 15° rather than the present 25°, and each expanded vertically to maintain the vertical acceptance. Major efforts are
being made to produce a monochromator system for use at $2\theta_m$ ~ 90° which
would further improve the resolution at large angles. We anticipate ~ 0.2%
resolution over the most important part of the angular range in this configuration.

Eventually, as MURR is upgraded to ~ 30 MW, the resolution of this instrument would be further improved to ~ 0.1-0.15%. At the same time intensity improvements, through focusing monochromators, expansion of the PSD
solid angle and enhanced source flux, should lead to higher data acquisition rates and an expansion in the experimental capabilities. These improvements parallel those contemplated at other neutron scattering centers,
although the details vary considerably from one facility to another.

Coupled with these are developments underway in analysis which may lead to more direct structure solution from powder data. We believe that problems of 200-300 parameters will be tractable with the contemplated instrument and anticipate that both the variety and complexity of powder diffraction applications at MURR will continue to grow.

Acknowledgments

We would like to acknowledge NSF grant DMR-81-14975 for help in the construction of the MURR PSD as well as significant contributions from C. W. Tompson, D. F. R. Mildner, J. Sudol and M. Mehregany.

References

1. H. M. Rietveld, J. Appl. Cryst. 2:65 (1969).
2. C. W. Tompson, D. F. R. Mildner, M. Mehregany, J. Sudol, R. Berliner and W. B. Yelon, J. Appl. Cryst. 17:385 (1984).
3. R. G. Teller, J. F. Brazdel, R. K. Gasselli and W. Yelon, Faraday Transactions 81:1693 (1985).
4. J. M. Newsam, R. H. Jarman and A. J. Jacobson, Mat. Res. Bull. 20:125 (1985).
5. J. J. Croat, J. F. Herbst, R. W. Lee and F. E. Pinkerton, Phys. Letter 44:148 (1984). M. Sagawa, S. Fujimura, N. Tagawa, H. Yamamoto and Y. Matsuura, J. Appl. Phys. 55:2083 (1984).
6. J. F. Herbst, J. J. Croat, F. E. Pinkerton and W. B. Yelon, Phys. Rev. B 29:417 (1984).
    J. F. Herbst, J. J. Croat and W. B. Yelon, J. Appl. Phys. 57:4086 (1985).
7. W. B. Yelon and J. F. Herbst, J. Appl. Phys., in press.
8. N. T. Littlewood, W. J. James and W. B. Yelon, Int'l Conference on Magnetism, J. Magn. and Magn. Mtls. (1985), in press.
9. R. Berliner and S. S. Werner, Int'l Conference on Neutron Scattering, Physica (1985), in press.

NEUTRON STRESS MEASUREMENTS WITH A POSITION SENSITIVE DETECTOR

A. D. Krawitz*, P. J. Rudnik**, B. D. Butler*, and J. B. Cohen**

* Dept. of Mechanical and Aerospace Engineering, University of Missouri, Columbia, MO 65211.
** Dept. of Materials Science and Engineering, Northwestern University, Evanston, IL 60201.

INTRODUCTION

The feasibility of measuring residual and applied stresses with neutrons employing a position sensitive detector (PSD) is demonstrated. Measurements with both a small beam, to probe internal regions, and a large beam, for bulk sampling have been made. With such a detector collection of data is rapid compared to ordinary neutron collection methods. This detector allows more detailed sampling, smaller probe regions and possibly even study of time-dependent processes.

EXPERIMENTAL ASPECTS

The spectrometer geometry is shown in Fig. 1a. The detector consists of a three-counter array of linear PSD's spanning a usable angular range of $25°2\theta$ at a distance of 1220 mm from the sample position. The detectors are stacked in a vertical plane which is normal to the horizontal diffraction plane. The front of the detector housing has an oscillating radial collimator which greatly reduces background and off-axis scattering. The data is rebinned into $0.1°$ 2θ increments from the basic channel width of $\approx 0.03°$ and the intensity of the three detectors is added. Further details of the PSD may be found in ref. 1.

For probing small internal regions a special slit geometry is required. In order that a probe region be defined and the beam allowed to reach the PSD, the diffracted beam slit must be placed as close as possible to the sample. The detailed geometry is shown in Fig. 1b. The basic constraints are the needs to define a probe region of a specified dimension and to allow sufficient divergence through the slit to record the entire peak of interest, including shifts due to elastic strain. The parameters of interest are related by:

$$\tan(\alpha/2) = (w_d/2)/l_d = (w_s/2)/l_s$$

where w_d is the distance along the detector to subtend $\alpha°$ (= 21 mm/° at the center), l_d is the distance from the goniometer center to the

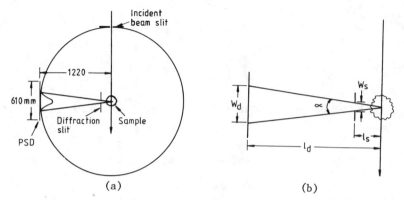

Fig. 1   (a) Spectrometer geometry for creation of internal probe region
             using PSD.
         (b) Detail of diffraction path showing parameters for relating
             size of probe region and usable area of PSD.

detector (1220 mm), w_s is the width of the diffraction slit, and l_s is
the distance from the goniometer center to the diffraction slit. The
slits in the incident and diffracted beams define the internal probe
region, or the "diffraction volume." Rectangular slits define a
parallelpiped while square slits define a rhomboid. The probe region
is fixed over the center of the diffractometer circle, and the sample
must be translated in order to examine different interior volumes.

For most detectors, the measured intensities are due only to the
particular peak of interest. However, because of the large angular
range of the PSD, it is quite easy to obtain additional Bragg peaks at
the same time due to crossfire through the diffracted beam slit. This
is illustrated schematically in Fig. 2a for the U-shaped aluminum
sample described below. The PSD is centered on the 311 peak at 63.5°
2θ for 0.1285 nm neutrons. As the sample is translated through the
probe region at the center of the goniometer, first the 222 then the
220 can pass through the diffracted beam slit, in addition to the 311.
Actual results are shown in Fig. 2b for a U-shaped Al alloy bar.
Probably the most important consequence of the geometry required for
measurements using a PSD is that if the peak of interest is not
precisely in the center of the PSD, then displacement of the
diffraction slit will lead to shifts in peak position. Such an error
in positioning may be due either to inaccurate placement of the
detector or mechanical error in positioning of the sample stage or
goniometer. This means that the width and position of the diffraction
slit should be sufficient to encompass the entire peak of interest
over the full range of stress under investigation and that the
diffraction slit should not be moved during the measurement.

All of the data was collected on the automated powder diffractometer at the University of Missouri Research Reactor (MURR) using a
neutron wavelength of 0.1285 nm. All scans were run under computer
control. Data acquisition with the PSD was performed for a preset
number of monitor counts which depended on the scattering
cross-section of the sample, slit size and desired counting
statistics. For example, in small beam experiments, the Al 311 peak
was counted for about 12 minutes while the broader 511/333 required
about 54 minutes. The peaks were fit to a Gaussian function, $y = a(\exp[-(x-c)^2/2b^2])$, where a is the peak intensity, b the breadth
(FWHM) and c the peak position (in ° 2θ).

SMALL BEAM (INTERNAL PROBE REGION) EXPERIMENTS

Blade Fixture Experiment

The ferritic iron blade fixture sample consists of alternating
blades and gaps; see Fig. 3a. It was designed to help evaluate the
resolution of internal probe experiments. The sample was mounted on
the diffractometer and moved, using a precision x-y translation stage,
in 0.508 mm increments through the incident beam. As the specimen is
translated through the probe region, the diffracted intensity from the
211 Fe peak (at 67° 2θ for the 0.1285 nm neutrons) should oscillate
with the period of the fixture. In particular, the blade/gap
interfaces should be located at the positions of 50% of the maximum
intensity. The results are shown in Fig. 3b. The half-maximum points

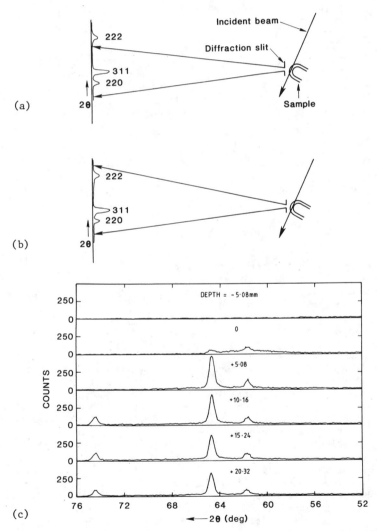

Fig. 2   Crossfire effects using U-shaped Al alloy bar sample. Schematic
         illustration of (a) presence of 220 and 311 peaks only for shallow
         beam penetration and (b) inclusion of 222 for deeper beam
         penetration. Actual results are shown in (c).

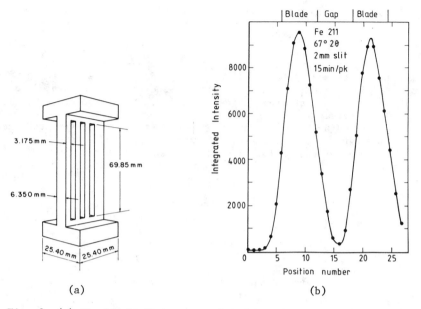

(a)   (b)

Fig. 3   (a) The blade fixture sample.
         (b) Results of internal probe scan through blade fixture.

of intensity are found to be slightly more than six steps apart, in
good agreement with the actual gap widths and with results obtained
using a standard detector[2].

Al Bar Experiment

     The second internal probe sample was a 2024-T86 Al alloy bar cut
to a "U" shape and having a 25.4 x 25.4 mm cross-section; see Fig. 4a.
Stress was applied by tightening the bolt shown in the figure. The
stress level was controlled via a strain gauge on the inner surface of
the central cross-section; a compressive stress of -413.7 MPa was set.
This sample was previously studied in a similar way but using a
standard detector[3]. Two experiments were performed in order to
evaluate the PSD relative to the conventional detector.

     First, the absorption profile of the bar was obtained, using the
311 peak. The sample was displaced into, and eventually through, the
probe region. Initially, when the sampled volume is in front of the
specimen, only background scattering is detected. The intensity
increases as the specimen is moved into the beam, and rapidly reaches
a maximum when the probe region is just entirely within the specimen.
Beyond this position the intensity decreases due to absorption. This
procedure enables the position of the surface of the sample relative
to the center of the diffraction circle to be determined since the
position of maximum intensity should be equal to one-half the depth of
the sampled volume. The results are shown in Fig. 4b. At the point
of maximum intensity, the depth was 2.04 mm. Since 4 mm slits were
used, the total penetration depth of the probe region was 4.53 mm at
the required diffraction angle, implying a depth of 2.26 mm and
revealing an error of 0.22 mm in the mechanical positioning of the
sample.

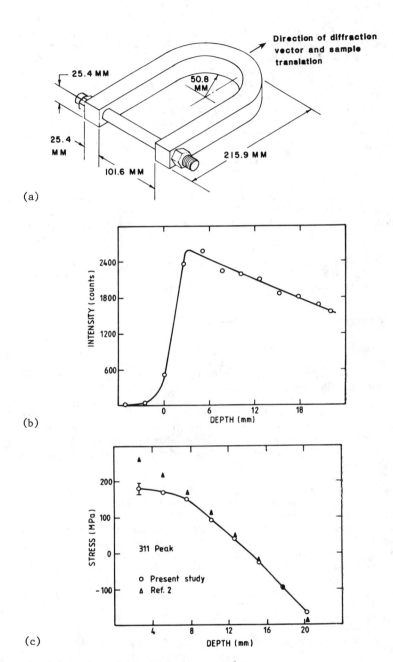

Fig. 4 (a) U-shaped Al alloy bar sample.
        (b) Absorption profile.
        (c) Results of measurement of stress gradient for present study
            and prior measurement using conventional detector. Maximum
            statistical error indicated on one data point.

Second, the bar was loaded and the resulting stress gradient through the bar was sampled using the two-tilt process[3] for psi values of 0 and 60°. Since the stress is proportional to the peak shift measured at two different psi angles[3], $\sigma = K(hkl)\Delta 2\theta$, stresses at varying depths through the material can be measured by repeating the procedure at different sample positions. The through-thickness stress measurement of the 311 peak, using two psi tilts and eight different depths, required less than 5 hours to complete. The actual collection of data represented only 100 minutes of this total; a similar period was needed to rebin the collected data, with the remainder used in changing the measurement parameters, i.e., the psi angle and probe region. The rebinning time has recently reduced to a matter of seconds so that it would be possible now to do these measurements in 2-3 hours.

The results obtained here and those obtained by Schmank and Krawitz [3] are compared in Fig. 4c. As is apparent, the values are generally very similar, although they deviate as the outer, tensile surface is approached. This is attributed to plastic deformation due to overload of the sample during testing and prior use though it is possible that the small difference in widths of the slits employed is partially responsible. The stress gradient for the 511/333 peak was also measured and found to be of the same shape.

LARGE BEAM (BULK SAMPLING) EXPERIMENTS

Experiments were conducted using conventional powder diffraction geometry and a tensile test device for in situ studies of specimens under load. The device can apply 1380 MPa to a 6.35 mm diameter sample having an 88.9 mm gauge length. Load is applied through a worm shaft and worm gear. Each revolution of the worm gear elongates the sample 7.03×10^{-5} mm; the maximum achievable strain is 8%. The applied strain is monitored using a strain gauge attached to the center of the samples. The beam size was 25.4 mm high, centered on the gauge length, and 12.7 mm wide so that the entire cross-section of the tensile sample was included.

Measurements of peak position vs. applied stress were performed on a 17-7PH austenitic stainless steel. Stress increments of 138 MPa (20 Ksi) were employed and the 211, 220, 310 and 222 peaks were measured. As these span 35°2θ, two settings of the PSD were required; a counting time of about 30 minutes per setting was used. A summary of information is presented in Table I. Figure 5(a) shows a plot of applied stress vs. peak shift for the 211 and 310 peaks. It clearly indicates the elastic anisotropy of the material. An example of a Gaussian peak fit is shown in Fig. 5(b), for the 310 peak under a stress of 276 MPa.

Table I. Some data for the 17-7PH Steel Sample
Used in the In Situ Tensile Device

hkl	Unstressed 2θ (°)	Average Peak Intensity (Cts)	Average Error from Gaussian Fits of Peak Position (°2θ)
211	66.6	5700	0.002
220	78.7	1200	.005
310	90.3	2000	.00
222	101.9	900	.008

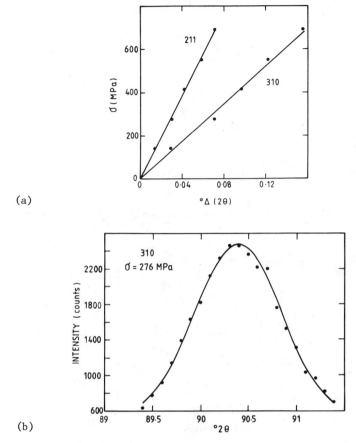

Fig. 5   (a) Applied stress vs. peak shift for the 211 and 310 peaks of the
             austenitic stainless steel sample using tensile device.
         (b) Example of Gaussian peak fit for 310 peak under stress of 276
             MPa.

The relative response of the four peaks is shown on a plot of
ν/E values vs. the crystallographic orientation parameter [4] $(h^2k^2 +$
$k^2l^2 + h^2l^2)/(h^2 + k^2 + l^2)$; see Fig. 6. The ν/E values were obtained
from least-mean-square fits of the slopes of the σ vs. $\Delta(2\theta)$ plots.
Since the measurements were made transversely to the tensile axis,

$$\varepsilon^{hkl} = (\nu^{hkl}/E^{hkl})\sigma ,$$

where ε^{hkl} is the transverse strain for the hkl planes, ν^{hkl} and E^{hkl}
are the Poisson ratio and Young's modulus of the hkl planes and σ is the
applied stress. Included are the Reuss and Voigt limits. The results
appear to follow the more realistic Kroner behavior, which is close to
the simple average of the Reuss and Voigt cases.

A related experiment was also performed using this sample. It
was loaded well into the plastic region. The material has a yield
strength of about 760 MPa, corresponding to a strain of about 0.37% or
3700$\mu\varepsilon$. The results for the 211 peak are shown in Fig. 7. The
plastic response lies between the limiting cases of continued elastic
and ideal elastic-plastic responses. This is a strengthened material

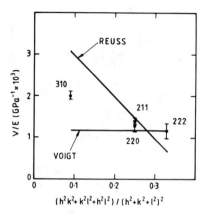

Fig. 6   Plot of ν/E vs. the crystallographic orientation parameter for
the austenitic stainless steel sample in elastic region.

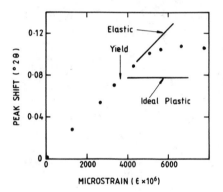

Fig. 7   Response of austenitic stainless steel sample (211 peak) in
plastic region.

for which the tensile strength is only about 10% greater than the
yield strength so that a rather sharp bend in the stress-strain curve
beyond yield is expected and observed. However, the actual yield point
seems to be delayed for the particular physical and crystallographic
orientations measured compared to the mechanically determined yield
point using the tensile test device and another sample. These
exploratory data suggest the potential of in-situ studies of two-phase
or composite systems with regard to mechanical aspects such as load-
sharing, constraint effects and in-situ yield points.

CONCLUSIONS

From these experiments we have shown that the PSD in use at the
University of Missouri Research Reactor Facility is an important tool
for the application of neutron scattering to engineering materials,
in addition to its use in more traditional neutron scattering experiments. This is because of the significant savings in data collection
time and the increased range of feasible experiments. The resolution
and behavior of this instrument has been demonstrated to be on a par
with standard detectors.

ACKNOWLEDGEMENTS

Mr. Cliff Holmes constructed and helped design the slit system and translation stage for the small beam experiments. Dr. W. Yelon also contributed greatly to the design. The tensile test device was designed by John Hollander and Rich Juergens and constructed by Rich Juergens. PJR thanks his associates at UMC and MURR. This work was performed with the support of MURR, Office of Naval Research contract No. N00014-80-C-116 with Northwestern University and the Research Council of the Graduate School, UMC.

REFERENCES

1.   C. W. Tompson, D. F. R. Mildner, M. Mehregany, J. Sudol, R. Berliner and W. B. Yelon, A Position Sensitive Detector for Neutron Powder Diffraction, J. Appl. Cryst., 17(1984)385-394.

2.   A. D. Krawitz, J. E. Brune and M. J. Schmank, Measurements of Stress in the Interior of Solids with Neutrons, in "Residual Stress and Stress Relaxation", Eds. E. Kula and V. Weiss, 1982, Plenum Press, New York, pp. 139-156.

3.   M. J. Schmank and A. D. Krawitz, Measurement of a Stress Gradient Through the Bulk of an Aluminum Alloy Using Neutrons, Metall. Trans. A, 13A(1982)1069-1076.

4.   F. Bollenrath, V. Hauk and E. H. Muller, Zur Berechnung der Vielkristallinen Elastizitatskonstanten aus den Werten der Einkristalle, Z. Metalkde., 58(1967)76-82.

LATTICE THERMAL EXPANSION EFFECTS IN PURE AND DOPED

CORDIERITE BY TIME-OF-FLIGHT NEUTRON DIFFRACTION*

P. K. Predecki and J. Haas
University of Denver, Denver, CO 80208

J. Faber, Jr. and R. L. Hitterman
Argonne National Laboratory, Argonne, IL 60439

ABSTRACT

The thermal expansion behavior of pure, Ge-doped and Li-doped hexagonal cordierites with respective compositions: $2MgO\ 2Al_2O_3\ 5SiO_2$, $2MgO\ 2Al_2O_3\ 4SiO_2\ GeO_2$, and $2MgO\ (2+x)Al_2O_3\ (5-2x)SiO_2\ xLi_2O$ with x = .174, was investigated using time-of-flight neutron powder diffraction at temperatures from 22 to 750°C in vacuum. The data were refined in space group P6/mcc using the Rietveld method. The lattice thermal expansion curves of all 3 samples were quite similar. The negative $\underline{c}$ axis expansion is associated with (1) displacement of the T_2 cations generally toward the $\underline{c}$ axis channels and (2) changes in the distortion of the coupled T_1/M tetrahedra/octahedra in the structure. Both contributions were present in all 3 samples but the first was more dominant in the Ge doped sample. The nature and origin of the distortions in T_1 and M are discussed.

INTRODUCTION

Cordierite is a magnesium-aluminum-silicate with the formula $2MgO$ $2Al_2O_3\ 5SiO_2$. It exists in two stable structural forms: (1) the high temperature hexagonal form with space group P6/mcc[2] isostructural with beryl, stable from 1450-1460°C (the melting point),[1] also known as indialite and (2) the slightly distorted and ordered orthorhombic form with space group Cccm,[3],[4] stable below 1450°C. The hexagonal form can also be formed at lower temperatures by devitrifying a glass of cordierite composition at temperatures near 1000°C.[5] The transformation of this hexagonal form to the orthorhombic follows a TTT type of behavior and is very slow below 1100°C: after 200 hrs at 1100°C the structure is still hexagonal.[6] On devitrifying the glass at 980°C the first phase to form is a metastable stuffed β-quartz structure of cordierite composition[5] also referred to as μ-cordierite.[7] After about 2 hrs at 980°C this transforms to the hexagonal form which is then stable for an indefinitely long period of time.[5]

In this study, only the low temperature hexagonal form was investigated. The objective was to gain an understanding of the negative $\underline{c}$ axis

─────────
*Work supported by Division of Basic Energy Sciences, DOE, Washington, D.C. on grant # DE-FG02-84ER45053.

173

expansion exhibited by this structure and the effect of cation dopants on the expansion mechanism.

EXPERIMENTAL

Samples were prepared from the component high purity oxide powders (mostly > 99.999%) correcting for moisture present in these powders. Compositions were, pure: $2MgO\ 2Al_2O_3\ 5SiO_2$, Li-doped: $2MgO\ (2+x)Al_2O_3$ $(5-2x)SiO_2\ xLi_2O$ with x = .174 (about the limit of Li solubility) and Ge-doped: $2MgO\ 2Al_2O_3\ 4SiO_2\ GeO_2$ (nominal). The Li was added as $LiAlO_2$. The well mixed powders were melted in Pt crucibles, homogenized at 1650°C, quenched and devitrified, using procedures described elsewhere.[8] Devitrification temperatures were: 1000°C for 20 hrs for the Li and Ge doped samples and 1000°C for 24 hrs followed by 1100°C for 20 hrs for the pure sample. Samples were hand ground wet to $\sim$ -325 mesh. The Li-doped sample was ground in 1% HF solution for 15 mins to remove a small amount of $Li_2Al_2Si_3O_{10}$ (PDF #25-1183), then washed and filtered. Samples were checked for phase content, homogeneity and chemical composition by X-ray diffraction and SEM. All samples were homogeneous and essentially single phase hexagonal cordierite with no evidence of the orthorhombic form, i.e. no splitting or broadening of the (211) hexagonal peak at $\sim$ 29.5°2θ $(CuK\alpha)$.[6]

For the neutron work, approximately 12 gms of sample powder were packed into a thin walled vanadium can $\sim$ 5 cms long x 1.2 cms in diameter placed inside a Ta resistance furnace and mounted at the center of the general purpose powder diffractometer (GPPD) at the Intense Pulsed Neutron Source (IPNS) at Argonne National Lab. Details of this instrument and its operation are given elsewhere.[9,10,11] In brief, the source is a spalla-tion type emitting pulses 30 times/sec. During each pulse, the neutrons arriving at the detectors of the GPPD have wavelengths, λ, which increase linearly with time, t. Equating the de Broglie relation to the Bragg law we obtain:

$$\lambda = \frac{ht}{mL} = 2d\ Sin\theta \qquad\qquad (1)$$

Here h is Planck's constant, m the neutron rest mass, L the total neutron flight path from source to detector (21.46m) and d is a particular inter-planar spacing. Fixed banks of detectors at 2θ near 90° and 159° are used and the neutron counts measured as a function of time from the start of each pulse (time-of-flight method), rather than as a function of angle. Thus from eq (1), d = C(θ)*t. Since the scattering angle is slightly different for each detector in a bank (the constant, C(θ) is slightly different), the counts for a particular d from all the detectors in a bank can be summed in real time and normalized to a particular scattering angle (90° or 159°), a technique known as time focussing. The range of d values covered in each pulse for the 159° bank was .38 to 2.86Å (4 to 30 msec). The resolu-tion, Δd/d of the GPPD (measured as the FWHM of a peak in terms of d) is .25% independent of d, compared with $\sim$ 1% (at 2θ = 20°) and $\sim$ .2% (at 2θ = 80°) for conventional X-ray diffraction. An important advantage of the IPNS is that the incident beam intensity increases rapidly with decreasing time, i.e. with decreasing d, up to the top of the Maxwellian distribution of neutron energies in the incident beam. Furthermore, the neutron scatter-ing lengths are constant and do not depend on sin θ/λ as scattering factors do for X-rays.

These features resulted in diffraction peaks being identifiable down to d ≃ .54Å, whereas the corresponding limit for cordierite with a con-ventional Guinier X-ray diffractometer and $CuK\alpha_1$ radiation was d ≃ 1.27Å

$(2\theta \approx 75°)$. Runs were made in vacuum ($< 10^{-5}$ torr) at temperatures from
21°C to 750°C. Each run was terminated when the number of counts at the
top of the Maxwellian distribution for the time-focused 159° detectors
reached 2000 (about 3 to 3.5 hrs with the accelerator beam current at
$\sim$ 12 μA). The 159° diffraction spectra at each temperature were fitted
to calculated patterns using the Rietveld method[12] modified for time-
of-flight neutron data.[13,14]

The structural refinements were carried out in the P6/mcc space
group, using the structural data of Meagher and Gibbs[2] as starting
values. The final refinement at each temperature utilized 36 variables:
lattice parameters (2), background (2), scale factor (1), profile
halfwidth (2), coordinates not fixed by the space group of the 5 atoms in
the asymmetric unit (7), extinction (1), occupancy of the T_1 and T_2 sites
constrained by stoichiometry (1), and anisotropic temperature factors (20).
The refinements done in several successive steps all converged to
weighted profile discrepancy factors, R, of less than 5%. R is defined
in eq (2):

$$R^2 = \sum_i w_i [Y_i(obs) - S*Y_i(calc)]^2 / \sum_i w_i [Y_i(obs)]^2 \qquad (2)$$

where $Y_i(obs)$ and $Y_i(calc)$ are the observed and calculated counts in the
i^{th} time channel, S is the overall scale factor for the calculated pattern and w_i a weighting factor for the ith channel.

The refined atom coordinates and lattice parameters were used to
calculate all bond distances ≤ 4Å and all bond angles involving bond distances ≤ 3Å in the unit cell.

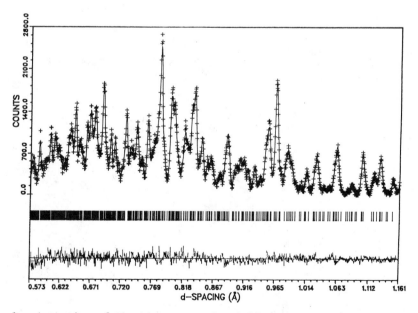

Fig. 1.   A portion of the neutron powder diffraction pattern from d =
          .573 to 1.161Å for pure hexagonal cordierite at 22°C. (+) are
          the measured data, the solid line is the best fit profile and the
          vertical marks are the calculated positions of the reflections.
          A difference curve is shown at the bottom. Background was sub-
          tracted before plotting.

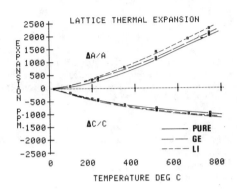

Fig. 2. Lattice thermal expansion of pure, Li-doped and Ge-doped hexagonal
         cordierites. Standard error bars on the individual points are 20
         to 40 ppm.

RESULTS AND DISCUSSION

     A portion of the neutron diffraction pattern for the pure sample
illustrating a typical fit obtained is shown in Fig. 1. Lattice thermal
expansion data are shown in Fig. 2. The structure of hexagonal cordierite
has been described elsewhere.[15,8] Briefly, the structure consits of
6-membered rings of T_2 (Si rich) tetrahedra stacked one above the other,
forming c axis channels (Fig. 3). Adjacent rings are rotated $\sim$ 30° relative to each other about the c axis. Rings are linked to each other
laterally and vertically by T_1 (Al rich) tetrahedra and M (Mg) octahedra.
Viewed normal to the c axis, T_2 and M polyhedra alternate, as shown in
Fig. 4.

     Data from the refinements were examined to see which structural
features were changing significantly with temperature. The complete data
are listed elsewhere.[16] The $M-O_1$ bond lengths increased with a thermal
expansion coefficient of $9.8 \pm 1 \times 10^{-6}$/°C for all 3 samples. The $O_1(1)-$

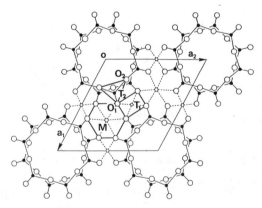

Fig. 3. Projection along the c axis of the hexagonal cordierite (beryl)
         structure in space group P6/mcc. The five atoms in the asymmetric
         unit: T_1, T_2, M, O_1 and O_2 are shown together with the 3 basic
         polyhedra: T_2, T_1 and M from which the structure is built. Each
         tetrahedron T_1 shares 2 edges with adjacent M octahedra. Each M
         shares 3 edges with adjacent T_1's. Origin is at 0.

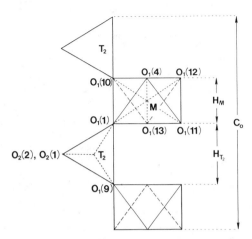

Fig. 4.   Schematic representation of a portion of the hexagonal cordierite
          structure viewed normal to the c axis. Octahedra M and tetra-
          hedra T_2 alternate along the c direction. Tetrahedra T_1 (which
          share edges with M) are not shown for clarity.

T_2-O_1(9) bond angle(defined in Fig. 4) decreased from room temperature to
750°C for all 3 samples (107.82° to 106.79° pure, 107.74° to 106.95° Ge
doped, 108.22° to 107.5° Li doped). Similar changes were reported by
Hochella et al.[17] in orthorhombic cordierite. The O_1-T_1-O_1 angles in
tetrahedron T_1 changed as shown in Fig. 5. The T_1 and T_2 site occupancies did not change much for the Ge and Li doped samples (Table I) and
became slightly more random with temperature for the pure sample. For
the Ge doped sample, two models were tried: Ge associated with the Si
(Si model) or with the Al (Al model). The Si model gave slightly better
R values.

Expansion Behavior

        The expansion of the structure was resolved into components parallel
and normal to the c axis.[8] Expansion parallel to c consisted of expansion of heights H_M and H_{T_2} (defined in Fig. 4) as shown in Fig. 6. H_M
decreased with temperature to a shallow minimum around 400°C for all 3

Table I: Fractional Al Occupancies in T_1 and T_2
           (Si model).   Standard deviations in the
           last digit are given in parentheses.

	Pure		Ge Doped		Li Doped	
Temp. °C	22	750	26	750	21	750
T_1	.70(7)	.52(10)	.53(3)	.55(5)	.85(8)	.85(10)
T_2	.32(3)	.41(5)	.40(2)	.39(5)	.24(4)	.25(5)

For the Li-doped sample, the Li content was too low to affect the refinement. A model with the Li in the c axis channels gave excessively large
temperature factors for the Li; the Li was therefore omitted from the
refinements other than to increase the Al/Si ratio.

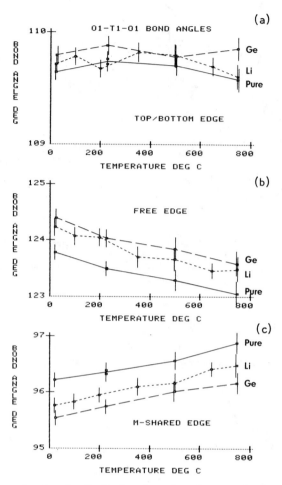

Fig. 5. Changes in the O_1-T_1-O_1 bond angles in tetrahedron T_1 with temperature in pure and doped samples containing Ge and Li. (a) Angle subtended by the top or bottom edges of T_1 (the diagonal lines on T_1 in Fig. 3). (b) Angle subtended by the two free (unshared) edges of T_1. (c) Angle subtended by the two M-shared edges of T_1.

samples while H_{T_2} decreased with temperature above about 400°C. Thus both H_M and H_{T_2} are involved in the negative $\underline{c}$ axis expansion. The sum H_M + H_{T_2} agrees well with half the $\underline{c}$ lattice parameter as shown in Fig. 6(c).

Expansion normal to $\underline{c}$ was conveniently expressed as the expansion of polygons in the basal plane of the unit cell, defined in Fig. 7. Changes in the areas of these polygons with temperature are shown in Fig. 8. It is evident that areas M and P expand with temperature for all 3 samples, while the area H of the six-membered rings stays constant or decreases slightly.

An alternate way of showing the expansion effects is shown in Fig. 9. Here displacements of the 750°C atom coordinates (in the basal plane) relative to coordinates at 22°C were magnified 20x to reveal the expansion effects. If expansion in the basal plane were isotropic, on the atomic scale, displacements of all atoms in Fig. 9 would be radial from the

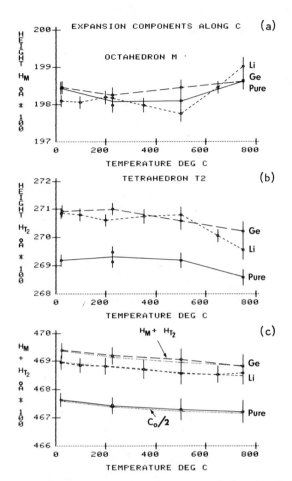

Fig. 6. Expansion components along c̲ in pure and doped samples. (a) The
        height, H_M, of octahedron M along c̲. (b) The height, H_{T_2}, of the
        $O_1(1)-O_1(9)$ edge (parallel to c̲) in tetrahedron T_2. (c) $H_M + H_{T_2}$
        compared with $c_o/2$ (dotted).

origin and would increase with increasing distance from the origin. This
is clearly not the case.

Distortion of Polyhedra

From Fig. 9 it is evident that with increasing temperature the T_2
atom moves significantly inward towards the c̲ axis channels for all 3
samples, rather than radially outward from the origin. The O_2 atoms also
move inward towards the origin while the O_1 atoms in tetrahedron T_2 move
small amounts approximately circumferentially about the origin. These
movements produce changes in the bond angle distortion index, σ, for
tetrahedron T_2, as shown in Fig. 10(b). σ is defined in eq. 3:[18]

$$\sigma^2 = \sum_{i=1}^{n} (\theta_i - \theta_o)^2/(n-1) \tag{3}$$

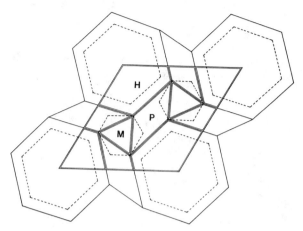

Fig. 7.   Division of the basal plane area into hexagons, H, formed by the
          outermost (O_1) oxygens of a ring of T_2 tetrahedra, triangles M
          formed by the top or bottom surfaces of octahedra and parallelo-
          grams P bounded by H and M. The total area of one unit cell
          base = H + 2M + 3P.

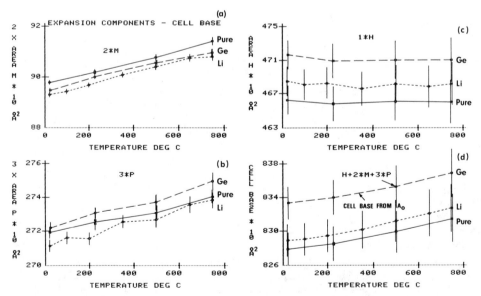

Fig. 8.   (a),(b),(c). Contributions of the polygons shown in Fig. 7 to
          the expansion of the unit cell base for the pure, Li-doped and
          Ge-doped samples. The sum: H + 2M + 3P of the contributions is
          compared with the cell base area calculated from the a_0 lattice
          parameter in (d).

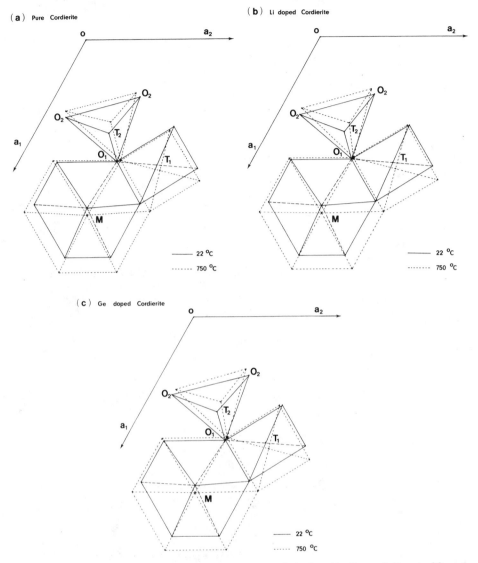

Fig. 9. c Axis projection of the 3 basic polyhedra M, T_1 and T_2 at 22 and
         $\overline{750}°C$ for (a) pure, (b) Li-doped, and (c) Ge-doped hexagonal
         cordierite. Displacements of the 750°C atom positions from the
         22°C positions have been magnified 20X. Origin is at O.

where θ_i is the bond angle subtended by an edge of the distorted polyhedron
at the interior atom, θ_o is the corresponding angle for the undistorted
polyhedron (109.47° for a tetrahedron, 90° for an octahedron) and n is
the total number of edges. σ passes through a shallow minimum around
400°C for the 3 samples. With increasing temperature above ∿ 500°C, the
increasing σ correlates for all 3 samples with the decreasing H_T (Figs.
6(b) and 10(a)). One can envisage that the inward movement of T_2^2 length-
ens the T_2-O_1 bonds causing the $O_1(1)-T_2-O_1(9)$ bond angle to decrease and
H_{T_2} to decrease. The movement of T_2 with increasing temperature in Fig.
7 is approximately away from the Mg atom at M in all 3 cases. Possible
reasons are discussed elsewhere.[8,16]

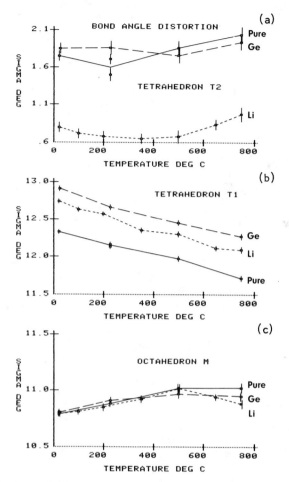

Fig. 10.   Variation of the bond angle distortion index, σ, (degrees) with
temperature for (a) tetrahedron T_2, (b) tetrahedron T_1 and
(c) octahedron M.

The distortion of the T_1 tetrahedron is much more severe than that
of the T_2, and decreases with increasing temperature as shown in Fig. 10(a)
and (b). The distortion is torsional about an axis parallel to $\underline{c}$ and
passing through the T_1 atom such that one pair of $O_1-T_1-O_1$ bond angles
(the pair subtended by the free edges of T_1) is much larger, at $\sim 124°$ than
the second pair at $\sim 96°$, subtended by the M-shared edges (Fig. 5). The
angles subtended by the top and bottom edges are relatively undistorted.
The distortion of the M octahedra is also severe and increases with increasing temperature (Fig. 10(c)).

The cause of the M and T_1 distortions appears to be that the M and T_1
polyhedra form rings surrounding the rather rigid 6-membered rings of T_2
tetrahedra (Fig. 3). The circumference of the M-T_1 rings would be too
small to fit around the 6-membered rings if the M and T_1 polyhedra were
undistorted; to fit around the 6-membered ring the M and T_1 must be stretched
circumferentially into their distorted shapes. Hence these polyhedra are
extended in the circumferential direction and shortened in the $\underline{c}$ direction.
M and T_1 are constrained to have the same heights in the $\underline{c}$ direction by
the space group.

With increasing temperature, the $M-O_1$ bonds expand, relieving some of the torsional distortion in T, but increasing the distortion in M. Possible reasons for the negative expansion of H_M up to 400°C are discussed elsewhere.[8,16]

Effect of Dopant Cations

The Ge and Li doped samples had larger T_2 tetrahedra and larger H_{T_2} values (Fig. 6(b)) than the pure sample, due to the larger Ge^{+4} ion size and the increased Al/Si ratio in the Li doped sample. The average T_2-O bond lengths at room temperature were: 1.648Å pure, 1.655Å Li doped and 1.658Å Ge doped. The size of the 6-membered T_2 rings was also larger in the doped samples (Fig. 8(c)) for the same reason. As a result of the larger T_2 ring size, the distortion of the T_1 polyhedra was also larger in the doped samples (Fig. 10(b); however, the H_M values for all 3 samples were about the same (Fig. 6(a)). Evidently the expected decrease in H_M from the increased distortion has compensated for the expected increase in H_M from the larger T_1 cation size in the doped samples.

The changes in H_M and H_{T_2} with temperature are quite similar for the 3 samples. The Ge doped sample shows a slightly smaller negative contribution from H_M and a slightly larger one from H_{T_2} than the pure sample (Fig. 6). Experiments with dopants producing larger effects are in progress.

CONCLUSIONS

Lattice thermal expansion coefficients for the samples investigated are all very similar: pure; α_a = 2.781, α_c = -1.351 x $10^{-6}c^{-1}$, Li doped; α_a = 3.155, α_c = -1.228 x $10^{-6}c^{-1}$, Ge doped; α_a = 2.943, α_c = -1.545 x $10^{-6}c^{-1}$ from room temperature to 750°C.

The negative c axis expansion results from two contributing effects: (1) displacement of the T_2 cations generally inward towards the c axis channels and (2) changes in the distortion of the coupled T_1 and M polyhedra. Both contributions are present in all 3 samples with the first more dominant in the Ge doped sample.

The severe bond angle distortion of the T_1 and M polyhedra apparently results from the fact that these polyhedra form rings which are stretched around the circumference of the 6-membered rings of T_2 tetrahedra in the structure.

ACKNOWLEDGEMENTS

We are grateful to Deane Smith of The Pennsylvania State University, M. F. Hochella of Stanford and R. N. Kleiner of Coors Ceramics Co. for helpful discussions. One of us (JF) acknowledges the support of the U.S. Department of Energy, BES-Materials Sciences, under Contract W-31-109-Eng-38.

REFERENCES

1. W. Schreyer and J. F. Schairer, Compositions and Structural States of Anhydrous Mg-Cordierites: A Reinvestigation of the Central Part of the System $MgO-Al_2O_3-SiO_2$, J. of Petrology 2:324 (1961).
2. E. P. Meagher and G. V. Gibbs, The Polymorphism of Cordierite II: The Crystal Structure of Indialite, Can. Mineralogist 15:43 (1977).

3. G. V. Gibbs, The Polymorphism of Cordierite I: The Crystal Structure
    of Low Cordierite, Am. Mineralogist 51:1069 (1966).
4. J. B. Cohen, F. K. Ross, and G. V. Gibbs, An X-Ray and Neutron Dif-
    fraction Study of Hydrous Low Cordierite, Am. Mineralogist 62:67
    (1977).
5. K. Langer and W. Schreyer, Infrared and Powder X-Ray Diffraction
    Studies on the Polymorphism of Cordierite, $Mg_2(Al_4Si_5O_{18})$, Am.
    Mineralogist 54:1442 (1969).
6. A. Putnis and D. L. Bish, The Mechanism and Kinetics of Al,Si Ordering
    in Mg-Cordierite, Am. Mineralogist 68:60 (1983).
7. A. Miyashiro, T. Iiyama, M. Yamasaki, and T. Miyashiro, Am. Jour. Sci.
    253:185 (1955).
8. P. K. Predecki, J. Haas, J. Faber, Jr. and R. L. Hitterman, Structural
    Aspects of the Lattice Thermal Expansion of Hexagonal Cordierite,
    submitted to J. Am. Ceram. Soc.
9. J. Faber, Jr., Sample Environments at IPNS: Present and Future Capa-
    bilities, Revue Phys. Appl. 19:643 (1984).
10. S. R. MacEwen, J. Faber, Jr., and A. P. L. Turner, The Use of Time-of-
    Flight Neutron Diffraction to Study Grain Interaction Stresses,
    Acta Met. 31:657 (1983).
11. J. Faber, Jr. and R. L. Hitterman, High Resolution Powder Diffraction
    Studies at a Pulsed Neutron Source, to be published in Vol. 29,
    Advances in X-Ray Analysis.
12. H. M. Rietveld, J. Appl. Crystallogr. 2:65 (1969).
13. J. D. Jorgensen and F. J. Rotella, J. Appl. Crystallogr. 15:27 (1982).
14. R. B. Von Dreele, J. D. Jorgensen and C. G. Windosr, J. Appl. Crystal-
    logr. 15:581 (1982).
15. E. P. Meagher and G. V. Gibbs, The Polymorphism of Cordierite II: The
    Crystal Structure of Indialite, Can. Mineralogist 15:43 (1977).
16. J. Haas, Mechanism of Lattice Thermal Expansion in Pure and Doped
    Cordierite, M.S. Thesis, Physics Dept., University of Denver,
    Denver, CO.
17. M. F. Hochella, Jr., G. E. Brown, Jr., F. K. Ross and G. V. Gibbs,
    High Temperature Crystal Chemistry of Hydrous Mg- and Fe-Cordierites,
    Am. Mineralogist 64:337 (1979).
18. K. Robinson, G. V. Gibbs, and P. H. Ribbe, Quadratic Elongation: A
    Quantitative Measure of Distortion in Coordination Polyhedra,
    Science 172:567 (1971).

XRDQUAL--A PROBABILISTIC AND LEAST-SQUARES SEARCH/MATCH PROCEDURE

William R. Clayton

Research Chemist
International Fertilizer Development Center
P.O. Box 2040
Muscle Shoals, Alabama 35662

INTRODUCTION

A series of computer programs have been written (collectively called XRDQUAL) for automated identification of XRD patterns. The algorithms used in XRDQUAL have been selected for maximum compatibility with computer-based calculations and decisions. XRDQUAL relies heavily on probabilistic logic and calculations. The overall sequence used can be divided into three general steps:

1. A subset of the JCPDS[1] file is selected by using the program XRDSRCH to scan the PDF file and calculate the probability of a line match between the PDF pattern and the observed powder pattern. The members of this subset are those patterns that have lines matching the observed pattern's d-spacings with a high probability value.

2. The subset generated by XRDSRCH is evaluated more rigorously (programs XRDSORT or XRDLSQ) and sorted in a descending order based on probability of a match between the PDF pattern and observed powder pattern. Chemistry may be used, if known, to bias the sorted order of the PDF subset.

3. The sorted PDF subset is used to synthesize an XRD pattern that matches the observed pattern as closely as possible. Least-squares and statistical procedures in program XRDLSQ are used to select the PDF patterns that will best reproduce the observed powder patterns.

Short descriptions of the program and algorithms that constitute XRDQUAL follow. The XRD data used in XRDQUAL examples are for the JCPDS/NBS Round Robin sample 7 (NBS-7) as reported by Edmonds and Henslee.[2]

XRDSRCH

XRDSRCH is designed to find a small subset of PDF files containing all phases actually present in the observed XRD powder pattern. A probabilistic approach is taken in order to make the acceptance/rejection simple and reliable. A Lorentzian function is used for simplicity of calculation. The probability function for matching single pairs of peaks has been discussed by Schreiner et al.[3] and has the form

$$P_i = \frac{1}{1 + (d/q)^2} \ .$$

Here, d is the difference between a close pair of lines when the PDF pattern is compared with the observed XRD powder pattern. The value of q is somewhat arbitrary but is defined as the difference value between pairs of lines corresponding to a probability of 0.50. In practice, this value should be determined experimentally. A satisfactory working value for data of good quality is usually the FWHM (full width at half maximum) value of resolved peaks. The FWHM is usually a slowly varying parameter in terms of angles (two-theta). A constant width probability window is adequate for determining P_i. A d-space representation of the probability window has a variable width and will shrink as d-space values become smaller.

The overall probability of matching a PDF pattern to the observed XRD pattern is the weighted probability

$$P_1 = \frac{\sum_i I_i P_i}{\sum_i I_i} \ .$$

The weighting function is the intensity of the PDF line of each pair. By trial and error, 0.80 has been established as a satisfactory cutoff point for an acceptance test. This corresponds to finding a match within the half-height points of a peak when the FWHM is used as a probability window.

XRDSRCH was run for NBS-7 against a JCPDS subfile of inorganic phases containing only common elements. Only 40 PDF patterns were retained for further consideration after some 12,200 PDF data sets were screened. The number of retained patterns will vary depending on the value of the probability window and the composition of the JCPDS subfile searched. A significant reduction in the number of incorrect "hits" can be achieved by using a small subfile.

XRDSORT

The list of PDF files produced by XRDSRCH is evaluated by XRDSORT and ranked in descending order of probability. The overall score is the continued product of all separate probability tests. Table 1 shows XRDSORT results for PDF patterns with total probabilities greater than 0.10. The four correct phases are found in the seven most probable PDF data sets. The overall probability may be considered as the probability that the

Table 1. XRDSORT Results for NBS-7. Data Sets Are Sorted in Descending Order of Probability of Being a Subset of NBS-7. JCPDS Phases Actually Present in the Mixture Are Marked With an Asterisk

JCPDS	Formula	Probability x 100				
		P1	P2	P3	P4	P
*251135	$BaCl_2 \cdot 2H_2O$ (Barium chloride hydrate)	89	52	78	75	28
*50664	ZnO (Zinc oxide/Zincite syn)	96	82	44	44	15
*40471	KI (Potassium iodide)	91	84	47	40	14
170385	$(ZrO_{0.35})$ (Zirconium oxide)	93	61	52	44	13
180394	$Cr_2(SO_4)_3$ (Chromium sulfate)	82	82	45	42	13
230215	$Cu_3Fe(CN)_6 \cdot 4H_2O$ (Copper iron cyanide)	92	84	41	33	10
*50565	(Si) Silicon	97	82	37	35	10
10433	$Zn_2Fe(CN)_6 \cdot 3H_2O$ (Zinc iron cyanide)	86	59	50	40	10
271402	(Si) Silicon	98	50	37	35	10
	Plus 31 additional phases with P<0.10					

reference pattern and observed pattern are identical. Because there is no distinct break in probability, the phases actually present cannot be separated from those not present.

The first test, P_1, is the same one used in XRDSRCH for the initial screening of the full PDF file. It is not sensitive to interference effects caused by multiple component mixtures nor to intensity distortions caused by preferred orientation. Thus it is a good test for initial screening.

It is, however, not discriminating enough for a complete analysis of the observed XRD powder pattern. Given enough lines (30-50), even a random collection of lines will find matches in the PDF file, especially among those PDF patterns containing only a few lines. The most important feature of the P_1 test is that it seldom fails to find a major phase that is actually present in the observed XRD pattern, assuming the phase has a PDF entry.

The second test, P_2, is the common "R^2" statistical test. It measures the linear correlation between observed XRD intensities and PDF pattern intensities. It is a natural measure of prediction accuracy and strength of linear association.

$$P_2 = \frac{[\Sigma XY]^2}{\Sigma X^2 \Sigma Y^2} \ .$$

The third test, P_3, estimates the information content of the PDF pattern relative to the observed intensities of the XRD powder pattern.

$$P_3 = [\Sigma I_{PDF}/\Sigma I_{XRD}]^{\frac{1}{2}} \ .$$

This test favors phases that are major contributors to the observed intensities. Subsequent least-squares analysis is aided by the analysis of the larger intensities first. Least-squares analysis becomes more robust as unexplained residuals become smaller.

The fourth test, P_4, estimates the information content of the PDF pattern relative to the number of lines in the observed XRD powder pattern.

$$P_4 = [\Sigma N_{PDF}/\Sigma N_{XRD}]^{\frac{1}{2}} \ .$$

This test favors phases that have many lines in common with the crystalline phases being analyzed. Subsequent least-squares analysis is aided by this test's discrimination against random matches from the PDF file. Random matches usually mimic minor phases with only one or two intense PDF lines apparently matching small observed intensities.

The last test, P_5, is an arbitrary attempt to consider the known chemistry, if any, of the sample. Each element in the PDF formula is compared with the chemistry given for the XRD sample and assigned a probability. The final value of P_5 is the product of all the individual elemental probabilities. Elemental probabilities are assigned as

$P_{element}$ = 1.00 element present;

$P_{element}$ = 1.00 unconcerned element;

$P_{element}$ = 0.70 undetermined element; or

$P_{element}$ = 0.00 element absent.

Unconcerned elements are light elements, such as carbon or oxygen, which may be present but are not usually included in an elemental analysis.

Undetermined elements are those that could have been analyzed but were
not. The presence of an element known to be absent removes that PDF
pattern from the subfile. In the absence of an analysis, chemistry may be
implied by the assignment of default probabilities for each element:

$P_{element}$ = 1.00 common element,

$P_{element}$ = 0.10 less common element, or

$P_{element}$ = 0.01 uncommon element.

The designations "common," "less common," and "uncommon" follow the
recommendations found in the Frequently Encountered Phases, Search Manual.[4]
Because of the arbitrary nature of its value, usually 0 or 1, P_5 is not
listed in the report.

XRDLSQ

XRDLSQ performs the evaluations described in XRDSORT and sorts the
PDF subset in the same manner. The sorted subset is then evaluated by
least-squares and statistical procedures. The basic concept is that the
observed XRD pattern is the simple sum of the individual patterns of its
component crystalline phases. The mathematical formulation of this is

$$I_i = a_1 I_{i,1} + a_2 I_{i,2} + \ldots + a_n I_{i,n} \ .$$

The least-squares evaluation (Table 2) involves finding the coefficients, a_n, for each of the PDF patterns in the subset generated by
XRDSRCH. The coefficients may seldom be determined directly from the full
subfile because least-squares evaluations require more data (intensities)
than the determined coefficients. A procedure must be followed which
will, ideally, result in the generation of a trial model containing only
patterns whose coefficients are non-zero. There is no general agreement
on a best procedure,[5] but one of the most reliable is the "forward addition, backward elimination" procedure. A model is formed by considering
the most probable patterns from the sorted XRDSRCH PDF list as candidates
for possible inclusion in a least-squares model. Candidates are added to
the trial model one at a time and evaluated by least-squares. If all
tests are favorable, the candidate is added to the model and an additional
candidate is selected for consideration. If the candidate is rejected, a
new candidate is evaluated. If a member of the model is found unacceptable, the evaluation process resumes at that point, and the next candidate
is the one immediately following the rejected member on the sorted PDF
list. This procedure is very conservative and can involve a large number
of cycles of calculation. The order of the sorted PDF candidate patterns
becomes very important in reducing the number of cycles necessary to find
the best solution.

The acceptance/rejection tests are the most critical aspects of the
"forward addition, backward elimination" procedure. The number of tests
to be performed have been held to a minimum. The most reliable tests have
been performed first. Failure of any test results in the rejection of the
PDF pattern from the least-squares model.

The first test checks the calculated coefficient for negativity. The
calculated coefficient is directly related to the weight fraction of a
component present in a mixture. The minimum weight fraction of a component
in a real mixture is "none." Therefore, if the coefficient is negative,
the phase cannot be present in the XRD powder pattern. This test is the
most powerful and reliable test performed by XRDLSQ.

Table 2.   XRDLSQ Final Results, NBS-7

R1 = 0.220
R2 = 0.934

Scale	Sigma	JCPDS	
0.577	0.034	251135	$BaCl_2 \cdot 2H_2O$ (barium chloride hydrate)
0.641	0.057	50664	ZnO (zinc oxide/zincite syn)
0.826	0.062	40471	KI (potassium iodide)
0.629	0.069	50565	Si (silicon)

			PDF Intensities			
d-Space	I_{obs}	I_{calc}	251135	50664	40471	50565
5.7200	19	7	7			
5.4520	66	49	49			
4.9380	54	35	35			
4.8440	19	12	12			
4.4960	29	32	32			
4.4270	61	58	58			
4.3330	15	10	10			
4.0790	38	35			35	
3.6610	29	26	26			
3.6220	13	10	10			
3.5330	100	91	9		83	
3.3890	29	32	32			
3.3600	15	14	14			
3.1350	62	63				63
2.9490	29	29	29			
2.9290	20	38	38			
2.9080	43	52	52			
2.8620	24	26	26			
2.8150	40	46		46		
2.7110	31	29	29			
2.6040	35	36		36		
2.5470	36	43	43			
2.5280	8	10	10			
2.4980	46	58			58	
2.4760	72	70	6	64		
2.4090	23	24	24			
2.3870	16	10	10			
2.3650	14	11	11			
2.2550	23	14	14			
2.2270	11	14	14			
2.2090	7	17	17			
2.1310	24	24			24	
2.1180	13	8	8			
2.0870	23	29	29			
2.0620	6	20	20			
2.0400	20	30	8		22	
1.9990	15	11	11			
1.9210	30	38				38
1.9110	17	21	2	19		
1.7670	11	12			12	
1.6380	29	22				22
1.6250	26	31		26	6	
1.5790	24	20			20	

(Continued)

Table 2.   XRDLSQ Final Results, NBS-7 (Continued)

d-Space	I_{obs}	I_{calc}	PDF Intensities			
			251135	50664	40471	50565
1.5570	9	0				
1.5000	9	0				
1.4780	25	22		22		
1.3790	30	18		18		
1.3580	22	16		9	2	5
1.3070	7	0				
1.2470	15	10			2	8
1.1090	7	11				11
1.0930	11	6		6		
1.0460	35	6				6
1.0420	6	6		6		
1.0160	6	3		3		

The second test is a statistical test to eliminate highly correlated pairs of PDF patterns. Such patterns are usually multiple observations of the same crystalline phase, isostructural compounds, or random matches. The best statistical factor for the determination of correlated pairs[6] is the partial correlation coefficient.

The partial correlation coefficient, $r_{12/3...p}$, is the correlation between the variables X_1 and X_2 when the other $p-2$ variables remain fixed. It is a measure of the direct linear dependence of variables X_1 and X_2. If B is the inverse of the matrix of normal equations, then

$$r_{12/3...p} = \frac{-b_{12}}{(b_{11}b_{22})^{\frac{1}{2}}} \ .$$

To test the hypothesis

$$r_{12/3...p} = 0,$$

compute

$$t = \frac{[(n - p)r^2_{12/3...p}]}{[1 - r^2_{12/3...p}]}$$

and test as Student's t with n-p degrees of freedom.[7] If a highly correlated pair of PDF patterns is found, one must be rejected while the other is retained. An absolutely reliable algorithmn for selecting the "correct" pattern has not been determined. The most reliable method is simply to accept the PDF pattern with the highest overall match probability. The use of known chemistry to restrict the search will resolve most ambiguities in this test.

The third test accepts or rejects the PDF pattern on the basis of the hypothesis

$$a_n > 0 \ .$$

The hypothesis is tested at a preset confidence level using the least-squares coefficient and its estimated error from the least-squares calculations. If the hypothesis is accepted, the PDF pattern is retained in the trial model. The acceptance level is held fairly high if no chemistry is used but can be relaxed considerably if the sample is well characterized and chemistry is used. The confidence level used for acceptance determines the lower detection limit of the procedure.

The final test checks to determine the number of unique matches between the observed XRD pattern and the PDF powder pattern. If the number is zero, the PDF pattern is rejected. This test has proved to be quite powerful in resolving multiple correlations. Such correlations can arise when parts of one PDF pattern systematically match parts of two or more other patterns. This is fairly common in inorganic compounds where molecular packing considerations determine the crystal structure.

Each cycle of refinement has two indicators for determining the quality of the trial model. R1 is the conventional crystallographic factor:

$$R1 = \frac{\Sigma \ |I_{obs} - I_{calc}|}{\Sigma \ I_{obs}} \ .$$

R1 approaches zero as the calculated model gets closer to the observed values. Little statistical information is gained from an examination of R1, but a low value of R1 is a reasonable indication that the model is correct. The R1 value, 0.220, in Table 2 is an indication of some misfit between the observed and calculated values. There is a systematic difference between the observed and calculated values at high d-spacings, probably due to differences in intensity calculation methods. If one set of data represents peak heights while the other represents integrated intensities, the differences are reasonable. There are many recognized problems in comparing data from different sources.

The second indicator, R2, measures the linear correlation between observed XRD intensities and the calculated intensities. Identical to the "r_{xy}^2" test,[8] it is a natural measure of prediction accuracy and strength of linear association. It is the ratio of explained variation in the dependent variable Y to the total variation in Y.

$$R2 = \frac{SS_y - SS_{residual}}{SS_y} \ .$$

This ratio is sometimes referred to as the coefficient of determination. The square root of this ratio is the Pearson product-moment correlation between variables X and Y.

The final trial model is used to generate a listing of d-space values and intensities from the sample, model, and individual PDF patterns used to create the model. The form of the listing is suitable for use as a manual worksheet if needed as such.

REFERENCES

1. JCPDS-International Centre for Diffraction Data, 1601 Park Lane,
       Swarthmore, Pennsylvania, 19081.
2. Edmonds, J. W., and W. W. Henslee. 1979. Advances in X-Ray Analysis,
       Vol. 22, p. 143, Plenum Press, New York and London.

3. Schreiner, W. N., C. Surdukowski, and R. Jenkins. 1982. J. Appl.
       Cryst., 15:524-530.
4. JCPDS. 1977. Frequently Encountered Phases, Search Manual.
5. Draper, N. R., and H. Smith. 1968. Applied Regression Analysis, 1st
       Edition, John Wiley and Sons, New York, New York.
6. Hamilton, W. C. 1964. Statistics in Physical Science, The Ronald
       Press Co., New York, New York.
7. Koehler, K. J. 1983. Technometrics, 25(1):103-105.
8. Nie, N. H. SPSS: Statistical Package for the Social Sciences, 2nd
       Edition, p. 324, McGraw-Hill Book Company, New York, New York.

RAPID COMPOSITION-DEPTH PROFILING WITH X-RAY DIFFRACTION:

THEORY AND PRACTICE

Karl E. Wiedemann and Jalaiah Unnam

Analytical Services and Materials, Inc.
28 Research Drive
Hampton, VA 23666

ABSTRACT

High precision composition-depth profiles can be determined quickly using a recent development. This method requires that noncompositional broadening arising from the instrument, crystal defects, and the radiation source be removed from the diffraction pattern before calculating the composition-depth profile. Effective deconvolution techniques, profiling theory, methodology of profiling, and the effect of residual noncompositional broadening on the profile are discussed. Examples include statistical analyses of error in the profile due to random counting errors and variance in the lattice parameter calibration.

INTRODUCTION

The diffraction peak will be broadened when x-rays are diffracted from a composition gradient within a single phase[1-5] (provided that there is a concomitant lattice-parameter gradient). This broadened peak is referred to as an intensity band.

The phenomenal intensity band (so called because it is the unmodified pattern produced by the diffractometer) has three fundamental constituents. They are (1) the ideal intensity band, (2) the noncompositional broadening, and, (3) random counting error. The ideal intensity band is pure compositional broadening; the noncompositional broadening is caused by crystal defects and instrumental limitations. The random counting error, on the other hand, is not a form of broadening at all; it is a random variable that represents the probability of determining the true counting rate after a given number of counts.

There are currently two approaches used for analyzing the phenomenal intensity band. The older method begins by assuming a profile and forming a simulated intensity band: the profile is then varied until the simulation and the phenomenal band are adequately matched. Unfortunately, this method required days of the researcher's time to analyze a single band, even when assisted by high-speed computers. The more recent method removes the noncompositional broadening and random counting error, and then treats the pure compositional broadening directly. This method requires only a few minutes for the analysis of a single band. This paper deals with various facets of this latter technique.

THE DERIVATIVES OF THE PHENOMENAL INTENSITY BAND

Before trying to remove noncompositional broadening, we want to quantify it, if possible, directly from the intensity band. If the edge of the ideal intensity band is like a step function and the noncompositional broadening is Gaussian, then, at tne step, the second derivative of the phenomenal band will be zero, and the ratio of the first and third derivatives will be $-\sigma^2$ where σ^2 is the variance parameter of the Gaussian broadening function.

The steps in the ideal intensity band mark solubility limits. In a simple intensity band one would expect only two steps: one marks the low solubility limit, and the other, the high solubility limit. In the example in figure 1 the second derivative of the phenomenal intensity band passes through zero four times. Whereas the outer two zeroes mark solubility limits, the inner two do not: they are a consequence of curvature in the ideal intensity band and are not key features. The Gaussian variance parameters determined at the solubility limits can be used in the deconvolution.

DECONVOLUTION THEORY

The phenomenal intensity band is the convolution of the ideal intensity band and noncompositional broadening to which has been added a random counting error. Its mathematical expression is

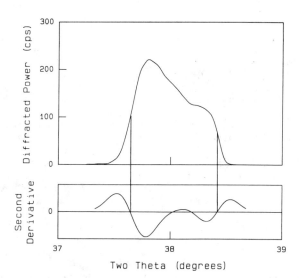

Fig. 1. A phenomenal intensity band (above) and its second derivative with respect to 2θ (below). The sample is an unalloyed titanium button exposed to air at 1400°F for 15 minutes. This is the (002) α-titanium reflection scanned with Cu $K\alpha_1$, 0.1°/min. The zero in the second derivative on the extreme left corresponds to a solubility limit of about 22 atomic percent oxygen. The zero to the extreme right corresponds to pure titanium.

$$\frac{dP_b}{d2\theta}(2\theta) = \int_{-\infty}^{\infty} \frac{dP}{d2\theta}(\zeta)\, g(2\theta-\zeta)\, d\zeta + \varepsilon(2\theta),\qquad(1)$$

where $dP_b/d2\theta$ is the distribution of power in the phenomenal intensity band, $dP/d2\theta$ is the distribution of power in the ideal intensity band, g is the normalized function representing the noncompositional broadening, and ε is the random counting error.

Because the random counting error is small, one might be tempted to neglect it and search for a function z such that

$$\int_{-\infty}^{\infty} z(2\theta-\xi) \int_{-\infty}^{\infty} \frac{dP}{d2\theta}(\zeta)\, g(\xi-\zeta)\, d\zeta\, d\xi = \frac{dP}{d2\theta}(2\theta).\qquad(2)$$

By virtue of Eq. (1) and the assumption $\varepsilon = 0$, Eq. (2) implies that the distribution of power in the ideal intensity band is simply

$$\frac{dP}{d2\theta}(2\theta) = \int_{-\infty}^{\infty} \frac{dP_b}{d2\theta}(\xi)\, z(2\theta-\xi)\, d\xi.\qquad(3)$$

But this results in an intensity band that will contain large, spurious oscillations: the neglected small random errors will have been amplified.

Instead of using Eq. (3), let us write the Lagrange functional

$$L\!\left(\frac{dP}{d2\theta}\right) = \int_{-\infty}^{\infty} H\!\left(2\theta,\frac{dP}{d2\theta}\right) d2\theta,\qquad(4a)$$

$$= \int_{-\infty}^{\infty} \cdot \left(\varepsilon^2\!\left(2\theta,\frac{dP}{d2\theta}\right) + \kappa_1\, \phi_1\!\left(2\theta,\frac{dP}{d2\theta}\right) + \kappa_2\, \phi_2\!\left(2\theta,\frac{dP}{d2\theta}\right)\right) d2\theta,\qquad(4b)$$

where ϕ_1 and ϕ_2 are constraints and κ_1 and κ_2 are Lagrange multipliers. These constraints are essential because they keep ε^2 from vanishing. There may be any number of constraints, but it is necessary that there be at least one.

The extrema of L are given by the set of $dP/d2\theta$ that satisfy

$$\frac{\partial H}{\partial(dP/d2\theta)}\!\left(2\theta,\frac{dP}{d2\theta}\right) = 0,\qquad(5)$$

which is the Euler equation for a singular problem in the calculus of variations.

Selection of the constraints is the crux of the deconvolution problem. One or more of the constraints may be used. The following two constraints are recommended as being particularly useful. The first is a smoothing constraint:

$$\phi_1\left(2\theta, \frac{dP}{d2\theta}\right) = \left(\frac{dP_s}{d2\theta}(2\theta) - \frac{dP}{d2\theta}(2\theta)\right)^2 - R_1^2(2\theta) = 0, \tag{6}$$

where $dP_s/d2\theta$ is the smoothing of the ideal intensity band (perhaps using a nine-point smoothing operation). Smoothing on each continuous segment of the ideal intensity band is done independently. The second constraint introduces the concept of the "current estimate" of the ideal band:

$$\phi_2\left(2\theta, \frac{dP}{d2\theta}\right) = \left(\frac{dP^*}{d2\theta}(2\theta) - \frac{dP}{d2\theta}(2\theta)\right)^2 - R_2^2(2\theta) = 0, \tag{7}$$

where $dP^*/d2\theta$ is the "current estimate." This is particularly useful when one adopts approximate numerical methods of solution (as one usually must).

For these constraints it is possible to show that the ideal band is given by:

$$\frac{dP}{d2\theta}(2\theta) = \frac{dP^*}{d2\theta}(2\theta) + \int_{-\infty}^{\infty} u(2\theta - \xi) \left(\frac{dP_b}{d2\theta}(\xi) + \kappa_1 \left(\frac{dP_s}{d2\theta}(\xi) - \frac{dP^*}{d2\theta}(\xi)\right)\right.$$

$$\left. - \int_{-\infty}^{\infty} \frac{dP^*}{d2\theta}(\zeta) \, g(\xi - \zeta) \, d\zeta\right) d\xi, \tag{8}$$

where u is defined as the function such that for any function y,

$$\int_{-\infty}^{\infty} u(2\theta - \xi) \left((\kappa_1 + \kappa_2) \, y(\xi) + \int_{-\infty}^{\infty} y(\zeta) \, g(\xi - \zeta) \, d\zeta\right) d\xi = y(2\theta). \tag{9}$$

There are many methods for determining u. Exact methods include use of Fourier Transforms for sectionally-continuous functions, and matrix inversion for discrete channels. An approximate numerical method, which relies upon iteration to achieve the precision of the exact methods, uses a series expansion of the matrix inversion for discrete channels. The method of tensors is a fourth method of deconvolution. It is predicated on a slightly different development for discrete channels but ultimately yields identical results.

Figure 2 shows a deconvoluted band. The broadening function, g, was assumed to be Gaussian and the variance parameters were computed from the analysis of derivatives in figure 1. The variance parameter was assumed to vary linearly between the end points.

PROFILING THEORY

The power diffracted by an infinitesimally thin volume element is given by[2,3]:

$$dP = P_o \, Q \, e^{\eta} \, dx/\sin\theta, \qquad (10)$$

where P_o is the power of the incident beam, Q is the fraction of power diffracted per unit of incident-beam-path length, $dx/\sin\theta$ is the incident-beam-path length within the element, and e^{η} is the fraction of power not absorbed within the specimen.

Substituting from Bragg's law ($\lambda = 2\delta \sin\theta$)

$$dP = 2 \, P_o \, Q \, e^{\eta} \, \delta \, dx/\lambda, \qquad (11)$$

where δ is the d-spacing and λ is the x-ray wavelength.

The fraction of power diffracted per unit of incident-beam-path length (pseudoreflectivity) is:

$$Q = \frac{r_e^2 \lambda^3}{V^2} \left(\frac{1 + \cos^2 2\alpha \, \cos^2 2\theta}{2 \sin 2\theta \, \sin\theta} \right) F^2 \, p \, e^{-2M}, \qquad (12)$$

where $r_e^2 = 7.94 \times 10^{-26} \, cm^2$, V is the volume of the unit cell, α is the Bragg angle of the monochromator, F is the structure factor, p is the multiplicity, and the exponential term is the Debye-Waller thermal factor.

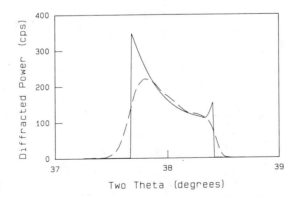

Fig. 2. An ideal intensity band (solid) extracted from a phenomenal intensity band (dashed). The phenomenal intensity band is the same as in Fig. 1. The position of the limits was determined from the zeroes of the second derivative (see figure 1).

The absorption term is given by:

$$\eta = \frac{-4\delta}{\lambda} \int_0^x \mu \, dx, \tag{13}$$

where μ is the linear absorption coefficient. The first derivative of η with respect to δ is:

$$\frac{d\eta}{d\delta} = \frac{-4\delta\mu}{\lambda}(1 + \gamma)\frac{dx}{d\delta}, \tag{14}$$

where γ is a small geometric term given by:

$$\gamma = \frac{1}{\mu\delta}\frac{d\delta}{dx}\int_0^x \mu \, dx. \tag{15}$$

Let us define a modified diffracted power by multiplying both sides of Eq. (11) by $2\mu(1+\gamma)/Q$ and integrating. From the left-hand side of Eq. (11)

$$\Psi = 2 \int_{\delta_0}^{\delta} \frac{dP}{d\delta}\frac{\mu}{Q}(1 + \gamma) \, d\delta, \tag{16}$$

where δ_0 is the d-spacing at the surface (x=0). From the right-hand side of Eq. (11)

$$\Psi = -P_0 \int_0^{\eta} e^{\eta} \, d\eta, \tag{17a}$$

$$= P_0 \left(1 - e^{\eta}\right). \tag{17b}$$

Hence,

$$\eta = \ln(1 - \Psi/P_0), \tag{18}$$

and, taking the derivative with respect to δ,

$$\frac{d\eta}{d\delta} = \frac{-2\mu\,(1 + \gamma)\,(dP/d\delta)}{Q\,(P_0 - \Psi)}. \tag{19}$$

Consequently, from Eqs. (14) and (19)

$$x = \frac{\lambda}{2} \int_{\delta_0}^{\delta} \frac{(dP/d\delta)}{\delta\, Q\, (P_0 - \psi)}\, d\delta. \tag{20}$$

The distribution of power $dP/d\delta$ is related to the distribution of power $dP/d2\theta$ through Bragg's law:

$$\frac{dP}{d\delta} = \frac{-4}{\lambda}\, \tan\theta\, \sin\theta\, \frac{dP}{d2\theta}. \tag{21}$$

From the Eqs. (16) and (17b) the constant P_0 is given by

$$P_0 = 2 \int_{\delta_0}^{\delta_\infty} \frac{dP}{d\delta} \frac{\mu}{Q}(1 + \gamma)\, d\delta, \tag{22}$$

where δ_∞ is the d-spacing at $x=\infty$.

The profile in figure 3, which was calculated from the intensity band in figure 2, is a composition-depth profile for oxygen in α-titanium. While the preceeding development has been for thick specimens without absorbing overlayers, a more general development is available elsewhere[5].

SYSTEMATIC ERRORS

Four sources of systematic error are (1) misalignment of the 2θ circle, which causes errors in composition; (2) sampling errors due to large grain size, which cause errors in depth; (3) nonuniform grain orientation with depth, which causes errors in depth; and (4) inept deconvolution. Inept deconvolution will cause errors in depth and composition

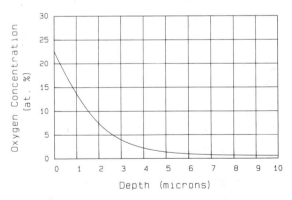

Fig. 3. A composition-depth profile of oxygen in α-titanium. The profile was constructed from the ideal intensity band in Fig. 2.

near the solubility limits; the center of the profile, however, will not be seriously affected (see figure 4).

Sampling errors and nonuniform grain orientation with depth can be tested for by collecting several intensity bands from the same specimen and then comparing the profiles. If the profiles all match then there is probably no problem. If sampling errors are present, it is a simple matter of reloading the specimen and of using oscillating/rotating specimen holders until the problem disappears. Nonuniform grain orientation, however, is a more difficult problem: one that is beyond the scope of this paper.

RANDOM ERRORS

The random counting error and imprecisions in the lattice parameter calibration will cause random errors in the profile. The variance due to the lattice parameter calibration, $var[C]$ can be estimated using linear regression. This variance will be with respect to composition.

In order to estimate the variance due to the random counting error, consider a scan that progresses from δ_0 to δ_∞. At any time τ the number of diffracted counts is N_D and the number of background counts is N_B. Because the arrival of x-ray photons at the detector follows a Poisson distribution[6], the variance of the modified diffracted power is given by:

$$var[\Psi] \cong 2 \int_{\delta_0}^{\delta} \frac{d}{d\delta}\left(\frac{N_D + N_B}{\tau}\right)\left(\frac{\mu}{Q}(1+\lambda)\right)^2 d\delta. \tag{23}$$

The net variance with respect to composition will then be

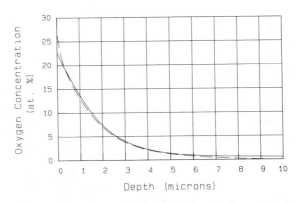

Fig. 4. The composition-depth profiles calculated from the ideal intensity band (solid) and the phenomenal intensity band (dashed) in Fig. 2. This figure demonstrates that removal of noncompositional broadening-- by deconvolution--strongly effects the ends of the profile while it does not seriously effects the center. Consequently, accuracy at the ends of the profile requires careful deconvolution.

$$\text{var}_{net}[C] \cong \text{var}[C] + \text{var}[\Psi]\left(\frac{d\delta}{d\Psi}\right)^2\left(\frac{dC}{d\delta}\right)^2, \tag{24}$$

where $dC/d\delta$ is determined from the lattice parameter calibration.

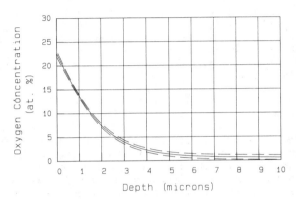

Fig. 5. A composition-depth profile (solid) and its estimated 90% confi-
         dence limits (dashed) for the ideal intensity band in Fig. 2.
         Random error arises from imprecisions in the lattic parameter
         calibration as well as the statistics of counting x-rays. The
         confidence intervals in this figure are based on the sum of the
         variances due to the lattice parameter calibration and the random
         counting error. For all practical purposes the entire width of the
         confidence bands can be attributed to imprecision in the lattice
         parameter calibration; the 90% confidence limits based just on the
         random counting error fall within the width of the solid line.

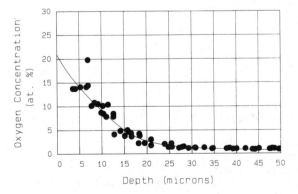

Fig. 6. Composition-depth profiles obtained by x-ray diffraction (line) and
         microhardness (circles). Specimen: unalloyed titanium exposed to
         air at 1100°F for 102 hours. X-ray data collected using Mo Kα.

Figure 5 is the profile in figure 3 with 90% confidence intervals calculated from the net variance. The intervals do not treat the systematic errors discussed in the previous section.

ACCURACY

Figure 6 is the comparison of composition-depth profiles determined by x-ray diffraction and Knoop microhardness. Microhardness has been used to obtain composition-depth profiles in the titanium-oxygen system[7]. The x-ray diffraction generated profile passes through the scatter in the Knoop microhardness profile. To obtain the diffraction data it was only necessary to load the sample in the diffractometer and scan; to obtain the Knoop microhardness data it was necessary to cut the sample, mount in cross section, polish, and make and measure dozens of indentations. Obviously, the profile obtained by x-ray diffraction is not only more precise, but easier to obtain as well.

SUMMARY

The phenomenal intensity band is composed of the ideal intensity band, noncompositional broadening, and random counting error. The ideal intensity band is pure compositional broadening which, before constructing the composition-depth profile, must be extracted from the phenomenal intensity band. The derivatives of the phenomenal band can be used to estimate the variance parameter of the noncompositional broadening and the location of discontinuities in the ideal intensity band. The deconvolution problem was formulated to remove the random counting error as well as the noncompositional broadening. The ideal intensity band extracted by this deconvolution technique is used to calculate the composition-depth profile. This profile is susceptible to analyses of random error due to imprecisions in the lattice parameter calibration and random counting errors in the phenomenal intensity band.

ACKNOWLEDGMENTS

The authors thank the National Areonautics and Space Administration's Langley Research Center for sponsoring this research and the Dr. Wilbur B. Ficter for helpful discussions of analysis techniques.

REFERENCES

1. B. Ia. Pines and E. F. Chaidouski, Dokl. Acad. Nauk. SSSR, 111, 1234 (1956).
2. D. R. Tenney, J. A. Carpenter, and C. R. Houska, J. Appl. Phys., 41, 4485 (1970).
3. J. Unnam, J. A. Carpenter, and C. R. Houska, J. Appl. Phys., 44, 1957 (1973).
4. K. E. Wiedemann, Thesis (unpublished), Virginia Polytechnic Institute and State University (1983).
5. K. E. Wiedemann and J. Unnam, J. Appl. Phys, 58, 1095 (1985).
6. B. D. Cullity, "Elements of X-Ray Diffraction," Addison-Wesley, Reading Massachusetts (1978).
7. C. E. Shamblen and T. K. Redden, Science and Technology and Application of Titanium (1968), p. 199.

QUANTITATIVE ANALYSIS OF THIN SAMPLES BY X-RAY DIFFRACTION

C.G. Brandt

Application Laboratory for X-ray Analysis
Philips I&E division, Almelo, The Netherlands

G.H. van der Vliet

I.C.T. Novotech
Nieuw Beijerland, The Netherlands

INTRODUCTION

For phase analysis of small amounts of material in an x-ray powder diffractometer the sample is often spread on a sample mounting plate or support plate. Different types of plates on which thin samples can be mounted are:
- silver membrane filters
- metal or other polycrystalline plates
- polymer membrane filters, for example PVC, polycarbonate, cellulose nitrate
- glass slides
- single crystals, cut so that no diffraction lines occur.

Filters are often used to collect environmental dust; these samples can be analysed "as received." Glass slides are inexpensive, flat, and readily available. Polycrystalline filters and plates have the disadvantage of possible line overlap with lines of the sample material (refer to figure 1). Glass plates or polymer filters have the high and curved background due to amorphous scatter as shown in figure 2.

Plates made of silicon or quartz single crystals have therefore become rather popular for phase identification of minute samples. They are cut so that no diffraction of the crystal is detectable. Consequently they show a very low and straight background (figure 3), which results in excellent detectability of weak diffraction lines of the sample.[1]

Quantitative analysis of a thin sample is somewhat more complicated than quantitative analysis of an "infinitely" thick sample. In the latter case the weight fractions of the phases in the mixture are unknown, together with the mass absorption coefficient of the sample. For thin samples the thickness and the density of the powder layer are further unknowns.

Until now quantitative analysis of thin samples has generally been done with samples deposited on polycrystalline plates, for example on

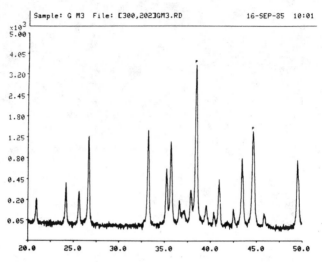

Figure 1. *Metal or other polycrystalline sample plates often cause line overlap.*

Ag filters. The attenuation of a line of the filter or plate material is used as a measure of the thickness and absorption of the sample layer.[2] A diffraction line from the plate is measured when the plate is clean and again after it has been covered by the sample material. Then the mass of the phase to be analysed can be calculated.

This method cannot be applied to samples on cellulose membrane filters or to thin samples mounted on single crystal plates. Therefore a new method has been developed for quantitative analysis of thin samples without use of a diffraction line of the mounting plate. In addition we will show that it is possible to calculate not only the mass but also the concentration or weight fraction of a compound. This gives better chances to meet the analytical requirements.

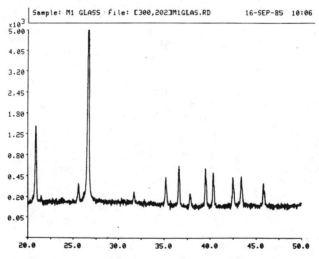

Figure 2. *Glass plates or polymer filters give high and curved background.*

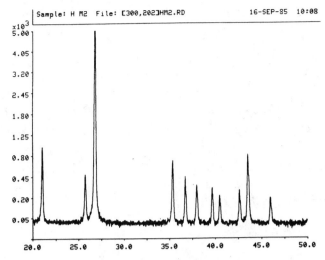

Figure 3. Obliquely cut single crystals are popular for the
low and straight background.

DERIVATION OF THE METHOD

The intensity of a diffraction line of a thin sample is

$$I_{it} = I_i \left(1 - e^{-\frac{2\mu\rho t}{\sin \theta_i}} \right) , \text{ refer to figure 4.}$$

In this formula I_i = intensity of the same line in an "infinitely"
thick sample
μ = mass absorption coefficient
ρ = density of the sample
t = thickness of the sample layer
θ_i = diffraction angle of the line "i".

The "thin sample" intensity can be converted into an intensity as if
measured at a thick sample.

$$I_i = \frac{I_{it}}{1 - T^{S_i}}$$

Here T $= e^{-2\mu\rho t}$ = "transmission factor" of the sample layer

$$S_i = \frac{1}{\sin \theta_i}$$

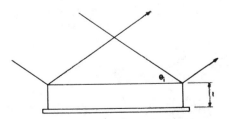

Figure 4. Thin sample on sample mounting plate.

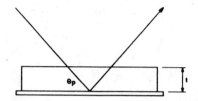

Figure 5. Thin sample on polycrystalline plate.

The transmission factor T can be determined in two ways. For a polycrystalline plate we measure the intensity of a line of the "clean" plate
(I_p0), at angle θ_p.   Then the same line of the dust-covered plate (I_p)
is measured.

From figure 5 we can see that

$$\frac{I_p}{I_p0} = e^{-\frac{2\mu\rho t}{\sin \theta_i}}$$

Therefore the transmission factor will be:

$$T = \left(\frac{I_p}{I_p0}\right)^{\sin \theta_p}$$

For amorphous or single crystal plates the previous method cannot be
applied. It also can not be used if the sample does not completely cover
the irradiated area. In these cases we propose the following solution.
Choose one of the components of the sample as the "reference compound."
The reference compound should be contained in all the samples in reasonable concentration (of course it can be the compound to be analysed).
Measure two lines of the reference compound $(I_{1t}$ and $I_{2t})$ in the thin
sample, see figure 6:

$$\frac{I_{1t}}{I_{2t}} \times \left(\frac{I_2}{I_1}\right) = \frac{1-T^{S1}}{1-T^{S2}}$$

Here $\frac{I_2}{I_1}$ is the intensity ratio of the two reference lines measured on
"infinitely" thick sample(s) for calibration.

And so   T   can be determined (by iteration) using the θ-dependence of
the transmission.

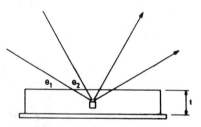

Figure 6. Thin sample on Si single crystal.

An intensity measured on a thin sample can be used for calculation of the mass of the compound, the method generally used up to now. The intensity measured on a thin sample can also be used for the calculation of the concentration (weight fraction) of the compound.

Mass can be determined in a thin sample because the mass of the sample is finite: the thickness of the sample layer is finite and independent of the mass absorption coefficient.

$$M_i = I_i \times B_{im} \times \ln\left(\frac{1}{T}\right)$$

I_i = intensity measured on thin sample, then converted to "thick" sample value

B_{im} = mass calibration constant of compound i

T = transmission factor of thin sample.

Concentration or "relative mass" can be calculated from the intensity measured on an "infinitely" thick sample, or measured on a thin sample and then converted.

$$C_i = I_i \times B_i \times \mu$$

B_i = calibration constant of compound i

μ = mass absorption coefficient of the sample.

EXPERIMENTAL

Twenty samples were prepared of α-quartz, α-corundum, two mixtures of quartz and corundum, and a mixture of quartz, corundum, and hematite. One bulk sample ("infinitely" thick) was prepared of each pure compound and of each mixture for calibration purposes. Eight thin samples of different well-defined weights of each of the pure materials and mixtures were prepared on aluminium sample plates, which had been measured previously ("clean" plate intensity I_p0). Seven such thin samples were prepared on Si single crystals cut in the "minimum background" way. The bulk samples and eight thin samples were used as calibration standard samples. The remaining seven samples were measured as "unknowns."

The samples were measured in a standard powder diffractometer, equipped with Cu LFF tube, automatic divergence slit, sample spinner (2 rev./second), secondary (diffracted beam) graphite monochromator, and proportional detector. Measuring times: 100 seconds per peak and 100 seconds on the background. A search for the maximum was performed before each peak measurement started.

RESULTS

The mass calculations give good results for both methods (refer to table 1): 2 lines of corundum as reference compound and the line of the Al plate as reference. In Table 1 "N.S." means: "not significant" because the intensity ratios of thin sample and thick sample do not differ significantly. In these cases no thin sample intensity conversion was done and mass calculation could not be done.
Note: the intensity ratios for the two corundum lines of the thin sample would differ more from the thick sample value if the difference in diffraction angle was larger. It is advisable to use higher order lines if possible, especially for heavily loaded sample plates.

For the Si-mounted samples the method of the plate line (Al) was not possible ("N.P.").

Table 1. Result of measurements

Sample Plate	Compound	Mass in mg 2 Ref.lines	Plateline	Concentration % 2 Ref.lines	Plate	Theoretical Mass	Conc.
BM1	Corundum	2.22	N.P.	15.33	N.P.	2.91	20.07
Si	Quartz	13.53		84.67		11.59	79.93
CM1	Corundum	11.70	13.91	19.99	19.96	13.0	20.07
Al	Quartz	50.61	59.33	80.01	80.04	51.8	79.93
DM3	Corundum	2.26	N.P.	29.76	N.P.	2.72	38.33
Si	Quartz	2.43		29.95		1.52	21.38
GM3	Corundum	N.S.	18.90	38.38	38.37	20.51	38.33
Al	Quartz		11.18	21.33	21.34	11.44	21.38
IM3	Corundum	N.S.	11.51	36.94	36.93	13.49	38.33
Al	Quartz		7.56	22.77	22.78	7.52	21.38
LC	Corundum	41.96	N.P.	99.96	N.P.	58.6	100
Si	Quartz	0.02		0.04			
MQ	Corundum	N.P.	0.31	N.P.	0.69		
Al	Quartz		48.08		99.31	38.5	100

One sample contained no corundum and mass calculation with the 2 reference lines method was not possible.

    The concentration calculations give, in general, more accurate results than mass calculations, because calibration has been done on thick samples and because the normalization of the calculated concentrations reduces the influence of the statistical error in the transmission factor used. Furthermore, if the intensity ratio of the thin sample does not differ significantly from the thick sample value, then the concentration calculation still can be done without using intensity conversion.

CONCLUSIONS

    1 Both polycrystalline plates and "minimum background" plates can be used for quantitative analysis of small amounts of sample material:
    - the method using the attenuation of the plate line works better on heavy loads for mass calculation
    - single crystals have the advantage of low background and do not cause overlap problems.

    2 Mass calculation as well as concentration calculation can be done using thin sample intensities:
    - mass determination is independent of other (even amorphous) components
    - concentration calculation is more precise especially for heavier loads, it works also on thick samples.

    3 Thick samples always give better results, are preferred for quantitative analysis:
    - they can have better homogeneity and particle statistics
    - no intensity conversion means one less source of statistical counting errors.

However: the fact that there is a choice of methods for small amounts of sample material makes x-ray diffractometry an even more useful analytical tool.

REFERENCES

1. B. Post, Norelco Reporter, 20, No. 1:8 (1973).
2. J. Leroux, A.B.C. Davey, and A. Paillard, Am. Ind. Hyg. Ass. J., 34:409 (1973).

QUANTITATIVE X-RAY DIFFRACTION ANALYSIS OF CALCIUM SULFATES AND QUARTZ IN WET-PROCESS PHOSPHORIC ACID FILTER CAKES

Kjell R. Waerstad

Tennessee Valley Authority, Muscle Shoals, Alabama

35660

Abstract

A quantitative X-ray diffraction method has been developed for rapid determination of the crystalline phases present in by-product filter cakes from wet-process phosphoric acid manufacture. A new technique to reduce preferred orientation of the crystallites by removing the top layer of the X-ray samples with adhesive tape is described. Statistical analysis of synthetic sample mixtures of $CaSO_4 \cdot 2H_2O$, $CaSO_4 \cdot 0.5H_2O$, $CaSO_4$, and $\alpha-SiO_2$ indicates that the calcium sulfate phases present in typical filter cakes can be determined with an accuracy of about 3% at the 95% confidence level; quartz can be determined at about 14%. The method has been applied to the analysis of 10 filter-cake samples from an experimental hemihydrate plant. The results are in good agreement with data from X-ray fluorescence and chemical analyses.

Introduction

The principal crystalline phases present in by-product filter cakes from wet-process phosphoric acid manufacture are $CaSO_4 \cdot 2H_2O$ (gypsum), $CaSO_4 \cdot 0.5H_2O$ (hemihydrate), $CaSO_4$ (anhydrite), and $\alpha-SiO_2$ (quartz). The calcium sulfates are formed when the mineral apatite reacts with H_2SO_4 during synthesis of H_3PO_4. The degree of hydration of the precipitating calcium sulfate salts depends primarily on the temperature and the P_2O_5 concentration of the acid but also may be influenced by other factors such as the SO_4^{2-} concentration and the level of impurities in the acid. Each calcium sulfate phase is either stable or metastable over wide and overlapping ranges of temperatures and P_2O_5 concentrations (1), permitting all phases to precipitate simultaneously under certain conditions. Therefore, it is important as a quality control measure and as a guide to the adjustment of process variables for optimum operation of any phosphoric acid plant to have access to rapid and reliable information about the state of hydration of the calcium sulfates in the filter cakes.

The present study was undertaken to develop a simple procedure for quantitative analysis of these phases using X-ray powder diffration and to assess the relative accuracy that can be expected from the analysis. The procedure is based on the Reference Intensity Method described by Chung (2).

211

The general equation for quantitative analysis based on this method is:

$$\frac{X_i}{X_r} = \frac{1}{k_i} \cdot \frac{I_i}{I_r} \tag{1}$$

In this equation, X_i is the unknown weight fraction of component i; X_r is the weight fraction of a reference phase; I_i and I_r are the integrated intensities of component i and the reference material, respectively; and k_i is the reference intensity ratio. This ratio is determined from a 1:1 weight mixture of component i and the reference material.

If all the phases present in a multicomponent sample are crystalline and identified (i.e., $\Sigma X_i = 1$), then the weight fractions of all phases present can be calculated from equation 1. The presence of amorphous material in the sample requires addition of an internal standard to determine the absolute concentrations of each phase present (2).

Materials

Calcium sulfate dihydrate and $CaSO_4 \cdot 0.5H_2O$ were prepared from reagent-grade H_2SO_4 and $CaCO_3$ by precipitation at 45° and 96°C, respectively. Anhydrite was prepared by heating gypsum at 400°C for 1 h. Seasand served as the standard quartz material.

Equipment

The intensity measurements were all acquired on a Philips diffractometer using Ni-filtered copper $K\alpha$ radiation. The operating conditions were as follows: constant potential voltage set at 35 kV, amperage set at 15 mA, goniometer speed set at 1/8 degree per minute, and a counting time of 400 sec. The integrated intensities were corrected for background by scanning on each side of the peaks.

Procedure for Sample Preparation

It became evident during preliminary investigations that the three calcium sulfates presented measurable degrees of preferred orientation, even after reducing particle size to less than 5 μm and sideloading the samples. The problem was particularly severe for the 020 reflection of gypsum. Therefore, some additional sample preparations were necessary to further increase the randomization of the crystallites.

Matrix absorption effects are essentially eliminated in X-ray quantitative analysis by the Reference Intensity Method; but the effects of microabsorption and preferred orientation depend to some extent on how the samples are prepared. These factors may cause serious experimental aberrations if not considered properly (3, 4). Microabsorption tends to decrease with increasing density of the sample, while the opposite is true for preferred orientation. The following procedure was developed to minimize the effects of these factors and to produce samples of uniform consistency.

Each sample was suspended in freon and milled in a grinding device to an average crystallite size of 5 μm. The sample then was sideloaded into a cavity-type sample holder according to the procedure outlined in NBS monograph 25, section 6 (5). Adhesive cellophane tape was placed on the surface of the sample. Then, with the tape supported by a smooth surface, such as a microscopic slide, the sample holder was tapped firmly on the counter top. The tape was removed before analysis, thereby stripping off the top sample layer and leaving a smooth, uniform sample surface.

Since the 002 reflection of CaSO₄ at 3.50 Å is very close to the 110 reflection of CaSO₄·0.5H₂O at 3.48 Å, it was necessary to integrate the intensities of both reflections and subtract the contribution of the CaSO₄·0.5H₂O–110 from the combined value to find the true intensity of CaSO₄–002. The intensity of the 110 reflection for the hemihydrate was determined to be 57% of the 100 reflection, which is well separated from other lines.

Results and Discussion

Table I shows how the intensities of some selected reflections for the calcium sulfates and quartz are affected by removing the top layer of the sample with adhesive tape, as described earlier. The data are based on five independent intensity measurements. The samples were emptied out of the sample holder and repacked before each intensity measurement. Using the Student t-test for comparing means (6), it was determined that the difference in the intensities before and after the tape treatment are statistically significant for all the reflections, except those for quartz.

Table II shows the intensity ratios that were determined from four separate measurements on 1:1 weight mixtures of the hemihydrate, anhydrite, and quartz relative to gypsum. These ratios were subsequently used to calculate the concentrations of the calcium sulfates and quartz in 14 synthetic sample mixtures consisting of these phases. The concentrations calculated from equation 1 were compared with the actual values to establish the mean error, standard deviation, and overall accuracy that can be expected by the method. The data are summarized in Table III.

In principle, any of the phases present in the samples may be used as the reference material since the intensity ratios between the various phases have been established. However, for the purpose of this presentation, the calculations were based on the 020, 12$\bar{1}$, and 14$\bar{1}$ reflections of gypsum as reference intensities. Notice that the smallest mean error and the smallest standard deviation are obtained by using the 12$\bar{1}$ reflection as the reference intensity. The accuracy of the method generally improved with increasing concentration of the phases being

Table I. Effect of tape treatment on intensities of selected reflections of calcium sulfate and quartz

		Intensity			
		Without tape treatment		With tape treatment	
Phase[a]	Reflection, hkl	Mean, cps	Standard deviation, %	Mean, cps	Standard deviation, %
G	020	2266	13.3	1199	8.0
G	121	655	2.0	698	1.9
G	141	631	2.8	597	1.6
H	100	827	3.3	624	2.7
H	110	411	3.4	354	2.1
A	002	1209	3.2	1062	2.4
Q	101	3331	1.9	3338	2.0

[a] G = CaSO₄·2H₂O, H = CaSO₄·0.5H₂O, A = CaSO₄, Q = α-SiO₂.

Table II. Mean values and standard deviations of intensity ratios for
1:1 mixtures of $CaSO_4 \cdot 0.5H_2O$, $CaSO_4$, and α-SiO_2 relative
to $CaSO_4 \cdot 2H_2O$

Group of reflections[a]	Intensity ratio, k_i (mean)	Standard deviation, σk	σk_i as % of k_i
H_{100}:G_{020}	0.762	0.033	4.3
H_{100}:G_{121}	1.136	0.015	1.3
H_{100}:G_{141}	1.333	0.047	3.5
A_{002}:G_{020}	1.232	0.084	6.8
A_{002}:G_{121}	1.982	0.051	2.6
A_{002}:G_{141}	2.417	0.037	1.9
Q_{101}:G_{020}	1.964	0.067	3.4
Q_{101}:G_{121}	2.609	0.106	4.1
Q_{101}:G_{141}	2.986	0.198	6.6

[a] $H = CaSO_4 \cdot 0.5H_2O$, $G = CaSO_4 \cdot 2H_2O$, $A = CaSO_4$, $Q = \alpha$-SiO_2.

Table III. Predicted accuracy of quantitative analysis at 95%
confidence level

Phase[a]	Reflection, hkl	Mean error, % of actual concentration, ε	Standard deviation of mean error, σ	Number of measurements, N	Accuracy of method[b]
G	020[c]	6.63	4.90	14	9.46
H	100	4.06	3.96	12	6.58
A	002	7.39	4.70	12	10.38
Q	101	13.72	9.77	7	22.42
G	121[c]	1.35	1.56	13	2.29
H	100	2.65	1.69	11	3.79
A	002	1.85	2.21	12	3.26
Q	101	11.09	3.19	6	14.44
G	141[c]	2.35	2.18	14	3.61
H	100	3.80	2.92	12	5.66
A	002	3.38	2.65	12	5.06
Q	101	9.65	3.99	7	13.34

[a] $G = CaSO_4 \cdot 2H_2O$, $H = CaSO_4 \cdot 0.5H_2O$, $A = CaSO_4$, $Q = \alpha$-SiO_2;
[b] $\varepsilon + (t_{95}^{N-1})\sigma/\sqrt{N}$ where t_{95}^{N-1} is Student's t test value from N
measurements at 95% confidence level; [c] Reference intensity reflections.

Table IV.   Composition of hemihydrate filter cakes as determined by
XRD, XRF, and chemical analyses

Sample No.	Method	Wt %						
		Ca	S	SiO₂	H₂O	Gypsum	Hemihydrate	Anhydrite
1	XRD	25.2	20.2	8.9	5.5	3.6	76.5	11.0
	XRF	25.9	19.2	5.2	–	–	–	–
	Chemical	25.2	–	6.4	6.6	–	–	–
2	XRD	25.1	20.1	7.0	7.8	18.1	64.5	10.4
	XRF	25.0	19.1	–	–	–	–	–
	Chemical	24.6	20.3	5.8	8.2	–	–	–
3	XRD	23.2	18.5	6.4	14.9	66.2	17.8	9.6
	XRF	24.8	18.3	–	–	–	–	–
	Chemical	23.2	–	5.6	14.6	–	–	–
4	XRD	22.7	18.2	7.1	15.8	74.5	2.7	15.7
	XRF	23.2	17.4	–	–	–	–	–
	Chemical	22.4	–	5.8	17.3	–	–	–
5	XRD	23.1	18.5	5.4	16.2	77.3	1.2	16.1
	XRF	23.4	17.4	4.6	–	–	–	–
	Chemical	22.2	18.5	5.6	17.5	–	–	–
6	XRD	22.5	18.0	6.5	17.1	81.9	0.0	11.6
	XRF	22.8	17.5	4.6	–	–	–	–
	Chemical	22.4	–	5.5	17.0	–	–	–
7	XRD	22.5	18.0	7.8	15.7	75.0	0.0	17.2
	XRF	23.1	17.7	5.1	–	–	–	–
	Chemical	23.1	–	5.8	16.9	–	–	–
8	XRD	24.0	19.2	7.4	11.1	45.7	25.2	21.7
	XRF	23.9	18.4	3.8	–	–	–	–
	Chemical	23.6	–	5.5	13.4	–	–	–
9	XRD	23.9	19.1	7.7	11.1	44.8	28.5	19.0
	XRF	24.1	18.5	5.6	–	–	–	–
	Chemical	23.6	–	6.4	11.6	–	–	–
10	XRD	25.4	20.4	7.3	6.3	12.2	˙60.5	20.0
	XRF	24.9	19.0	7.0	–	–	–	–
	Chemical	24.9	20.4	5.8	7.6	–	–	–

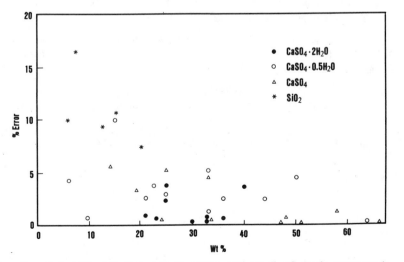

Fig. 1.   Relationship between % error and wt % of each component.

determined. Figure 1 shows how the relative error for each phase is
related to the concentration in the sample mixtures. The mean concentrations of gypsum, hemihydrate, anhydrite, and quartz in the 14 samples
were 35.9, 30.2, 37.7, and 11.7 weight percent, respectively. At these
concentrations the data in Table III indicate that the calcium sulfates
can be determined with an accuracy of about 3% at the 95% confidence
level, and quartz can be determined with about 14% accuracy (based on
gypsum $12\bar{1}$ as the reference reflection).

The present quantitative method was tried on 10 filter-cake
samples from an experimental hemihydrate phosphoric acid plant. The
compositions of the filter cakes, as calculated from phase determinations
by the X-ray diffraction (XRD) method, are compared with those from X-ray
fluorescence (XRF) and chemical analyses in Table IV. The data show good
agreement among the three methods.

Conclusions

X-ray powder diffraction provides a fairly accurate and rapid method
for quantitative analysis of by-product filter cakes from the synthesis of
H_3PO_4. Satisfactory precision is obtained without elaborate sample preparations. The method requires grinding the sample to an average crystallite
size of less than 5 μm, side-loading, and then applying adhesive tape to
reduce the degree of preferred orientation.

This procedure also may be successfully applied to the analysis
of other sample mixtures that are naturally prone to preferred orientation.

References

1. Dahlgren, S. E. "Calcium Sulfate Transitions in Superphosphate,"
    J. Agric. Food Chem. 8 (5), 411–412 (1960).

2. Chung, F. H. "Quantitative Interpretation of X-Ray Diffraction
    Patterns of Mixtures. III. Simultaneous Determination of a Set
    of Reference Intensities," J. Appl. Cryst. 8, 17–19 (1975).

3. Hubbard, C. R. and Smith, D. K. "Experimental and Calculated
    Standards for Quantitative Analysis by Powder Diffraction,"
    Advances in X-Ray Analysis 20, 27–39 (1977).

4. Hubbard, C. R., Evans, E. H., and Smith, D. K. "The Reference
    Intensity Ratio, I/I_c, for Computer Simulated Powder Patterns,"
    J. Appl. Cryst. 9, 169–174 (1976).

5. National Bureau of Standards, "Standard X-Ray Diffraction Powder
    Patterns," NBS Monograph 25, Section 6, Washington, D.C. 1968.

6. Dixon, W. G. and Massey, F. J. Introduction to Statistical Analysis,
    2nd ed., McGraw-Hill, New York, 1969, pp 114–124.

CLAY MINERAL ANALYSIS BY AUTOMATED POWDER DIFFRACTION ANALYSIS USING THE

WHOLE DIFFRACTION PATTERN

Deane K. Smith, Gerald G. Johnson, Jr., Clayton O. Ruud

The Pennsylvania State University

University Park, PA 16802

INTRODUCTION

Clay minerals are one of the most difficult classes of materials to analyze by x-ray powder diffraction, yet powder diffraction is the <u>only</u> technique which can yield the important crystallographic information necessary to identify and classify the minerals. The importance of clay minerals in industry and in the studies of rocks, due to their chemical properties and sensitivity to geological changes, often requires the analyses of large numbers of samples in short periods of time. Such sample throughput requires computerized analysis. Because definitive, meaningful d-I data are difficult to obtain from the broad diffraction peaks obtained from most clay samples, this problem has been approached by using the whole diffraction trace as the basis of a computerized analytical scheme.

Very elaborate procedures (based on characteristic d-spacings) have been developed to identify clay minerals using powder diffraction analysis. Samples prepared to enhance preferred orientation of the platy morphology of the clay particles yield diffraction patterns composed almost totally of the 00ℓ family of peaks. These oriented samples are more commonly employed than the usual random sample, because they markedly increase the intensity values from the clay components and allow definitive identification of most samples. Often a single x-ray sample is insufficient to confirm the identification and special chemical and thermal treatments are used prior to recording the x-ray pattern. Confirmation of the identification usually requires more than one x-ray run and may require as many as four different runs on the same sample.

Clay mineral identification may mean only the identification of the major clay mineral families present in a given sample or it may mean the determination of the specific clay specie. Sometimes the distinction of dioctohedral or trioctahedral character of the clay is sufficient. Occasionally, a very specific question may be asked such as the percent expanding clay present in the sample or the presence and nature of iron-containing species. Each level of identification requires a different approach to the set of samples needed for the analysis and to the analytical procedure to reach the required answer. All these options must be incorporated into any computer-based scheme. The scheme must be able to incorporate the multiple sample approach necessary to confirm some identifications.

The basis of the scheme described in this paper is the analysis of
the whole diffraction pattern from 2°-50°2θ (70° for random samples)
through comparisons with a reference database of diffraction traces
obtained from a set of carefully purified reference clay samples run
under the same conditions. Patterns are recorded for the specified
treatments for each clay sample. The comparisons leading to the
identification and quantification of clay minerals are user controlled
through interactive computer terminals. Both numerical and graphical
analysis are employed.

THE DATABASE

The heart of the clay mineral analysis scheme is the database of
diffraction patterns recorded from well characterized clay mineral
samples. This database is supplemented by calculated diffraction traces
using the techniques of Reynolds[1,2], Pevear[3], and Smith et al.[4]. All
the patterns used for the database were recorded on a Philips APD3600/01
system using a range of 2° to 50° 2θ for the oriented samples and 2° to
70° 2θ for the randomized samples. Data points were recorded using 1
sec. count at intervals of 0.02° 2θ. These traces are stored in digital
form along with other documentation including a d-I table of peaks
located by the Philips system. Graphical displays were recorded for each
sample as well.

The samples in the present working database include examples of each
of the principle clay mineral families, kaolinites, smectites, micas,
chlorites as well as mixed-layered samples were available. In addition
to the clays, the database also contains representative patterns of the
common sedimentary minerals associated with the clays. For each of the
clay samples, patterns were recorded for a randomized sample and for
oriented samples in the air-dried (50% R.H.) state, glycol-saturated
state and for thermal treatments of 375°C and 550°C. Provisions are
included in the database structure for selected ion-saturated samples,
but such samples are not collected to date. The present database
contains over 200 diffraction patterns.

The clay samples were purified by settling techniques to isolate the
appropriate size fraction for the particular clay mineral. These techniques are described in detail in several references including the Clay
Minerals Society.[4] Quartz was difficult to remove from the kaolinite and
chlorite samples due to the coarse size of the clay component, which is
usually around 1 μm. Kaolinite and illite traces were difficult to
remove from some of the smectites even when using a 0.2 μm. fraction.
For those samples which were impossible to purify completely, computer
techniques, to be described, were used to eliminate the unwanted peaks.

For quantitative estimation, it was necessary to record a pattern
for a 50:50 mixture of the clay sample with some reference materials.
For the random samples, α-Al_2O_3, corundum, was used for the reference;
for the oriented samples, $Mg_6Si_4O_{10}(OH)_4$, talc, was used. Talc was
selected as an easily obtainable layer mineral whose d-spacings did not
interfere with the minerals under study. X-ray samples of the oriented
samples were prepared by rapid suction and transfer to prevent any
separation of the components due to differential settling properties.
The resulting samples were used to obtain the reference-intensity-ratios.

The working database requires the elimination of the background
portion of the recorded data. This correction is described below. The
application of this correction produces a working database with an
identical structure to the raw data database.

INTEGRATED COMPUTER PROGRAMS

   Successful clay mineral analysis with the whole-pattern database
requires a complex family of closely integrated computer programs to
build the database, correct the unknown pattern and compare selectively
the unknown with the database reference patterns. The programs in the
present system are integrated through common file structures and program
menus which allow the user at a terminal to select interactively the
paths of data flow and analysis and to examine the progress of the
analysis. All the programs were developed on a DEC VAX780 using REGIS
graphics. The general program struture is shown in the flow diagram in
Figure 1.

   Three types samples are processed: 1) The purified reference clay
which yields the patterns that make up the working database. 2) A 50:50

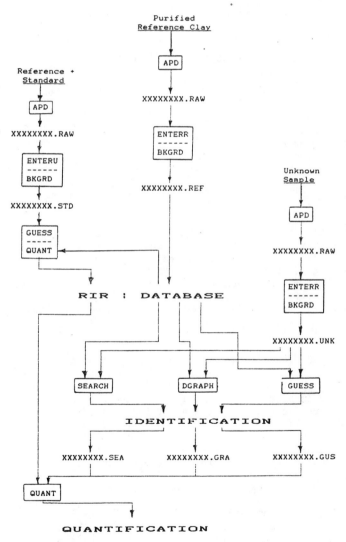

Figure 1. Flowchart for CLAY/APD

mixture of the reference clay with an intensity standard (either corundum or talc) which yields a pattern from which the reference-intensity-ratio is determined. 3) The unknown whose pattern is to be analyzed. All the samples are processed by the Automated Powder Diffractometer which yields a raw diffraction data file identified as X.RAW.

Patterns to be added to the database are processed through ENTERR. This routine calls BKGRD which allows the stripping of unwanted impurity peaks and finds and strips the background. ENTERR then enters the corrected reference pattern (X.REF) into the permanent working database.

BKGRD is a subroutine called during the processing of all X.RAW data files produced by the APD. It allows the user to display the X.RAW file either numerically or graphically and then allows the user to perform several optional procedures on these data prior to building the final data file. The first option which is usually implemented is the elimination of peaks of impurity phases. Any selected peak may be replaced by a smooth line between designated points. BKGRD then determines a background. Either the user or the computer selects low points along the diffraction trace, and then the program fits a chosen function to these points and subtracts this function from the raw data to create the desired X.REF, X.STD or X.UNK working data file.

ENTERU is used to process data in a manner similar to ENTERU. Patterns are returned as a separate datafile for further analysis. Both the reference + standard (X.STD) and the unknown (X.UNK) files are run through ENTERU, which calls BKGRD to make the background correction. The resulting file from the reference + standard sample is then examined by QUANT, which yields an estimate of the weight fractions of the components. The specific patterns in the database to be used by QUANT are identified by the user using GUESS. The scale factor for the intensity of the reference pattern is adjusted to yield the desired 50:50 weight percentage, and this scale factor is stored in the database as the reference-intensity-ratio, RIR.

The unknown pattern, after correction of ENTERU and BKGRD, may be examined by any of three routines to achieve an identification: 1) Routine GUESS allows the user to designate preconceived minerals for later processing by QUANT. 2) Routine SEARCH allows a user-controlled search/match/subtract sequence which is based on numerical comparisons of the database reference patterns with the pattern of the unknown. 3) Routine DGRAPH allows the user to examine the unknown graphically and to superimpose and strip reference patterns sequentially to achieve an identification.

Routine GUESS is employed when the user already knows which minerals are present in the sample. This approach may be used when a series of related samples are being examined in which the minerals have already been identified in an earlier run. It may also be used to force the selection of specific patterns in the database. The resulting choices are retained in a X.GUS file for use by QUANT.

Routine DGRAPH is a graphics-based program which may be used to display the unknown or any pattern in the database. The display can allow the user to select probable identifications by inspection or by successive superposition and subtraction of reference patterns. The intensity level of the known can be adjusted to achieve the best fit prior to subtraction. DGRAPH allows a 2θ shift to be made in the patterns to improve the fit. After each subtraction, the residual pattern is displayed for the next match. The pattern subtracted is remembered in a X.GRA file. for use by QUANT.

Routine SEARCH is used to find an identification when the minerals in a sample are unknown. It proceeds by a user-defined path to use the d-I information selecting patterns which match the first line (highest d) or the strongest line to find possible matches in the database. Matches are given figures-of-merit which guide the user in the selection of one pattern for subtraction. After the subtraction, the process is repeated. Each time an identification is accepted, the selected mineral is added to the X.SEA file for later use by QUANT. SEARCH can be controlled interactively by the user or can be set to run as a batch job.

Once the identification has been completed, the identification file (X.GUS, X.SEA or X.GRA) is passed to QUANT, which makes an estimate of the weight percentage of each selected mineral in the unknown sample. This estimate is based on a least-squares fit of all the points in the diffraction pattern with a weighted sum of the selected database patterns. Combining the pattern weighting factors and the reference-intensity-ratio values produces a weight ratio for each mineral. If the intesity standard is included in the sample, this weight-ratio can be converted to an absolute weight fraction indicating any unaccounted for intensity as an x-ray amorphous component. If the standard is not included, the weight ratios are converted to relative weight fractions whose sum is unity.

EXAMPLES

The programs are combinations of menu driven and sequential interactive question modes, so the user at a terminal can control the flow of analysis. Evoking the main calling program displays a master menu shown in Figure 2 from which the user selects the path to be followed. Option 4 allows ENTERU or ENTERR to process experimental patterns. Data are processed initially through ENTERU, and the processed data is stored as a file in the working directory. Data that proves suitable for inclusion in the database is reprocessed through ENTERR. Both these processing routines follow a path of interactive questions which includes options to remove extraneous peaks and the background.

Processed unknown patterns, including those with internal standards, may now be analyzed by one of three search/match options: 6. DGRAPH (graphical), 7. SEARCH (numerical), and 8. GUESS (user designation). DGRAPH allows the user to view the unknown pattern and to perform super-positions of reference patterns, shift and scale the reference patterns to best fit the unknown, and finally to subtract patterns and cycle the procedure. The menu for this option and one graph are shown as Figure 3.

SEARCH presently utilizes the matching of d-I sets employing the option philosophies of emphasizing the strongest lines or the lines with the largest d-values. This latter approach is analogous to techniques used in manual analysis of data from oriented clay samples. A third option is being implement, which "fits" the whole trace of the reference patterns to the unknown pattern and provides a fitting figure-of-merit to guide the user in mineral selection. SEARCH is sequential in that the routine provides the user with possible matches, and the user selects interactively one for subtraction. The process is then repeated using the residual pattern. The conversational dialogue is too long to show here.

Both DGRAPH and SEARCH provide an identification file of accepted matches for use by QUANT. The GUESS option is employed when the user knows the minerals in the unknown. One of the identification files is then designated as input to QUANT which establishes the reference

```
THIS IS THE CLAY/APD SYSTEM
============================

ADAPTED FOR THE CLAY/APD CONSORTIUM BY:
     MATERIALS RESEARCH LABORATORY
     THE PENNSYLVANIA STATE UNIVERSITY
     UNIVERSITY PARK,  PA 16802
             JUL-85                              TODAYS DATE:  16-SEP-85
----------------------------------------------------------------------------
        MAIN MENU FOR CLAY ANALYTICAL SOFTWARE
     ********************************************
     (0) EXIT PROGRAM
     (1) PERFORM VISSER'S PEAK SEARCH ON RAW DATA
     (2) PERFORM A VISSER-ITO'S INDEXING ("VISSER")
     (3) CREATE A CALCULATED PATTERN (POWDER OR REYNOLDS)
     (4) ENTER A PATTERN TO SOFTWARE SYSTEM
     (5) UPDATE DATA BASE ("INIT")
     (6) CLAY MINERAL GRAPHICS SEARCH ("DGRAPH")
     (7) AUTOMATED SEARCH/MATCH ("SM" OR JCPDS' "SM")
     (8) QUANTITATIVE ANALYSIS
     (9) ACCESSORY PROGRAMS              --   OPTION: 8
```

Figure 2. Main Menu

patterns to be used in the least-squares fitting of the whole unknown
pattern. Part of the QUANT dialogue, including GUESS, is illustrated in
Figure 4.

DISCUSSION

     The use of the whole diffraction pattern for clay mineral analysis
has many advantages over methods based solely on d-I data. Because of
the small crystallite sizes inherent in clays, all of the diffraction


```
GRAPHIC OPTIONS MENU

  (0)  Exit program
  (1)  Display a new pattern
  (2)  Superimpose a pattern on current display
  (3)  Erase a pattern on current display
  (4)  Expand or contract the scale of the graph
  (5)  Make a hard copy of the display
  (6)  Subtract one pattern from another pattern
  (7)  Show a list of displayed patterns
  (8)  Show a list of the subtraction log
  (9)  Search for a pattern              ENTER OPTION     1

PATTERN SELECTION MENU

  (0)  Exit this menu
  (1)  Reference pattern
  (2)  Unknown pattern
  (3)  Residual pattern

                                         ENTER OPTION     2
ENTER UNKNOWN FILE NAME      unktx2.dat
ENTER ONE LINE OF TITLE INFORMATION -- 79C
test
ENTER TWO THETA OFFSET       0
ENTER INTENSITY OFFSET       0
```

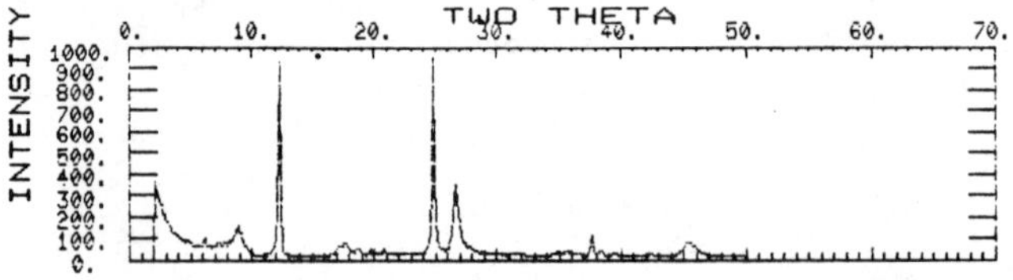

Figure 3. DGRAPH Menu and Example

profiles are broadened, and the precise position of the peaks are usually difficult to locate. In addition, the presence of 00ℓ stacking faults drastically affect hk0 and hkℓ type reflections, further complicating the pattern. All the clays show these effects to some degree with the smectites rarely showing any hkℓ reflections at all. The profiles, however, still contain considerable information about the sample. There is no way that a d–I table can represent this informaton, so it is logical that this profile information should be retained when doing identifications and quantitative estimations of clay minerals in rock samples.

Because of the poor hkℓ reflections for most clay minerals, the oriented sample technique enhancing the 00ℓ intensities is extensively employed and usually provides and adequate characterization of the clay component of a rock. Usually, determination of the relative proportions of each of the principle clay families is sufficient to satisfy the needs of the requestor. For this purpose, the 00ℓ reflections can distinguish smectites and the expanding component of mixed-layer clays from other 14Å clays, specifically the chlorites. Heat treatments are useful to distinguish Na and Ca-smectites, vermiculites and the presence of a kaolinite in a chlorite-containing sample.

The use of 00ℓ reflections are not without pitfalls. Properly treated samples, where the clays are all in an equilibrium state usually result in symmetric 00ℓ peaks which are easily interpreted. However, if the treatment is incomplete, the clay is in a mixed state, and the diffraction profiles are complex. Glycolation of expanding clays usually results in a more constant state than does dehydration through various drying techniques. Heat treatments need to be given sufficient time to complete the desired reaction or the interpretation of the diffraction pattern will be difficult. Sample preparation is still the most important step in QXRPD.

```
QUANTITATIVE PROGRAM REQUIRES TWO FILES TO OPERATE.

OUTPUT FROM THE QUANTITATIVE ANALYSIS CAN BE FOUND IN THE FILE QUANT.OUT; .
THIS FILE CAN BE PRINTED ON THE SYSTEM PRINTER AFTER
EXITING THE "XRAY" PROGRAM.

          QUANTITATIVE PROGRAM MENU
          *****************************************
          (0) RETURN TO MAIN MENU
          (1) BUILD GUESS FILE ("GUESS")
          (2) RUN QUANTITATIVE ANALYSIS ("QIN" AND "Q")
          ENTER SELECTION        : 1

PROGRAM GUESS---YOU THINK YOU KNOW THE ANSWERS
          VERSION SEPT 5, 1984

          The file name represented by XXXXXX, should be consistant from a
          logical point of view and should allow identification of the sample

ENTER THE NAME OF THE OUTPUT FILE (XXXXXX.GUS)  UNKSC2.GUS
ENTER ONE LINE OF TITLE INFORMATION    -- 79C
TEST FROM HOME
NOTE THAT NO CHECK IS MADE TO ASCERTAIN THAT THE
          PATTERN NUMBER THAT ARE ENTERED ARE VALID.
          PLEASE CHECK THE OUTPUT FROM THE PROGRAM "INIT" TO
          GET THE NUMBER VS NAME OF THE REFERENCES.
          YOU COULD ALSO CHECK YOUR NOTEBOOK FOR THESE NUMBERS

ENTER THE PATTERN REFERENCE NUMBERS , ONE PER LINE (ENTER 0 TO QUIT)
46
125
136
0
FORTRAN STOP
```

Figure 4. Quantitative Analysis Menu and GUESS Example

Even under ideal conditions, species within a family are indistinguishable using only 00ℓ reflections. Kaolinite, dickite and nacrite are stacking polytypes and have identical 00ℓ intensities. The width of the 00ℓ peaks may indicate the presence of smaller than usual crystallites typical of the more disordered forms, but still cannot distinguish the polytypes. To resolve this problem, it is necessary to look at the hkℓ intensities using the random sample or at the hk0 reflections using the oriented sample in transmission. This project has chosen the former approach, primarily because of the lack of availability of a transmission diffractometer.

The discussion above emphasizes the need to retain all the information in a diffraction pattern when possible. Clay mineral characterization, i.e. identification and quantitative estimation, is one of those instances. The success of the method depends on the completeness of the database and the reproductibilty of the sample preparation techniques. Consequently, half the effort in this project was the location of suitable clay samples and the purification of these samples for entry into the database.

ACKNOWLEDGEMENTS

This project was developed under the sponsorship by a Consortium of five oil and oil-related companies. R. C. Reynolds and the late G. W. Brindley were consultants. D. M. Melcher did much of the computer programming. D. E. Pfoertsch, J. A. Garland and K. E. Schultz prepared the samples and collected the x-ray patterns for the database. K. E. Schultz also calculated the theroretical patterns for the database.

SOFTWARE AVAILABILITY

Because of the nature of the sponsorship of this project, the software and experimental database remain proprietory to the sponsors and The Pennslvania State University for two years following the termination of activities. Because the project is still active, no date for availability can be announced.

REFERENCES

1. R. C. Reynolds, Interstratified Clay Systems: Calculation of
    the total one-dimensional diffraction functon, Am. Mineral.
    52:661-672 (1967).
2. R. C. Reynolds, Interstratified clay minerals Chapt 4 in Crystal
    Structures of Clay Mineral and their X-ray Identification G.
    W. Brindley, and G. Brown, Editors, Mineralogical Society,
    Great Britian (1982).
3. D. R. Pevear, (Program for calculating oriented clay mineral
    x-ray patterns FORTRAN Version) Private Communication (1982).
4. D. K. Smith, M. C. Nichols, and M. E. Zolensky, POWD10, A
    FORTRAN-IV program for calculating X-ray powder diffraction
    patterns. College of Earth and Mineral Sciences, Pennsylvania
    State University (1983).
5. The Clay Minerals Society, Oriented Sample Mounts for the
    Analysis of Clay Minerals by X-ray Diffraction. Workshop
    Syllabus, Annual Meeting, Honolulu Hawaii (1982).

THE USE OF CALCULATED PATTERNS AS AN AID IN PREPARATION OF POWDER DIFFRACTION STANDARDS: MINYULITE, $KAl_2(PO_4)_2(OH,F) \cdot 4H_2O$, AS AN EXAMPLE

Frank N. Blanchard

Department of Geology
University of Florida
Gainesville, FL   32611

INTRODUCTION

As part of a continuing project dealing with evaluation, testing, and preparation of new high quality X-ray powder diffraction standards for phosphate and arsenate minerals, new X-ray and supporting data for minyulite, $KAl_2(PO_4)_2(OH,F) \cdot 4H_2O$, have been obtained and interpreted with the aid of a calculated powder pattern.

ACKNOWLEDGEMENTS

I am indebted to the JCPDS-International Centre for Diffraction Data for partial support by repeated grants-in-aid, and to Dr. James L. Eades for obtaining the funds to modernize the X-ray diffraction laboratory in the Department of Geology at the University of Florida.

METHODS

Inspection of the current PDF card for minyulite and examination of preliminary X-ray data indicated that a substantially more complete and more accurate reference pattern could be obtained.

The minyulite used for this new reference pattern came from Dandaragan Wells, Western Australia, and consists of a white encrustation of aggregates of radiating acicular crystals on a matrix consisting of limonite and an iron phosphate mineral of undetermined identity. Individual crystals are roughly 0.05 by 1.0 mm, and are essentially free from contaminating phases, as evidenced by imaging and X-ray emission analysis under the scanning electron microscope (SEM), and by examination under the polarizing microscope (Figure 1). The only element detected which is not ideally present in minyulite was a trace of iron. The following chemical analysis for material from the same locality is given in Dana's System of Mineralogy (1951): K_2O 12.97, Al_2O_3 28.08, P_2O_5 39.10, H_2O 19.85.

In order to obtain a clean sample for preparation of an X-ray standard, part of the minyulite was broken off from the surface of the specimen; this material was sieved and the 30 to 100 mesh fraction was separated grain by grain under a binocular microscope to obtain material showing no visible contamination. This clean sample was ground under methanol, dried, and smeared onto a glass slide with amyl acetate and collodian as a glue. The "smear" was used for collection of X-ray powder

225

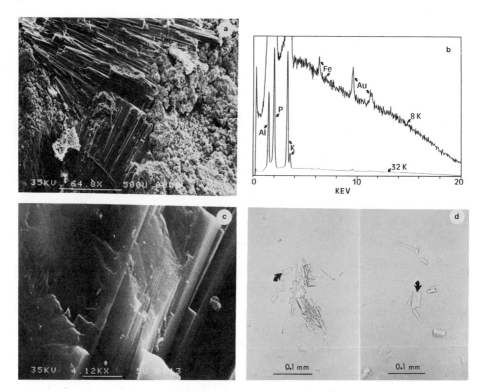

Figure 1. SEM photographs of minyulite (along with iron oxide and iron phosphate minerals) showing crystal habit, high degree of purity, and probable cleavage perpendicular to the length of the crystals (a and c); X-ray emission spectrum (b), showing K, Al, P, and a trace of Fe (Au is from the coating on the sample); and polarized light photomicrographs (d) showing high degree of purity and cleavage perpendicular to the length of the crystals. Arrows point out cleavages.

data with a Philips' APD 3600 with the following experimental conditions: Cu X-rays at 45 KV and 30 MA, a 6 degree take-off angle, theta-compensating slit (18 mm irradiated length), a graphite monochromator, a step size of 0.01 degree and count times of one second (4 different scans) and 4 seconds (one scan), and a step size of 0.02 degree with a count time of 0.5 second (one scan).

   Peak searching (second derivative method), alpha-2 stripping, correction of 2-theta angles (with an external standard of quartz prepared in the same way as the minyulite), and determination of peak intensities were accomplished using the Philips' analytical software. The final d-values were obtained from the scan made with a step size of 0.01 degree and a count time of 4.0 seconds. Final intensities for the smear preparations were obtained by averaging values from 6 scans, followed by conversion to intensities that would have been obtained using a diffractometer with a one degree fixed divergence slit.

DATA ANALYSIS AND DISCUSSION

   Indexing and lattice parameters. The 2-theta angles from the first 30 reflections were used as input to the program written by Visser (1969) in order to find the unit cell. Only one of the "4 most probable solutions" had a high figure of merit (32.6), and the parameters

corresponding with this solution were used as input for the program by
Appleman and Evans (1973), which indexes the reflections and performs a
least-squares refinement of lattice parameters. With reasonably
successful results from this computation, the presence criteria for space
group Pba2 (assigned by Kampf, 1977) were added and the computation was
repeated with good success. Although the results of the Appleman and
Evans program were essentially acceptable, there were 24 of the 106 input
reflections which were marked "rejected". Rejection may occur because the
difference between a calculated and observed 2-theta angle (delta
2-theta) exceeds the tolerance permitted by the investigator (0.04 degree
2-theta in this instance), or because there are two or more possible
reflections with 2-theta angles too close together for the computer to
"decide" which set of hkl planes is responsible for the reflection. In
the minyulite data, all of the rejections were due to the second cause.
Correct assignment of hkl's to all of these 24 rejected reflections was
accomplished by comparison of the data with a calculated pattern for
minyulite, as described in the next paragraph.

A calculated pattern for minyulite was obtained from the structure
determination of Kampf (1977), aided by the program POWD10 (Smith et al.,
1983) for calculation of powder diffraction patterns. The calculated
pattern may be presented in tabular form or as a simulated diffractometer
chart. The latter is shown, along with an experimental pattern obtained
on a G.E. XRD-5 diffractometer, in Figure 2, and the close similarity
between the two is evident. Detailed comparison of the calculated
intensities with the observed pattern, in the vicinity of each of the 24
unindexed (rejected) reflections, made it possible to unequivocally index
all of the reflections which were rejected by the Appleman and Evans
program (Table 1). In some instances the "closest hkl" proved to be the
correct one, in other instances an hkl with a slightly greater delta
2-theta than the closest fit was correct, and in still other instances
the correct assignment consisted of multiple hkl's.

Analysis of intensities Comparison of intensities of the observed
patterns (average of 6 scans) and the calculated pattern shows a good
correlation, but with some noticeable discrepancies. Except for the 100%
line nearly all of the observed intensities are considerably less than the
calculated intensities (Figure 2). In order to adjust the observed

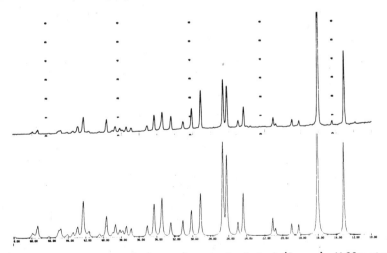

Figure 2. Experimental (upper) and simulated (lower) diffractograms
of minyulite, showing close similarity between observed and calculated
patterns. Right end is 10 degrees and left end is 50 degrees 2-theta.

Table 1. Examples of the use of a calculated pattern to resolve problems in indexing. "R" indicates a rejected reflection from the Appleman and Evans output. Arrows and brackets show correct indexing, while "OK" indicates a confirmed correct indexing of a rejected line.

I_{calc}	hkl		I_{obs}	d_{obs}	$2\theta_{calc}$	$2\theta_{obs}$	$\Delta 2\theta$
Calculated Intensities			Observed Data and Calculated 2θ Angles				
3.2	421		1.2	1.9684	46.0756	46.0750	0.00064
0.6	340	R	8.9	1.9190	47.3264	47.3320	-0.00555
29.8	232				47.3424		0.0103
0.1	150				47.6157		
6.7	322		1.9	1.9026	47.7680	47.7640	0.00400
8.2	430		4.7	1.8966	47.9128	47.9260	-0.00424
4.8	233		2.0	1.5152	61.1104	61.1120	-0.00156
9.9 (OK)	252	R	2.4	1.5074	61.4622	61.4600	0.00223
0.2	323				61.4687		
14.1	601		3.5	1.4986	61.8424	61.8620	-0.01965
6.2	442		2.6	1.4393	64.7161	64.7130	0.00314
10.4	62$\bar{1}$				65.0435		
15.2	541	R	6.4	1.4324	65.0707	65.0640	0.00671
3.8	41$\bar{3}$				65.1658		

pattern to "good correspondence" with the calculated pattern, the scale factor of the observed pattern was doubled (Figure 3), which resulted in truncation of the 100% line (at half its height) on the intensity scale. With this adjustment the discrepancy between the observed and calculated patterns is reduced greatly. The need for the adjustment may be due in part to preferred orientation. The 100% line is the 001 reflection; if this reflection were enhanced due to preferred orientation, this would have the effect of reducing the intensities of all other reflections (except for higher orders of 001). This appears to be the case. Evidence from the SEM image (Figure 1a and 1c) and from polarized light microscopy (Figure 1d), suggests that there may be a cleavage

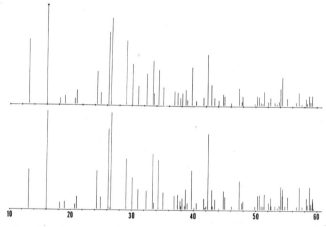

Figure 3. Stick patterns of observed (upper) and calculated (lower) patterns, showing close correlation between observed and calculated patterns. The observed pattern is from a smear preparation and the intensity scale is twice that of the calculated pattern.

Table 2. Evaluation of new data and current PDF data by AIDS83.

MINYULITE: PDF 27-371

TOTAL LINES INPUT = 39
NUMBER INDEXED   = 38
NUMBER UNINDEXED = 1

FOR INDEXED LINES

Average 2-theta difference = -0.019
with Diff > +0.05(2-theta) = 7
with Diff < -0.05(2-theta) = 13

F(30) = 10.19 (Delta 2-theta = 0.0545, # Possible = 54)

MINYULITE: NEW PATTERN

TOTAL LINES INPUT = 106
NUMBER INDEXED   = 106
NUMBER UNINDEXED =   0

FOR INDEXED LINES

Average 2-theta difference = -0.000
with Diff > +0.05(2-theta) = 0
with Diff < -0.05(2-theta) = 0

F(30) = 62.22 (Delta 2-theta = 0.0127, # Possible = 38)

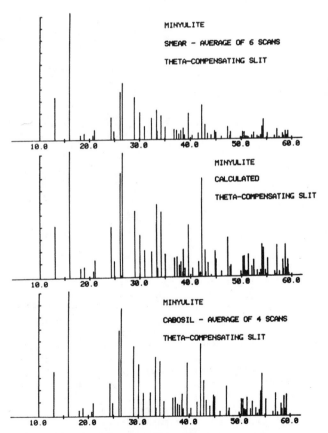

Figure 4. Stick patterns of "smear", calculated, and "cabosil" patterns of minyulite. The "cabosil" pattern shows the best total intensity correspondence with the calculated pattern, while the "smear" pattern shows the most complete correlation of reflections.

Table 3. Summary of powder pattern for minyulite. I_1 = intensity from cabosil preparation (more nearly random orientation of particles) and I_2 = intensity from smear preparation. Intensities recalculated from theta-compensating slit (18 mm irradiated length) to 1° divergence slit.

d A	I_1	I_2	hkl	d A	I_1	I_2	hkl	d A	I_1	I_2	hkl
6.745	35	35	110	1.9190	9	3	232	1.4393	2	1	442
5.519	100	100	001	1.9026	1	1	322	1.4324	5	2	541,621
4.871	4	3	020	1.8966	2	2	430	1.4246	2	1	333
4.674	5	4	200	1.8270	1	<1	042	1.4180	1	<1	352
4.277	3	2	111	1.8135	5	2	341	1.4044	1	<1	630
4.215	8	6	210	1.8033	4	2	151	1.3998	1	<1	062
3.653	18	12	021	1.7937	2	<1	431	1.3809	3	1	004
3.569	7	4	201	1.7836	1	<1	402	1.3610	<1	<1	631
3.402	45	25	121	1.7766	5	<1	113	1.3527	1	<1	114
3.349	55	30	211	1.7546	2	1	412	1.3436	3	1	612
3.068	35	20	130	1.7435	4	1	332	1.3342	3	1	270
2.967	23	12	310	1.7225	1	1	023	1.3250	2	1	153
2.878	10	6	221	1.7114	<1	<1	251	1.3066	1	<1	622,542
2.762	10	9	002	1.7015	1	1	242	1.2998	2	<1	513
2.681	25	12	131	1.6938	6	3	123	1.2965	2	1	461,271
2.669	3	3	230	1.6863	11	5	440,213	1.2780	2	<1	224
2.612	20	9	311	1.6642	4	1	521	1.2663	1	<1	523
2.557	6	5	112	1.6250	1	1	060	1.2594	2	1	134
2.437	7	3	040	1.6125	5	2	441	1.2520	7	<1	314
2.405	7	3	022,231	1.6003	1	<1	160	1.2432	1	<1	443
2.378	4	2	202	1.5832	3	1	351	1.2388	1	<1	371
2.358	4	3	140	1.5785	1	<1	133	1.2346	2	<1	730
2.328	8	4	122	1.5700	5	2	152	1.2260	1	<1	560
2.311	2	1	212	1.5644	2	1	313	1.2171	1	<1	650,063
2.271	18	9	410	1.5578	5	1	600,061	1.2078	1	<1	163,180
2.248	<1	<1	330	1.5362	1	<1	161,260	1.2045	3	<1	731
2.229	2	1	041	1.5285	2	1	512	1.2007	1	<1	462
2.169	4	2	141	1.5152	1	1	233	1.1914	1	<1	144
2.136	25	11	222	1.5074	2	1	252	1.1887	2	<1	404,603+
2.108	11	4	420	1.4986	4	1	601	1.1800	2	<1	181,414
2.084	5	2	331	1.4824	2	1	540,611	1.1683	1	<1	800,471
2.0530	3	1	132	1.4754	4	1	522	1.1610	<1	<1	453
2.0220	6	2	312	1.4688	2	1	043	1.1550	2	<1	543,623
2.0120	6	2	241	1.4511	1	1	143	1.1350	1	<1	811
1.9684	1	<1	421	1.4455	2	1	403,451	1.1240	1	<1	660
								1.1128	1	<1	821

perpendicular to the length of the crystals. Fleischer et al. (1984) describe the elongation of minyulite as parallel with the c direction; cleavage perpendicular to the length would be consistent with the (001) plane. Dana's System of Mineralogy (1951) lists a probable cleavage parallel with the elongation, while Fleischer et al. indicate no cleavage. Neither of these are consistent with the observation of a probable 001 cleavage as noted above. The possibility of preferred orientation (parallel with (001)) is also indicated by the fact that the only reflection other than 001 which has an observed intensity close to the calculated intensity is 002.

In order to further clarify the discrepancy between observed and calculated intensities, a sample of minyulite was prepared to reduce preferred orientation by mixing the mineral powder with cabosil and vegetable oil. Four scans with steps of 0.01 degree and count times of 3, 4, 5, and 10 seconds were made and the intensities from these were averaged. Comparison of these values ("cabosil") with the calculated intensities showed a much closer correlation than obtained from the "smear" slide (Figure 4). Disadvantages of the "cabosil" preparation include the high background, the very broad amorphous peak of the cabosil between 14 and 28 degrees 2-theta, and the failure to record many of the very weak reflections even with very slow scanning.

Identification of minyulite using the PDF. As an additional test of suitability of the current PDF pattern for minyulite, the data collected in this study were used as input to a search/match program (SANDMAN (Jenkins et al., 1983)) which attempts to identify an unknown from its observed pattern using the PDF as a data base. The program SANDMAN found and identified minyulite from the observed data, but, in addition, identified seidozerite as also present. Furthermore, 13 out of 52 common d-range lines are not shown on the PDF card, and two lines on the current PDF card have no counterpart on either the observed or calculated patterns. The two lines apparently are due to an impurity.

Evaluation by AIDS83. The powder data described in this report, as well as the data on the current PDF card, were evaluated by the program AIDS83 (used by the NBS and by the JCPDS for evaluation of crystallographic data, Mighell et al., 1981); part of the output is given in Table 2. The results indicate that the new data are a substantial improvement over the current PDF card and will be submitted to the JCPDS as a recommended replacement for card #27-371. The new data (Table 3) are more accurate and more complete, closely spaced reflections are better resolved, all lines are indexed, all delta 2-thetas are less than 0.05 degree, and intensities correlate well with calculated intensities based on the crystal structure. Comparison of the figure of merit (Smith and Synder, 1979) of the new data with the old, and with the distribution of values for essentially all of the phosphate minerals through set 31 of the PDF (Blanchard, 1984), is shown in Figure 5. The new pattern is in the top several percent in quality of all of the phosphate minerals.

SUMMARY AND CONCLUSIONS

This particular report is taken from a continuing project dealing with improvement of powder diffraction standards for the phosphate minerals.

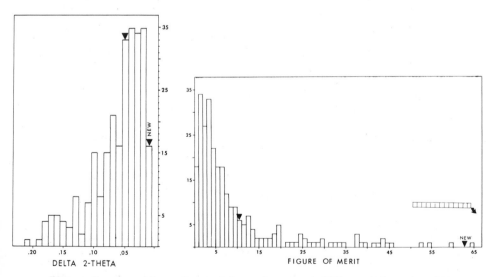

Figure 5. Location of new data and current PDF card for minyulite on diagrams of average delta 2-theta (left) and figure of merit (right), for all of the phosphate minerals through set 31 of the PDF.

Inspection of the current PDF card for minyulite and comparison of the data with an experimental pattern indicated that a new reference pattern would be desirable. Problems with the current PDF data include low accuracy of the d-values, many missing reflections, the presence of two lines which do not belong to minyulite, less than satisfactory results in computer identification, and incomplete optical properties.

Preparation of powder diffraction standards is greatly facilitated by the use of calculated patterns, based upon known crystal structures, and, according to Moore (1984), crystal structures have been determined for about 120 of the 300+ known phosphate minerals. Calculated patterns are useful for the following two reasons. (1) They aid in alleviating the problem of ambiguities in indexing, which are almost always present in patterns from low symmetry phases. (2) Calculated patterns may disclose conditions of preferred orientation in sample preparation, which result in X-ray intensities inconsistent with the ideal random orientation of crystals expected in the powder method. Both of these advantages are illustrated in the example of minyulite presented in this report.

The new data, which will be submitted as a recommended replacement for the current card, have been evaluated by the program AIDS83 and the results indicate that the new pattern meets the criteria for "star" quality. This pattern ranks in quality in the top several percent of all of the phosphate minerals in the PDF.

REFERENCES CITED

Appleman, D. E. and Evans, Jr., H. T., 1973, Indexing and least-squares refinement of powder diffraction data. Geol. Survey Computer Contribution, No. 20, 62 p.

Blanchard, F. N., 1984, Evaluation of existing X-ray powder diffraction standards for phosphate minerals. Advances in X-ray Analysis, 27, 61-66.

Dana's System of Mineralogy, 1951, 7th Ed., Vol. 2, 1124 p.

Fleischer, M., Wilcox, R. E., and Matzko, J. J., 1984, Microscopic determination of the nonopaque minerals. U.S.G.S. Bull. 1627, 453 p.

Jenkins, R., Hahn, Y., Pearlman, S., Schreiner, W. N., 1983, A computer aided search/match system for qualitative powder diffractometry. Norelco Reporter, 30, 16-18.

Kampf, A. R., 1977, Minyulite: its atomic arrangement. Am. Mineral., 62, 256-262.

Mighell, A. D., Hubbard, C. R., and Stalick, J. K., 1981, NBS*AIDS80: a FORTRAN program for crystallographic data evaluation. NBS Technical Note 1141, 54 p.

Moore, P. B., 1984, Crystallochemical aspects of the phosphate minerals, in Phosphate Minerals, ed. Nriagu, J. O., and Moore, P. B., (Springer-Verlag), p. 155-170.

Smith, D. K., Nichols, M. C., and Zolensky, M. E., 1983, A FORTRAN IV program for calculating X-ray powder diffraction patterns - version 10. The Pennsylvania State University, University Park, PA, 72 p.

Smith, G. S. and Snyder, R. L., 1979, F(N): a criterion for rating powder diffraction patterns and evaluating the reliability of powder pattern indexing. J. Appl. Cryst., 12, 60-65.

    Visser, J. W., 1969, A fully atomatic program for finding the unit
cell from powder data. J. Appl. Cryst., 2, 89-95.

ADDENDUM

    After completing this manuscript, a single crystal rotation photograph
was obtained from an elongate crystal of minyulite. Interpretation of the
results indicates that the morphological elongation is coincident with the
c unit cell dimension. This confirms the orientation of cleavage parallel
with {001} and the associated likelihood of preferred orientation of
powder diffraction preparations. I am indebted to Dr. Gus Palenik
(University of Florida) for the single crystal pattern and its
interpretation.

APPLICATION OF QUANTITATIVE X-RAY DIFFRACTION ANALYSIS COMBINED WITH OTHER

ANALYTICAL METHODS TO THE STUDY OF HIGH-MAGNESIUM PHOSPHORITES

Frank N. Blanchard, Robert E. Goddard and
Barbara Saffer

Department of Geology
University of Florida
Gainesville, FL 32611

INTRODUCTION

Phosphorite is a sedimentary rock with a high enough content of phosphate minerals to be of economic interest. Most phosphorites are composed predominantly of microcrystalline to cryptocrystalline carbonate fluorapatite (henceforth in this report referred to simply by the mineral group name apatite). Florida produces roughly 1/3 of the world's supply of phosphate rock, most of which is used in the fertilizer industry.

Long term continuation of phosphorite mining in Florida will require exploitation of the extensive high-magnesium phosphorite deposits south of the present mining district in central Florida, and this will require new technology in order to produce beneficiated concentrates with less than 1% MgO, a limit imposed by fertilizer processing technology. In order to develop benefication methods applicable to these ores, it is essential to know how Mg occurs in phosphorites. Dolomite, $CaMg(CO_3)_2$, is the chief host of Mg in phosphorites from Florida. Magnesium may also be present, however, as a substituent in apatite (the chief phosphate mineral in these deposits), as a minor substituent in calcite ($CaCO_3$), in certain clay minerals (particularly palygorskite and to a lesser extent in some smectites), and/or in organic matter within the apatite particles.

DOLOMITE IN PHOSPHORITES FROM FLORIDA

General statements. The presence of dolomite in Florida phosphorites containing appreciable Mg is well known and is best established by X-ray diffraction. Polarizing microscope examination of the phosphorite known to contain dolomite clearly shows the dolomite in contrast to other phases, such as apatite and various other minor minerals. Evidently, dolomite is responsible for most of the Mg in "high-Mg phosphorites".

Polarized light microscopy. Under the polarizing microscope it is possible to distinguish, in Mg-bearing phosphorites, (1) particles of apatite which are essentially free from dolomite, (2) particles of dolomite which are free from apatite, and (3) particles of apatite which contain inclusions of dolomite (Figure 1). Even if beneficiation techniques can be developed to separate pure dolomite from pure apatite, particles of apatite which contain inclusions of dolomite will still constitute a problem.

Apatite with inclusions of dolomite. In samples of high-Mg phosphorite, it is common to find a wide range of percentage of apatite particles which contain dolomite inclusions. Furthermore, within any one

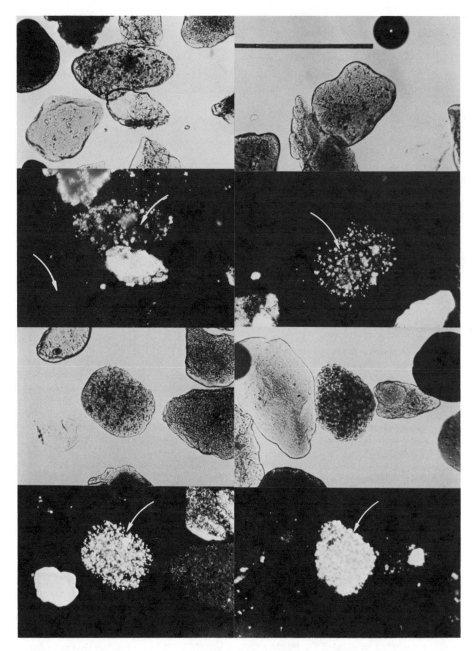

Figure 1. Photomicrographs in plane light (upper) and between crossed polars (lower) showing particles of apatite with varying amounts of dolomite inclusions. Dolomite inclusions show up with strong birefringence between crossed polars in contrast with the near absence of birefringence in the apatite. Apatite grains of special interest are marked with arrows. A. Lower left grain with no dolomite, center grain with minor amount of dolomite. B. Grain of apatite with moderate amount of dolomite inclusions. C. Grain of apatite with large amount of dolomite inclusions. D. Grain of apatite dominated with dolomite inclusions. Bar scale is 1.0 mm.

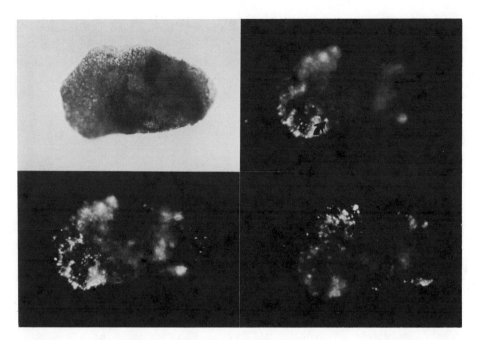

Figure 2.   Photomicrographs of apatite particle in plane light upper
left, and between crossed polars. The three photos with crossed polars
were focused at different levels to show dolomite inclusions (strongly
birefringent) at various depths beneath the surface. This is the same
particle (and in the same orientation) as shown in Figure 3 (upper left)
under the SEM. The arrow on the upper right photo points to position "A"
on the upper left SEM photograph of Figure 3.

sample the abundance of inclusions of dolomite in apatite particles
ranges from almost nil to well above 50%. In phosphorites with an
abundance of apatite grains with dolomite inclusions, reduction of the Mg
content of the beneficiated product depends either on (1) selective
separation of dolomite-bearing apatite particles from dolomite-free
apatite particles, or (2) grinding of the phosphorite to a size such that
the dolomite is liberated from the apatite. The latter is not practical
because observations indicate that the dolomite inclusions are in the
sub-micron to several micron size range. The possibility of selective
separation of the dolomite-bearing from the dolomite-free particles of
apatite may depend on whether or not appreciable amounts of dolomite
occur on the surface of the particles of apatite. In order to answer
this question several samples were examined with both the polarizing
microscope (to determine the abundance of dolomite inclusions) and the
scanning electron microscope (to search for dolomite on the surface of
particles).
       Scanning electron microscopy. Because of the extreme birefringence of
dolomite, it shows up clearly between crossed polars under the polarizing
microscope; however, in a grain roughly 0.5 mm in diameter, it is
difficult to determine whether the dolomite is completely enclosed by
apatite or whether it is abundant on the particle surface. In order to
answer this question, grains of apatite from the "sink fraction" of a
flotation experiment in which dolomite was floated, were selected under
the polarizing microscope for large, medium, small, and no percentages of
dolomite inclusions; the same grains were then separately glued to small
glass coverslips, and were photographed under a polarizing microscope

(plane light and crossed polars); the glass cover slips containing the grains were then transferred to a stub, coated with gold-palladium, and examined under the scanning electron microscope (SEM). In this way a single apatite particle could be studied under the polarizing microscope (Figure 2), under the SEM (Figure 3), and chemically by X-ray emission

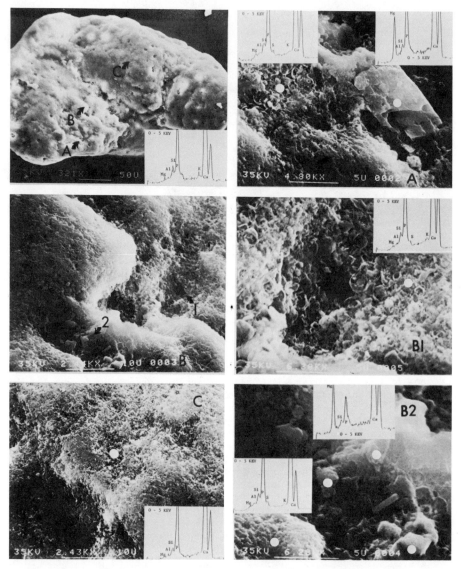

Figure 3. SEM photographs and X-ray emission spectra of the same particle of apatite as shown in Figure 2. Areas "A", "B", and "C" from the upper left are enlarged in the other illustrations as labeled. "B1" and "B2" are enlarged from the two areas of "B" marked "1" and "2". Locations of emission spectra are marked by white or black dots. Locations where dolomite is on the surface of the particle are indicated by strong emission of Mg K-radiation. In spite of the fact that much dolomite can be seen in abundance with the polarizing microscope (Figure 2), it is rare and requires careful searching to find it on the surface of the particle.

under the SEM (Figure 3). The presence of substantial Mg in the X-ray
emission spectrum under the SEM was used to identify the presence of
dolomite on the surface of the particles (in passing through less than 7
microns of apatite, Mg K-radiation is 99% absorbed). Limited success was
achieved in locating areas which showed dolomite under the polarizing
microscope, by using X-ray emission analysis under the SEM. Examination
of many grains known to contain varying amounts of dolomite indicates
that the SEM image and X-ray emission analysis fail to indicate as much
dolomite as would be expected from the view under the polarizing
microscope. It is concluded, therefore, that much of the dolomite is
contained within, and is completely surrounded by, apatite. This suggests
that some grains which may behave as apatite in flotation experiments may
contain a very appreciable amount of dolomite as inclusions, and this
partially explains the excessive amount of chemically determined Mg in the
experimentally derived flotation fraction which should contain apatite.

It is not uncommon to find that X-ray emission analysis of apatite
grains shows the presence of a very minor Mg peak from particles which,
under the polarizing microscope, show almost no visible dolomite. This
may be due to Mg-substitution in the apatite structure or the presence of
dolomite at a depth of several microns beneath the surface. In other
instances areas of an apatite grain fail to emit any detectable
Mg-radiation.

In rare instances, surface fibers (possibly palygorskite, a Mg-bearing
clay mineral) were seen under the SEM. In other rare instances, the X-ray
emission spectra indicated abnormally high Si and Al, which could be
accounted for by clay minerals, Al-hydroxide phases along with quartz, or
quartz and such minerals as wavellite or crandallite (both known as common
alteration phases in Florida phosphorites). The presence of substantial
amounts of these other possible phases on the surface of apatite particles
may have a bearing on flotation experiments.

MAGNESIUM AS A SUBSTITUENT IN CARBONATE FLUORAPATITE

General statements. It is generally accepted that Mg may substitute
for Ca in the structure of apatite. The amount of this substitution is
difficult to estimate and is critical to the attempted reduction of Mg
content during beneficiation of phosphorites. Even if beneficiation can
remove all of the dolomite (both pure particles and those with inclusions)
the Mg substituting in apatite will still be present. If this amount can
be determined, then the maximum allowable dolomite after beneficiation
could be ascertained.

The method of estimating Mg substitution in apatite presented in this
report consists of chemical analysis for MgO contained in phosphorite,
combined with quantitative X-ray diffraction analysis for dolomite. From
these analyses, the amount of Mg contributed by dolomite to the total Mg
content of the phosphorite can be calculated. If this amount is less than
the chemically determined Mg, the difference could be accounted for by Mg
substitution in apatite (and to a lesser extent by Mg contained in minor
amounts of other Mg-bearing minerals).

Methods of quantitative XRD. Quantitative X-ray diffraction analysis
for dolomite in phosphorite was based on 26 standards prepared by mixing
appropriate proportions of Steinhatchee dolomite (1) with NBS SRM-120b (a
phosphorite from Florida). Chemical analysis of the Steinhatchee dolomite
indicates that it consists of 0.77% impurities, and the certified chemical
analysis of SRM-120b (along with X-ray diffraction data) indicates that it

---

(1) The term Steinhatchee dolomite is used in this paper to refer to a
particular bulk sample of dolomite from the (dolomitized) Suwannee
limestone (Oligocene age) collected along the Steinhatchee River in FL.

contains 0.725% dolomite. These two figures were used to correct the proportion of dolomite actually contained in each of the 26 prepared standards and the corrected percentages range from 3.90 to 20.43% dolomite.

For each of the 26 standards, 10 X-ray intensity ratios were calculated and averaged; each ratio consists of the measured intensity of the dolomite 104 peak in the standard, divided by the measured intensity of the 101 peak of an external quartz reference (used as a drift monitor). Because of differences between intensity measurements for dolomite from the phosphate mining district, compared with Steinhatchee dolomite in the standards, a correction was made based on comparative intensity measurements and on the purity of the dolomite from the phosphate mining district. The intensity ratio for Steinhatchee dolomite was 0.929. This was corrected for the 0.77% non-dolomite impurities to a ratio of 0.936. The intensity ratio for the dolomite from the phosphate district was 0.781 and this was corrected for the 8.245% non-dolomite impurities to a ratio of 0.851. The final intensity correction factor, used to adjust the dolomite of the standards (Steinhatchee) to dolomite from the phosphate mining district is 0.851/0.936 = 0.909.

The corrected intensities for the standards were plotted against the corrected weight percentages of dolomite, and a linear regression for the data points yielded the equation $y = 0.00634x + (-0.00642)$, with a correlation coefficient of 0.997. The negative y intercept is due to the location of the dolomite peak on the shoulder of the strongest apatite peak. Because of this, the background trace for the dolomite peak is a straight line which lies above the actual diffraction trace where no dolomite (and even a small amount) is present.

The corrected intensity measurement for each standard along with the corresponding mass absorption coefficient (computed from a complete chemical analysis) permitted calculation of the constant K in the Klug and Alexander equation as follows (for each standard):

$$K = \frac{((I+0.00634)0.909)\rho\mu^{*}t}{x}$$

The 26 standards yield an average value of $K = 45.9\rho$ (ρ = the density of dolomite), with a standard deviation of 2.53. With the value for K established, the concentration of dolomite in an unknown phosphorite can be computed from the formula

$$x = \frac{(I+0.00642)\mu^{*}t}{45.9}$$

Use of equation 1 requires measurement of an intensity (compared to the quartz drift monitor) and calculation of a value for the mass absorption coefficient. Chemical analyses were obtained for all unknowns so that appropriate mass absorption coefficients could be computed.

Results and interpretations of quantitative XRD. Measurements of dolomite content were carried out for 18 samples of phosphorite and the results were used to calculate the amount of MgO contributed to each sample from the dolomite. In this calculation a 2.2% atomic deficiency of Mg in the dolomite was assumed for all samples. The 2.2% Mg deficiency is based upon refinement of lattice parameters for samples of phosphorite containing "sufficient dolomite", "sufficient dolomite" meaning enough dolomite to yield X-ray diffraction peaks for all of the 18 peaks used in the least-squares refinement of lattice parameters. The lattice parameters were used to estimate Mg deficiency based upon the relationship elucidated by Goldsmith et al. (1961).

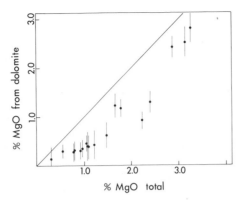

Figure 4. Plot of chemically determined MgO against MgO contributed by dolomite (determined by quantitative XRD) for 18 samples of phosphorite. Error bars represent the 95% confidence limits. The line represents a 1:1 ratio. The fact that all the points lie to the right of the line indicates that there is more MgO than can be accounted for by dolomite contained in the phosphorite. Most of this excess (an average of 0.57%) is believed to represent Mg substituting in the apatite structure.

Figure 4 shows a plot of chemically determined MgO against MgO contributed by dolomite, as determined by quantitative X-ray diffraction. The "error bars" represent the 95% confidence limits based on the standard deviations associated with the 10 measurements made on each sample. The average excess MgO content above that accounted for by dolomite is about 0.57% ($\pm$0.17, at 95% confidence level). This is considered a reasonable estimate of the amount of Mg substituting in the apatite structure plus any very minor amount of magnesium occurring in other phases.

According to McClellan (1980) the a unit- cell dimension for carbonate fluorapatite can be used as an estimate of the Mg substitution for Ca. McClellan and Van Kauwenbergh (written communication, 1985) find that for the southern extension of the central Florida phosphate field the average a cell dimension is 9.330(3), corresponding to a MgO content of 0.49 to 0.41% (by weight), depending upon the model used. The average a dimension for 9 samples in the present study for which the cell parameters were refined is 9.322(3). According to Van Kauwenbergh (International Fertilizer Development Center, written communication, 1985) this value would correspond to 0.55% (by weight) MgO in the apatite, and to 0.13 molecular proportion of Mg. This shows a good correspondence with the average of 0.57% MgO interpreted from the results of this study. The reasonably close similarity of the results reported in this study with those of McClellan suggest that a combination of chemical analysis and quantitative X-ray diffraction may provide a new and fairly rapid alternative to other methods which have been used to estimate the amount of Mg substitution in apatite.

ACKNOWLEDGEMENTS

    This material is based upon work supported by the Florida Institute of Phosphate Research under Award #82-02-023. Any opinions, findings, and conclusions or recommendations expressed in this publication are those of the author(s) and do not necessarily reflect the views of the Florida Institute of Phosphate Research.

REFERENCES

Goldsmith, J. R., Graf, D. L., and Heard, H. C., 1961, Lattice constants of the calcium-magnesium carbonates. Am. Mineral., 46, 453-457.

McClellan, G. H., 1980, Mineralogy of carbonate fluorapatites. J. Geol. Soc. London, 137, 675-681.

INSTRUMENTATION FOR SYNCHROTRON X-RAY POWDER DIFFRACTOMETRY

W. Parrish, M. Hart,[*] C. G. Erickson, N. Masciocchi,[†] and T. C. Huang

IBM Research Laboratory

San Jose, California 95193

ABSTRACT

The instrumentation developed for polycrystalline diffractometry using the storage ring at the Stanford Synchrotron Radiation Laboratory is described. A pair of automated vertical scan diffractometers was used for a Si (111) channel monochromator and the powder specimens. The parallel beam powder diffraction was defined by horizontal parallel slits which had several times higher intensity than a receiving slit at the same resolution. The patterns were obtained with 2:1 scanning with a selected monochromatic beam, and an energy dispersive diffraction method in which the monochromator is step-scanned, and the specimen and scintillation counter are fixed. Both methods use the same instrumentation.

INTRODUCTION

The increasing use of storage rings as synchrotron X-ray sources has opened new fields of X-ray analysis.[1] High energy X-ray sources are available at three major laboratories in the U.S.A. (Stanford University, Cornell University and Brookhaven National Laboratory) and several more in Europe, U.S.S.R. and Japan. They attract a large number of users in various fields. The most important differences of the synchrotron X-rays and X-ray tubes are the higher intensities, the availability of any wavelength over a wide range of wavelengths, the absence of line spectra so that each reflection is a single peak, the parallel rays and the polarized nature of the beam.

The instruments used for X-ray tube focussing geometry had to be modified to take full advantage of the properties of the synchrotron X-ray source. This paper describes the instruments developed for parallel beam geometry powder diffractometry. Experiments were conducted at the Stanford Synchrotron Radiation Laboratory in two two-week sets of runs in 1984 and 1985. The white X-ray beam in hutch No. 4 located 17 meters from the storage ring on bending magnet line II was used.

---

[*]Department of Physics, The University, Manchester M13 9PL, U.K.
[†]Permanent address: Istituto Di Chimica Strutturistica Inorganica, Universita Degli Studi, 20133 Milano, Italy.

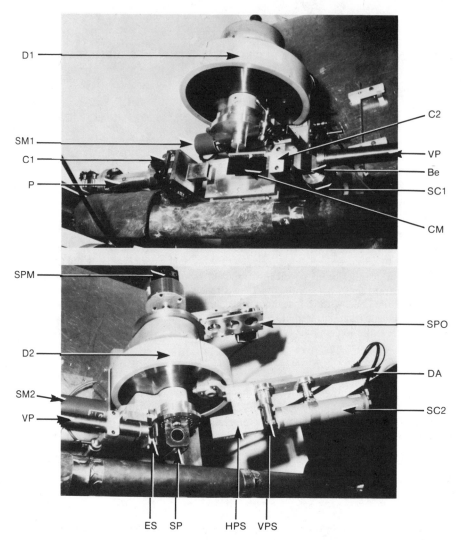

Fig. 1. Diffractometers with shielding removed. White X-ray beam from storage ring enters hutch in pipe P. C1 adjustable horizontal and vertical slits to limit beam striking Si channel monochromator CM mounted on diffractometer D1; C2 adjustable position radiation shield with fixed opening; Be foil inclined 45° and vertical scintillation counter SC1 used for monitor; VP vacuum pipe; SM1 and SM2 stepper motors operated from computer; SPM synchronous motor for rotating specimen; SPO stepper motor for oscillating specimen; SP specimen in rotating device on diffractometer D2; HPS horizontal parallel slits; VPS vertical parallel slits; DA detector arm; SC2 scintillation counter for powder measurements; D1 and D2 are in line.

DIFFRACTOMETERS

Two diffractometers were used in line: D1 (with detector arm removed) for the incident beam monochromator CM and D2 for the powder specimen, Fig. 1.   Both diffractometers scanned in the vertical plane because of the beam orientation and to avoid problems arising from the polarization of the beam in the horizontal plane.[2] The instruments were enclosed in a hutch and operated remotely because of the extreme radiation safety requirements. The white beam entering the hutch through pipe P was 22 mm wide and 6 mm high. Adjustable slits C1 limited the size of the beam striking the monochromator. The beam shield C2 had a fixed opening whose position could be adjusted during the alignment procedure. The detector arm of the first diffractometer was removed and the θ-axis could be set manually or driven with a precision stepper motor SM1 (Compumotor Model M57-102). The smallest repeatable angle step was $0.0005°\theta$. The powder diffractometer was driven by another Compumotor SM2 with the same characteristics.

The intensity of the primary beam was not constant because of the gradual decrease of current over a period of several hours between the electron refillings, and the occasional orbital shifts.   Since scanning methods were used it was essential to monitor the monochromatic beam striking the powder specimen to obtain accurate relative intensities. The beam passed through a 0.05 mm thick beryllium foil Be inclined 45° and the down-scattered X-rays were recorded with a shielded vertical scintillation counter SC1. The count rate was reduced to about 20,000 c sec^{-1} with an aperture over the window. The average number of counts for each 100 diffractometer steps was recorded with the powder data and stored in the computer to later correct the observed powder intensities. The tracking of the ring current with the monitor is shown in Fig. 2.

The beam from the monochromator (described below) was limited by an entrance slit ES mounted on D2 which determined the illuminated area of the specimen and could be

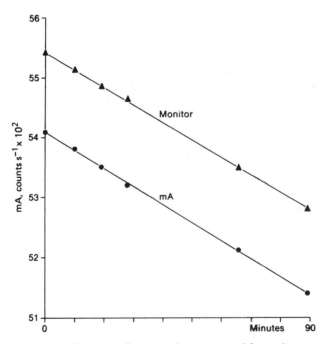

Fig. 2. Tracking of storage ring current with monitor.

changed for various angular ranges. Using ES width=2.7 mm, the irradiated specimen length was 18 mm at 17.2° 2θ and the irradiated specimen width was 16 mm at all angles. Unlike the X-ray tube focusing diffractometer the parallel beam remains centered at all angles and there is no flat specimen aberration.

The detector arm DA of D2 was modified by adding a rectangular aluminum arm with threaded holes to mount the various slits and detector. Close fitting holders could be fastened to the arm in any desired position. When using a receiving slit to define the diffracted beam, it was mounted at the front of the VPS housing 200 mm from the D2 axis of rotation. A 105 mm long rectangular vacuum pipe with thin Mylar windows was mounted between the specimen and receiving slit to reduce the air path. To use the horizontal parallel slits HPS, the rectangular pipe and receiving slit were removed and the front end mounted 65 mm from the D2 axis of rotation.

The horizontal parallel slits, obtained from Rigaku Corp., were made of 0.05 mm thick steel foils, 100 mm long, with 0.15 mm spacers which transmitted 67% of the beam. The full aperture was 0.17° and the usable area 19.8 mm wide and 11.1 mm high. The axial divergence was limited by vertical parallel 0.025 mm thick molybdenum foils VPS with 2.3° aperture. They were 12.5 mm long (parallel to the beam), 20 mm wide and 5.7 mm high stacked in a rigid frame and sets with different foil spacings could be inserted to change the aperture.

The X-ray path from the end of pipe P to SC2 was 969 mm. A 316 mm vacuum pipe VP with thin Mylar windows reduced the air absorption. It was not possible to use a vacuum specimen chamber because of interference with a beam pipe line passing through the hutch (lower portion of Fig. 1). The energy dispersive patterns had to be corrected for the varying air path transmission with wavelength. It was essential to carefully design the radiation shielding around the X-ray beam path and to use anti-scatter slits to avoid background scattering.

Because manual adjustments could be made only with the beam off, the alignment procedure was more involved than in an X-ray laboratory. Mechanical procedures using squares, height gauges and Polaroid films were used to align the diffractometers, monochromator and slits. The zero-angle calibration of D2 was made with a rapidly rotating pin-hole in the specimen holder, a narrow receiving slit and step scanning.[2] The ES height was set with a height gauge, the receiving slit removed and the centering checked with the pin-hole and step-scanning. Manual adjustments were made to bring the position of the ES to a height which gave the same zero-angle as the receiving slit. The 2:1 setting was made with the diffractometer at 0° and turning the specimen post until a spirit level indicated it was horizontal. The accuracy of the calibration could then be checked by recording profiles of standard specimens.

The automation was controlled with an IBM Personal Computer to provide selection of the angle increment for the step scan, count time and scan range and to record pertinent data for each run. The completed pattern was displayed on a Tektronix 4013 terminal with paper copy. The pattern and monitor data were recorded on diskettes and later transferred to our laboratory host computer for data reduction, graphics and analysis.

Single channel pulse height analyzers were used with the scintillation counters. When using monochromatic radiation the window transmitted about 95%. Only lower level discrimination was used for the energy dispersive runs.

CHANNEL MONOCHROMATOR

The monochromator was silicon (111) cut from a high quality boule with a channel to form a pair of parallel plates, Fig. 3. An important advantage of this type of monochromator

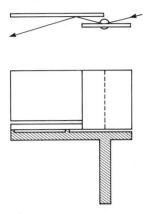

Fig. 3. Silicon channel monochromator.

is that one diffractometer alignment satisfied all wavelengths in the range 0.55 to 2Å, and rotating the crystal to change wavelengths did not change significantly the direction of the diffracted monochromatic rays. For example, changing the wavelength from 1.5Å to 1Å gave the same zero-angle. The first wafer was 30 × 25 mm, and the second 57 by 25 mm. A strain-free mount was obtained by cutting a slot in the lower portion of the crystal which was mounted on an aluminum holder with a few drops of epoxy cement. The middle of the front surface of the first wafer was centered on the rod which was inserted in a flange on the rotation axis of D1. This design provided a rigid mount which was stable in the primary beam heating. The wavelength calibration was made by scanning the absorption edges of elements in the specimen or pure element foils placed in the beam.

SPECIMEN PREPARATION

The specimen sizes and holders were the same as in conventional diffractometry. The powders were packed in cylindrical aluminum holders exposing 22 mm diameter powder surface, 0.5 to 1 mm deep, using 1:6 collodion: amyl acetate as a binder. The number of particles correctly oriented to reflect the parallel rays is much smaller than in divergent beam focussing diffraction.[3] Experiments using different particle size specimens with the present beam and specimen sizes showed it was necessary that the particles be <10 μm and that it was also essential to rotate the specimen. The powders were carefully prepared with acoustically driven sifters and Lektromesh and rotated 70 rpm. A device to simultaneously oscillate and rotate the specimen was built but there was not sufficient time to test it.

PARALLEL BEAM X-RAY OPTICS

Several methods were used to scan the patterns: a) 2:1 scanning (2θ detector: θ specimen) using a selected wavelength as in X-ray tube focusing; b) Specimen fixed at small θ with 2θ detector scan, the parallel beam equivalent of Seemann-Bohlin focusing geometry; c) Detector and specimen fixed in a selected 2:1 position and the monochromator step scanned for energy dispersive diffraction; d) Same as c) but the detector and specimen not in 2:1 relationship.

The specimen could be set in either the reflection or transmission positions simply by rotating the specimen $90^{\circ}\theta$. This was useful in detecting preferred orientation, but of course required an X-ray transparent substrate for mounting the powder.

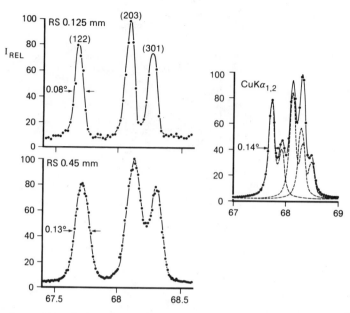

Fig. 4. Dependence of resolution of quartz triplet on receiving slit width RS, 1.54Å X-rays; (Right) Cu X-ray tube pattern with profile fitting showing relative intensities of the Kα doublets.

The first method used a receiving slit to define the diffracted beam as in conventional focusing diffractometry.[3] The profile widths increased with increasing receiving slit width as shown in Fig. 4 for a quartz triplet. (The intensities also increased but were normalized to compare the profile shapes). Each reflection is a single peak and hence it is much easier to determine the correct relative intensities and to do profile fitting than with the X-ray tube pattern in which each reflection is a $CuK\alpha_{1,2}$ doublet and the fourth peak is an overlap of $(203)\alpha_2$ and $(301)\alpha_1$.

By replacing the receiving slit with a nearly perfect single crystal plate the profile widths were reduced from $0.08° - 0.20°$ to $0.02° - 0.05°$ (2θ).[4] This method is useful for high resolution but the intensity is also greatly reduced.

The diffracted beam could also be defined by long horizontal parallel slits HPS whose aperture determines the profile width.[5] This method has above five times higher intensity than the single receiving slit at the same resolution and was used for most of the runs.

An improved method of energy dispersive diffraction (EDD) was also used with the HPS geometry.[5] The detector and specimen were set in a selected 2:1 position and the monochromator step-scanned to change the wavelength. Unlike conventional EDD each step of the monochromator produces a different wavelength which strikes the powder specimen and is measured with a conventional scintillation counter. The d-range of the pattern decreases with increasing 2θ-setting and there is only a small decrease at the higher 2θs. The wavelength range used was 0.55 to 2.04Å; for $2\theta=10°$ d=3.1−11.7Å, and for $2\theta=90°$ d=0.39−1.4Å.

    Patterns of tungsten powder using 2:1 scanning and EDD illustrate the different
aspects of the methods. Figure 5(a) was recorded with a 0.75 mm wide receiving slit,
$\lambda = 0.68883$A (Zr K-absorption edge) and contains reflections from d=2.2 to 0.37Å. The
background is low across the entire pattern and each reflection is a single peak. Figure 5(b)
was recorded with the horizontal parallel slits, step scanning monochromator and $2\theta = 135°$;
the reflection range was d=1.0 to 0.3A. The higher background in the short wavelength
region is due to W L fluorescence which decreases sharply as the incident wavelengths cross
the W L-I, L-II and L-III absorption edges. The fluorescence background could be reduced
without reducing the peak intensities by adding a vacuum or He pipe to increase the distance
between the specimen and detector, by increasing the length of the parallel slits, or by using
a solid state detector. The abscissa scales show the monochromator Bragg angle θ_M, the
wavelength λ and energy E. The computer scale numbers are related to θ_M. The intensity
of the highest peak in this range (321) was 27650 c sec^{-1}. Figure 5(c) is an enlargement of
the small-d range in which the intensities were low but the profiles retained good shape and
could be precisely measured. Patterns of tungsten recorded with conventional X-ray tube

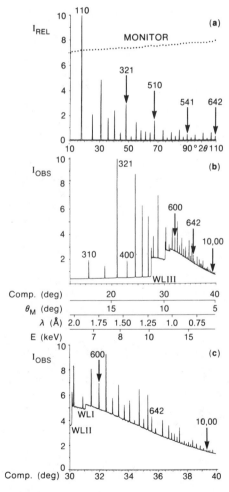

Fig. 5. Patterns of tungsten powder. (a) 2:1 scan with 0.75 mm wide receiving slit,
$\lambda = 0.68883$Å; monitor points are averages of 100 scanning steps; (b) energy
dispersive scan with horizontal parallel slits, $2\theta = 135°$, $\lambda = 2.04$ to 0.55Å,
$\Delta\theta_M = 0.005°$; (c) expansion of $30° - 40°$ range of (b) to show low intensity peaks.

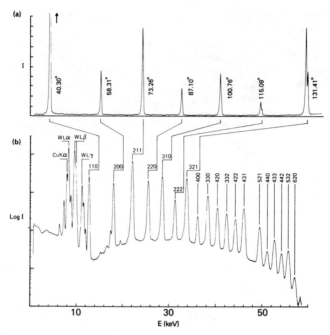

Fig. 6. Patterns of tungsten powder recorded with conventional methods.   (a) CuKα
focusing diffractometer; (b) Si(Li) solid state detector, $2\theta=25°$, Cu tube 60 kV,
log intensity scale.

focusing diffractometer and solid state detector methods are shown in Fig. 6 for
comparison.[6] The reflections in the diffractometer pattern have the well-known asymmetry
due to the Kα doublet and instrument aberrations, and those in the solid state detector
pattern are more than an order of magnitude broader than the X-ray tube or synchrotron
profiles and have much lower counting rates.

ACKNOWLEDGMENT

We are indebted to the staff of the Stanford Synchrotron Radiation Laboratory for
their aid in carrying out this program. Gerald L. Ayers of IBM was responsible for the
personal computer automation program.

REFERENCES

1. H. Winick and S. Doniach, eds., "Synchrotron Radiation Research," Plenum Press,
    New York, (1980).
2. W. Parrish, "X-Ray Analysis Papers," Centrex Publishing Co., Eindhoven (1965).
3. W. Parrish, M. Hart and T. C. Huang, "Synchrotron X-Ray Polycrystalline
    Diffractometry," J. Appl. Cryst., 19 : in press (1986).
4. J. B. Hastings, W. Thomlinson and D. E. Cox, "Synchrotron X-Ray Powder
    Diffraction," J. Appl. Cryst., 17:85 (1984).
5. W. Parrish and M. Hart, "Synchrotron Experimental Methods for Powder Structure
    Refinement," Trans. Am. Cryst. Assoc., 20 : in press (1985).
6. M. Mantler and W. Parrish, "Energy Dispersive X-Ray Diffractometry," Adv. in X-Ray
    Anal., 20:171 (1977).

SPECIAL APPLICATIONS OF THE DEBYE MICRODIFFRACTOMETER

P. A. Steinmeyer

Rockwell International
Rocky Flats Plant
Golden, CO 80401

INTRODUCTION

The Debye microdiffractometer, a recently developed instrument,
enables diffraction patterns to be obtained from very small samples or
from selected areas of samples. The instrument discussed here, a Rigaku
Model 2870E1 microdiffractometer, features an annular sealed proportional
counter which is moved along the primary beam axis for two theta scanning, as seen in Figure 1. This configuration allows most of a Debye
cone to be detected, making the use of a primary beam typically less than
100 microns in diameter practical. An integral microscope facilitates
sample positioning for diffraction from selected areas of a specimen.
The capability of the instrument is apparent from the diffractometer
traces of Figure 2, in which intermetallic compounds in a braze joint are
identified.

Referring to Figure 3, the specimen is mounted on the end of a
sample rod, which is inclined from the vertical by 25 to 35 degrees.
This specimen tilt allows a two theta scanning range of 30 to 150 degrees. In addition, the specimen tilt may be combined with a 360-degree
rotation about the sample rod axis to give an integrating effect for
coarse-grained samples. By employing both of these rotations to increase
the number of crystal orientations, a diffraction pattern can be obtained
from a single crystal sample.

SPECIAL METHODS AND APPLICATIONS

The instrument may be used for nondispersive XRF analysis by
operating it as a simple energy-dispersive spectrometer, and a technique
is proposed for XRD texture analysis enabling complete pole figure
coverage from a spherical sample. In addition, the application of a
simple differential filtration technique for attainment of monochromatic
conditions is described.

BALANCED FILTERS

Because Debye-Scherrer x-ray optics are used, monochromating methods
are confined to incident beam monochromators and filter techniques. The

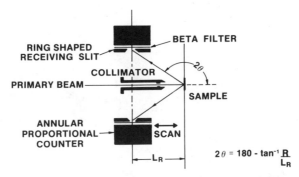

FIGURE 1.   Microdiffractometer schematic. Two theta scanning is conduc-
            ted by moving the detector along the primary beam axis.

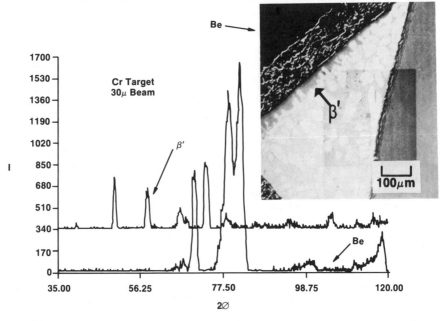

FIGURE 2.   SEM Cross-section and microdiffractometer traces of failed
            cusil braze. The brazed parts are monel (60 Ni/39 Cu/L Fe)
            on the right side of the micrograph and beryllium on the upper
            left. Fracture has occurred in the Be/Cusil interface at the
            Be-Cu intermetallic compound denoted as beta prime.

beta filter/PHA combination gives reasonably good results, but for some
applications, a simple method for improving P/B ratio is desired. A
significant background reduction and slightly increased sensitivity
result from the balanced filter technique, which involves isolation of a
characteristic line by bracketing the line between the absorption edges
of two appropriate filter materials.[1] [2] [3] The Cu K alpha lines lie
between the K absorption edges of cobalt and nickel, and foils of these
two elements may be used together to give a monochromating effect. In
use, two separate diffractometer scans are conducted, one using each
filter. At each two theta positions, the intensity of the cobalt filter
signal is subtracted from the nickel filter signal, the net signal
corresponding only to a narrow passband centered on the K alpha line.

This condition is realized only when the filters are balanced so that they exhibit identical transmission characteristics outside of the passband.

In addition to the monochromating effect, another advantage also results from the use of balanced filters. Diffracted beam intensities vary strongly with counter position in the Debye geometry, sensitivity being maximum at approximately 90 degrees two theta. By placing filters in the incident beam, the angle-dependent intensity variation is reduced and diffractometer sensitivity is improved.[4]

An example of the application of primary beam balanced filters is seen in Figure 4. A diffractometer trace taken on a uranium sample with

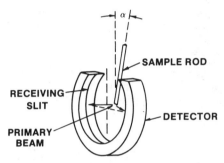

FIGURE 3. Configuration of microdiffractometer with Debye-Scherrer X-ray optics. The annular proportional counter receives a large fraction of the diffracted rays, which take the form of a Debye cone.

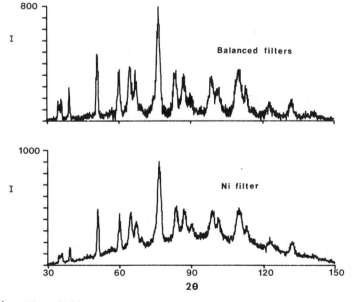

FIGURE 4. Microdiffractometer traces for a uranium specimen taken with a diffracted beam nickel beta filter (bottom) and with primary beam balanced filters (top).

balanced filters is contrasted with the corresponding trace obtained with
a standard diffracted beam beta filter, showing the increase in P/B and
improved linearity with balanced filters.

NONDISPERSIVE XRF

Although extremely useful for diffraction purposes, it has been
found that the microdiffractometer can also be used for qualitative
chemical analysis by operating it as an energy-dispersive spectrometer.
Although conventional Bragg-Bretano diffractometers have been employed
for EDS fluorescence analysis,[5] their use is not common because of
their long x-ray path and attendant low count rate. The Debye micro-
diffractometer, however, does not have this drawback. Referring to
Figure 1, when the counter is positioned for a two theta angle near
90 degrees, the secondary beam (either diffracted or fluorescent) path
length is minimum. For the Rigaku microdiffractometer, the minimum
sample-to-counter distance is 1.3 cm. Thus, even relatively soft fluorescent radiation can reach the counter without being severely attenuated. Furthermore, no diffracted beam soller slits or other collimating
devices are between the sample and the detector. Finally, the receiving
slit may be widened if necessary. All of these factors contribute to a
high count rate for secondary fluorescence emitted by the sample. Rocky
Flats has recently investigated the practicality of the microdiffractometer for EDS analysis and has shown that elements as light as aluminum
can be detected.

In a preliminary experiment, a 100-micron x-ray beam (copper target,
50 kV, 40 mA) was directed onto the surface of seven elemental standards,
and pulse height analysis was conducted with a single-channel analyzer.
Only slight changes in the diffractometer setup were necessary for the
EDS experiments. The beta filter was removed, and a relatively wide
(0.4 mm) receiving slit was used. The detector was placed so that the
two theta position was near 90 degrees, resulting in minimum fluorescent
x-ray path length. (The 90-degree value is not critical, the only
requirement being that the instrument not be set at an angle corresponding to a reflection from the sample). When scanning a weak characteristic line, if a long count time cannot be tolerated, the count rate
can be increased by widening the annular receiving slit. By using shims
or washers, the receiving slit can be made several degrees wide. This
must be done carefully, however, to avoid detection of diffracted x-rays.

The pulse height analysis results are summarized in Figure 5, which
confirms the linearity of the detector/PHA combination and demonstrates
the detector sensitivity for soft x-ray lines. Although the highest
energy line checked was that of zinc, it should be possible to excite
fluorescent radiation having energies up to that determined by the short
wavelength cutoff limit, as dictated by the tube voltage. Naturally, an
x-ray tube having a short wavelength characteristic line (such as Mo)
would be preferable for excitation of high Z elements.

The primary limitation of EDS with a proportional counter is energy
resolution; it is not possible to completely separate the characteristic
lines of elements differing by two or less in atomic number. The situation might be improved in several ways. First, the PHA data might be
computer fitted to Gaussian energy distributions. Such a procedure
would, in principle, enable the determination of individual characteristic line contributions to the net energy peak. A more direct method
would entail an actual increase in spectrometer resolution by the use of
selective filtering methods. In particular, differential filtration[2 3 4] would greatly enhance selectivity while sacrificing very
little intensity.

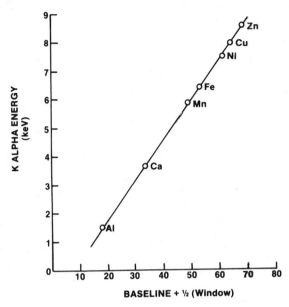

FIGURE 5.   Summary of nondispersive X-ray fluorescence results with
             proportional annular counter and single channel pulse height
             analyzer

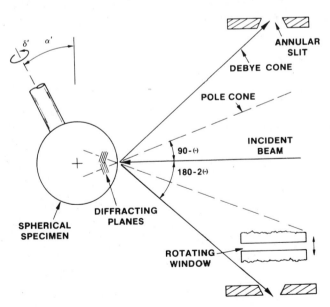

FIGURE 6.   Microdiffractometer configuration for texture measurement with
             spherical specimens

TEXTURE ANALYSIS

By inserting a rotating azimuth ring inside the annular detector, the variation in Debye ring intensity can be measured, yielding information about preferred orientation. Using a spherical texture specimen and combining the two rotations of the microdiffractometer with the rotation of the azimuth ring, a complete pole figure can be plotted from intensity data obtained in this manner.

The proposed arrangement is shown in Figure 6. Note that there exists a combination of sample orientations and rotating window positions which correspond to the same point on the pole figure. Thus for each location on the pole figure, a number of intensity readings can be taken and be summed to give an integrating effect, resulting in improved counting statistics. This feature is not available when spherical specimens are used in a conventional Bragg texture diffractometer.

SUMMARY

The microdiffractometer described here has been successfully adapted for basic EDS analysis, resulting in a combined XRD/XRF technique. A simple, inexpensive means for monochromatism and a novel procedure for texture analysis has also been suggested.

REFERENCES

1. P. A. Ross, Phys. Rev. 28:425 (1926).
2. C. H. Macgillavry and G. D. Riech, in: "International Tables for X-Ray Crystallography," Vol. III, 78, Kynock Press, Birmingham, United Kingdom (1962).
3. P. Kirkpatrick, Rev. Sci. Inst. 10:186 (1939).
4. P. A. Steinmeyer, Advances in X-Ray Analysis, 28:233 (1985).
5. B. D. Cullity, in: "Elements of X-Ray Diffraction," p 419, Addison-Wesley, Reading, Massachusetts, (1967).

CHANGES IN THE UNIT CELL DIMENSIONS OF ZSM-5 PRODUCED BY

THE ADSORPTION OF ORGANIC LIQUIDS*

S.S. Pollack, R.J. Gormley, and E.L. Wetzel

U.S. Department of Energy
Pittsburgh Energy Technology Center
P.O. Box 10940
Pittsburgh, PA 15236

INTRODUCTION

The adsorption of liquids in zeolites has been a widely studied phenomenon, and with the development of ZSM-5, considerable work has been carried out on this zeolite[1-4]. However, few articles have been published dealing with changes in the X-ray diffraction patterns when the zeolite pores contain adsorbed molecules. The purpose of this paper is to report changes in the X-ray diffraction pattern of ZSM-5 after mixing it with various organic solvents.

This work began after studies of both deactivated and partially deactivated ZSM-5 catalysts showed that their X-ray diffraction patterns changed in intensities and peak positions after use in microreactor studies[5]. If coke or hydrocarbons were causing the above changes, as postulated, would hydrocarbons adsorbed in ZSM-5 produce the same changes? Simpson and Steinfink showed that the low-angle lines of faujasite decreased in intensity after the adsorption of m-dichloro-benzene[6]. They concluded that the molecules form a liquid in the super-cages and achieved a best fit for the intensity data when the atoms con-stituting four molecules had a uniform distribution throughout a sphere of 0.58 nm inside each supercage. They didn't mention any peak shifts resulting from the adsorption of the liquid. Wu et al.[7] showed that upon calcination, as-synthesized ZSM-5 with very high SiO_2/Al_2O_3 ratios changes from apparent orthorhombic to monoclinic symmetry with a decrease in the intensity of the low-angle lines and with peak shifts at the higher angles. With the addition of ammonia, the monoclinic crystals appeared to return to orthorhombic symmetry. Wu et al.[7] men-tioned that although they worked with samples having SiO_2/Al_2O_3 ratios of 70 to 3000, the orthorhombic-monoclinic transformation is much easier with samples having the higher ratios. They showed data for one sample that had a ratio of 1600.

---

*The U.S. Government's right to retain a non-exclusive royalty-free license in and to the copyright covering of this paper, for governmental purposes, is acknowledged.

EXPERIMENTAL

Zeolite Materials

The ZSM-5 batch ZRG was prepared according to example 8 of Mobil European Patent Application 34,444 issued to W.O. Haag and R.M. Lago in 1981[8]. Silicalite batch Sil was prepared according to example 5 of Union Carbide patent U.S. 4,061,724 issued to R.M. Grose and E.M. Flanigen in 1977[9]. A zeolite with a structure similar to ZSM-5 is made by Union Carbide and called ELZ-105-5[10]. The term pentasil is used to refer to all of these high silica zeolites[11].

Adsorption

The various organic liquids (97% pure or purer) were adsorbed into the zeolites by mixing approximately 0.15 g of zeolite with approximately 0.5 mL of liquid. Originally diffraction patterns were run on the moist pastes packed into a glass holder with a cavity 10 x 20 x 0.2 mm and also after the pastes had dried to the point that they flowed like powder. The patterns of the pastes were useful for rapid determination of whether or not peak shifts had occurred and for measuring relative intensities within a sample. Since the intensities of the peaks increased upon drying, the intensities of the pastes were not useful for comparing samples. For some liquids, the excess evaporated in an hour or less, while for others, such as tetradecane, evaporation of the excess took one day or longer.

X-ray diffraction measurements. The X-ray diffraction measurements were made on a Rigaku[#] horizontal goniometer using procedures described earlier[5].

RESULTS

Before proceeding with a discussion of the adsorption of liquids into the zeolite samples, a description of the crystallinity of the samples is presented. In earlier work, a series of ZSM-5 preparations with a range of crystallinities had been studied. Samples with SiO_2/Al_2O_3 ratios of 100 and 150 had crystal sizes of 10-30 microns and showed the highest X-ray diffraction intensities. Since there was no indication of other crystalline phases or amorphous material, these samples were considered to be 100%, or close to 100%, crystalline. The sample with the higher SiO_2/Al_2O_3 ratio was used as a 100% standard in both the earlier work and this work. X-ray diffraction intensity data for the samples used in this work are listed in Table 1. Sample number 150-C is the same one that was used by Pollack et al.[5]. The only treatment it had received after synthesis was calcination at 538°C, and Na⁺ was the main exchange cation. Other samples in the table are ZRG and Sil prepared at PETC. For the ZRG samples, data are shown for the calcined forms that have mainly Na⁺ as the exchangeable cation (designated with a Na⁺) and for the NH₄⁺ and H⁺ samples. Data for only the NH₄⁺ and Na⁺ forms of Sil are also shown. Theoretically, comparisons of intensities and crystallinities should only be made for samples having the same calcination and exchange cation, but the data in Table 1 show that in most cases, the change in exchange ion did not cause a large change in intensities. Data for an uncalcined sample of ZRG are included to show the difference in intensities of the peaks between 6.4° and 10.4° for calcined and uncalcined samples, first described by Wu et

---

[#]Use of brand names facilitates understanding and does not necessarily imply endorsement by the U.S. Department of Energy.

TABLE 1.   Integrated Intensities of Samples ZRG and Sil

Integrated Intensities, Arbitrary Units
2θ, Degrees

Sample	6.4-10.4	21.4-25.2	44.1-46.0	6.4-10.4 + 21.4-25.2 + 44.1-46.0	Δ2θ
150-C - Na⁺ (Ref. 5)	87,300	93,600	7,900	188,800	
ZRG - U	39,100	102,500	6,900	148,500	0.30
ZRG - Na⁺	78,000	86,700	5,900	170,600	.33
ZRG - NH₄⁺	84,000	92,500	6,800	183,300	.36
ZRG - H⁺	87,600	86,000	5,700	179,300	.38
Sil-NH₄⁺	117,600	98,300	7,900	223,800	.52
Sil-Na⁺	101,000	83,500	6,400	190,900	.53

al.[7]. The standard, sample 150-C, has a ratio of 1.07, while the ratios for the $21.4°-25.2°/6.4°-10.4°$ regions for all the ZRG samples are one or slightly higher except for the uncalcined sample, which has a ratio of 2.6. Comparison of the intensities for each angular range indicate that the samples all have about the same crystallinity.

Table 2 contains integrated intensity and Δ2θ values showing the changes produced by the adsorption of organic liquids for selected ZRG-H⁺ samples. Mesitylene is the only liquid used that produced no significant changes when mixed with the ZRG samples (compare samples 1 and 2). Anderson et al.[2] were able to adsorb mesitylene on ZSM-5, but their samples had been outgassed in vacuum at 720-770K before the sorption measurements. Since mesitylene is about the largest molecule that can enter ZSM-5, a treatment at a higher temperature than the one used is probably required before adsorption will occur using the mixing procedure described earlier. Other liquids, such as toluene (sample 3), cause a decrease in the intensity of the $6.4°-10.4°$ range but no significant change in the Δ2θ value. Some of the halogenated aromatics, such as o- and m-dichlorobenzene, produce both intensity and Δ2θ shifts. These shifts occur for the ZRG samples in the H⁺ form (samples 4 and 5) and also in the NH₄⁺ and Na⁺ forms (not shown). With octane and tetradecane, the ZRG-H⁺ shows no significant change in Δ2θ (samples 6 and 7).

Based on the knowledge that numerous organic and some inorganic liquids adsorb in ZSM-5, we conclude that the intensity and Δ2θ changes described above are due to the presence of organic liquids in the channels of the pentasil samples. The Δ2θ measurement provides a rapid method for detecting the changes mentioned above, but determination of the unit cell dimensions is required to reveal the nature and magnitude of the changes. The lattice constants were estimated by measuring peaks as described earlier[5] and using the peaks shown in Table 3.

In that work, indexing of the powder diffraction lines had shown that the two lines between 45° and 46° were produced by a number of superimposed lines but that the line at about 45° was due mainly to the (804) and (10 0 0), while the line at about 45.5° was the (0 10 0)[5]. Therefore, a shift of these lines to higher angles means the a and b axes are getting smaller, while a shift to lower angles indicates they are increasing. This interpretation is valid as long as there is no change in crystal systems. The peak at about 24.3° never split into

TABLE 2.   Selected Intensities and $\Delta 2\theta$ Values for ZRG
with Adsorbed Liquids

Integrated Intensities, Arbitrary Units
2θ, Degrees

Sample Number	Sample	6.4-10.4	21.4-25.2	44.1-46	$\Delta 2\theta$ Degrees
1.	ZRG - H^+	87,600	86,000	5,700	0.38
2.	ZRG - H^+ + mesitylene	90,600	92,300	6,400	.38
3.	ZRG - H^+ + toluene	72,300	94,600	6,500	.36
4.	ZRG - H^+ + o-dichlorobenzene	64,400	95,500	7,200	.18
5.	ZRG - H^+ + m-dichlorobenzene	47,900	96,300	7,200	.24
6.	ZRG - H^+ + octane	54,700	106,900	7,500	.40
7.	ZRG - H^+ + tetradecane	43,200	109,500	7,400	.39

two peaks except for sample Sil, indicating that with all the liquids, the other samples maintained apparent orthorhombic symmetry.   In cases such as mesitylene, where there are only small or no changes in intensity and $\Delta 2\theta$ values, we believe the liquid does not enter the channels or enters in very small amounts.

The assignment of indices was checked by measuring accurately the peaks used for the intensity measurements in the region of 22.5° to 25.2° and the (352) peak at about 29.3° and by carrying out a refinement using the method of Main and Woolfson[12].   Table 3 contains the indices and the expected relative intensities of the reflections used in the refinement.   The intensities were calculated from the data of Olson et al.[13], which were obtained from an uncalcined ZSM-5 crystal with a Si/Al ratio of 86.   Figures 1 through 4 show some of the intensity and peak position shifts produced by calcination and the adsorption of liquids. In Figures 1 and 2 are shown peak shifts for the patterns with the sharpest lines from sample Sil.   This sample is monoclinic in the calcined (not shown) and NH_4^+ forms.   The diffraction pattern of the latter shows a (133) and (313) doublet at 24.31° and 24.53° (see Figure 1).   After adsorbing o-dichlorobenzene or tetradecane, the doublet is replaced by a single peak denoting apparent orthorhombic symmetry.   Wu et al.[7] were the first to report such a change when they treated a ZSM-5 having a high SiO_2/Al_2O_3 ratio with NH_4Cl.   Figures 3 and 4 show the peak position and intensity changes for the ZRG sample, which has broader lines.   The SEM studies of these two samples showed that the Sil sample contains crystals one to three microns large, while the ZRG particles are about 0.1 micron but do not show the rectangular forms seen in the Sil sample and may be aggregates rather than individual crystals.

Table 4 contains all the unit cell dimensions measured.   Peak shifts similar to those produced by o-dichlorobenzene have also been observed for p-dichlorobenzene and the xylenes, but the unit cell dimensions have not been measured.   Also shown in Table 4 are the C, H, and Cl contents for some of the samples after adsorption of the liquids.

TABLE 3. Relative Intensities of Selected Reflections
Calculated from Work of Olson et al.[13]

h	k	l	Intensity
5	0	1	1.0
0	5	1	0.38
1	5	1	0.27
5	1	1	0.03
3	0	3	0.51
0	3	3	0.04
3	1	3	0.07
1	3	3	0.13
3	5	2	0.14
5	3	2	0.01
8	0	4	0.10
10	0	0	0.09
0	10	0	0.16

These can only be used to estimate the approximate amounts of the
adsorbed liquids, since some of these adsorbates may not be in the pore
channels.

Peak shifts similar to those produced by the aromatic hydrocarbons
were reported earlier, but in that work, the changes in the unit cell
dimensions were not measured. At the bottom of Table 4 are shown the
unit cell dimensions for catalyst BOM when fresh and after use in a
dual-stage reactor. The used sample had shown the most deactivation of
any sample that has been studied. Aromatics in the C_5+ fraction had
been reduced from 80% initially to 50% at the end of the run. The used
sample contained 14.7% C and 1.4% H, two to four times greater than the
amount of adsorbed aromatics. After use, the a axis had decreased and
the b axis increased approximately the same amount, as observed when o-
dichlorobenzene was added to pentasils. The c axis did not change
significantly.

The changes in lattice constants produced by tetradecane (increases
in all three axes) and o-dichlorobenzene (decreases in a and c, and
increase in b) for both pentasils were the same. However, the magnitude
of the changes was larger for the Sil, and smaller for the ZRG. While
discussing these changes, the relationships between the axes and the
channels should be kept in mind[14]. The sinusoidal pore system runs
parallel to a and perpendicular to b, and the straight channels are
parallel to b and perpendicular to a.

DISCUSSION

The most significant conclusion of this study is that organic molecules adsorbed on pentasils produce small changes in their unit cell
dimensions and that the changes produced by aromatics such as o-
dichlorobenzene and chloronaphthalene are different from those produced
by paraffins.

A comparison of untreated ZRG-H[+] with that treated with paraffins
shows that the three unit-cell dimensions increase after treatment, but
the changes are larger with the tetradecane. This indicates that the
nature of the adsorbed molecule determines what changes will occur, but
the size of the paraffin chain determines the amount of change. Toluene
is the only aromatic used in this work that entered the channels but

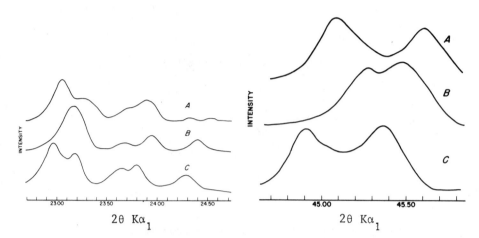

Figure 1.   Left Sample Sil.   Portions of diffraction patterns from
             22.7°-24.7°.   (A) Sil-NH$_4^+$. (B) Sil-NH$_4^+$ + o-dichlorobenzene.
             (C) Sil-NH$_4^+$ + tetradecane.
Figure 2.   Right Sample Sil.   Portions of diffraction patterns from
             44.6°-45.8°.   (A) Sil-NH$_4^+$.   (B) Sil-NH$_4^+$ + o-dichlorobenzene.
             (C) Sil-NH$_4^+$ + tetradecane.

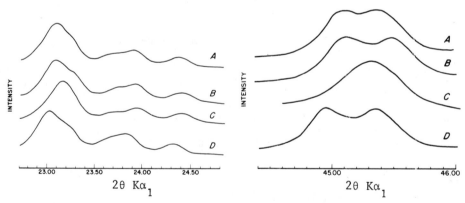

Figure 3.   Left Sample ZRG.   Portions of diffraction patterns from
             22.7°-24.8°.   (A) ZRG-uncalcined.   (B) ZRG-NH$_4^+$.   (C) ZRG-NH$_4^+$
             + o-dichlorobenzene.   (D) ZRG-NH$_4^+$ + tetradecane.
Figure 4.   Right Sample ZRG.   Portions of diffraction patterns from
             44.4°-46.0°.   (A) ZRG-uncalcined.   (B) ZRG-NH$_4^+$.   (C) ZRG-NH$_4^+$
             + o-dichlorobenzene.   (D) ZRG-NH$_4^+$ + tetradecane.

TABLE 4. Lattice Constants of Pentasils with Adsorbed Organic Liquids

Sample	a (nm) ± 0.002	b (nm) ± 0.002	c (nm) ± 0.002	α^{o}*	Weight % C**	Weight % H**	Weight % Cl**
1. Sil-NH₄⁺	2.011	1.989	1.341	90.5	0.7	0.1	
2. Sil-NH₄⁺ + o-dichlorobenzene	2.002	1.994	1.337		3.7	0.3	2.8
3. Sil-NH₄⁺ + tetradecane	2.018	1.998	1.344		9.8	1.7	
4. ZRG-uncalcined	2.008	1.996	1.339		6.4	1.8	
5. ZRG-NH₄⁺	2.002	1.993	1.339		0.3	0.6	
6. ZRG-NH₄⁺ + o-dichlorobenzene#	2.002	1.994	1.337		5.6	1.0	2.9
7. ZRG-NH₄⁺ + tetradecane	2.014	1.998	1.344		8.5	1.7	
8. ZRG-H⁺	2.009	1.993	1.338		0.4	0.4	
9. ZRG-H⁺ + chloronaphthalene	1.998	1.998	1.339		6.8	0.8	1.7
10. ZRG-H⁺ + hexane	2.011	1.994	1.341		0.6	0.8	
11. ZRG-H⁺ + octane	2.011	1.995	1.341		2.1	1.0	
12. ZRG-H⁺ + tetradecane	2.015	1.999	1.344		10.0	1.9	
13. ZRG-H⁺ + toluene	2.010	1.994	1.340		2.5	0.7	
14. ZRG-H⁺ + mesitylene	2.011	1.994	1.341		1.6	0.5	
15. BOM	2.009	1.993	1.340				
16. BOM - used	2.003	1.999	1.341		14.7	1.4	

* - For all samples except NH₄-Sil, α had a value of 90°, and β and γ were 90° for all samples.

** - Analysis performed by Huffman Laboratories, Wheatridge, Colo.

produced no lattice constant changes. With aromatics, the effect of size seems more important than the nature of the molecule, since all the aromatics larger than toluene caused decreases in $\underline{a}$ and $\underline{c}$ and increases in $\underline{b}$. The change in the $\underline{a}$ axis produced by the addition of aromatics is similar to that observed when deactivation of pentasils occurs. Deactivated pentasils or calcined pentasils with adsorbed aromatics have a shorter $\underline{a}$ axis and longer $\underline{b}$ axis than the untreated pentasils. The use of X-ray diffraction studies combined with the adsorption of aromatics on pentasils may provide a useful technique for studying pentasil deactivation. Chen and Weisz[15] have indicated that shape-selective catalysis is due to geometric factors, coulombic field interactions, and/or diffusion rates. Although it is not known why organic molecules distort the pentasil structures or why the paraffins produce changes different from the aromatics, it seems likely that the geometric factors of size and shape of the channels, and coulombic interactions all are involved in the changes produced by the organic molecules.

REFERENCES

1. Breck, D.W., Zeolite Molecular Sieves, John Wiley and Sons, New York, 1974 (p. 493).
2. Anderson, J.R., Foger, K., Mole, T., Rajadhyaksha, R.A., and Sanders, J.V., J. Catal., 1979, **58**, 114.
3. Gabelica, Z., Gilson, J.P., Debras, G., and Derouane, E., Therm. Anal., Proc. Int. Conf., 7th, 1983, Vol. 2, 1203.
4. Kulkarni, S.J., Kulkarni, S.B., Ratnasamy, P., Hattori, H., and Tanabe, K., Appl. Catal., 1983, **8**, 43.
5. Pollack, S.S., Newbury, D.E., Wetzel, E.L., and Adkins, J., Zeolites, 1984, **4**, 181.
6. Simpson, H.D., and Steinfink, H., J. Am. Chem. Soc., 1969, **91**, 23.
7. Wu, E.L., Lawton, S.L., Olsen, D.H., Rohrman, A.C., and Kokotailo, G.T., J. Phys. Chem., 1979, **83**, 2777.
8. Haag, W.O., and Lago, R.M., European Patent Applications, 34444 (1981).
9. Grose, R.M., and Flanigen, E.M., U.S. Patent 4 061 724 (1977).
10. Grose, R.W., and Flanigen, E.M., U.S. Patent No. 4 257 885 (1981).
11. Kokotailo, G.T., and Meier, W.M., Chem. Soc. Special Pub., 1980, **33**, 133.
12. Main, P., and Wolfson, M.M., Acta Crystallogr., 1963, **16**, 731.
13. Olson, D.H., Kokotailo, G.T., Lawton, S.C., and Meier, W.M., J. Phys. Chem., 1981, **85**, 2238.
14. Flanigen, E.M., Bennett, J.M., Grose, R.W., Cohen, J.P., Patton, R.L., Kirchner, R.M., and Smith, J.V., Nature, 1978, **217**, 512.
15. Chen, N.Y., and Weisz, P.B. Chem. Eng. Prog., Symp. Ser., 1967, **63**, 86.

X-RAY MEASUREMENT OF RESIDUAL STRESS NEAR FATIGUE FRACTURE SURFACES

OF HIGH STRENGTH STEEL

Yukio Hirose

Faculty of Education, Kanazawa University
1-1 Marunouchi, Kanazawa 920, Japan

and

Keisuke Tanaka

Faculty of Engineering, Kyoto University
Yoshida Honmachi, Sakyo-ku, Kyoto 606, Japan

INTRODUCTION

The residual stress left on the fracture surface is one of the important parameters in *X-ray fractography* and has been used to analyse fracture mechanisms in fracture toughness, stress corrosion cracking and fatigue tests especially of high strength steels.

In this study, the distribution of residual stress near fatigue fracture surfaces made in air and in 3.5% NaCl solution was measured by the X-ray diffraction method. The effect of aqueous environment on the plastic deformation near fatigue fracture surfaces was discussed on the basis of the residual stress distribution.

EXPERIMENTAL PROCEDURE

The material used for experiments is AISI 4340 steel, which had the following chemical compositions (wt%) : 0.39C, 0.74Mn, 1.38Ni, 0.78Cr, 0.23Mo. The heat treatment was made prior to the experiments, in which the specimens were held at 880°C for 1 hour for normalizing, then at 850°C for 1 hour austenizing and then oil quenched. The tempering processes were conducted at 200°C and 600°C. The yield strengths of 200°C and 600°C tempered steels are 1530 MPa and 951 MPa, respectively. Figure 1 shows the test specimen which was subject to four point bending. A servo-hydraulic closed loop testing machine was used for fatigue tests and the crack length was measured with a traveling microscope. The fatigue tests were conducted under constant range of stress intensity factors (constant ΔK) and stress ratios R of 0.1 and 0.5. For the constant ΔK fatigue tests, the load was decreased step-wise to maintain the ΔK value within 3 % deviation of the prescribed value. The fatigue tests were conducted both in air and 3.5% aqueous solution.

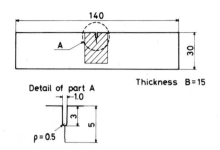

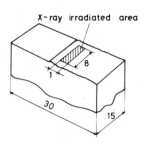

Fig. 1. Dimensions of test specimen   Fig. 2. Schematic illustration
         (in mm).   of X-ray irradiated area.

The solution was circulated by a vane pump made of synthetic resin
between a corrosive chamber and corrosion reservoir. The temperature
of corrosive liquid was kept at 16°C ± 2°C by a thermo-regulator. A
test specimen was sandwiched between two corrosion cells.

    The distribution of residual stress on the fatigue fracture surface
was measured with the X-ray diffraction method. The method adopted is
the standard $\sin^2 \psi$ method by using the pararell beam of Cr-Kα X-rays as
described in a previous paper.[1] The area irradiated by X-rays was 1 mm
width and 8 mm length at the middle of the thickness of fracture
surfaces as illustrated in Fig.2. The direction of residual stress
measured is along the crack extension. The residual stress in the
subsurface was measured by electro-polishing the surface successively.

EXPERIMENTAL RESULTS

    Figure 3 shows the relation between the crack growth rate and the
range of the stress intensity factor under R=0.1. The crack growth
rate is higher for the 200°C tempered specimen than that for the 600°C
tempered specimen both in air and in 3.5% NaCl solution. An aqueous
environment causes an enhancement of crack growth especially at low ΔK
values for both intervals. The experiment was also conducted for R=0.5.
For the 200°C tempered specimens, the effect of R was negligible while
a slight increase in crack growth rate was observed for the 600°C
tempered specimens.

    X-ray stress measurements were conducted on fracture surfaces ob-
tained for constant ΔK tests. Figure 4 shows the relation between the
residual stress measured on the fracture surface and the maximum stress
intensity factor K_{max}. The residual stress, σ_R on the fatigue fracture
was tensile. σ_R for the 200°C tempered specimen was higher than that
for the 600°C tempered specimen. σ_R increases with K_{max} for the 200°C
tempered specimen, while it has a maximum value at about K_{max}=30 MPa√m̄
for the 600°C tempered specimen. σ_R for R=0.5 was larger than that
for R=0.1, but the difference is small. The σ_R value on corrosion-
fatigue surfaces is clearly smaller than that on air-fatigue surfaces.

    The distribution of the residual stress beneath the fatigue fracture
surface for K_{max}=22 MPa√m and 44 MPa√m. Figures 5 and 6 show the
results. σ_R for R = 0.5 is slightly larger than that for R = 0.1.
There is no effect on the size of the region with residual stress. The

tensile residual stress diminished with depth for the 600°C tempered specimen, while it decreased at first and then increased and shows compressive residual stress for the 200°C tempered specimen.   The plastic zone size determined from the distribution of residual stress for the 600°C tempered specimen was larger than that for the 200°C tempered specimen when compared at the same ΔK.   The effect of the plastic zone size on environment was observed and will be discussed next.

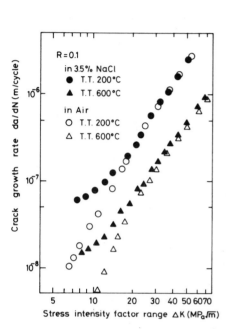

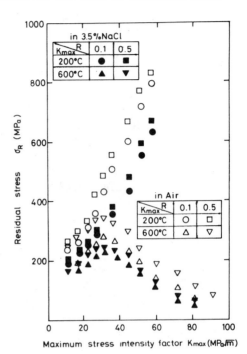

Fig. 3.   Relation between crack growth rate and stress intensity factor range.

Fig. 4.   Relation between residual stress and maximum stress intensity factor.

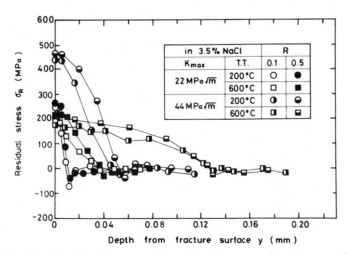

Fig. 5.   Residual stress distribution near fracture surface. (in 3.5% NaCl solusion)

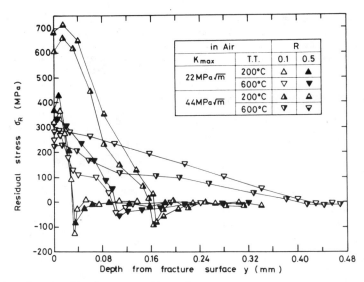

Fig. 6.   Residual stress distribution near fracture surfaces
          (in air)

DISCUSSION

     The measured distribution of residual stress near fatigue fracture
surfaces can be obtained as a superposition of two types of residual
stress distributions. In Fig.7, the distribution of Type I is caused by
the plastic deformation near the crack tip which is extension in the
loading direction and contraction in the perpendicular direction.   The
contraction yields tensile residual stress on the fracture surface owing
to the spring back by the surrounding elastic part.   The distribution
of Type II is, opposite to that of Type I, compression near the surface.
This corresponds to the relieved component of Type I residual stress
due to the roughness of fracture surfaces.   The compressive plastic
deformation and crack closure near the tip of fatigue cracks will also
give the Type II distribution of residual stress.

     The aqueous environment will modify the aforementioned distribution
in two ways.   The roughness of corrosion-fatigue surfaces may be less
than that of air-fatigue surfaces because of the more brittle nature of the
fracture process.   The drop of the tensile residual stress in the close
vicinity of fracture surfaces is small for the case of corrosion-fatigue.
The magnitude of the tensile residual stress for the case of corrosion-
fatigue is lower than that for the air-fatigue fracture surface when
compared at the same values of ΔK and R.   This may be related to the
hydrogen occlusion near the corrosion-fatigue fracture surface, which
gives rise to the reduction of tensile stress.

     The depth of the residual stress zone from the fracture surface
means the extent of the plastic zone.   This size ω_y was measured and
plotted against the maximum stress intensity factor divided by the
yield strength in Fig.8.

For each environment, the relation is approximated by the second power
equation:

$$\omega_y = \alpha \ (\ K_{max} \ / \ \sigma_y \ )^2 \tag{1}$$

where α is 0.19 for the case of air-fatigue and 0.06 for the case of corrosion-fatigue.   There is no influence of tempering temperature on this relation.

In Table I, the values of α obtained in a similar way for the cases of fracture toughness test,[2] stress corrosion cracking test[3] and fatigue test[4] are summarized.   The α value for the stress corrosion cracking test is smaller than that for fracture toughness test.   Levy et al[5] derived $\alpha = 0.15$ on the basis of the elastic-plastic finite element method for perfectly plastic material.   The α value different from 0.15 is now assumed to be caused by the difference of the yield strength in the plastic zone from that in simple tension test.   The yield strength in the plastic zone $\sigma_{\tilde{y}}$ is evaluated from the following equation:

$$\omega_y = 0.15 \left(K_{max} / \sigma_{\tilde{y}} \right)^2 = \alpha \left(K_{max} / \sigma_y \right)^2 \tag{2}$$

or

$$\sigma_{\tilde{y}} = \left(0.15 / \alpha \right)^{\frac{1}{2}} \cdot \sigma_y \tag{3}$$

The calculated values of α are given in Table I.   The cyclic softening is seen in the case of air-fatigue.   It is very interesting to note that the yield strength is increased both in stress corrosion cracking and in corrosion-fatigue.   The mechanism of hardening in aqueous solution will be studied in the future.

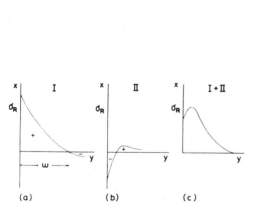

(a)   (b)   (c)

Fig. 7.   Schematic illustration
of residual stress
distribution on
fatigue fracture
surface.

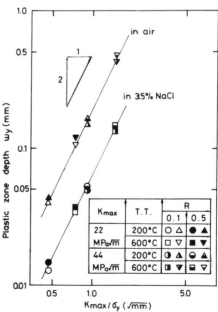

Fig. 8.   Relation between
plastic zone depth
and stress intensity
factor divided by
yield strength.

Table I　Values of α and $\sigma_{y}^{\,\prime}$.

		Tensile test	Fracture toughness fracture surface	Stress corrosion cracking fracture surface	Air-fatigue fracture surface	Corrosion-fatigue fracture surface
α		——	0.14	0.08	0.19	0.06
Yield strength $\sigma_{y}^{\,\prime}$ (MPa)	T.T. 200°C	1530	1584	2095	1359	2419
	T.T. 600°C	951	984	1302	845	1503

CONCLUSIONS

The main results obtained in the present study are summarized as follows:

(1)　The residual stress on fatigue fracture surfaces in air and 3.5% NaCl solution were tensile. The residual stress increased with K_{max} for the 200°C tempered specimen, while it had a maximum value at about K_{max} = 30 MPa√m for the 600°C tempered specimen both in the cases of air-fatigue and corrosion-fatigue. The aqueous environment reduced the tensile residual stress.

(2)　The distribution of residual stress beneath the fatigue fracture surface in air and 3.5% NaCl solution can be decomposed into two parts: 1) the distribution caused by monotonic plastic deformation near the crack tip, 2) that relieved by surface roughness or that resulting from the compressive deformation due to crack closure.

(3)　The maximum depth of plastic zone ω_y determined from residual stress was correlated to K_{max} and the yield strength σ_y as follows

$$\omega_y = \alpha \left(K_{max} / \sigma_y \right)^2$$

where α is 0.19 for the case of air-fatigue and 0.06 for the case of corrosion-fatigue. There is no influence of tempering temperature on this relation.

REFERENCES

1.　Y. Hirose, Z. Yajima, and T. Mura,　"X-Ray Fractography on Fatigue Fracture Surfaces of AISI 4340 Steel", To be published in Advances in X-Ray Analysis,　29:　(1986).
2.　Y. Hirose, Z. Yajima, and K. Tanaka,　"X-Ray Fractographic Approach to Fracture Toughness of AISI 4340 Steel", Advances in X-Ray Analysis,　28:289 (1985).
3.　Y. Hirose, Z. Yajima, and K. Tanaka,　"X-Ray Fractography on Stress Corrosion Cracking of High Strength Steel", Advances in X-Ray Analysis,　27:213 (1984).
4.　K. Tanaka, and N. Hatanaka,　"Residual Stress near Fatigue Fracture Surfaces of High Strength and Mild Steels Measured by X-Ray Method", J. Soci. Mat. Sci. Jap.,　31:215 (1982).
5.　N. Levy, P. V. Marcal, W. J. Ostergren, and J. R. Rice,　"Small Scale Yielding near a Crack in Plane Strain: A Finite Analysis", Int. J. Frac. Mech.,　7:143 (1971).

SINGLE-LINE PROFILE ANALYSIS OF SUPERPLASTICALLY DEFORMED ALUMINIUM ALLOYS

A. W. Bowen

Materials and Structures Department
Royal Aircraft Establishment
Farnborough, Hants GU14 6TD, UK

INTRODUCTION

Superplastic deformation involves the controlled processing of materials in order to produce extensive deformation, the longest recorded elongation being nearly 5000%[1]. Components of very complex shape can be formed in this way[2], thus offering substantial savings over the use of conventional manufacturing routes. Typically, these savings can be 50% in cost (through reduced man-hours required for component manufacture, from improved structural efficiency and from reduced material costs) and 25% in weight (through more efficient use of materials and the reduced requirement for joints and fasteners).

The success of the superplastic process will depend, in large part, on achieving an adequate balance between the deformation input and the relief of work hardening by thermally activated relaxation processes. An attempt has therefore been made to establish whether the residual strain in superplastically deformed materials is related to deformation parameters, and if so, to interpret this strain data in terms of the mechanisms of superplasticity.

The technique of line profile analysis has been used to extract information on lattice strain from the shape of X-ray diffraction lines. In the work reported here only the analyses for a single high-angle diffraction line (the (422) reflection) using the rapid method proposed by Langford[3] are presented. Although this introduces some degree of inaccuracy[3,4] into the actual strain values, general trends will nevertheless be true[3,4]. Furthermore, it was considered more important at this stage to establish the feasibility of the approach, and only then to follow it up with a more detailed multi-line analysis. The work forms part of a longer-term study of structural materials by line profile analysis.

BACKGROUND TO THE ANALYTICAL METHOD

An X-ray diffraction line profile can be defined in terms of three parameters[3]: the peak height I_{max}; the peak area A; and the full width (2w) at $I_{max}/2$. The first two parameters can be combined to give the integral breadth (β) which is the width of a rectangle with the same height as the peak ie $\beta = A/I_{max}$. The broadening of diffraction lines can arise from a finite crystallite size (which will not be considered further at the present time) and strain in the lattice due, eg, to dislocations and stacking faults. The contribution of the former can be expressed by a Cauchy function whilst the latter more nearly fits a Gaussian function. The rapid method due to Langford[3] involves the fitting of a Voigt function (which is a convolution of a Cauchy and a Gaussian function) to experimental profiles. The shape of a Voigt function (although not its absolute intensity) can be expressed by just two parameters: 2w and β. Thus from the 2w and β for the g (standard) and the h (measured) profiles it is possible to extract the 2w and the β values for the f (specimen) profile, since h = g*f. To calculate the lattice strain, use has been made of the following empirical equations[4]:

$$\frac{\beta_G}{\beta} = 0.6420 + 1.4187 \left(\left(\frac{2w}{\beta} \right) - \frac{2}{\pi} \right)^{\frac{1}{2}} - 2.2043 \left(\frac{2w}{\beta} \right) + 1.8706 \left(\frac{2w}{\beta} \right)^2 \quad (1)$$

where β_G is the Gaussian component of the Voigt function and:

$$\left[\beta_G (f) \right]^2 = \left[\beta_G (h) \right]^2 - \left[\beta_G (g) \right]^2 \quad (2)$$

Mean relative strain values can then be calculated using the equation[4]:

$$\text{Strain (E)} = \frac{\beta_G (f)}{4 \tan \theta} \quad (3)$$

BACKGROUND TO SUPERPLASTIC DEFORMATION MECHANISMS

Whilst the many models[5] describing superplastic deformation mechanisms differ in degree, they are all in general agreement regarding the predominant mechanism, namely grain boundary sliding where grains maintain an approximately equiaxed shape and permit extension by sliding and rotating past one another during deformation. There is also general agreement that there must be an additional contribution from material transport, either by diffusion or dislocation motion, in the immediate vicinity of the grain boundary in order to maintain material continuity. A consequence of this grain boundary sliding is that in some materials, including aluminium alloys, cavities can develop in the boundaries[6]. These are undesirable because they can grow in size, coalesce and lead to premature failure. They also have an adverse effect on subsequent properties. Since a local tensile stress is necessary for cavitation, any compressive stress is beneficial. A successful way of combating cavitation is therefore to superimpose a hydrostatic pressure during deformation[6].

MATERIALS AND EXPERIMENTAL PROCEDURE

Three aluminium alloys have been studied[7,8] (Table 1), although most of the work reported here will be for the Lital A alloy. The testing conditions used are given in Table 2.

After stressing specimens were cut from the gauge lengths of uniaxial test pieces and from the bottoms of biaxially formed boxes (Fig 1). They were polished to mid-section with a 0.25 μm finish. All specimens were analysed in an automated Philips diffractometer operating in step-scan mode (0.02° 2θ steps with counting times of 60s per step) using Cu K$_\alpha$ radiation. A specimen spinner was employed. An annealed silver powder sample was used as a standard in order to measure the instrumental contribution to the line broadening ie to measure the g profile. The results of texture analyses on these specimens are reported elsewhere[7,8].

Fig. 1

Typical superplastically deformed box for evaluation.

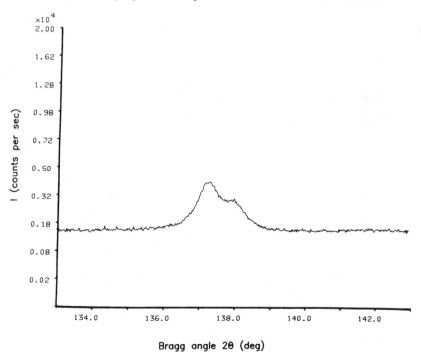

Fig. 2a

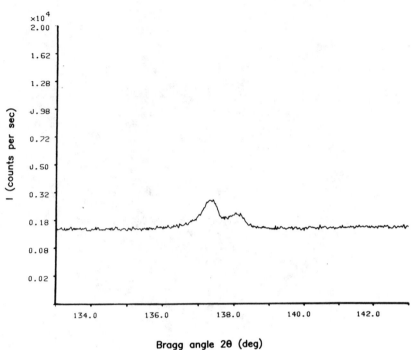

Fig. 2b

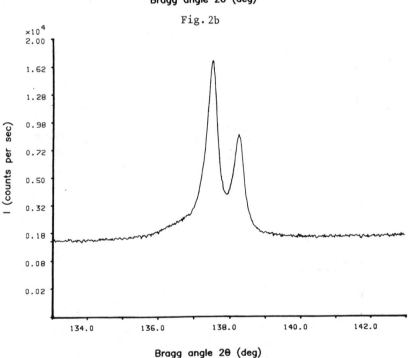

Fig. 2c

Diffractometer traces for the (422) reflection of Lital A (a) as-received
(b) thermally-cycled and (c) superplastically deformed to a strain of
1.63.

Table 1.   Chemical Compositions

Alloy	Typical compositions (wt %)									
	Li	Cu	Mg	Zr	Zn	Fe	Si	Cr	Ge	Al
Lital A*	2.45	1.3	0.75	0.12	–	0.15	0.04	–	–	rem
Supral 220*	–	5.9	0.35	0.4	0.07	0.18	0.12	–	0.1	rem
7475E**	–	1.55	2.25	–	5.7	0.12	0.1	0.21	–	rem

* As-rolled sheet.
** supplied in thermo-mechanically processed condition.

(All nominally 2 mm thick sheet except for Supral 220 which had been chemically milled from 3 mm thickness.)

Table 2.   Testing Conditions

Alloy	Strain rate (s^{-1})	T°C	Imposed hydrostatic pressure
Lital A	5×10^{-4}	530	Yes
Supral 220	10^{-3}	480	Yes
7475E	5×10^{-4}	516	No

The following correction procedures were applied to the the line profiles of the measured patterns:

1.   Background removal by a linear interpolation between the low and the high angle ends of the profile.
2. α_2 stripping assuming an α_1/α_2 ratio of 0.5 and using the method of Delhez and Mittemeijer[9].
3.   Correction for the high angle dependence of the Lorentz-Polarization factor[10].
4.   Determination of the peak position and I_{max}.
5.   Determination of $2w$ at $I_{max}/2$.
6.   Determination of β_G (f) using equations (1) and (2).
7.   Calculation of mean relative strain values using Equation (3).

RESULTS

The effects of thermal cycling and superplastic strain on peak shape are shown in the examples of diffractometer traces in Fig. 2. The as-received sheet showed a poorly resolved $\alpha_1\alpha_2$ doublet (Fig. 2a), which

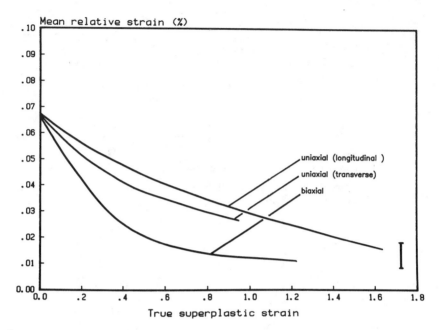

Fig. 3   Mean relative strain values for superplastically deformed
          Lital A under uni- and bi-axial conditions. (Values for the
          (422) reflection.) (I - error bar.)

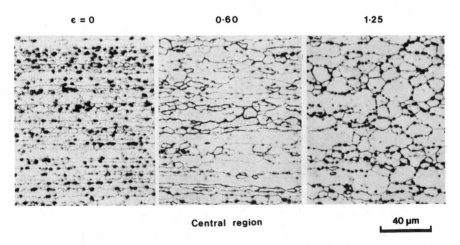

Fig. 4   Breakdown in aligned microstructure and increase in grain size
          with increasing superplastic strain in Lital A.

changed little on thermal cycling (Fig. 2b). On the other hand, the
conjoint action of stress and temperature had a considerable effect on $\alpha_1\alpha_2$
resolution (Fig. 2c).

The mean relative strain for the Lital A alloy in the as-received
condition was 0.075%, which reduced to 0.069% after thermal cycling.
Increasing superplastic strain in longitudinal uniaxial tests progressively
reduced these values of internal strain (Fig. 3). Somewhat lower values
were obtained for transverse uniaxial tests (Fig. 3).

In biaxially deformed material the reduction in mean relative strain
was greater than in the uniaxial tests (Fig. 3).

Less extensive data are available on the other two alloys. Nevertheless, the results are not in disagreement with the trends for Lital A. For
7475E after a uniaxial strain of 0.8 the internal strain was 0.034%, and
after a biaxial strain of 0.9 the internal strain was 0.035%. For
Supral 220 a uniaxial strain of 1.05 produced an internal strain of 0.041%,
and a biaxial strain of 1.27 produced an internal strain of 0.032%.
Interestingly, the results for the 7475E alloy (deformed at nearly the same
temperature and strain rate but with no imposed hydrostatic pressure) are
almost identical to those for Lital A but those for Supral 220 (deformed at
a lower temperature and a faster strain rate) are distinctly higher. This
may be related to a more stable grain size in this alloy[7].

Changes in microstructure and texture[7,8,11,12] with deformation
parameters can be summarized as follows:

- there is a reduction in dislocation density due to the removal of low
  angle boundaries as a result of thermal cycling
- the original aligned microstructure is progressively removed
  (Fig. 4)
- the grain size increases (Fig. 4)
- cavitation is prevented by increasing hydrostatic pressure
  (cf Figs. 5a and 5b)
- there is a gradual loss in texture intensity under uniaxial conditions
  but an initial increase, followed by a decrease, under biaxial
  conditions. This decrease did not fall below the intensity of the
  starting condition
- there is a much greater degree of rotation of poles under uniaxial
  compared to biaxial conditions.

DISCUSSION

Values of internal strain in three superplastically deformed aluminium
alloys, calculated by a rapid single-line method[3], vary in a systematic way
with deformation parameters, particularly for the Lital A alloy. It is
possible to correlate these changes in internal strain with the observed
microstructural and textural changes, in the following way.

The reduction in internal strain as a result of thermal cycling can be
attributed to the decrease in overall dislocation density as a result of the
removal of the low-angle boundaries.

The progressive reduction in internal strain with increasing superplastic strain, in both uni- and bi-axial tests, can be attributed to a
reduction in grain boundary dislocations as a consequence of the decrease
in total grain boundary areas as the grain size increases.

Fig. 5a   50 μm

Fig. 5b

Effect of superimposed hydrostatic pressure on the extent of cavitation
in Lital A (a) 2.06 MPa (b) 3.45 MPa.

Although slip must occur to cause the observed texture changes[7,8,11], the dislocations so generated do not remain in the grains, since transmission electron microscopy investigations[13-17] have shown that after superplastic deformation, grains are essentially dislocation-free.

The effect of superimposed hydrostatic pressure is to prevent cavitation and its associated localised plastic flow (cf Figs. 5a and 5b). The higher internal strain after uniaxial tests (Fig. 3) can therefore be interpreted in terms of a higher residual dislocation configuration assocated with this localised flow. This flow must resemble that in a hot tensile test ie it must occur at an effectively faster local strain rate. Under these non-superplastic conditions greater dislocation activity occurs[13-17] with the result that these may be retained in cell form after deformation.

Analyses of higher accuracy would now seem to be justified, particularly to study both lattice strain and crystallite size. Multi-line analyses are therefore underway: to provide a more detailed correlation between deformation parameters and changes in internal strain, crystallite size, and alloy microstructure.

CONCLUSIONS

Rapid single-line analysis of peak broadening in superplastically deformed aluminium alloys has shown that there is:

1. A progressive reduction in internal strain during superplastic deformation.
2. These reductions in internal strain seem to be due to the removal of dislocations, in both low- and high-angle boundaries.
3. Localised plastic flow as a result of cavitation results in a measurable increment in residual internal strain. This increment is attributed to dislocations retained as a result of slip in these local regions.

ACKNOWLEDGEMENTS

The author would like to thank D. S. McDarmaid and A. Shakesheff for providing Figures 1, 4 and 5. The paper is published by permission of the Controller, HMSO, London 1985.

REFERENCES

1. M. M. Ahmed and T. G. Langdon, Exceptional ductility in the super-plastic Pb-62Sn eutectic, Met. Trans. A 8A:1832 (1977).
2. See several papers in Conference Proceedings on Superplastic Forming of Structural Alloys, N. E. Paton and C. H. Hamilton, Eds, AIME, NY (1982).
3. J. I. Langford, A rapid method for analysing the breadths of diffraction and spectral lines using the Voigt function, J. App. Cryst. 11:10 (1978).
4. Th. H. de Keijser, J. I. Langford, E. J. Mittemeijer and A. B. P. Vogels, Use of the Voigt function in a single-line method for the analysis of X-ray diffraction line broadening, ibid, 15:308 (1982).
5. B. P. Kashyap and A. K. Mukherjee, On the models for superplastic deformation, in Int. Conf. on Superplasticity p.4-1, B. Baudelet and M. Suery, Eds, CNRS, Paris (1985).
6. N. Ridley and J. Pilling, Cavitation in Superplastic alloys - experimental, ibid p.8-1.

7. P. G. Partridge, A. W. Bowen and D. S. McDarmaid, The relationship
    between texture, r-values and superplasticity in aluminium alloy sheet,
    in Int. Conf. on Superplasticity in aerospace-aluminium, R. Pearce,
    Ed., Cranfield Inst. Tech., UK (in the press).
8. P. G. Partridge, A. W. Bowen, C. D. Inglebrecht and D. S. McDarmaid, in
    Reference 5 p.10-1.
9. R. Delhez and E. J. Mittemeijer, An improved α_2 elimination, J. App.
    Cryst. 8:609 (1975).
10. R. Delhez, E. J. Mittemeijer, Th. H. de Keijser and H. C. F. Rozendaal,
    Corrections for the angle dependence of Lorentz, polarization and
    structure factors in X-ray diffraction line profiles, J. Phys. E: Sci.
    Inst. 10:784 (1977).
11. A. W. Bowen et al, Residual internal strain in superplastically
    deformed aluminium and titanium alloys, to be published.
12. A. J. Shakesheff and D. S. McDarmaid, Effect of processing parameters on
    the superplastic behaviour of an Al-Li-Cu-Mg-Zr alloy, in Reference 7.
13. H. W. Hayden, R. C. Gibson, H. F. Merrick and J. H. Brophy, Super-
    plasticity in the Ni-Fe-Cr system, Trans ASM 60:3 (1967).
14. R. Z. Valiev, O. A. Kaibyshev, M. M. Myshlayev and D. R. Chaleev, The
    observation of intragranular slip during in-situ deformation in the HVEM,
    Scripta Met. 14:673 (1980).
15. S-A. Shei and T. G. Langdon, A microstructural examination of the flow
    behaviour of a superplastic copper alloy, J. Mat. Sci. 16:2988 (1981).
16. L. C. A. Samuelsson, K. N. Melton and J. W. Edington, Dislocation
    structures in a superplastic Zn-40 wt % Al alloy, Acta. Met. 24:1017
    (1976).
17. K. Matsuki, Y. Uetani, M. Yamada and Y. Murakami, Superplasticity in an
    Al-6 wt % Mg alloy, Met. Sci. 10:235 (1976).

COMPUTER CAPABILITY FOR THE DETERMINATION OF

POLYMER CRYSTALLINITY BY X-RAY DIFFRACTION

Andrew M. Wims, Mark E. Myers Jr.,
Jack L. Johnson, and Julia M. Carter*

Analytical Chemistry Department
General Motors Research Laboratories
Warren, MI

INTRODUCTION

Polymer Crystallinity

The physical and mechanical properties of many industrially important polymers are profoundly influenced by their degree of crystallinity[1]; such properties include flex modulus, tensile strength, percent elongation, and impact strength. Commonly used polymers influenced by their crystallinity level include polyethlene, polypropylene, polyesters, and nylons. Many of these materials are above their glass transition temperature at room temperature and would be useless were it not for their crystalline phase which typically has a melting point far above room temperature. The crystalline regions (domains) in these materials are frequently very small, typically in the nanometer range in diameter. These crystalline domains act as reinforcing fillers (in somewhat the same manner as carbon black in rubber) and give strength to the polymer.

Crystallinity is also important because too much leads to brittle fracture and failure of polymers, measured as low impact strength and increased susceptibility to stress cracking. Furthermore, the percent crystallinity of a polymer is strongly influenced by its thermal history from pellet extrusion through molding into a useful product, and through subsequent heat treatment and environmental aging. Certainly, the measurement and correlation of polymer crystallinity with mechanical properties is very important. This information is also useful in the analysis of field failed manufactured components.

Polymer Crystallinity Methods

Several methods have been successfully used to determine polymer crystallinity[2]: x-ray diffraction (XRD), density, differential scanning calorimetry (DSC), nuclear magnetic resonance (NMR), and infra-red (IR) spectroscopy. The major requirement or assumption of each of these methods is shown in Table 1.

* Churchill College, Cambridge CB3-ODS, England

Table 1. Principal Methods of Crystallinity Measurement

Method	Requirement
XRD	Resolve the diffractogram into contributions from the crystalline and amorphous phases.
Density	Theoretical or experimentally determined density of a single crystal of the polymer and the use of standards.
DSC	Heat of fusion of a polymer standard of known percent crystallinity.
NMR	Resolve crystalline and amorphous phases spectroscopically.
IR	Distinct differences in the spectra of amorphous and crystalline phases.

X-RAY DIFFRACTION (XRD) METHOD

General Principles

The atoms in amorphous materials such as glasses, resins, and liquids do not have the symmetrical repeating arrangement characteristic of crystalline materials. Thus, the x-ray patterns of amorphous materials at low diffraction angles appear as one or two weak, broad bands superimposed upon a continuous background. However, a highly crystalline substance has very explicit peaks at specific diffraction angles. The terms crystalline and amorphous are important to define because in a polymer the two states can be intertwined in many ways yielding states described as quasi-crystalline, para-crystalline, semi-ordered, or even randomly ordered. Thus, polymers consist of various regions, each with a characteristic degree of internal order, ranging continuously from something close to the ideal crystalline state to the completely amorphous state[3].

Crystallinity is defined as the weight fraction of the crystalline portion of the polymer. The higher the degree of ordering, the more crystalline is the polymer. For real systems, virtually no amorphous substance will be completely without order. There will always be some succession of chemical bonds with fixed lengths and angles that will give rise to a particular radial density distribution function $P(r)$ that we view as a diffractogram. As structural features coalesce in a more ordered fashion, the x-ray diffractogram moves away from a diffuse halo to explicit peaks whereas the total integrated area (halo plus explicit peaks) remains constant. This fact can be used to calculate a percent crystallinity by defining the crystallinity to be the area of these explicit peaks and the amorphous content to be the area of the diffuse halo.

Advantages of XRD

The XRD method has several advantages over the other methods in that it provides a rather rapid, convenient, and accurate method which can be implemented without standards for determining percent polymer crystallinity. Modern XRD systems have a major advantage in that the instrumentation can be automated and computer methods can be used for data analysis providing flexibility to use special data evaluation schemes when required. Further, with modern systems, XRD is particularly attractive because we do not need to use time-consuming film techniques which require a densitometer. Modern x-ray detectors, monochromators, and pulse-counting electronics render the effective x-ray radiation monochromatic, thereby avoiding the necessity for applying other corrections to the data.

An explicit assumption in the XRD method is that the diffractogram represents the sum of the amorphous and crystalline regions which can be graphically or mathematically separated[2][4]. Fortunately, the resolution of the two regions is readily accomplished for many polymers[2][5]. Examples do exist, notably nylon-6, where the separation is difficult because of the presence of para-crystalline regions[6][7].

A number of XRD techniques[2][5][7] have been used to determine polymer crystallinity, some specific for certain polymers. Some of these techniques admittedly measure only relative crystallinity between samples and do not purport to measure absolute polymer crystallinity[2]. The method we used is an expansion of an earlier method by Farrow[8] and Farrow and Preston[9]. The basic feature of this method is that a diffractogram of an amorphous polymer is scaled until it overlaps the diffractogram of a partially crystalline sample of the same polymer at diffraction angles (2-theta values) outside the region of the crystalline peaks.

Hindeleh[7] pointed out in his method that one should not overlook the effect of overlap from adjacent broad peaks and the difficulty of separating long tails from the amorphous background. When these effects are present, the percent crystallinity determined by the Farrow and Preston method will be lower than the true value. Because we are primarily interested in following changes in percent crystallinity with thermal and mechanical treatment of industrial polymers and copolymers which may contain fillers, we have not added the complexity of separating overlapped peaks.

Thus, our method lies somewhere between "relative" methods (which almost always give very low values for percent crystallinity) and "absolute" methods which often come very close to the true value. The most primitive method of processing x-ray diffraction data utilizing the two phase assumption is to simply draw in the amorphous curve by hand and graphically subtract it from the total diffractogram. This procedure always produces low values for polymer crystallinity[2].

In our application of the Farrow and Preston x-ray method, we first used a manual process whereby a polymer diffractogram was first recorded on a strip chart recorder and integrated with a planimeter followed by an appropriate scaling of the amorphous diffractogram to the sample diffractogram. This paper presents the design and advantages of a computer method (SHOPOL) that we developed to determine the percent crystallinity of polymeric materials.

XRD Calculations

The calculation of percent crystallinity first requires a numerical integration of the sample and amorphous XRD data over a specified two-theta range. In addition, a scaling factor must be determined using intensity data at a two-theta value outside of the crystalline peak region. The resulting mathematical formula for percent crystallinity takes the form

$$\% \, C = \frac{\int I_S \, d\theta - \left[\int I_A \, d\theta \right] \left[K \right]}{\int I_S \, d\theta} \times 100 \qquad (1)$$

Where I(S) and I(A) are the sample and amorphous intensities. The scaling factor K is the ratio of the sample and amorphous intensities at a particular two-theta value outside of the crystalline peak region. All of the intensities are baseline corrected, and the integrals are taken over the two-theta range which encompasses the crystalline peaks and the amorphous halo. For most polymeric materials, a range of 5 to 40 degrees two-theta is optimum. The biggest advantage of Eq.1 is that the percent crystallinity is independent of the absolute areas of the sample and amorphous diffraction curves. Thus, one does not have to be concerned about measurement and normalization of the incident x-ray intensity I, sample thickness, size, shape, or keeping track of the efficiency of the counting-system pulse-height analyzer and other electronics.

EXPERIMENTAL

X-ray Instrumentation

An automated, Siemens D-500 x-ray diffractometer system equipped with a carbon monochromator was used for the polymer crystallinity analyses. The x-ray generator was operated at power settings of 40 KV and 30 mA with a copper target x-ray tube, and the diffractometer was controlled by a Digital PDP 11/03 computer from a Tektronix 4010-1 graphics terminal. The diffractometer collimators used were: three 1 degree beam apertures, one 0.15 degree detector aperture, and one 0.15 degree diffracted beam aperture. Samples were scanned on the automated diffractometer from 5 to 40 degrees two-theta, with counts taken at intervals of 0.05 degrees for two seconds. The diffractograms can be displayed on a Tektronix terminal and preserved using a Tektronix hard copier unit.

Polymer Samples

Three types of polymer systems were studied: nylon 6,12 copolymer, polyethylene terephthalate (PET) blends, and nylon-6 homopolymers. Some of the semi-crystalline samples had been annealed at various temperatures and lengths of time to produce maximum crystallinity. The amorphous reference samples had been quenched in liquid nitrogen after heating to melting temperature under pressure to keep crystallinity to a minimum.

COMPUTER METHOD

Implementation of the SHOPOL Program

The SHOPOL program described in this paper was written using parts of the Siemens software plotting program SHOTEK which uses the Tektronix terminal's graphing capability to display raw data files. Like SHOTEK, SHOPOL is structured to receive raw data over any scan range for graphing. However, the SHOTEK subroutines that overlay a standard XRD pattern on the raw data pattern have been replaced by several subroutines that deal with polymer crystallinity. In SHOPOL, consistency with SHOTEK in file structure and floppy disk input-output has been maintained; however, temporary file data storage has been eliminated to reduce space demands on the system disk. The SHOPOL source program is written in FORTRAN IV (version 2) for the RT11/03 RSTS/E V06B operating system.

The new SHOPOL subroutines and their functions are as follows:

1) INPOLY: Inputs the amorphous and sample raw data files.

2) POLCRY: Does baseline and amorphous data correction; calculates the scaling factor (K-value) and percent crystallinity; and generates data for corrected and raw data graphs.

3) PKSUB: Subtracts extraneous peaks (e.g., inorganic filler crystalline peaks) from raw data and draws a straight line through the subtracted area's endpoints.

Use of the SHOPOL Program

The SHOPOL program has a conversational dialog that prompts the user for necessary information. In addition, several pre-selected default values are assumed for convenience unless overridden by the analyst. After issuing the "RUN SHOPOL" command, the user is asked whether a long or short dialog is preferred. The long dialog gives detailed explanations of several questions and advises the user about input formats. However, the short dialog is sufficient for most purposes and will also reduce execution time. Next, the user is given a chance to change the radiation wavelength from the default copper K-alpha value. Then a choice is made between a raw data plot and a polymer crystallinity calculation. If the first is chosen, SHOPOL transfers control to the INPARM subroutine (from Siemens) which will input graphing parameters and data for any raw data file. If polymer crystallinity is chosen, the INPOLY subroutine is entered.

INPOLY has a similar function to INPARM except that it is modified for polymer crystallinity calculations. It first prompts for the amorphous and the first sample raw data file names, and the number of samples. Up to fifteen samples can be run at one time, and the format for data file specifications can be: DEV:MEASXX.DAT, MEASXX.DAT, MEASXX, or simply XX, where XX refers to the number of the raw data file. Next, the user is given the chance to alter the default two-theta range for graphing (which is originally set to the sample range) and to input the number of degrees per screen (no default value). A printout for verification of the plot parameters follows, and the amorphous and first sample raw data files are read in from the floppy disk. Program execution then returns to SHOPOL which transfers control to the POLCRY subroutine.

The POLCRY subroutine is responsible for the polymer graph data generation and percent crystallinity calculations. It is organized around a menu consisting of ten options. Three of these are necessary for each percent crystallinity calculation and are indicated by an asterisk "*". The options and their explanations are as follows:

1) VIEW RAW DATA: Graphs out either the amorphous or sample raw data, compressing it to fit the chosen graph configuration.

2) * BASELINE CORRECTION: Prints out the default baseline points (6 and 36 degrees two-theta) and permits the user to change them if necessary; the linear baseline is then calculated using a standard line-generation procedure, and the equations are printed for inspection.

3) VIEW BASELINE CORRECTED DATA: Generates baseline-subtracted amorphous or sample data.

4) * K-VALUE CALCULATION: Prompts the user to change the default K-value point set at 14 degrees two-theta.

5) VIEW-AMORPHOUS CORRECTED DATA: Generates a graph of the K-value scaled, baseline-corrected amorphous halo subtracted data from the baseline-corrected sample data.

6) * PERCENT CRYSTALLINITY CALCULATION: Integrates the baseline-corrected amorphous and sample peaks between the baseline points and then calculates the percent crystallinity using Equation 1.

7) NEW RAW DATA FILE: Gets the next consecutive sample raw data file by reentering the INPOLY subroutine.

8) RESTART PROGRAM.

9) EXIT PROGRAM.

10) SUBTRACT PEAKS: Transfers control to the PKSUB subroutine which asks for the number of peaks to be subtracted and then prompts for the start and end angle of each peak.

     The graphical output from the SHOPOL program is produced by the GRAPH and TICS subroutines which call subroutines in the Siemens software graph package TKPLOT. TKPLOT resides in the main program overlay segment of the SHOPOL program and is used for enabling the Tektronix graphics capabilities. Examples of the graphs produced during SHOPOL crystallinity calculations are shown in the APPLICATIONS section of this report.

Advantages of the Computer Method (SHOPOL)

     The new computer-based polymer crystallinity method can make full use of an automated diffractometer system since both data acquisition and re- duction will be handled by the computer. Another advantage of the computer method is that polymer x-ray diffraction data can be stored and later re- called to apply other data reduction schemes. In addition, the computer method improves the accuracy and precision of the data and significantly reduces the time spent in data analysis.

APPLICATIONS

Polyethylene Terephthalate (PET)

     Five annealed PET samples were analyzed for percent crystallinity using both our manual method and the SHOPOL computer program. K-values were calculated at 14 degrees two-theta, and a summary of the crystallinity results for the polymer samples is presented in Table 2. The values agree rather well; however, where differences occur, one would expect to have more confidence in the computer method since slight errors in manual measurement can have significant effects on the calculated percent crystallinity.

     Figures 1a through 1f show the output data that can be obtained from the SHOPOL program at various stages of its operation in determining the crystallinity of a polymer sample. These figures are, respectively, for a semi-crystalline PET sample raw x-ray diffractogram, an amorphous PET sample

Table 2. Percent Crystallinity of Five Annealed PET Samples.

Sample	% Crystallinity (Manual Method)	% Crystallinity (Computer Method)	% Diff.
PET 1	44	38	-16
PET 2	49	46	- 7
PET 3	52	53	+ 2
PET 4	53	51	- 4
PET 5	51	53	+ 4

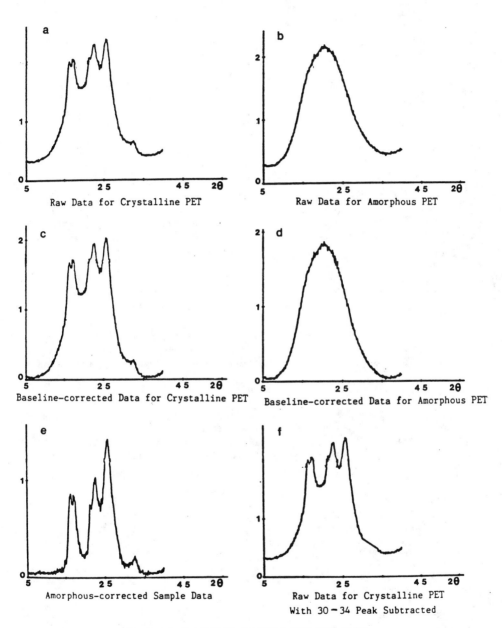

Figure 1 Graphical Results For PET Data.

Table 3. Comparison of Percent Crystallinity Results

	% Crystallinity	
Nylon 6,12 Ratios	XRD	DSC
100/0	47*	49.7
90/10	54*	36.5
80/20	37*	18.9
70/30	12	---
60/40	12	12.7
50/50	15	12.6
40/60	10	11.4
30/70	16	15.7
20/80	19	18.4
10/90	19*	---
0/100	31*	29.1

* The amorphous halo was estimated.

raw x-ray diffractogram, a baseline corrected semi-crystalline sample, a baseline corrected amorphous sample, an amorphous corrected semi-crystalline sample, and finally, the semi-crystalline sample with the peak between 30 and 34 degrees two-theta removed.

Nylon 6,12 Copolymer System

This copolymer system demonstrates an example of data processing using our manual (planimeter) method. The x-ray diffractograms were integrated with a planimeter and numerical results were obtained via equation 1. Table 3 shows a comparison of results obtained from x-ray diffraction and DSC data for a series of these nylon copolymers of varying composition.

The measured percent crystallinity is strongly and systematically influenced by the copolymer composition with minimum crystallinity observed for the samples with nylon 6,12 ratios in the 50/50 composition region. (The percent crystallinity was found to correlate strongly with the reciprocal of the impact strength, which is a key physical property governing the usefulness of a thermoplastic material for automotive applications.)

Nylon-6 Homopolymer

As mentioned in the Introduction, some polymers are not described well by the two-phase model. Nylon-6 is one of these, and the determination of its crystallinity is rendered even more difficult because of the problem in producing an amorphous standard. (Nylon-6 tends to crystallize at room temperature even if it has been heated to its melting point and quenched in liquid nitrogen.) Usually this problem forces one to estimate the amorphous halo.

We have recently overcome some of these difficulties. Amorphous standards produced from very thin films can be quenched in liquid nitrogen effectively without undergoing inadvert annealing. If such a sample is run immediately in the x-ray diffractometer a diffractogram is observed which shows essentially zero residual crystallinity. Using such an amorphous standard with the SHOPOL program, we obtain amorphous corrected semi-crystalline nylon-6 diffractograms which appear to contain almost entirely crystalline peaks, and for which the SHOPOL program should produce fairly good numerical values for absolute crystallinity. Figure 2 is an example of such a curve.

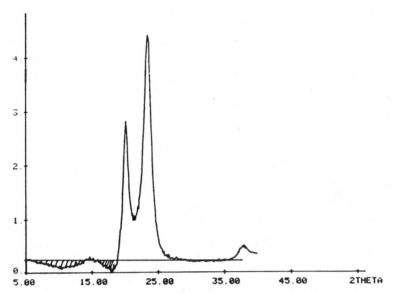

Figure 2.   Nylon-6 Homopolymer (Amorphous-Corrected Data). Hatched Area
represents negative area caused by over subtraction of
amorphous curve.

The Siemens TKPLOT subroutine has the feature of not plotting negative numbers. TKPLOT adds a positive constant to a data file generated
by SHOPOL so that the smallest number in the TKPLOT file becomes zero.
By definition, the output of the SHOPOL program is zero at the two-theta
value where K is evaluated. The departure of the ordinate (intensity)
from zero at this two-theta value indicates that there are negative numbers in the SHOPOL data file. Two effects will cause this; 1) an impre-
cise value for K which produces over-subtraction of the amorphous curve
(causing negative area), and 2) an amorphous standard with residual crystallinity which will again cause over-subtraction and negative area. The
negative area subtracts from the actual positive area and this effect
leads to an under-estimation of the percent crystallinity. The magnitude
of these effects can be observed directly in Figure 2 where they are seen
to be rather small. The largest negative number in the SHOPOL data file
is at the two-theta value where the difference curve touches the x-axis.
The cross hatched area in Figure 2 represents negative area which is, of
course, causing error in the percent crystallinity determination. The
value of 37 % crystallinity obtained for this unannealed nylon-6 sample
is reasonable[6] and gives us confidence that we can make measurements on
nylon-6.

The overall workings of the SHOPOL program and the accuracy of the
crystallinity determination can be gleaned by examining the amorphous
corrected plots, such as shown in Figure 2, observing how far the ordinate value at the normalization angle departs from zero, and how much
negative area is entering into the calculation. Thus, the operator can
make a judgment as to quality of the data (i.e, validity of the % crystallinity data) that the computer program is calculating.

ACKNOWLEDGEMENT

The authors thank Drs. Zack Gardlund, Dave Garner, and Ismat Abu-
Isa for preparing the polymers used in this study and for the DSC data.

REFERENCES

1. F. W. Billmeyer, Jr., <u>Textbook of Polymer Science</u>, Wiley-Inter-
    science, New York, 1971, p.235-7.

2. M. Kakudo and N. Kasai, <u>X-Ray Diffraction by Polymers</u>, Elsevier
    Publishing Company, New York, 1972, p.359-366.

3. M. Kakudo and N. Kasai, <u>op. cit.</u>, p.111.

4. R. L. Miller, "Crystallinity", in <u>Encyclopedia of Polymer Science and
    Technology, Vol 4.</u>, H. F. Mark, N. G. Gaylord, and N. M. Bikales
    editors. Interscience Publishers, New York, 1966, p.480-1.

5. R. L. Miller, <u>op. cit.</u>, p.477-484.

6. R. L. Miller, <u>op. cit.</u>, p.509-510.

7. A. M. Hindeleh and D. J. Johnson, "Crystallinity and Crystallite Size
    Measurement in Polyamide and Polyester Fibres", <u>Polymer</u>, <u>19</u>, 27
    (1978).

8. G. Farrow, "The Measurement of Crystallinity in Polypropylene Fibres
    by X-ray Diffraction", <u>Polymer</u>, <u>2</u>, 409 (1961).

9. G. Farrow and D. Preston, "Measurement of Crystallinity in Drawn
    Polyethylene Terephthalate Fibres by X-ray Diffraction", <u>British
    Journal of Applied Physics</u>, <u>11</u>, 353 (1960).

DETERMINATION OF CRYSTALLINITY IN GRAPHITE FIBER-REINFORCED
THERMOPLASTIC COMPOSITES

M.R. James
Rockwell International Science Center, Thousand Oaks, CA 91360

D.P. Anderson
University of Dayton Research Institute, Dayton, OH 45469

Interest in advanced thermoplastic composites for use in high performance structures stems from their order of magnitude improvement in
fracture toughness and delamination resistance over epoxy based composites, their strong solvent resistance, and the possibility of dramatically lower fabrication costs through processing flexibility. The chemical and mechanical properties of semicrystalline thermoplastics depend on
the morphology of the material, such as the crystallinity content and
spherulite size. We describe here the use of x-ray diffraction to characterize the degree of crystallinity of the polyetheretherketone-graphite
composite system, a leading thermoplastic candidate for use in aerospace
vehicles. In reflection, diffraction from the microcrystalline graphite
fibers dominates the scattered signal and must be adequately accounted
for. The technique is useful on large samples and for quality control.
In transmission, the graphite signal is weak, thus simplifying data
analysis; however, sample thickness must be limited.

The main elements of data reduction, profile fitting and crystallinity calculations are described. Data on the dependence of crystallinity content on cooling rate and matrix grade show that high crystallinity contents are obtainable in fiber-reinforced PEEK.

1.0 INTRODUCTION

The application of advanced composites in the construction of aircraft and aerospace structures has advanced significantly during the past
two decades. During this period, composite development activity has
evolved from material and fabrication process development through structural demonstration and production utilization. Current development
activities are being prioritized toward the reduction of manufacturing
cost, although improved mechanical properties are always desired. Thermoplastic composite systems have shown much promise lately because of
their adaptability to high rate production processes, ease of handling
and storage, capability to be thermally formed subsequent to curing, and
ease of reprocessing. Thermoplastic composites offer many of the advantages of thermosetting resin composites with the added advantages of
dramatically improved toughness, delamination and creep resistance, and
potential for unique processing capabilities, such as thermal forming.
Previously, application of advanced thermoplastic composite materials has
been limited by their susceptibility to chemical degradation, especially

during contact with certain aircraft fluids. An emerging thermoplastic system, semicrystalline polyetheretherketone (PEEK), developed by Imperial Chemical Industries (ICI), a British petrochemical company, has demonstrated the necessary solvent resistance for use in aircraft.

The dependence of strength and fracture properties and the chemical resistance of semicrystalline PEEK polymers on microstructural factors are well documented. The degree of crystalline content of the neat resin and of graphite reinforced laminates is quite dependent on processing parameters such as cooling rate. To date, a nondestructive technique to monitor the crystalline content has not been developed, yet the need for characterizing crystallinity and other morphological features to investigate different processing procedures clearly warrants such a capability. This paper describes the use of x-ray diffraction to determine the crystalline content of PEEK/graphite laminates. The addition of a third constituent, graphite, complicates the analysis as compared to a neat resin system, and techniques to overcome this problem are described. Finally, the effects of different cooling rates on the crystallinity content are given to highlight the ability to tailor the morphological properties of the PEEK/graphite system.

2.0 BACKGROUND

2.1 Morphology of Polyetheretherketone

Thermoplastics are polymers having a linear two-dimensional structure which can be repeatedly softened by heat or dissolved; when the heat or solvent is removed, the material solidifies. Unlike thermosets, these polymers undergo only physical changes with no cross-linking. For amorphous thermoplastics, the glass transition is the point at which the polymer begins changing from a hard, relatively glassy state, to a viscous fluid. Since the polymer viscosity gradually decreases with increasing temperature, the processing of amorphous polymers is done at temperatures above the glass transition such that the viscosity is in a range commensurate with the processing technique. In semicrystalline polymers, such as PEEK, the melting temperature is 100-200°C or more above the glass transition. While the polymer has only limited use above the glass transition, processing must usually be done in the melt where the viscosity drops off very rapidly. Schultz[1] has described the morphology of semicrystalline polymers as three levels of microstructure. On the molecular level, polymer chains interact with one another in both the amorphous and crystalline phases. Crystal packing defects and dislocations occur at this level. The next morphological structure is the crystalline phase, made up of ribbon-like crystals (having a thickness of about 100Å) that grow from the nucleating site. Finally, spherulite formation results from the growth of the ribbon-like crystal structures from a common center to fill a spherical space. The space between the individual crystals and at the spherulite boundary is amorphous, giving an overall lamellar structure within the spherulite. Blundell and Osborn[2] have characterized the morphology of neat PEEK resin and find a spherulite radius of 13 to 22 μm and a lamella thickness of 20 to 60Å, depending on crystallization conditions. Dawson and Blundell[3] report that PEEK has an orthorhombic unit cell with lattice parameters of approximately a = 7.7Å, b = 5.9Å and c = 10Å.

The microstructure of PEEK resin in a graphite-reinforced composite is quite different. The presence of the fiber introduces a very large number of crystal nucleation sites. Cogswell[4] has reported a transcry-

stalline spherulite structure with the polymer chains extending parallel
to the fibers, and a spherulite size no larger than 2 μm.

Currently five grades of PEEK are available from industry, of which
three are reported on here. These include:

FGP – Film grade PEEK is an amorphous film material produced by
extrusion from pellets. Its amber color is characteristic
of its amorphous state, but when heated above its melt
temperature and cooled slowly, it transforms into an
opaque, semicrystalline state, gray in color.

XAS/MG-1 – This PEEK composite consists of Courtaulds Grafil XAS
graphite fiber, impregnated by ICI, with a first generation material resin, MG-1. The prepreg product is referred to as APC-1 by ICI.

AS4/MG-2 – This material grade contains Hercules AS4 graphite fiber
impregnated by ICI with a PEEK resin modified from MG-1.
According to ICI, this material grade (MG-2) is not as
sensitive to cooling rate after processing above the melt
temperature as MG-1, and it provides less variation in
crystalline structure.

2.2 X-ray Methods for Determination of Crystallinity

The use of wide-angle x-ray diffraction for characterizing the
degree of crystallinity in organic polymers and glass-ceramic materials
has achieved considerable sophistication. Even so, its application to
fiber-reinforced thermoplastic composites has not seen wide use because
these materials have only recently been available. Fundamentally, the
x-ray method is based on the fact that the ratio of the integrated intensities of crystalline diffraction lines with that of the broad halo
due to the amorphous regions is related to the volume fraction of crystallites. Experimentally, however, determination of the correct ratio is
not trivial. The ordered regions that are considered crystalline are
very small and often columnar in that they preferentially nucleate perpendicular to the fibers during cooling, giving a so-called transcrystal-
line structure. This causes them to be highly strained and to produce
very broad diffraction lines. Additionally, the graphite fiber used in
the laminates produces a microcrystalline diffraction halo in the same
scattering space as the crystalline polymer, making separation of the
diffracted intensities tedious. Fortunately, with modern computer-
automated x-ray diffractometers these problems can be surmounted. We
begin by reviewing some of the techniques used in the past.

A number of methods have been proposed to determine the crystalline
content of polymeric systems, or, in general, any two-phase system where
both phases have the same absorption coefficient. Wakelin et al.[5] have
described a technique that uses two standards (100% amorphous and 100%
crystalline) which they applied to determine the relative crystalline
content of cotton cellulose. The correlation method described by Wakelin
et al. has been modified by a number of investigators: Sotton et al.[6] on
polyethylene terephthalate (PET); Radhakrishnam and Nadkarni[7] on poly-
phenylene sulfide (PPS); and Chung and Scott[8] on PET. Another approach
involving simply dividing the integrated intensity under the diffraction
curve attributed to the crystalline phase by the total integrated intensity was used by Brady[9] on PPS and Kays and Hunter[10] in characterizing the
neat resins of four thermoplastics, including PEEK. Blundell and Osborn[2]
used this method on neat PEEK and found a linear correspondence between
the crystallinity index and the specific volume for samples ranging from

0 to 50% crystalline. Although the crystallinity index achieved in this manner does not correspond exactly to the volume fraction of crystalline material, the procedure is rapid and provides a valid measurement for qualitative purposes.

Lately, the method of Ruland[11] as modified by Vonk[12] has attracted the most attention.[13-17] This method has a sound theoretical basis, and although it requires a computer for efficient analysis, gives both an absolute degree of crystallinity and an isotropic disorder parameter reflecting the deviation of the atoms from their ideal positions in the crystalline phase. Assuming one can make the proper delineation of scattering due to the crystalline and amorphous parts of the sample, Ruland defines the weight fraction of crystallinity as:

$$X_{cr} = \frac{\int_0^\infty s^2 I_{cr} ds}{\int_0^\infty s^2 I \, ds} \cdot \frac{\int_0^\infty s^2 |f|^2 ds}{\int_0^\infty s^2 |f|^2 D \, ds} \tag{1}$$

where

I_{cr} is the crystalline coherent intensity

I is the total coherent intensity,

$|f|$ is the mean square scattering factor,

D is Ruland's disorder parameter,

$s^2 ds$ is defined as the volume element of integration in s-space ($2\sin\theta/\lambda$) where λ is the wavelength in Å.

The disorder parameter, D, is assumed to have the form $\exp(-ks^2)$ which includes thermal fluctuations and assumes the other lattice imperfections are generally isotropic.

A further assumption is that finite integration limits (s_0 and s_p) exist such that:

$$\int_{s_0}^{s_p} s^2 I(s) \, ds = \int_{s_0}^{s_p} s^2 |f|^2 ds \tag{2}$$

is independent of the sample crytallinity and:

$$\int_{s_0}^{s_p} s^2 I_{cr}(s) \, ds = X_{cr} \int_{s_0}^{s_p} s^2 |f|^2 D \, ds \quad . \tag{3}$$

Using the assumptions in Eqs. (2) and (3), Eq. (1) may be rewritten as:

$$w_{cr} = \frac{\int_{s_0}^{s_p} s^2 I_{cr} \, ds}{\int_{s_0}^{s_p} s^2 I \, ds} \cdot K(s_0, s_p, D, |f|) \tag{4}$$

where X_{cr} is a constant and

$$K = \cfrac{\int\limits_{s_o}^{s_p} s^2 \, |f|^2 \, ds}{\int\limits_{s_o}^{s_p} s^2 \, |f|^2 \, D \, ds} \qquad .$$ (5)

The integration limits s_o and s_p are determined experimentally, pre-
ferably with a common s_o. One then solves the equation by determining
the series of K's that yield a constant X_{cr}.

An important intensity correction for this analysis is the subtraction of incoherent (or Compton's) scattering. This correction
requires a detailed knowledge of the chemical composition of the sample.

Vonk[12] took the basic equations of Ruland's theory and defined the
ratio, $R(s_p^2)$:

$$R\left(s_p^2\right) \sim K/X_{cr}$$ (6)

where

$$R\left(s_p^2\right) = \cfrac{\int\limits_{s_o}^{s_p} s^2 \, I \, ds}{\int\limits_{s_o}^{s_p} s^2 \, I_{cr} \, ds}$$ (7)

and K is defined by Eq. (5). The symbol $\sim$ indicates that the left-hand
function oscillates about the right-hand function.

When D has the form:

$$D = \exp(-ks^2)$$ (8)

then K can be approximated by the first two terms of the exponential
series:

$$K = 1 + b \, s^2$$ (9)

where b is a constant equal to $k/2$. Combining Eqs. (9) and (6), a plot
of $R(s_p^2)$ vs s_p^2 should oscillate about a straight line defined by:

$$y = 1/X_{cr} + \left(k/2X_{cr}\right) s_p^2 \quad .$$ (10)

The crystalline content and disorder factor are obtained from the intercept and slope of a plot of $R(s_p^2)$ versus s_p^2.

3.0 EXPERIMENTAL ASPECTS

3.1 Material

The crystallinity of XAS/MG-1 and AS4/MG-2 laminates was studied on
samples having undergone various cooling rates. A 16-ply press molded
unidirectional laminate of XAS/MG-1 procured from ICI was analysed in the
as-received condition. ICI recommends a fast cooling rate (greater than

3°/min) which this material probably underwent. A similar laminate of AS4/MG-2 was bag-molded by North American Aircraft Operations of Rockwell International using a gas pressure consolidation technique in which the laminate was slow quenched from 400°C in a temperature programmed press. Faster cooling rates were obtained for both materials by careful cooldown of the laminates in a convection heated oven. Fast (~ 500°C/min) and very fast (~ 1000°C/min) cooling rates were obtained by quenching into water or liquid nitrogen, respectively. Neat resin was examined in both an as-received state (translucent amber color) and after slow quenching (opaque grey color).

3.2 Equipment

Two diffractometer setups were used in the experimental program. A Picker FACS-1 diffractometer with a primary beam crystal monochrometer was used at the University of Dayton Research Institute for the experiments done in symmetric transmission geometry. These data were taken using copper radiation and were analysed using a computer program entitled DIFF.[18] The symmetric reflection data were obtained at the Rockwell International Science Center using a modified G.E. XRD-5 diffractometer equipped with a Kevex energy dispersive detector and Data General MP/100 microcomputer for automated control. Chromium radiation was used. Data analysis was done on a DEC VAX 11/780 after downloading the data from the micro.

The two scattering geometries were explored because of the differing contribution of the graphite fiber scattering. The highest intensity diffraction peak from graphite fibers is from the (002) reflection. This results from diffraction between the graphitic planes in the graphite structure. The interplane distance of ~ 3.4Å producing the (002) reflection is present whether the graphite has perfectly ordered crystals or disordered layers such as the turbostratic structure. The graphite sheets in a fiber are aligned parallel to the fiber axis but are randomly oriented through the fiber cross-section.

In a typical composite structure, the reinforcing fibers are arranged parallel to the surface of the structure. As a result, the fibers all lie in a plane; when performing wide angle diffraction in the most common symmetric reflection mode, that plane is in the Bragg diffraction condition. Since the graphitic planes producing the (002) reflection are parallel to the fiber axes, they too are in the Bragg condition and their intensity will be present in the diffraction scan.

In order to move the graphitic planes out of the diffraction condition, the fiber axes must be tilted out of the plane in the Bragg θ to 2θ position. This is not possible unless one only looks down on the middle of a composite or if the composite is examined in the transmission mode. Absorption of the incident x-ray beam becomes important in transmission geometry and must be addressed in the analysis.

Figures 1 and 2 show typical raw data scans for a 1 mm thick AS4/MG-2 GPC sample in transmission and reflection geometries, respectively. The intensities were measured in a step-scanning mode with 300-1000 data points over the range shown for each geometry. In transmission, a relatively low intensity, broad peak from the graphite is seen at 44 °2θ. The intensity profile of the same sample, run in reflection (Fig. 2), is dominated by the much stronger graphite peak at 39 °2θ. Because the graphite is microcrystalline, with crystals in the 25Å - 50Å range, the diffraction peak is very broad and overlaps all the crystalline peaks in reflection.

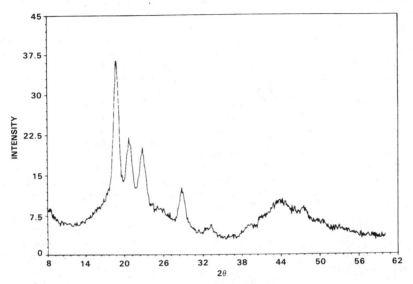

Fig. 1.   Raw diffraction scan of AS4/MG-2 gas pressure con-
          solidated sample taken with copper radiation in sym-
          metric transmission geometry.

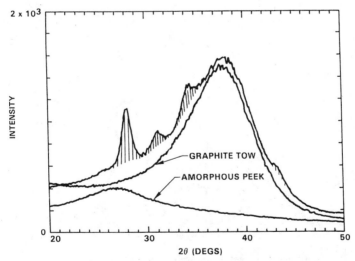

Fig. 2.   Diffraction scan of AS4/MG-2 gas pressure con-
          solidated sample taken with chromium radiation
          in symmetric reflection geometry. Relative
          contribution of graphite and amorphous PEEK have
          been added to illustrate the weak nature of the
          crystalline peaks in reflection geometry.

3.3   Data Reduction

Ruland,[11] Vonk[12] and Milberg[19] detail a number of elementary precautions that must be considered when analyzing polymers having low x-ray absorption coefficients. All raw data were corrected for the usual factors according to the appropriate geometry: Lorentz-polarization factor, absorption correction, background intensity, and incoherent scattering intensity. Air scatter in the background was accounted for by normalizing a background scan (no sample) to account for the absorption of the main beam by the sample. The incoherent scattering intensity was calculated by assuming that the intensity at the highest angle is the sum of the incoherent and coherent scattering of individual atoms (no crystalline diffraction). Both the coherent and incoherent scattering factors were calculated by interpolating between values at equally spaced $\sin\theta/\lambda$ calculated for the sample's composition from literature values.[20-21] The ratio of the calculated total scattering intensity at the maximum angle to the calculated incoherent intensity is then used to scale the calculated incoherent intensity at all other angles. The relative magnitudes of these corrections are shown by the appropriate curves in Fig. 3 for the transmission case.

After these corrections, the data from a PEEK/graphite composite represent the combined coherent intensities from the amorphous and crystalline regions of the polymer, and from the graphite. To subtract both the amorphous and graphite contributions, diffractometer scans over the same scattering space were taken of a graphite tow and an amorphous PEEK resin sample. After corrections, these curves were normalized to the composite data using the method proposed by Vonk.[12] By inspection,

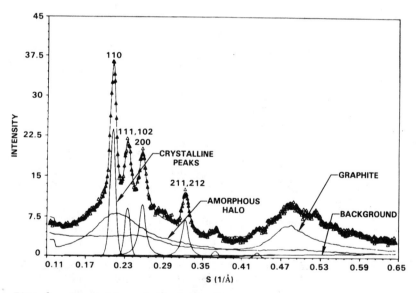

Fig. 3.   Separation of the contributing sources of intensity
to the diffraction scan of AS4/MG-2 sample in transmission. The relatively weak graphite peak at s = 0.48
(44 °2θ) is not only less difficult to ·subtract than
in reflection (see Fig. 2), but also is at high enough
angle that the Vonk analysis can be done below the
contributing region.

the regions of the composite data void of any crystalline peaks are fit
to the appropriate calibration curves, and the entire curve is normalized. This is obviously a critical step in determining the crystallinity
content and is best performed interactively with an operator viewing the
corrected curves on the terminal. This is especially true in the
reflection geometry where the graphite peak is so dominant. Typical
curves showing the separate contributions are given in Fig. 3.

To determine the integrated area under the diffraction peaks, we
have found that functional fitting of the profiles using a Pearson Type
VII distribution[22] works best. This is given by

$$y(x) = y_o \left[1 + (x - \bar{x})^2/(ma^2)\right]^{-m} \tag{11}$$

where

  y_o is the peak intensity maximum,
  $\bar{x}$ is the center of the distribution,
  m is the Pearson VII exponent, and
  a is related to the peak width according to the following

$$\beta = 2a[m(2^{1/m} - 1)]^{1/2} \tag{12}$$

where

  β is the full width at half the maximum intensity.

When the exponent, m, is unity, the Pearson VII distribution reduces to a
Cauchy, while as m approaches infinity the Pearson VII distribution approaches the Gaussian form. At the University of Dayton, the Gauss-
Newton method of least-squares is used to best fit the 4 parameters to
the data; whereas at the Rockwell Science Center, a Levenburg-Marquardt
routine available in the IMSL library[23] is used. A plot of fitted data
is given in Fig. 4 for a reflection scan. Using a functional fit such as
Eq. (12) eliminates some of the problems associated with intensity stripping because the tails of the crystalline peaks can be properly accounted
for. It was not found feasible to fit the amorphous or graphite profiles
because these are not symmetric enough.

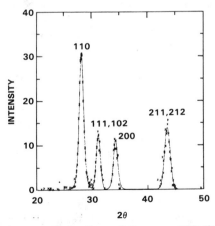

Fig. 4. Typical plot of Pearson VII distri-
         bution to crystalline data taken in
         reflection geometry.

4.0   RESULTS AND DISCUSSION

The results of crystallinity measurements obtained by Ruland's method are presented in Table 1. Typical plots of $R(s_p)$ as a function of s_p^2 are shown in Figs. 5 and 6. The straight lines, extrapolated to zero to give $1/X_{cr}$ in Eq. (10), are determined using a linear least-squares fit to a specified range in s_p^2. For the neat resin sample shown in Fig. 5, the choice of the range is straightforward, although choice of the lowest s_p^2 values is hampered by the considerable oscillations. These are expected to be large as discussed by Vonk,[12] and the lower limit is usually set to the value of s^2 corresponding to the second peak. Figure 6 shows a plot of $R(s_p)$ as a function of s_p^2 for the AS4/MG-2 gas pressure consolidated sample analyzed in transmission. The plot has been carried out to rather high s_p^2 to make an important point; the considerable oscillations at high s_p^2 make the choice of the least-squares fit region very difficult to determine. This results from the difficulty of

Table 1.   Crystallinity of PEEK/Graphite

Quench Rate (°F/min)	Crystallinity Index	
	XAS/MG-1	AS4/MG-2
As-received from ICI	38	–
Gas Pressure Consolidated	–	59
3	45	56
10	43	53
150	31	51
Fast	–	44
Very Fast	–	42

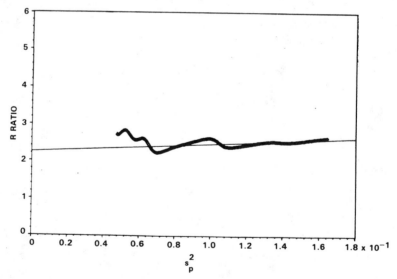

Fig. 5.   Vonk plot for PEEK resin slow cooled from 400°C. The crystallinity is 44% and the disorder parameter k is 1.89.

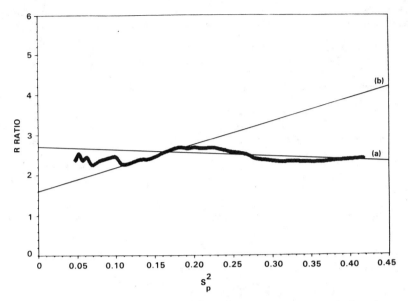

Fig. 6 Vonk plot for AS4/MG-2 gas pressure consolidated
        sample in transmission. Line (a) represents a
        linear least-squares fit from 0.05 - 0.4 s_p^2
        and gives X_{cr} = 37%. Line (b) was fit from 0.1
        to 0.8 s_p^2 which gives X_{cr} = 62%.

subtracting the contribution from the graphite fiber at s_p^2 greater than
0.18 (>38 °2θ). A crystalline content of 62% is obtained when the s_p^2
range is set to 0.10 - 0.18 and a value of 37% is obtained when the
range is extended to 0.42. Obviously, this places considerable doubt as
to the absolute nature of the results, although the relative variations
should be precise if identical ranges are used in the data analysis. The
shorter appropriate range for chromium radiation was used to obtain the
results in Table 1.

The data in Table 1 support two hypotheses concerning the crystalline morphology of graphite/PEEK composite. One is that the crystallinity of AS4/MG-2 does not appear to vary significantly with quench rate.
Even the fast and very fast quenches did not lower the crystallinity
drastically. This contrasts with the behavior of XAS/MG-1, where the
crystallinity drops sharply at a quench rate of 150°F/min. If maintaining a higher degree of crystallinity is desirable, AS4/MG-2 has an advantage over XAS/MG-1 for use in processing methods where rapid cooling is
involved. A second conclusion from data in Table 1 is that AS4/MG-2 appears to be more crystalline than XAS/MG-1 composite. Both PEEK composite resin materials show a higher degree of crystallinity than film grade
neat resin, which is ~ 32% crystalline. This may be due to the high
nucleating ability of the graphite fiber surface, which induces a higher
degree of crystallinity in the composite.

The chromium radiation used in these experiments has a penetration
depth of only about 30 μm in graphite/PEEK composite. To determine if
measurements made at the surface were representative of the bulk, approximately 125 μm of material was removed from the sample quenched at
150°F/min, and the analysis was repeated with no change in crystallinity

content observed. Tests of this type are important for characterizing variations in crystallinity which may be present through the thickness of fabricated test parts. The inhomogeneity between the surface skin and bulk microstructure of injection-molded semicrystalline polymers is well documented.[24] Fortunately, the large number of nucleation sites provided by the graphite fibers probably acts as a randomizing factor and improves the homogeneity between the surface and the bulk.

On completion of this work, an article was published by Blundell et al.[25] on crystallinity measurements in PEEK APC-2 (referred to as MG-2 here) composites. They developed a simple x-ray procedure by measuring the ratio of the 110 peak intensity of PEEK to that of the graphite signal and calibrated the procedure with infrared reflection spectroscopy. Their measured values show a maximum of 48% crystallinity achieved at the slowest cooling rate of ~ 1°F/min. Our results show a slightly higher crystalline content (~ 56%) at this rate. The Ruland/Vonk procedure is expected to give higher crystalline contents since effects of disorder are accounted for.

ACKNOWLEDGEMENTS

The authors kindly acknowledge the support of this work by Rockwell International IR&D (MRJ) and the Materials Laboratory, Air Force Wright Aeronautical Laboratories (DPA).

REFERENCES

1. J.M. Schultz, Microstructural Aspects of Failure in Semicrystalline Polymers, Polymer Eng. and Sci., 24:770 (1984).
2. D.J. Blundell and B.N. Osborn, The Morphology of Poly(aryl-ether-ether-Ketone), Polymer, 24:953 (1983).
3. P.C. Dawson and D.J. Blundell, X-ray Data for Poly(aryl ether ketones), Polymer, 21:577 (1980).
4. F.N. Cogswell, Microstructure and Properties of Thermoplastic Aromatic Polymer Composites, in 12th National SAMPE Symposium, April 12-14, p. 528 (1983).
5. J.H. Wakelin, H.S. Virgin and E. Crystal, Development and Comparisons of Two X-Ray Methods for Determining the Crystallinity of Cotton Cellulose, J. Appl. Phys., 30:1654 (1959).
6. M. Sotton, A. Arniaud, and C. Rabourdin, Crystallinity and Disorder in PET Fibers, J. App. Polymer Sci., 22:2585 (1978).
7. S. Radhakrishnan and V.M. Nadkarni, Modification of Surface Structure and Crystallinity in Compression-Molded PPS, Polymer Eng. and Sci., 24:1383 (1984).
8. F. H. Chung and R. W. Scott, A New Approach to the Determination Of Crystallinity of Polymers by X-Ray Diffraction, J. Appl. Cryst., 6:225 (1973).
9. D.G. Brady, The Crystallinity of PPS and its Effect on Polymer Properties, J. Appl. Polymer Sci., 20:2541 (1976).
10. A.O. Kays and J.D. Hunter, Characterization of Some Solvent-Resistent Thermoplastic Matrix Composites, in Composite Materials: Quality Assurance and Processing, ASTM STP 797, C.E. Browning, ed., American Society for Testing and Materials, p. 119 (1983).
11. W. Ruland, X-Ray Determination of Crystallinity and Diffuse Disorder Scattering, Acta. Cryst., 14:1180 (1961).

12. C.G. Vonk, Computerization of Ruland's X-Ray Method for Determination of Crystallinity in Polymers, J. Appl. Cryst., 6:148 (1973).

13. F. Fontaine, J. Ledent, G. Groeninckx, and H. Reynaers, Morphology and Melting Behavior of Semi-Crystalline PET: 3. Quantification of Crystal Perfection and Crystallinity, Polymer, 23:185 (1982).

14. F.J. Balta Calleja, J. Martinez Salazar, H. Cackovic and J. Loboda-Cackovic, Correlation of Hardness and Microstructure in Unoriented Lamellar Polyethylene, J. Mater. Sci., 16:739 (1981).

15. M. Ahtee, T. Hattula, J. Mangs and T. Paakkari, An X-Ray Diffraction Method for Determination of Crystallinity in Wood Pulp, Paperi ja Puu - Papper och Tra, 8:475 (1983).

16. D. Hlavara, J. Baldrian and J. Plestil, Quantitative Characterization of the Structure Changes in Low Crystalline PET Fibers by X-Ray Diffraction, Acta Polymerica, 33:256 (1982).

17. Kh.M. Mannan and L.B. Lutfar, Effects of Grafted Methyl Methacrylate on the Microstructure of Jute Fibers, Polymer, 21:777 (1980).

18. D.P. Anderson, X-Ray Analysis Software: Operation and Theory Involved in Program "DIFF", AFWAL-TR-85-4079, June 1985, Materials Laboratory, Air Force Wright Aeronautical Laboratories.

19. M.E. Milberg, Transparency Factors for Weakly Absorbing Samples in X-Ray Diffraction, J. Appl. Phys., 29:64 (1958).

20. International Tables for X-Ray Crystallography, Kynoch Press, Birmingham, England (1968).

21. A.H. Compton and S.K. Allison, X-Rays in Theory and Experiment, D. Van Nostrand Co., Inc., New York, 1935, pp. 780-782.

22. S.U.N. Naidu and C.R. Houska, Profile Separation in Complex Powder Patterns, J. Appl. Cryst., 5:190 (1982).

23. Problem-Solving Software System for Mathematical and Statistical FORTRAN Programming, IMSL®, Library Edition 9.2, IMSL, Inc. Houston, TX.

24. S.S. Kati and J.M. Shultz, The Microstructure of Injection-Molded Semicrystalline Polymers: A Review, Polymer Eng. and Sci., 22:1001 (1982).

25. D.J. Blundell, J.M. Chalmers, M.W. MacKenzie and W.F. Gaskin, Crystalline Morphology of the PEEK-Carbon Fiber Aromatic Polymers Composite I. Assessment of Crystallinity, SAMPE Quarterly, 16:22 (1985).

ADJUSTMENT OF SOLID-SOLID PHASE TRANSITION TEMPERATURE

OF POLYALCOHOLS BY THE USE OF DOPANTS*

D. Chandra and C. S. Barrett, University of Denver,
Denver, CO 80208

D. K. Benson, Solar Energy Research Institute, Golden,
CO 80401

ABSTRACT

   The transition temperatures of solid-solid phase changes in selected
polyalcohols, "plastic crystals," can be adjusted by using interstitial
and substitutional dopants. An investigation is under way of the structural changes in these during heating and cooling, and of the thermodynamic
properties such as the transition temperatures and enthalpy changes, as a
function of the percent of dopant. The purpose of the investigation is to
find and evaluate materials having potential value in thermal storage
applications. Dopants for pentaerythritol discussed in this report are
trimethylolpropane (TMP), ammonia, boron trifluoride, pentaglycerine (PG)
and neopentylglycol (NPG).

INTRODUCTION

   Solid-state phase-change materials are being considered as potential
candidates for thermal storage of energy.[1] These materials can store
large amounts of (latent) heat per unit mass at a constant transition temperature well below their melting points. At this transition temperature
these materials undergo a phase transformation which is associated with a
reversible storage of latent heat and is, in general, orders of magnitude
larger than the conventional sensible heat-storage materials. Organic
molecular crystals such as polyalcohols are potential candidates for
thermal storage of heat in which such solid-state phase changes occur; in
general crystallographic changes from a lower symmetry to a higher one are
observed during the heating cycle. Some of the polyalcohols which have
tetrahedral globular molecules are termed "plastic crystals" after
Timmermans;[2] these are characterized by a large solid-solid entropy of
transition and very low entropy of fusion, making them attractive as
thermal storage materials.

   These polyalcohol plastic crystals generally have layered structures
in the low temperature phase (II) and isotropic structures in the high

*Work supported by contract No. DE-AC03-84SF12205 of the Dept. of Energy,
 San Francisco Operations Office, Oakland, CA.

temperature phase (I). Some important examples of these energetic materials are pentaerythritol (PE), pentaglycerine (PG) and neopentylglycol
(NPG) whose transition temperatures are at 188°C, 89°C and 44°C,
respectively. The structure and lattice parameters of phase II and I are
listed in Table 1.

Table 1. Structure and lattice parameters of
polyalcohol plastic crystals above and
below transition temperatures

Plastic Crystal	Structure and Lattice Parameters of the Room Temp. Phase II	Structure and Lattice Parameters of the High Temp. Phase I
Pentaerythritol (PE) ($C_5H_{12}O_4$)	Tetragonal[3] $a_o = 6.0756Å$ $c_o = 8.7798Å$ @ 26°C	Cubic[3] $a_o = 8.999Å$ @ 189°C
Pentaglycerine (PG) ($C_5H_{12}O_3$)	Tetragonal[4] $a_o = 6.2609Å$ $c_o = 8.5354Å$ @ 26°C	Cubic[5] $a_o = 8.86Å$ @ 100°C
Neopentylglycol (NPG) ($C_5H_{12}O_2$)	Monoclinic[4] $a_o = 6.0773Å$ $b_o = 10.9642Å$ $c_o = 10.1125Å$ $\beta = 99.68°$ @ 21°C	Cubic[4] $a_o = 8.838Å$ @ 46.5°C

The fixed transition temperature of the plastic crystals limits their use
in thermal storage applications; consequently adjustment of the phase
transition temperature is required in practice. This paper describes
the adjustment of solid-state phase transition temperatures in plastic
crystals by the introduction of substitutional and interstitial dopants
in the lattice of the host crystal of pentaerythritol. The phase transformation in pure pentaerythritol will also be discussed. Research on
adjustment of transition temperatures of other plastic crystals is in
progress.

STRUCTURE OF PENTAERYTHRITOL

     The crystal structure of pentaerythritol has been determined by Nitta
and Watanabe[6] and Llewellyn, Cox and Goodwin[7] independently. Eilerman
and Rudman[8] refined the structure and determined the atom positions with
precision. Using these refined data the (001) and (100) projections of
the unit cells are plotted in Figure 1. There are two molecules per unit
cell in the body centered tetragonal pentaerythritol, in which the central
carbon atoms of the molecules are at 0,0,0 and 1/2,1/2,1/2, each molecule
having a 4 fold symmetry. This PE crystal has space group symmetry I$\bar{4}$
(#82). The (100) projection shows the layering in the structure with
a period $c_o/2 = 4.469Å$. The neighboring molecule linkage within a sheet
is mainly hydrogen bonding via interaction of hydroxyl groups. In this
linkage four hydroxyl oxygen atoms, one from each adjoining molecule,
constitute a square bridge by hydrogen atoms (dashed lines in Fig. 1).
The length of these hydrogen bonds is 2.71Å. The intermolecular hydrogen
atoms, not shown in Figure 1, are located on or near these dashed lines.
Interatomic distances given by Eilerman and Rudman[8] for the hydrogen bond
are: O...O = 2.710Å, O-H = 0.752Å, H...O = 1.983Å, angle O-H...O = 163°.
The C-C distance in one arm of the molecule is 1.527Å and the C-O distance
is 1.422Å, one C-C-C angle in a molecule is 107.33° and the other is 110.55°;

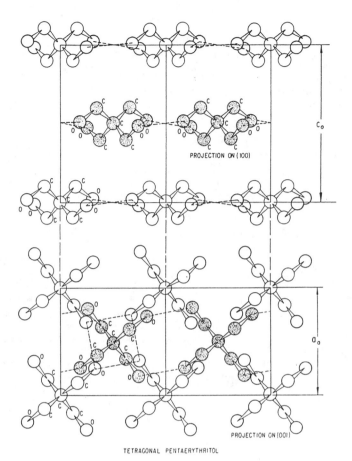

Figure 1. The layered structure of tetragonal pentaerythritol. Unit cells are projected parallel to the c-axis in the lower drawing and two cells are projected parallel to the a-axis in the upper drawing. The dashed lines represent the O-H...O hydrogen bond (H atoms are not plotted).

compared with ideal tetrahedral angles there is slight compression within the H-bonded sheets and slight elongation perpendicular to these sheets.

A high temperature crystal structure study of pentaerythritol was first performed by Nitta and Watanabe[9] who reported the structure of the high temperature transformed phase of PE to be face centered cubic. They measured the lattice parameter to be a_o = 8.963Å at 230°C. More recently Chandra, Fitzpatrick and Jorgenson[3] confirmed the Nitta and Watanabe[9] structure and also performed detailed high temperature studies using Guinier X-ray diffractometry and film techniques. The thermal expansion coefficients in the low temperature tetragonal phase were evaluated; in the a-axis α_a = 2.644 x 10^{-5}/°C and in the c-axis α_c = 13.967 x 10^{-5}/°C between 26 and 180°C. This anisotropy, due to the large differences in thermal expansion coefficients along the a and c axes, suggests that there is strong hydrogen bonding in the (001) layer and that the inter-layer bonds are weak due to weak Van der Waal forces. The thermal expansion coefficient for the high temperature cubic phase was reported to be α_a = 1.57 x 10^{-4}/°C.

EXPERIMENTAL

Finely ground recrystallized pentaerythritol was ground in a glove box and batches of these materials were used to mix and melt the dopants in various concentrations. The following dopants were used for transition temperature adjustment: a) trimethylol propane (TMP), b) pentaglycerine (PG), c) neopentyl glycol (NPG), d) sodium acetate, e) Lewis acids (BF$_3$) and bases (NH$_3$) and f) 2-amino-2-methyl propanol.

Evaluation of thermal properties of these materials in undoped and doped conditions is of great importance in thermal storage research. A Perkin-Elmer Differential Scanning Calorimeter (DSC I) was used to evaluate the transition temperatures and the enthalpies of transition. A few milli-grams of the samples were placed in a crimped aluminum pan and, in general, a heating rate of 10°C/min was used for heating.[1]

The structural properties of the undoped and doped materials were evaluated at room temperature using an automated Philips diffractometer using Bragg-Brentano focusing. A diffracted beam monochromator (CuKα) was used on the diffractometer. The diffractometer was interfaced with a PDP-11/23 DEC computer. The data were reduced by using the following Nicolet programs: 1) D-SPACE for listing 2θ and d's of reflections and assigning indices of reflections for structures with given cell parameters, 2) PEAK for peak fitting by least squares to a 5, 7 or 9-point cubic polynomial, 3) ELST for refining unit cell dimensions from a selected list of reflections of known hkl indices, and for computing differences between observed and calculated d's for each reflection. The scans covered a 10°-65° range in steps of 0.02° with 4 sec counts at each step, and with the Cu X-ray tube operating at 35kv, 20Ma.

RESULTS AND DISCUSSION

The solid-state transition temperatures of pentaerythritol (PE) and homologous compunds have been adjusted by the use of substitutional and interstitial dopants. In general the goal was to decrease the transition temperature for thermal storage purposes, but it was found that increases in transition temperature were also possible. The substitutional dopants can be added in large amounts so as to form solid solutions. Heating of crystals in some cases results in order-disorder transformations which allow drastic decreases in the transition temperture. Small amounts of interstitial dopants can have significant effects on the thermal proper-ties. The effect of the dopants on thermal properties will be discussed first, and correlation of thermal properties with the structural data will be discussed later in this section.

The phase transformations in organic plastic crystals are generally of a first order transformation type,[1] where a low temperature phase II changes into a high temperature phase at transition temperature T_t. Equilibrium at the phase transition temperature T_t is represented by $\Delta G = \Delta H - T_t(\Delta S)$, where ΔG, ΔH and ΔS are the changes in Gibbs free energy, enthalpy and entropy. But since $\Delta G = 0$, $T_t = \Delta H/\Delta S$ at equilibrium. It is this entropy term that influences the enthalpy by adjusting the transition temperature. Perhaps increases in entropy in materials are obtained by increased molecular movement at elevated temperatures in plastic crystals. In general, the plastic crystals by definition[2] have a large enthalpy change of solid-solid transition and a low enthalpy change of melting, which is demonstrated by plotting the ΔH vs T_t for pentaerythritol (PE), penta-glycerine (PG), and neopentylglycol (NPG). This is shown in Figure 2. The low enthalpies of PG and NPG compared with PE are due to the smaller number of O-H bonds per molecule. The slope of ΔH vs T_t gives the

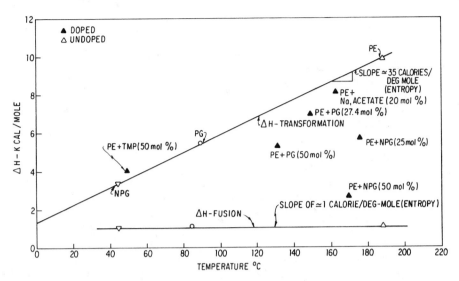

Figure 2. The enthalpies ΔH (Kcal/mole) of transition and fusion for undoped pentaerythritol (PE), pentaglycerine (PG) and neopentylglycol (NPG) plastic crystals as a function of their transition temperatures. The average entropy of solid-solid transition is estimated to be ≃ 35 cal/mol-deg and that of the fusion to be ≃ 1 cal/mol-deg given by the slope of the line. Also shown are PG, NPG and trimethylol propane (TMP) doped samples of PE which, in general, lower the transition temperature of PE (of 188°C). In the case of TMP as a dopant, decreases in T_t as much as 140°C have been observed.

entropy for the plastic crystals, which is ≃ 35 cal/deg-mole for the solid-solid transition compared with ≃ 1 cal/deg-mole for fusion; this makes the plastic crystals important for heat storage materials. Also shown in Figure 2 are the transition-temperature-adjusted doped PE materials, which show small as well as large losses in enthalpies of transition depending on the type of the dopant.

The changes in thermal properties due to addition of branched alcohol substitutional dopants such as pentaglycerine (PG), neopentyl glycol (NPG) and trimethylol propane (TMP) are shown in Table 2. Some of the interstitial dopants which yielded very interesting results were straight chain alcohols, ethers, Lewis acids (borontrifluoride) and bases (ammonia) and di- and tri-hydroxy alcohols (ethylene glycol and glycerol), some of which are listed in Table 2.

The most effective dopant was trimethylolpropane (TMP) which lowered the transition temperature from 188 to 48.2°C with a resultant enthalpy of ΔH 30 cal/gm. An order-disorder type of transformation mechanism is suggested. In some cases the changes in enthalpies of the doped materials are almost proportionate to the decreases in transition temperatures. This PE-TMP is the best example of a temperature adjusted solid-state phase change material in the 25-48°C temperature range. The type and amount of dopant are also critical if one is seeking high resultant enthalpies after doping, but in general transformation temperatures can be varied with other substitutional dopants such as PG and NPG. The interstitial dopant such as the Lewis acid (BF_3) has the opposite effect on the transition temperature and enthalpy from the effect of the base (NH_3), which is of significance from the standpoint of molecular interactions in the lattice. In fact, introduction of NH_3 in the lattice increased the transition temperature by 6.9°C with an enthalpy increase of ΔH by 10.91 cal/g. Although the goal is to decrease the transition temperature, increases give more latitude in the optimization of thermal storage material properties. Thus, if appropriate dopants which provide steric hindrances are found for polyalcohols whose transition temperatures are well below the desired temperature, then there may be a possibility of increasing the transition temperature of the materials. This may allow more heat-storage capacity than would be realized in the undoped condition.

CORRELATION OF STRUCTURE AND THERMAL PROPERTIES

The thermal properties and lattice parameters of pentaglycerine (PG) doped PE crystals were studied. Substitutional doping of PE with PG and neopentyl glycol (NPG) resulted in decreases in transition temperatures and enthalpies, but the enthalpy reductions were more than expected. This is attributed to differences in the number of O-H bonds per molecule; four in the case of PE and three in the case of PG and two in the case of NPG. The relative decreases in enthalpies associated with decreases in transition temperatures are more than expected, perhaps because of broken hydrogen bonds or severe localized distortions in the lattice due to the differences in the number of O-H bonds per molecule in these materials. The thermal data in Figures 3a and 3b show the changes in transition temperature (T_t) and the enthalpies (ΔH_t), respectively, as a function of concentration of PG in the host PE crystal.

The structural data for the same samples in Figures 3c and d show changes in the lattice parameters as a function of concentration of PG measured at room temperature. The contractions and expansions of the lattice in different directions correlate with the decreases in transition temperature of the doped samples. It appears that there is a complete but non-ideal solid-solution series between PE-PG with gradual increase in a_o and decrease in c_o, an indication of shrinkage in lattice between the layers of molecules with increase in the dopant concentration.

Also shown in Figures 3c and d are differences in lattice parameters in the undoped PE powders obtained from various sources as well as values for the purified PE powders obtained in our and other laboratories. This

Table 2. Results of Thermal Properties of Selected
Pentaerythritol in Undoped and Doped Conditions

Dopant		Transition Temperature (T_t)	Solid-Solid Enthalpy (ΔH) cal/gram	Change in Transition Temperature (ΔT_t)	Increase/ Decrease in Transition Temperature
Material	Amount Mole %				
No dopant	0%	188°C	68.0	–	–
TMP	50%	48.2°C	30.0	140.2°C	Decrease
NH_3	–	194.9°C	78.9	6.9°C	Increase
BF_3	–	180.5°C	50.2	8.5°C	Decrease
PG	50%	108.2°C	41.5	80°C	Decrease
NPG	50%	169°C	21.4	19°C	Decrease
Na-Acetate	20%	155°C	41.4	33°C	Decrease

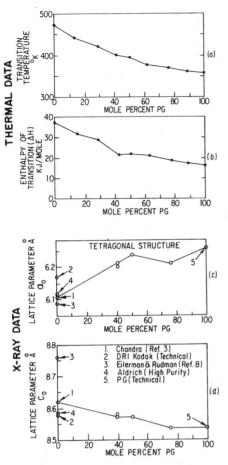

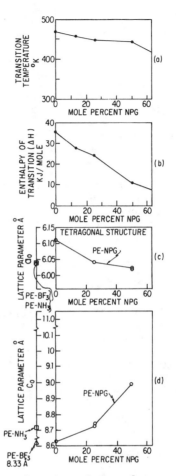

Figure 3. Thermal data for penta-erythritol (PE) doped with penta-glycerine (PG) showing variation in a) transition temperature (T_t); b) enthalpies of transition (ΔH) as a function of concentration of PG. X-ray data for this PE–PG series in c) and d) show complete solubility of PE in PG.

Figure 4. Thermal data for penta-erythritol (PE) doped with neo-pentylglycol (NPG) showing variation in a) transition temperature (T_t); b) enthalpies of transition (ΔH) as a function of concentration of NPG. X-ray data for this PE–NPG in Figures c) and d) show solubility up to ≃ 40% NPG. Interstitial Lewis acid and base dopants which cause swelling or concentration normal to the layer plane (001) are also shown in this figure.

suggests that either small amounts of impurities or some sample preparation techniques can also alter the lattice dimensions.

The changes in transition temperatures, enthalpies of transition and lattice parameters associated with the addition of NPG to the host PE crystal are shown in Figures 4a through 4d for the PE–NPG series. It was observed that complete solid solutions were not obtained in this series beyond 50% PG additions, where a second phase began to appear with its crystal structure of the monoclinic type, quite different from the tetrag-

onal structure of PE. However, the structure of PE is retained with up
to 50% NPG addition.

In comparing the thermal data obtained from the PE-PG series discussed above, the decreases in transition temperatures caused by the
addition of NPG to PE in the same concentrations as the PG additions,
were not as high as with the PG additions. Also the decreases in enthalpies of transition were greater for a given concentration of NPG than for
a similar concentration of NPG, possibly due to distorted or broken
hydrogen bonds. More interestingly, the axial ratios (c/a) increased
with the increase in the NPG concentration which is an effect quite the
opposite of that observed in the case of PG.

The addition of Lewis acids as interstitial dopants has shown that
there is swelling of the c-axis lattice with the addition of ammonia, and
shrinkage along the c-axis due to borontrifluoride. But there is negligible change in the a-axis of the tetragonal cell of PE. Preliminary
X-ray diffraction data on the TMP-doped PE samples have been obtained
recently, but have not yet been indexed.

DISCUSSION ON THE PHASE TRANSFORMATION IN PENTAERYTHRITOL

The changes in the lattice dimensions and crystal structure due to
heating through the transition temperature have been studied to evaluate
the possible crystallographic relationships between the low and high temperature phases. But the possible alterations in these changes accompanying doping have yet to be studied. When PE is heated through 188°C where
the body centered tetragonal transforms to face centered cubic the distances between neighboring molecules increase slightly and suggest epitaxy.
There is no shrinkage in the volume of the cell in spite of the fact that
there is closer packing in the FCC structure. If the tetragonal (001)
plane begins to transform epitaxially to face centered cubic, the molecules could reorient and shift short distances so that the (001) plane of
the tetragonal may become (001) of the cubic phase, with c_o tetragonal,
8.939Å becoming c_o cubic, 8.999Å (the dimensions are for 180° tetragonal
and 189° cubic[3]), an expansion of 0.69%. In the (001) tetragonal plane
the nearest neighbors are at a_o = 6.100Å, which may become the face
diagonal distances in the face centered cubic, $a_o \sqrt{2}/2$ = 6.364, an expansion of 4.3%. These dimension changes accompanying the transformation
may account for the crystal breakup and reorientation that was shown by
the Laue photographs of Nitta and Watanabe[9] when a tetragonal crystal was
heated into the cubic temperature range. Crystal breakup was also
observed in our project by optical microscopy when crystals were heated
through the transformation temperature, and the outgrowth of identically
oriented pyramids of the high temperature phase were seen on the cleaved
surface (001) of a crystal of PE.

The amplitudes and complexities of libration and translation movements
of molecules are undoubtedly very large in the cubic phase compared with
those of the tetragonal phase. Layering is not maintained; the cubic
phase is "plastic" not layered, and evidence has indicated that the
molecules are not rigid. The statistical preferences of molecular orientations in the cubic cell were addressed by Nitta and Watanabe,[9] but these
have not been determined, and perhaps cannot be, because of the very few
reflections available at these high temperatures (the temperature factor
is very high). As tetragonal crystals approach the transformation temperature, there is increasing molecular rotation due to progressive breaking
and rearrangement of hydrogen bonds and there may even be small cubic
nuclei forming and dissolving. Proton magnetic resonance studies[10] and

some Fourier transformation infrared spectroscopic experiments and calorimetric work in the present project have suggested these changes.

Possible models of the cubic PE structure have been considered in which dopant molecules can be inserted interstitially without drastic displacements of neighboring molecules. In one 3-dimensional model the interstitial molecule is placed in the octahedral void that is centered at 1/2 1/2 1/2, distant from the neighboring molecular centers by $a_0/2$ where it might be thought to fit even if it were a sphere of radius 2.249Å. In studying possible molecular orientations in this model, we suggest that rotation of two nearest neighbor host molecules on the cubic lattice can readily occur without altering the hydrogen bond length of 2.71Å, provided the rotations in these molecules are synchronous.

REFERENCES

1. D. K. Benson, J. D. Webb, R. W. Burrows, J. D. McFadden, and C. Christensen, SERI Report, Task Nos. 1275.00 and 1464.00 WPA304 (Mar. 1985) (available through NTIS, SERI/TR 255-1828 Category 62e).
2. J. Timmermans, J. Phys. Chem. Solids 18(1):1-8 (1961).
3. D. Chandra, J. J. Fitzpatrick, and G. Jorgensen, Adv. in X-Ray Anal. 28:353-360 (1985).
4. D. Chandra and C. S. Barrett, Effect of Interlayer and Substitutional Dopants on Thermophysical Properties of Solid State Phase-Change Materials, Topical Report to Dept. of Energy, Contract DE-AC03-84SF12205, Oct. 1, 1984 to Oct. 15, 1985.
5. N. Doshi, M. Furman and R. Rudman, Acta Cryst. B-29:143 (1973).
6. I. Nitta and T. Watanabe, Science Paper I.P.C.R. 34:1669 (1938).
7. F. J. Llewellyn, E. G. Cos and T. H. Goodwin, J. Chem. Soc. 1883 (1937).
8. D. Eilerman and R. Rudman, Acta Cryst. B35:2458-2460 (1979).
9. I. Nitta and T. Watanabe, Bull. Chem. Soc. Japan 13:28-35 (1938).
10. G. W. Smith, J. Chem. Phys. 50:3595-3605 (1969).

ANALYSIS OF THE THERMO-STRUCTURAL BEHAVIOR OF POLYETHYLENE

USING SIMULTANEOUS DSC/XRD

C. E. Crowder, S. Wood, B. G. Landes, R. A. Newman,
J. A. Blazy, R. A. Bubeck

Michigan Applied Science and Technology Laboratories
The Dow Chemical Company
Midland, Michigan

INTRODUCTION

Over the past 25 years, numerous studies of polymers utilizing both
X-ray diffraction (XRD) and differential scanning calorimetry (DSC) have
been reported in the literature. These studies have suffered because the
two techniques must be performed on separate samples and under conditions
that are often dissimilar. By combining the two techniques into one
instrument, typical problems encountered with variations in sample preparation and thermal and atmospheric environment are eliminated. This is quite
important in the study of polymers since one must match not only temperatures between the two techniques, but also heating rates as well. Matched
thermal conditions are necessary because polymer properties such as
crystallinity and crystallite size depend on both the temperature and
thermal history of the sample under study.

A device capable of performing simultaneous XRD and DSC experiments
has been described in a previous report[1]. The use of a position sensitive
detector for the X-ray analysis allows the two techniques to be performed

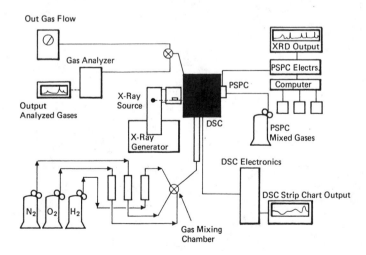

Figure 1. The DSC/XRD instrument.

315

on similar time scales. In this paper are described the capabilities of
this instrument in characterizing polymers, with polyethylene used as an
example.

EXPERIMENTAL METHOD

    A schematic of the instrument is shown in Figure 1. The design consists
of three basic systems: (1) an XRD system utilizing Guinier geometry and a
position sensitive detector, (2) a DSC oven with computerized data collection
and control, and (3) a gas intake manifold and gas detection system. A more
detailed description exists in a previous report[1]. Recent modifications have
been made such that the oven temperature control is now within 0.1°C, with
an upper temperature limit of 600°C.

    The polyethylene samples were hot-pressed films, 5-10 mg in mass.
Before taking data, samples were heated slowly to 160°C then cooled slowly
($<2^{\circ}$C/min) to eliminate effects of thermal history, strain, and orientation
in the samples. X-ray diffraction scans of the 10-35° 2θ range (Cu $K_{\alpha 1}$
radiation) were typically obtained in 1 to 2 minutes depending on the
crystallinity of the sample. These scans were taken at 2-4 minute intervals
during the second heating and cooling sequence. DSC heating and cooling
rates were 1-2°C/min depending on the time required for the XRD scans. The
sample chamber was continually purged with nitrogen during the experiments.

DATA ANALYSIS

    X-ray diffraction scans for a typical heating/cooling cycle are shown
in Figures 2 and 3. Each pattern was collected over a temperature interval
of 2°C. The patterns show three major features - two crystalline peaks,
namely the 110 and 200 Bragg reflections, and a broader amorphous signal.
The integrated area of each of these was determined by profile fitting using
a Split Pearson VII algorithm[2]. The result was used to determine the percent
crystallinity of the polyethylene for any pattern via the following formula:

$$\text{Percent crystallinity} = \frac{\text{Area of 110 \& 200 crystalline reflections}}{\text{Total area of 110, 200, and amorphous signals}}$$

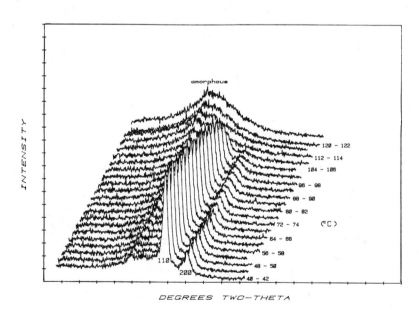

Figure 2. XRD data taken during heating of polyethylene.

The 110 and 200 reflections comprise more than 95% of the crystalline diffraction intensity for polyethylene; therefore the sum of the areas from these two reflections is a good approximation of the total diffraction from the crystalline portion of the sample.

The d-spacings of the 110 and 200 crystalline reflections were monitored with temperature to indicate the rate of thermal expansion in the crystalline part of the samples. The full width at half maximum (FWHM) for the diffraction peaks was used as an indication of the average crystallite size in the material as a function of temperature.

The DSC data for the melting (and subsequent crystallization) was digitally collected and subsequently plotted (Figure 4).

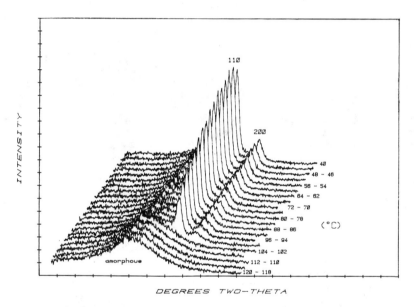

Figure 3. XRD data taken during cooling of polyethylene.

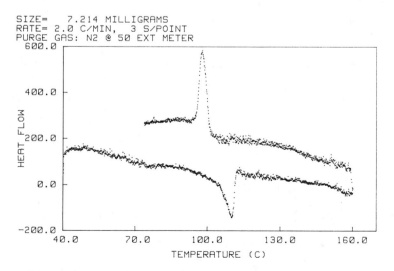

Figure 4. DSC data taken simultaneously with XRD data in Figures 2 and 3.

RESULTS

Out of many polyethylene samples that have been studied using this technique, two have been chosen to illustrate the differences that can be characterized via DSC/XRD. Sample A is a linear high-density polyethylene (LHDPE) and sample B is a linear low density polyethylene (LLDPE). The diffraction patterns shown in Figures 5 and 6 illustrate the differences in thermo-structural behavior for these two polyethylene samples. Figure 7 illustrates the XRD crystallinity as a function of temperature for the two samples. Sample A has an initial crystallinity of ~60% and shows little melting until a temperature of ~124°C is attained. Above this temperature, melting occurs rather suddenly as evidenced by the abrupt drop in the intensity of the crystalline diffraction peaks and the increase in the amorphous scattering. Sample B has an initial crystallinity of ~18% and shows gradual melting from room temperature to ~90°C.

On cooling, sample A begins to crystallize at about 10 degrees below the temperature that marked the end of the melting. This lag can be explained in terms of nucleation time. The crystallinity during cooling quickly matches the crystallinities obtained at the same temperatures during heating, indicating that crystallization proceeds at a relatively rapid rate. Sample B shows no corresponding lag in crystallization temperature on cooling. It appears that at the lower temperature, nucleating occurs as soon as crystallization is thermodynamically favorable. The percent crystallinity below this temperature duplicates the values found during heating. This would indicate that the crystallinities measured in this sample are very close to equilibrium values at the given temperatures.

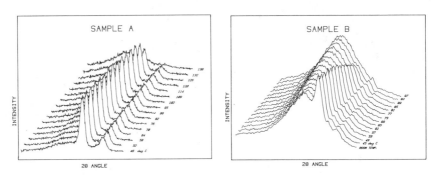

Figure 5.   Comparison of XRD patterns for heating of two
polyethylenes that differ widely in crystallinity.

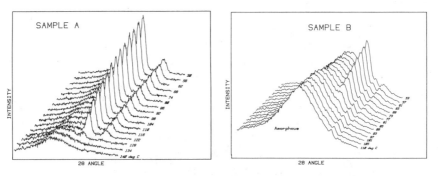

Figure 6.   Same as Figure 4, except during cooling.

Figure 8 illustrates the change in the 110 peak position as a function of temperature for samples A and B. The d-spacings corresponding to the observed 2θ's were used to determine a linear coefficient of thermal expansion along the 110 direction in the crystalline regions of this sample. For sample A, the d-spacing is increasing at a rate of ~0.00057 Å/deg which corresponds to a coefficient of ~0.000081°C^{-1}. For sample B, there is little expansion observed in the crystalline regions. This result is not fully understood; however it is recognized that the crystalline regions make up only 18% of the bulk. It is not to be inferred that this polymer sample does not expand with increasing temperature, but rather that any expansion that does occur must be attributed to the amorphous component of the material.

Figure 9 shows how the FWHM of the crystalline reflections changed as a function of temperature for the two samples. Little change is seen in the FWHM of the crystalline peaks for sample A indicating that the average crystallite size is not changing during the experiment. Sample B initially shows a larger FWHM corresponding to a smaller crystallite size. However, above ~75°C, the FWHM becomes narrower, indicating a larger average crystallite size exists at higher temperatures. Since the percent crystallinity is continually decreasing as the temperature is raised, this result is most likely due to a selective melting of the smaller crystallites at lower temperatures. This would mean that the average size of unmelted crystallites at the higher temperature is greater even if no crystallite growth is taking place.

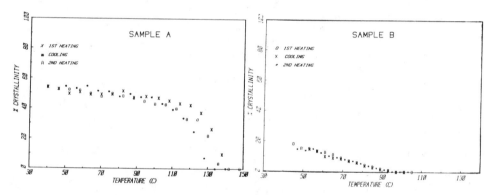

Figure 7.   Comparison of XRD crystallinities for samples A & B
             as functions of temperature.

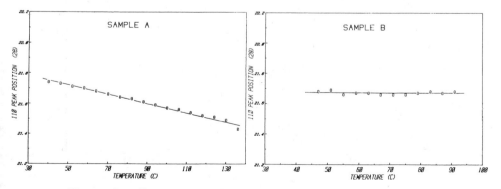

Figure 8.   Observation of crystalline unit cell expansion
             in samples A & B.

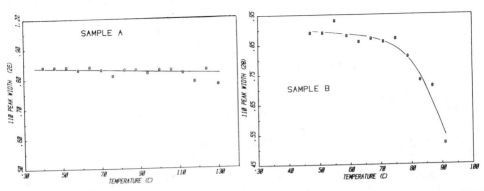

Figure 9.   FWHM's for the 110 reflections of samples A & B as functions of T.

       Results of crystallinities vs. temperature for other polyethylene
samples characterized by this technique are shown in Figure 10.   These
curves characterize each polymer according to its melting behavior and
indicate the state of the polymer at any given temperature.   This type of
information is valuable for many types of polymer fabrication processes.

ANOMALOUS MELTING BEHAVIOR IN A POLYETHYLENE SAMPLE

       A sample of LLDPE polyethylene that exhibited two distinct DSC melting
points above 120°C was studied using DSC/XRD in an attempt to explain this
behavior.   A trace of the DSC curve for this material when heated slowly
(0.3°C/min) is shown in Figure 11a.   In an effort to characterize the
material as it existed between the two melting endotherms, the sample was
heated at the same slow rate in the DSC/XRD cell to a temperature between
the two endotherms.   At this point, the material was held isothermal for 70
minutes for characterization before resuming the heating.   The resulting DSC
curve is shown in Figure 11b.   The second endotherm is obviously greater in
proportion to the first endotherm when compared to the original DSC.   The
X-ray diffraction scans show decreasing crystallinity up to the time that

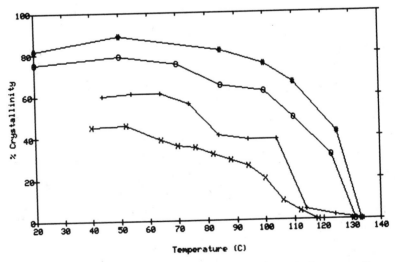

Figure 10.   Crystallinities of various polyethylene samples
              as functions of temperature.

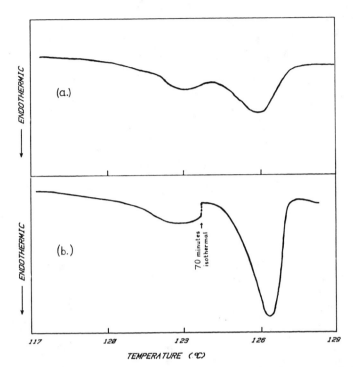

Figure 11. DSC traces for a polyethylene sample having two melting points.
(a) obtained at a constant 0.3°C/min, (b) obtained at 0.3°C/min
with a 70 minute isothermal wait at 124°C.

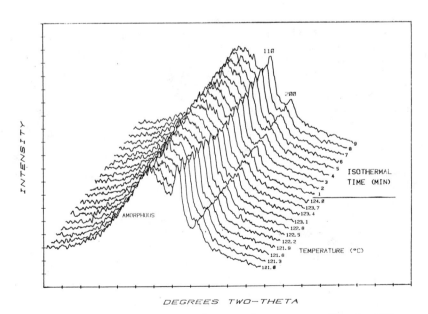

Figure 12. XRD patterns for sample corresponding to DSC
in Figure 11b.

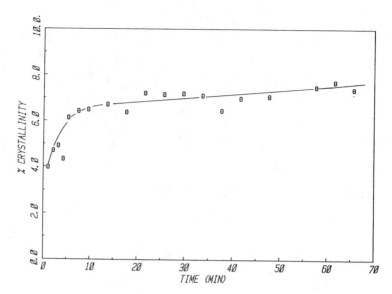

Figure 13.   Crystallinity vs. isothermal time at 124°C
             corresponding to Figure 11b.

isothermal conditions began (see Figure 12). During the isothermal period,
crystallinity was observed to increase from ~4% to nearly 8% as illustrated
in Figure 13. This evidence indicates that an isothermal recrystallization
of the polyethylene occurs at this temperature. Thus at least part of the
second endotherm represents melting of material which recrystallized from
the original melt.

CONCLUSIONS

     Thermo-structural analysis of polymers is one very powerful application
of the DSC/XRD. It has been shown that several structural variables of
crystalline polyethylene can be monitored as a function of temperature in a
single experiment. These include crystallinity, thermal expansion, and
crystallite size. Once these are known as functions of temperature, the
information can be used in optimizing plant processes to provide for more
efficient operation and easier fabrication of products. The information is
also useful in basic research for probing the microstructure of polymers.
Although this report has been concerned with only polyethylene, the technique is equally useful in studying other crystalline polymers (polypropylene, polyethylene terephthalate, etc.).

REFERENCES

1. T. G. Fawcett, et.al., The Rapid Simultaneous Measurement of Thermal
        and Structural Data by a Novel DSC/XRD Instrument, Advances in
        X-ray Analysis, Vol. 28, Barrett, Predecki, and Leyden, eds.,
        Plenum, 1985, p. 227.
2. A. Brown, J. W. Edmonds, The Fitting of Powder Diffraction Profiles to
        an Analytical Expression and the Influence of Line Broadening Factors,
        Advances in X-ray Analysis, Vol. 23, J. Rhodes, ed., Plenum, 1985,
        p. 361.
3. J. L. Matthews, H. S. Peiser, and R. B. Richards, The X-ray Measurement
        of the Amorphous Content of Polythene Samples, Acta Cryst., Vol. 2,
        (1949), p. 85.

ANALYSES OF MULTI-PHASE PHARMACEUTICALS USING SIMULTANEOUS

DIFFERENTIAL SCANNING CALORIMETRY AND X-RAY DIFFRACTION

T. G. Fawcett, E. J. Martin, C. E. Crowder, P. J. Kincaid,
A. J. Strandjord, J. A. Blazy, D. N. Armentrout, R. A. Newman

Michigan Applied Science and Technology Laboratories
The Dow Chemical Company
Midland, Michigan

INTRODUCTION

The analysis of multi-phase pharmaceuticals, particularly when similar structures are involved (i.e. polymorphs, salts or hydrates), can often be a difficult task. Historically, x-ray powder diffraction (XRD) and differential scanning calorimetry (DSC) have been utilized to study pharmaceutical samples. Relative to other materials, diffraction data for pharmaceuticals are often complex due to the large number of diffraction maxima caused by the size of the molecule and/or the molecular symmetry. Multi-phase mixtures tend to have a large number of overlapping peaks which can hinder the diffractionist's ability to identify phases and interpret the data. When similar structures are analyzed calorimetrically, their thermal events may severely overlap (as will be shown), preventing accurate interpretation of the data. In addition there are several types of thermal events which may not be related to structural transitions. A common one in pharmaceuticals is the loss of solvent or absorbed (versus molecular) water.

One way to study these complex problems is to utilize both high temperature x-ray diffraction and calorimetric techniques. However when comparing results from these two methods, questions usually arise concerning sampling methods, homogeneity, and instrumental differences, in particular with respect to heating and sample environment.

In a previous publication[1], we discussed the development of an instrument at The Dow Chemical Company which combines x-ray diffraction capabilities with simultaneous thermal analysis using differential scanning calorimetry. The calorimeter and diffractometer co-operate and co-act in real time on a single sample undergoing analysis. In the past year, a second generation DSC/XRD has been built. The high temperature range (600°C), sensitivity, resolution, and data handling facilities have been substantially improved for both the DSC and XRD components of the instrument.

This instrument has been used to look at the thermo-structural behavior of multi-phase samples obtained from Merrell-Dow Pharmaceuticals. A single experiment has been used to identify several structures and delineate their thermal characteristics.

EXPERIMENTAL

Three sets of samples were analyzed. Set I is a series of hydrates from a pharmaceutical intermediate. Set II contains two close melting polymorphic structures, and set III is a pharmaceutical hydrate. All three sets of samples were analyzed as lightly ground powders, approximately 5-10 mg per experiment. The samples were ground in a mortar and pestle for 3-5 seconds. Short grinding times were used to minimize atmospheric dehydration of the samples.

The DSC/XRD instrument is shown schematically in Figure 1 of the previous paper by C. E. Crowder et. al. and has been described previously[1]. In the second generation unit, the original DSC oven has been replaced with a custom designed sample holder which performs the calorimetric experiment while allowing the sample to be exposed to an x-ray beam. The change in the sample holder design has increased the operative temperature range to 600°C, and improved the sensitivity of both the x-ray diffraction and calorimetric components of the instrument. The x-ray diffraction data is recorded via a position sensitive proportional counter (PSPC) which has been interfaced to a PDP 11/34 minicomputer. The x-ray diffraction optics are in a transmission-subtraction Huber-Guinier geometry. Previous publications by this group have shown the utility of using Guinier optics to improve resolution[2], and the combination of Guinier optics and high temperature capabilities to thermally deconvolute complex samples[3]. Interfacing these systems to position sensitive detectors provides the powerful combination of rapid data collection and high resolution[4]. Recent work using an improvised slit system has shown that the position sensitive detector loses only about 20% resolution relative to the film detector data shown in reference 4. The high resolution and focusing optics allow the diffractionist to use a relatively short focusing radius (57 mm) compared to a diffractometer. In turn, the short radius enables the collection of data for approximately 25 degrees two theta ($CuK_{\alpha 1}$ radiation) simultaneously with the position sensitive detector (PSPC) in these experiments. Once the data is transferred to the PDP 11/34, the data can be fitted, calibrated, reduced, plotted and refined using several internally developed programs and those outlined in Figure 2 of reference 3.

The second generation DSC/XRD has replaced the DSC strip chart recorder and DSC electronics of the first unit with a new computerized controller which displays the DSC output on a CRT in real time. The controller/data analysis system uses a sophisticated internally developed program that not only controls the DSC operation but also analyzes the data including scaling, plotting and report outputs.

The entire system is gas tight, and a wide variety of gases have been used in the experiments. The system is currently arranged so that up to three gasses can be blended simultaneously. This manifold system allows samples to be exposed to inert, oxidizing or reducing atmospheres. A mass spectrometer has been used to monitor reaction gasses at the gas outlet of the sample holder. All gas lines from the sample to the mass spectrometer are controllably heated in the second generation unit. In the experiments described here, all except for one were run in a nitrogen environment with a flow rate of 18 mL/min. The mass spectrometer was not interfaced for these experiments.

Most experimental parameters on the DSC/XRD are completely programmable and can be varied interactively by the user. For instance, the XRD data collection times usually vary from 3 seconds to 10 minutes. In practice the XRD scan time usually depends on the sample crystallinity and the type of information which is desired from the experiment. The DSC has been run isothermally and also with scan rates from $0.5-10^\circ$/minute; once again the practical time used is generally related to the type of information desired.

RESULTS AND DISCUSSION

Sample Set I
     Figure 1 shows three DSC scans of a pharmaceutical intermediate. The
samples were taken from different lots of the same intermediate, all of
which were shown to be pure by HPLC; however they show several differences
in thermal behavior. These include the exotherm at approximately 105°C
observed only in the middle sample, the four endotherms observed in the
bottom sample, and the 3 degree melting point difference (133°C vs 136°C)
between the final melt in the bottom sample and the top two samples. These
data suggest that the chemistry of the intermediate may have changed from
lot to lot, and the 3 degree final melt difference suggests the possibility
of a polymorphic compound (similar to sample set II).

     A sample from the bottom lot of Figure 1 was analyzed using the DSC/-
XRD. The sample was run in air with a heating rate of 2 degrees/minute and
an XRD data collection rate of 5 minutes/scan. Therefore each XRD scan
covered a temperature range of 10 degrees. Figure 2 shows the resulting XRD
and DSC data. Even at a first glance it is apparent that several structural
transitions have occurred along with their corresponding thermal events
during the analysis. The peak at 3.80 Å steadily decreases from 20-70
degrees, while peaks at 4.76 and 5.27 Å increase from 40-70 degrees then
rapidly disappear in the next scan. To help delineate the scan to scan
changes, adjacent scans were subtracted from each other as shown in Figure 3.
In these data positive peaks are those which have disappeared due to a melting while negative peaks are those formed upon crystallization. These data
allow the user to assign both strong and weak peaks to a particular structure. In this manner, this single experiment allows the identification of
four distinct structures. Two of the structures exist at room temperature,
one is formed (and later dehydrates) between 70 and 90 degrees while the
final structure is the anhydrous form.

     From approximately 60-130 degrees there are 7 thermal events, four
endotherms and three exotherms. Many of these events strongly overlap as
shown by the DSC/XRD data in Figure 2. The weak endotherm observed at
approximately 75°C is actually a simultaneous endotherm and exotherm.
Knowing the thermo-structural chemistry of this sample allows one to explain
all the DSC scans shown in Figure 1 by using different concentrations of the
four observed structures. These data can also explain the different profiles given for the two samples from the same lot (bottom scan Figure 1
versus Figure 2). The sample used in the bottom scan of Figure 1 was in a
closed pan environment with a pinhole opening as compared to the open
environment of the DSC/XRD experiment. The closed pan environment
apparently altered the equilibria between various hydrates.

     Additional DSC/XRD scans of different samples show that one always
obtains the same final anhydrous form of the pharmaceutical intermediate.
Combining weight loss data with the DSC/XRD data, the three observed structures are identified as 1.5, 0.5 and 0.25 hydrates. Room temperature
samples are usually combinations of the 1.5, 0.5 and anhydrous forms. The
0.25 hydrate has only been found in some experiments, and usually exists in
a narrow 20°C temperature range.

     Figure 4 shows DSC/XRD data from another lot of the same intermediate.
At room temperature this material contains the 0.5, 1.5 and 0.0 hydrate
forms in order of decreasing concentration (identifications were made using
the characteristic data from the first DSC/XRD experiment). The XRD data
were particularly complex; therefore a whole series of difference plots similar to those shown in Figure 3 were used to outline the chemistry. Figure
4 shows the DSC data and the difference plot data for this experiment. The
difference plot data is simpler since it only records the changes between

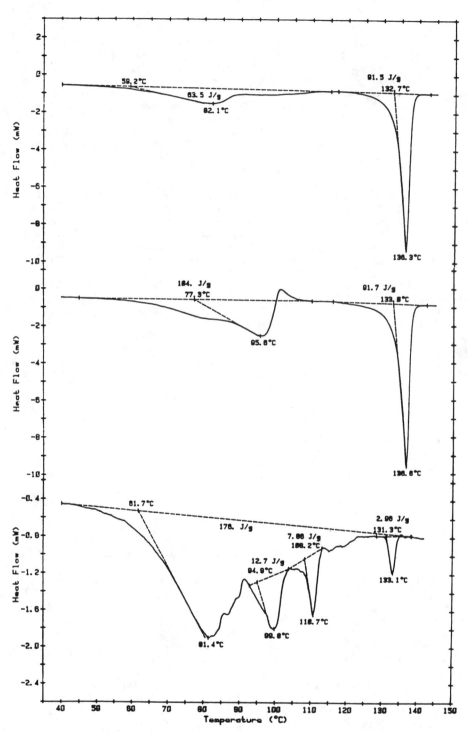

Figure 1.   Three DSC scans of different lots of the same pharmaceutical
            intermediate taken on a commercial DSC.

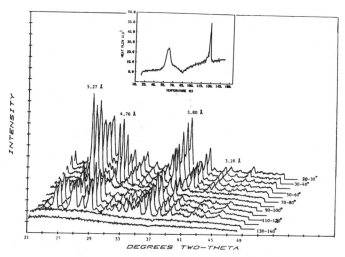

Figure 2.   DSC and XRD data of the pharmaceutical intermediate taken on
the DSC/XRD instrument.   In this DSC scan endotherms point up,
in all other data endotherms point down.

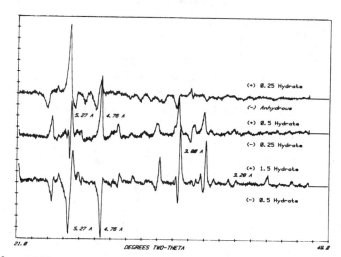

Figure 3.   Difference plots where adjacent XRD scans from Fig. 2 were
subtracted to highlight structural changes.

adjacent XRD scans such as melts (shown as positive peaks) and crystallizations (shown as negative peaks).   In this experiment the heating rate was
1 degree per minute with XRD scans every 5 minutes.   Therefore the difference plots show the changes taking place from one 5 degree interval to the
next.   These data closely reflect the DSC data.   The first scan shows the
melt of small amounts of the 1.5 hydrate.   Then some 0.5 hydrate crystallizes, adding to the 0.5 hydrate already present at room temperature.   The
0.5 hydrate has a broad melt as shown both in the XRD and DSC data from
40-80°C.   At approximately 80°C to 100°C the anhydrous form crystallizes,
but the exotherm is masked by both the endotherm of the 0.5 hydrate melt and
the small crystallization and dehydration of the 0.25 hydrate (90-100°C).
Finally, the data show the melt of the anhydrous form.   These data, both DSC

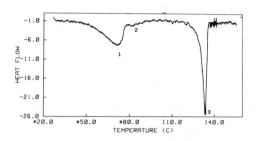

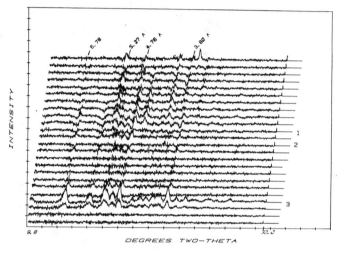

Figure 4. DSC (top) and difference XRD data (bottom) taken on another sample of the pharmaceutical intermediate. Numbers show data taken at equivalent temperatures.

and XRD, outline the thermo-structural behavior of these hydrates as well as their crystallization and dehydration kinetics.

Sample Set II

The second sample set demonstrates the interactive abilities of the DSC/XRD instrument. This pharmaceutical exists as two anhydrous polymorphic structures which melt at 146 and 150°C. DSC and XRD data were used at different laboratories to quantitate the polymorphic forms in various lots. The DSC quantitation used partial peak integration of the endothermic doublet to quantify the polymorphs. In addition, DSC scans must be conducted at relatively slow rates (2.5 deg/min or slower) to better separate the two endotherms. XRD quantitation was done using standard addition methods on room temperature samples. The DSC and XRD results were similar but did not agree in several cases.

A DSC/XRD experiment was conducted using the first generation DSC/XRD[1]. In this experiment the lower melting polymorph was heated until the start of its characteristic endotherm (145°C), at which point a 2 minute XRD scan was taken. The sample was cooled to 110°C, heated to 148°C, cooled to 110 and heated to 146 again, taking a 2 minute XRD scan at each interval.

The XRD data are shown in Figure 5. The characteristic diffraction maxima of the low melting polymorph is designated with a II, while characteristic peaks of the higher melting polymorph are designated I. The data show that the sample underwent a polymorphic interconversion with 80% of the lower

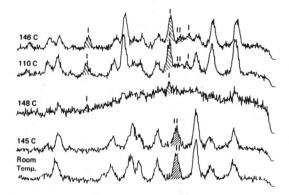

Figure 5: XRD data from an interactive DSC/XRD experiment showing a polymeric interconversion.

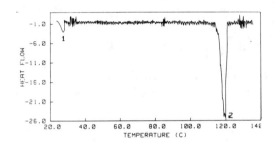

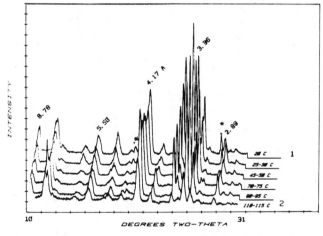

Figure 6: DSC (top) and XED (bottom) data from a hydrated pharmaceutical. Numbers show data taken at equivalent temperatures.

melting polymorph being converted to the high melting polymorph. The important result of this experiment is that slow heating of polymorph II can cause melting and partial recrystallization to polymorph I, causing erroneous quantitation of the polymorphs by DSC methods. The degree of error depends on the heating conditions and the amount of polymorph II originally present at room temperature.

Sample Set III

In the third set of samples a hydrated pharmaceutical was analyzed to delineate its complex hydration chemistry. This pharmaceutical exhibited lot to lot variations in melting behavior, crystal habit, and solubility when crystallized from aqueous solutions. In the first DSC/XRD experiment an anhydrous sample was dissolved in water and subsequently evaporated (not to dryness) at approximately 60°C until crystals formed. The entire procedure took 10 minutes. Figure 6 shows the DSC and XRD data from this experiment. The sample was heated at a rate of 1 degree per minute from room temperature (23°C) to 140°C. An XRD scan was taken every 5 minutes which corresponded to a 5 degree temperature interval. In Figure 6 only selected XRD scans are shown where the structure of the pharmaceutical changed.

As shown in the DSC there was a small endotherm at approximately 25°C. The XRD data show this endotherm as a structural transition: diffraction peaks at 4.3 and 3.0 Å disappear and the anhydrous peaks at 4.14 and 3.36 Å increase in intensity from 20-50 degrees. These data show one of the advantages to the DSC/XRD in that the XRD data identifies the early endotherm as a structural event and not an instrumental anomaly. After 50°C, the structure is in the anhydrous state until melting at 119°C, which is shown by both the large DSC endotherm and the corresponding XRD data.

A second sample of the same pharmaceutical was dissolved in water and allowed to slowly crystallize over a period of several days. The resulting elongated needles were crushed and lightly dried on a filter paper. The DSC/XRD data are shown in Figure 7. Both the DSC and XRD data show multiple events as the pharmaceutical dehydrates. Previous single crystal analyses and powder diffraction data performed at room temperature enable us to identify two anhydrous structures, a monohydrate, and a 1.5 hydrate by their characteristic d-spacings in the XRD data of this DSC/XRD experiment. Hygroscopicity data and density gradient data suggest that this pharmaceutical may also crystallize as a 2.0 or higher hydrate. Crystals of this pharmaceutical have been found containing approximately 18 weight percent water. At room temperature at least two structures are present; the 1.5 hydrate and most probably a higher hydrate. The DSC data show an endothermic doublet from 20-45°C. The XRD data indicate the disappearance of at least three structures and the recrystallization of two others over this 25°C temperature range. The 1.5 hydrate is characterized by a strong peak at 6.4 Å, which reaches maximum intensity at 35°C; it melts from 35-50°C (between points 1 and 2 on Figure 7). The hydration level of the other observed phases has not been identified. At 50°C, the DSC data are near baseline and the XRD data show the sample to contain a single monohydrate structure. At approximately 56°C, the monohydrate melts as shown by the loss of the characteristic XRD peaks and the observed DSC endotherm. The dehydration of the monohydrate coincides with the crystallization of one of the three known anhydrous structures. The background is elevated indicating the presence of an amorphous compound(s) in addition to the anhydrous structure. (The raised background is easily observed in the data even though it is not easily seen in the reduced graphical presentation of Figure 7.) At approximately 77°C (point 3 in Figure 7) an endotherm is observed in the DSC data and simultaneously a new phase crystallizes with a reduction in the amorphous scatter. This new phase melts at approximately 83°C as shown by a DSC endotherm and loss of the characteristic XRD peaks (point 5 in Figure 7). The anhydrous form increases in concentration with the melting of this phase. In the final stages of the experiment the anhydrous form melts over a broad temperature range from 100-120°C. Previous experiments show that the melt may also be accompanied by a partial sublimation. From 110-120°C there is a small amount of polymorphic interconversion as the initial anhydrous structure partially converts to another anhydrous

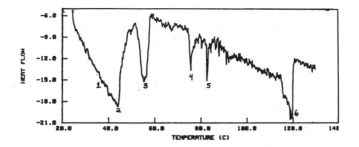

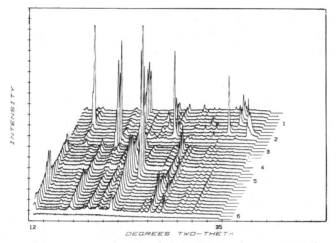

Figure 7: DSC (top) and XRD (bottom) data
showing the multi-stage dehydration of a
hydrated pharmaceutical.

structure. Both of these structures have been isolated and characterized
in previous experiments.

This experiment also illustrates some of the difficulties in analyzing
multi-phase pharmaceuticals. The number of thermo-structural transitions
make it hard to identify individual phases. For the above case there was
additional analytical data which was used to identify the different structures. For these experiments a mass spectrometer interfaced to the gas
outlet stream of the DSC/XRD could have aided in monitoring water and gases
from decomposition and sublimation processes. This data could have been
used to delineate dehydrations from polymorphic phase transitions. It is
interesting to note that in the dehydration processes shown in the first
and third sample sets, the DSC data are dominated by endotherms despite the
XRD evidence of multiple crystallizations (exothermic). There are additional
endotherms in this data due to the vaporization of water during the dehydrations. Therefore the endotherms of melting and vaporization overwhelm the
exotherm of crystallization during the simultaneous processes.

There is a large variation in the signal to noise levels for the
various DSC data shown in this report. Three factors contribute to the
variation. First the DSC/XRD experiments were run in an open pan configuration to simulate process conditions, second the large number of
thermal processes involved in the multi-stage dehydrations make the baseline
appear to be variable when in reality there are overlapping thermal events.
Finally, during the course of these experiments the DSC oven controller

design was altered. Initially temperatures could be controlled to within 1°C, currently temperatures are controlled to better than 0.1°C.

CONCLUSION

A second generation DSC/XRD instrument has been developed at The Dow Chemical Company. The instrument provides better data handling facilities, a higher temperature range (600°C) and increased sensitivity for both XRD and DSC data relative to the initial instrument. The DSC/XRD has been used to analyze the complex thermo-structural behavior of pharmaceuticals. In a single experiment the characteristic diffraction and thermal data for several structures can be elucidated. The instrument is especially powerful in analyzing multi-component mixtures since thermal transitions (or overlapping transitions) can be directly assigned to one or more structural events.

REFERENCES

1. T. G. Fawcett, C. E. Crowder, L. F. Whiting, J. C. Tou, W. F. Scott, R. A. Newman, W. C. Harris, F. J. Knoll, V. J. Caldecourt, The Rapid Simultaneous Measurement of Thermal and Structural Data by a Novel DSC/XRD Instrument, Adv. in X-Ray Analysis, 28, 227 (1985).
2. A. Brown, J. W. Edmonds, The Fitting of Powder Diffraction Profiles to an Analytical Expression and the Influence of Line Broadening Factors, Adv. in X-Ray Analysis, 23, 361 (1980).
3. T. G. Fawcett, P. Moore Kirchhoff, R. A. Newman, The Measurement of Thermally Induced Structural Changes by High Temperature (900°C) Guinier X-Ray Powder Diffraction Techniques, Adv. in X-Ray Analysis, 26, 171 (1983).
4. R. A. Newman, T. G. Fawcett, P. Moore Kirchhoff, Evaluation of Straight and Curved Braun Position-Sensitive Proportional Counters on a Huber-Guinier X-Ray Diffraction System, Adv. in X-Ray Analysis, 27, 261 (1984).

STRUCTURAL EXAMINATION OF IRIDIUM-BASED SINGLE-CRYSTAL PREPARATIONS

K. M. Axler, Materials Science and Technology Division

R. B. Roof, Chemistry Division

Los Alamos National Laboratory
Los Alamos, New Mexico

INTRODUCTION

A high-temperature crystal growth experiment produced discrete single-crystal products of AlIr and IrSi. The preparation and examination of these phases is described within. This project is part of a materials compatibility study relating to radioisotopic heat sources. These heat sources are comprised of a PuO_2 fuel pellet encapsulated in an Ir alloy containment shell. Thorium is introduced as an additive within the Ir to maintain ductility.[1] Si and P are picked up inadvertently in the fuel processing. The compatibility of the heat sources with Al is of interest because of potential interactions with the Al alloy hardware associated with the heat source environment.

MATERIALS PREPARATION

Single crystal products were obtained utilizing a technique in which the reactants were dissolved in a molten metal flux. A similar technique is described by Meisner[2] in a previously reported work. The starting materials were Th metal with a minimum purity of 99.9%, red amorphous P with a minimum purity of 99.7%, hydrogen-reduced Cu powder with a purity of 99.36%, and Ir metal with a minimum purity of 99.9%.

The Th, Ir, P, and Cu were combined in a molar ratio of 1:1:1:15 with additional P added to achieve an 8.4 wt. percent with the Cu. The additional P extended the liquid range of the flux by achieving the eutectic[3] composition within the Cu-P system. This provided a larger range of temperature for crystal growth. The materials were mixed thoroughly and then pressed into a pellet @ 5000 PSI. The pellet was then placed into an Al_2O_3 tube with the end packed with quartz wool and then sealed under vacuum. The sample was allowed to equilibrate at a maximum temperature of 1150°C for 24 hours. The furnace was then cooled at a rate of 1°C/hr until the sample was down to 660°C. When the heating cycle was completed, the tube was opened and the Cu-rich matrix appeared as a dark friable material which was dissolved in nitric acid. Microscopic examination of the remaining material revealed discrete metallic crystals with two distinct habits. One was cubic in appearance and the other was a rectangular prism with faceted faces on each end.

PRODUCT ANALYSIS

X-ray precession protographs of cube-shaped crystals indicate that they are indeed cubic. The space group is Pm3m. The refined lattice constant obtained from a Gandolfi film is 2.9867(1) Å. The lattice constant and space group suggest a metallic CsCl structure type. This structure has an Ir atom at the corner of the cubic unit cell and an Al atom in the center of the cube at ½, ½, ½. Based on the crystallographic data obtained, this crystal was identified as AlIr. Previously published information[4] on this compound confirms our identification. Elemental Al was made available in situ by the reduction of the Al_2O_3 crucible by Th. This

Figure 1. SEM photo of a crystal of AlIr viewed approximately along the cube diagonal. Surface contamination is ThO_2. See text for further details. 250 X.

Table 1. Comparison of observed and calculated intensities for AlIr. See text for details.

d_{obs}, Å	d_{calc}, Å	I_{obs}	I_{calc}	hkl
2.976	2.987	10	11	100
2.107	2.111	20	22	110
1.723	1.724	5	6	111
1.492	1.493	5	6	200
1.334	1.336	15	12	210
1.218	1.219	20	17	211
1.056	1.056	5	8	220
0.995	0.996	10	12	300,221
0.944	0.945	15	17	310
0.900	0.901	10	12	311
0.862	0.862	10	8	222
0.828	0.828	30	20	320
0.798	0.798	100	100	321

Figure 2.   SEM photo of a crystal of IrSi viewed approximately
along the orthorhombic unit cell diagonal.   In contrast to Fig. 1
the surfaces are clean.   See text for further details.   350 X.

is evident because of the relative stability of ThO$_2$ in respect to Al$_2$O$_3$ at
the elevated temperatures of the experiment.   The Gandolfi film clearly
reveals weak lines of a second material which is identified as ThO$_2$.
Figure 1 is a SEM photograph (250 X magnification) of an AlIr crystal
viewed approximately along the cube body diagonal.   The surface contamination is ThO$_2$.

A comparison between the observed and calculated intensities for the
Gandolfi powder pattern of an AlIr crystal is given in Table 1.   The observed intensities were obtained from visual estimates of the Gandolfi
film.   The calculated intensities for AlIr include the effects of absorption for a crystal approximately 0.1 mm in size.

X-ray precession photographs of the prismatic-shaped crystals revealed
orthorhombic symmetry.   Characteristic extinctions indicated space group
Pnma.   Refined lattice constants obtained from a Gandolfi film are a =
5.4996(7), b = 3.3052(9), and c = 6.1857(13) Å.   The lattice constant
ratios and space group suggest the IrSi structure as reported in the literature.[4]   X-ray fluorescence techniques applied to the single crystal revealed only the presence of Ir.   Si cannot be detected by this method as
fluorescence is generally only valid for elements heavier than Ca in
the periodic table.   The important point is that no heavy elements other
than Ir were found.   Elemental Si was made available in situ by the dissolution of the quartz wool into the metal flux and the subsequent reduction of
the quartz by the Th metal.   Figure 2 is a SEM photograph (350 X magnification) of a crystal of IrSi viewed approximately along the unit cell diagonal.   It is of interest to note that the crystal faces are clean and well
defined in contrast to Fig. 1 where the surface contamination was identified as ThO$_2$.

The IrSi crystal has the B31 (MnP) structure type with the Ir atoms in
the position set (4c) with x = 0.005 and z = 0.20.   The Si atoms are also
in the position set (4c) with x = 0.19 and z = 0.57.   A comparison between
observed and calculated intensities for the Gandolfi pattern of the mater-

Table 2.   Comparison of observed and calculated intensities
           for IrSi.   Only the low angle lines and intensities
           greater than 2% are listed.   Intensities for $\underline{d}$
           spacings smaller than 1.5 Å are in general quite
           weak.

d_{obs}, Å	d_{calc} Å	I_{obs}	I_{calc}	hkl
4.12	4.11	15	19	101
3.09	3.09	40	38	002
2.906	2.915	100	100	011
2.742	2.750	40	43	020
2.575	2.576	5	3	111
2.087	2.089	30	40	112
2.053	2.055	30	37	202
1.997	2.000	100	88	211
1.931	1.931	30	27	103
1.760	1.758	30	$\left\{ 5 \right.$	301
1.749	1.749		$\left. 16 \right\}$	013
1.650	1.653	15	15	020
1.530	1.533	--	3	121

ial is given in Table 2.   The observed intensities were visually estimated
from the Gandolfi film.   The calculated intensities are for position sets
fully occupied.   Corrections for absorption were not deemed appropriate for
the small range of d spacings listed.

Although the lattice constants obtained in the current work are slightly
different than the previously reported values[4] we maintain that the material
is IrSi.   Other known compounds of Ir (aluminides, silicides, and phosphides)
do not have the proper combination of axial ratios and space group to match
our observed data.

ACKNOWLEDGMENT

We would like to thank Dean Peterson for providing valuable technical
guidance in these studies.

REFERENCES

1. C. L. White, C. T. Liu, Acta Metallurgica $\underline{29}$, (1981), pp. 301-310.
2. G. P. Meisner, "The Superconductivity, Structure, and Magnetism
    of Some Ternary Metal Phosphides and Arsenides," Ph.D. Thesis,
    University of California San Diego (1982).
3. M. Hansen, "Constitution of Binary Alloys, Second Edition," (McGraw-
    Hill, New York, 1958), pp. 607-609.
4. W. B. Pearson, "Handbook of Lattice Spacings and Structures of
    Metals," Vol. 2, (Pergamon Press, New York, 1957), p. 121,
    p. 297.

X-RAY DOUBLE CRYSTAL DIFFRACTOMETRY OF MULTIPLE AND VERY THIN

HETEROEPITAXIAL LAYERS

B.K. Tanner and M.J. Hill

Department of Physics, University of Durham,
South Road, Durham, UK

INTRODUCTION

The intense interest in production of heteroepitaxial quaternary structures of $Ga_xIn_{1-x}As_yP_{1-y}$ on InP for electro-optical tele-communications systems has stimulated development of non-destructive techniques for their analysis. One of the most important is double axis X-ray diffractometry, a technique originally developed in the 1920s but only now coming into widespread use as a routine assessment tool. The basic theory is well treated by James[1] and discussion of alignment errors are found in references cited by Fewster[2] in a paper describing alignment procedures for the automated diffractometer manufactured by Bede Scientific Instruments of Durham. The application to III-V systems has been discussed by Tanner, Barnett and Hill[3].

In the (+−) parallel setting, the rocking curve width is the convolution of the plane wave reflecting ranges of the specimen and reference crystal. For a single epitaxial layer two peaks are observed, one each from the layer and substrate and the effective lattice mismatch obtained directly from the peak separation. However, injection lasers often contain an active quaternary layer sandwiched between two other layers of a different composition. Rocking curves from such structures are not so simple and, as we show here, a full simulation is necessary to interpret unambiguously the data. Naive assignment of one layer composition to each peak can lead to serious error.

ROCKING CURVE SIMULATION

Several workers have developed theoretical models for simulation of rocking curves from crystals in which the lattice parameter varies with depth as for example in an ion implanted crystal or an epitaxial layer whose chemical composition varies with depth. References are cited in our recent paper[4] describing the application of a model[5,6] based on solution of the Takagi-Taupin equation to graded epitaxial layers and multiquantum well structures. The model is quite general and is here applied to simulate rocking curves from typical multiple layer structures.

337

MULTIPLE LAYERS

With epitaxial layers greater than about 1μm in thickness interpretation is <u>normally</u> straightforward[b] but for sub-micron layers dynamical diffraction effects seriously affect peak shape and position.   An extreme example is shown in Fig. 1 of successive addition of layers A and B of mismatch -500 and -860 p.p.m.. Fig.1(a) is a symmetric 004, CrKα rocking curve of just two layers 0.6 μm (A) and 0.24 μm (B).   (InP 004 was used for the reference crystal reflection).   Here the positions of the peak maxima do yield the correct mismatch but the peak corresponding to the thinner layer (B) shows subsidiary maxima, X, almost as intense as the central peak. These are Bragg case Pendellosung fringes from a thin crystal. Similar oscillations are seen in the tail of the A peak.   Addition of a third layer A' 0.16 μm thick changes the shape of the A composition markedly although with little change in position of the peak maximum (Fig.1(b)).   The interference fringe structure Y becomes more pronounced, this now corresponding to 'gap fringes' which have been treated analytically in the Laue case but not, to our knowledge, in the present Bragg geometry.   The most dramatic change is in the peak B

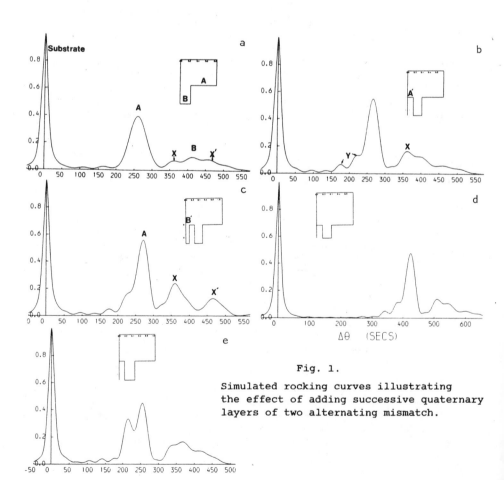

Fig. 1.

Simulated rocking curves illustrating
the effect of adding successive quaternary
layers of two alternating mismatch.

where the interference between layers A and A' skews the peak shape. Now, the heighest peak is at the position of the subsidiary maximum X in Fig. 1(a). Determination of mismatch from $m^* = \cot \theta_B \, \Delta\theta$ (where $\Delta\theta$ is the separation of this peak from the substrate peak) would thus lead to an error of about 15%. Note that the Fig.1b layer structure is similar to some real laser structures studied experimentally.

Addition of a very thin 0.1 μm layer B' has a dramatic effect on the B peak (Fig.1(c)). The maximum X, in Fig.1(b) increases in size and the other subsidiary peak X' also increases dramatically. At the same time, the central peak intensity falls, resulting in a rocking curve of three peaks A, X and X'. A naive interpretation of this curve would give three quaternary compositions not the two known to be present. Use of a second set of reflecting planes, (necessarily an asymmetric reflection), is needed to establish whether 2 or 3 compositions are present. Only if there are three compositions will the simple mismatch measurement be consistent between reflections. Use of the same reflection with a different wavelength does not give unambiguous identification. In the cases where we have altered the wavelength, the peak structures simply scale without changing shape.

Changes in the layer mismatches lead to similar rocking curve structures but on a different scale. Fig. 1(d) shows the rocking curve corresponding to Fig.1(b) but with mismatches −885 p.p.m. and

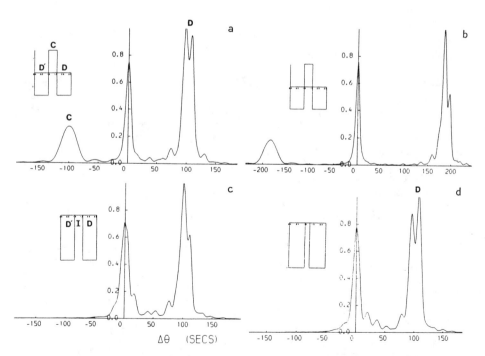

Fig. 2. Simulated rocking curves showing the effect of varying the separation between two quaternary layers of equal mismatch.

-1240 p.p.m. All structures visible in Fig.1(b) can be identified.
The interference peak structure is determined both by the layer
thicknesses and the X-ray structure factors. Over the range of
mismatch parameters used in devices the scattering factors change
relatively little and thus the layer thickness is the dominant factor.
As the layer thicknesses are identical in Figs.1(b) and 1(d), the
structures have the same form. A very small change in the thickness
of the layer B from 0.24 to 0.30 μm changes the rocking curve
dramatically (Fig.1(e)).

        We thus see why different reflections, with very different
structure factors, give contrasting rocking curve shapes. Fig.2 shows
an example of a three layer quaternary structure of mismatch −400,
+400, −400 p.p.m, 1 μm, 0.6 μm and 1 μm thickness respectively. The
peak D corresponding to the two −400 p.p.m. layers shows a clear
splitting in the 004 reflection with CuKα radiation and an 004 InP
reference crystal reflection (Fig.2(a)). It has a very different shape
in Fig.2(b) for the 115 reflection. Comparison of these two curves
enables "interference" peaks to be distinguished from "composition"
peaks. Substitution of an InP layer I for the 0.6 μm thick quaternary
layer C gives significant change in the peak D associated with the
other two layers (compare Figs.2(a) and 2(c)). The refractive index
for an InP layer is significantly different from that of the
quaternary and thus the phase difference across the "gap" of layer I
is different to that across C. Reduction of the thickness of the InP
layer I to 0.3 μm (Fig.2(d)) gives a return to a similar structure to
Fig.2(a).

SINGLE VERY THIN LAYERS

        We have seen how small changes of only a few thousand Angstroms
in the thickness of layers in a multiple structure can influence
markedly the rocking curve structure. However, for a single layer
less than about 0.2 μm thick, the peak profiles become extremely wide
and of low intensity (see Fig. 1(a) for an example). Use of grazing
incidence should enable sharp peaks of such layers to be obtained
permitting much more precise measurement of mismatch. Lyons and
Halliwell[7] have suggested that use of an asymmetrical reflection where
the Bragg angle θ_B is larger than the angle between Bragg planes and
crystal surface is a way of implementing this in the laboratory. By
rotation of the sample, a geometry can be chosen to give a grazing
incident beam. We have already shown experimentally how the layer to
substrate peak ratio rises as the wavelength of the radiation is
tuned to give a small angle of incidence[3]. The 044 reflection with
CuKα radiation is ideally suited to such low angle diffraction
experiments[3].

        In Fig.3(a) we show an experimental rocking curve of a layer
approximately 0.2 μm thick taken using the 004 reflection at a
wavelength of 1.5 Å. Here, both incident and diffracted beams make
large angles to the surface, and we see that the layer peak is very
broad and of low intensity. Fig.3(b) shows the low incidence angle
224 rocking curve at 1.54 Å of the same layer and we note now the
sharpness and relative height of the layer peak. Use of such highly
asymmetric reflections reduces the extinction distance and hence the
depth penetration of the X-ray wave.

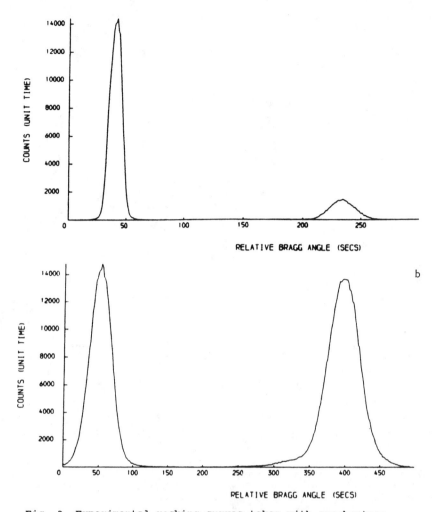

Fig. 3. Experimental rocking curves taken with synchrotron
radiation of a very thin quaternary layer on InP.
(a) 004 high incidence angle reflection 1.5 Å.
(b) 224 low incidence angle (4.74°) reflection 1.54 Å.

Although these experiments were performed with synchrotron radiation
at the SRS at Daresbury Laboratory, use of the 224 reflection and CuKα
radiation makes them transferable directly to the laboratory.   Lyons
and Halliwell's[7] technique, as yet experimentally unproven, gives even
greater flexibility.   The simulated rocking curves in Fig.4 of InGaAs
layer on InP suggests that experimental data on layers as thin as 350 Å
should be obtainable with CuKα radiation and the 044 reflection.

CONCLUSIONS

    We have shown that use of rocking curve simulation is essential
for the correct interpretation of double crystal X-ray rocking curves
of multiple layer heteroepitaxial structures.   Experimentally, we have

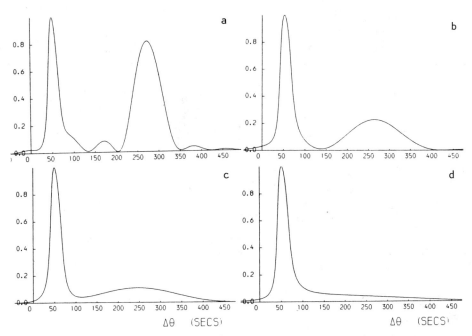

Fig. 4. Simulated rocking curves of very thin InGaAs layers taken
        with the 044 reflection at 1.54 Å where the incidence
        angle is very low.   (a) 1000 Å, (b) 500 Å, (c) 350 Å,
        (d) 200 Å thick layers.

demonstrated that the quality of III-V epilayers is such that sub-
micron layers can be studied in highly asymmetrical settings where an
X-ray beam grazes the specimen surface.   In such settings diffraction
peaks sharpen up dramatically enabling accurate mismatch and layer
thickness measurements to be made. We suggest that such X-ray rocking
curve measurements may be the most reliable method for measurement of
thickness of sub-micron layers.

ACKNOWLEDGEMENTS

     Financial support from the Science & Engineering Research Council
and British Telecom p.l.c. are gratefully acknowledged.

REFERENCES

1.   R.W. James, "The Optical Principles of the Diffraction of X-rays"
     Bell, London (1962).
2.   P.F. Fewster, Alignment of Double-Crystal Diffractometers, J.Appl.
     Cryst., 18 (1985), in press.
3.   B.K. Tanner, S.J. Barnett and M.J. Hill, X-ray Double Crystal
     Topographic Studies of III-V compounds, in "Microscopy of
     Semiconducting Materials, Oxford 1985", Inst. Phys. Conf. Ser.,
     Inst. Phys., London, Bristol (1985), in press.
4.   M.J. Hill, B.K. Tanner, M.A.G. Halliwell and M.H. Lyons,
     Simulation of X-ray Double Crystal Rocking Curves of Multiple and
     Inhomogeneous Epitaxial Layers, J. Appl. Cryst., 18 (1985), in
     press.

5. M.A.G. Halliwell, J. Juler and A.G. Norman, Measurement of Grading
    in Heteroepitaxial layers, in "Microscopy of Semiconducting
    Materials, Oxford 1983", Inst. Phys. Conf. Ser. 67. Inst. Phys.,
    London, Bristol, p.65 (1983).
6. M.A.G. Halliwell, M.H. Lyons and M.J. Hill, The Interpretation of
    X-ray Rocking Curves from III-V Semiconductor Device Structures,
    J. Crystal Growth 68: 523 (1984).
7. M.H. Lyons and M.A.G. Halliwell, Double Crystal Diffractometry of
    III-V Semiconductor Device Structure, in "Microscopy of Semi-
    conducting Materials, Oxford 1985", Inst. Phys. Conf. Ser., Inst.
    Phys., London, Bristol (1985), in press.

ASSESSMENT OF MULTIPLE EPILAYER III-V COMPOUND SEMI-CONDUCTORS

BY SYNCHROTRON RADIATION DIFFRACTOMETRY

D.K. Bowen, S.T. Davies and S. Swaminathan*

Department of Engineering
University of Warwick
Coventry, U.K.

INTRODUCTION

An extensive programme of characterization of optoelectronic device material has been performed at the Synchrotron Radiation Source, Daresbury Laboratory, in collaboration with Plessey Research, Caswell. The material was grown by Plessey Research by liquid phase epitaxy on InP substrates and had quaternary active layers with, usually, four epilayers in total. Some specimens had graded epilayers. This paper reports use of the methods of double crystal topography, rocking curve analysis and simulation, selective etching and Talysurf measurement in order to develop and assess non-destructive methods of evaluation. The destructive methods above were therefore used in order to test and verify the non-destructive X-ray techniques.

EXPERIMENTAL TECHNIQUES

Specimens were grown on indium phosphide substrates by liquid phase epitaxy at Plessey Research, Caswell. Many configurations were studied, but a typical specimen would consist of an InP substrate, n-InP buffer layer a few microns thick, a quaternary (Q) active layer around 0.2 μm thick, a p-InP layer about 2 μm thick, followed by a thin ternary (T) or Q cap in order to reduce the contact resistance. The use of Q layers permits the device designer, in principle, to maintain independent control over both the band gap and the crystal lattice parameter. Thus, tuned solid state lasers may be made that are free from interfacial dislocation networks, which are associated with both poor device performance and with device degradation.

Nevertheless, many problems can occur in device fabrication. The difference in thermal expansion coefficients between the substrate and the Q or T layer or layers means that the lattice parameter can only be matched exactly at one temperature, normally chosen as the growth temperature, leaving some residual strain at room temperature. The growth of a sub-micron layer by liquid phase epitaxy may be completed in a matter of seconds, leading to difficult control problems and some uncertainty in layer thicknesses. Selective loss of certain components from the solid or

---

*Now at Dept. of Electronics, Roorkee University, Roorkee, India.

melt may occur, leading to non-stoichiometry, and convective, turbulent and diffusive effects in the melt may lead to a variation in stoichiometry across the surface of the material as well as an error in the absolute stoichiometry.

A hetero-epilayer is therefore mismatched, bent, misoriented with respect to the substrate and has variable stoichiometry in all three orthogonal directions. At some level, these defects will affect device performance, through their influence on carrier life times, band gap values, and on non-radiative recombination centres. It is therefore essential to characterise and to assess the materials in ways that are directly useful to the crystal grower and to the device maker, and it is non-destructive methods that enable the loop between growth, structure, properties and device performance to be closed. As has been shown by, for example, Petroff, Sauvage and Riglet[1] and by Tanner and Hill[2] double crystal X-ray topography and diffractometry is an important, sensitive and non-destructive method for device material characterisation. It is of course necessary to demonstrate that X-ray methods give reliable information, and in particular that the assessment is more accurate than the predictions of the crystal grower. In the work reported herein, quantitative comparisons using destructive but very accurate metrological methods have been used to substantiate this claim.

The Synchrotron Radiation Source at Daresbury Laboratory was used for the experiments. The double crystal camera at this facility has been described by Bowen and Davies[3], and was used either with an InP first crystal or with a double reflection silicon 111-order sorting beam conditioner as described in the above reference. For the topographs and rocking curves reported in this paper the 400 reflection from the InP or quaternary layer was used. Whilst many of the experiments could be performed on a laboratory generator, synchrotron radiation provides the advantage (apart from intensity) of tunable wavelengths, polarization, penetration, etc., which can be exploited in order to discover the optimum conditions. In the reflection case, the intensity of the wavefield inside the crystal falls off as $\exp(-2\pi/\xi)$ where ξ is the extinction distance. Absorption only becomes important at low incident angles to the surface, as illustrated in the paper by Tanner and Hill[2]. By choosing the radiation and the reflection (including its symmetry), the penetration may thus be varied between about 0.05 and 10 μm . This is ideally matched to device structures; of course, the geometric distortion of the image is not then under control, but this is unimportant.

The normal experimental procedure was to take a rocking curve of the specimen crystal, to repeat this after rotation through 90° and 180° in its plane in order to distinguish between mismatch and misorientation, then to take topographs with the specimen set at different positions on the rocking curve. This can be used to distinguish between defects in the substrate and those in the epilayers. When using the beam conditioner, the X-ray beam divergence seen by a point on the specimen is about 1 arc second. Since the narrowest rocking curve observed for III-V crystals is several arc seconds wide (full width half maximum) the specimen is, in effect, illuminated by a plane wave; the (+ , -) symmetrical double crystal setting with an InP first crystal will effectively also give plane wave contrast (strictly, a convolution of the plane wave reflection curve of the first crystal with that of the second crystal). The qualitative features of topographic contrast may therefore be understood by considering the diffraction of a plane wave from the specimen. It is clear that

1.   The substrate and an epilayer that is slightly mismatched will diffract at different angular positions; for example, a fairly typical

mismatch of 220 ppm gives a peak separation of 28 arc seconds for the 400 reflection at 1.5 A .

2. A misorientation of more than a second or so of the epilayer with respect to the substrate also gives a peak splitting effect, and both the magnitude and direction of the misorientation may be found and distinguished from that due to mismatch by repeating the measurements of the rocking curve at at least two other rotations of the specimen about its surface normal.

3. Curvature of the epilayer/substrate through the induced stresses caused by mismatch will broaden the rocking curve; if a small collimator is used so that the broadening is not appreciable then there will be an absolute shift in the angle of the rocking curve if different parts of the specimen are examined.

4. Since for a mismatch of 100 ppm or greater, the peaks from the substrate and epilayer are clearly distinct, the specimen may be set at one or other of these peaks for the exposure of a topograph. This allows distinction of the defects in the epilayer from those in the substrate, and thus allows one to see which of the substrate imperfections are transmitted to the epilayer.

For complex epilayer structures, however, as shown by Tanner and Hill[2], there may not be a simple correspondence between rocking curve peaks and individual layers. In these circumstances, which often in practice corresponds to the circumstances in which topographs show little contrast due to the quality of the layers themselves, the method of rocking curve simulation[2] is far more powerful. It can be used to determine stoichiometric fluctuations, epilayer compositions and gradings, and layer thicknesses.

The use of X-ray methods for determining layer thickness after verification by computer simulation of a complex multiple epilayer structure, is a new application and requires verification. This was performed by partial or selective etching followed by engineering metrology. Selective etchings are discussed by Phatak and Kelner[4], Komiya and Nakajima[5], Nelson et al.[6], Abrahams and Buiocchi[7] and by Akita et al.[8]. Those used in this case were:

3:1:1 $H_2SO_4:H_2O_2:H_2O$ for selectively removing a GaInAsP or GaInAs layers, and
4:1 $HCl:H_2O$ for InP layers

both these etchants remove material at approximately 1 μm per minute, but careful control of the time is not necessary if the whole of a layer is being removed. Lacomit varnish was used to inhibit the etching action on certain areas, in order that windows could be etched through to differentiate layers. This meant that the thickness of the layers could be measured by engineering metrology, after rocking curves had been taken from each windowed area.

The metrology was performed using a Rank Taylor Hobson Talysurf Mk.V. This is a stylus instrument with a measuring accuracy of about 1% for sub-micron surface features. The stylus was drawn across the surface of the specimen, enabling the measurement of the step height associated with the transition to a new layer.

EXPERIMENTAL RESULTS

Application of topography

Double crystal x-ray topography will show regions of strain in the lattice, with a strain sensitivity that might be as high as 10^{-8}, and will also image directly any dislocations that are present in the sample. Figure 1(a) shows the rocking curve of an edge emitting LED structure. The FWHM is about 10 times the theoretical width for a perfect specimen, yet there is no splitting of the peak. Whilst a perfect mismatch is unlikely, it is plausible that the splitting is masked by other strains in the crystal. The topograph, 1(b), shows one reason for this: there is a substantial cross-hatched dislocation network typical of interfacial dislocations. There may, of course, be other sources of strain or inhomogeneity present, but for a relatively simple device such as an LED this characterisation is all that is necessary. LED's made from the same wafer as this specimen showed degradation effects.

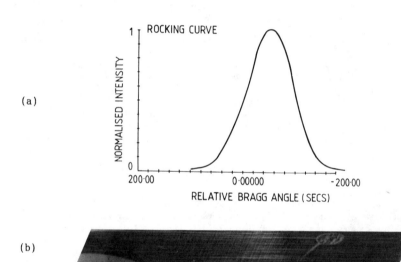

Figure 1. (a) Rocking curve for an edge emitting LED structure, (b) Topograph taken with the crystal set at the top of the peak. Silicon 111 Beam conditioner, 1.0A, slit size 100 x 500 microns.

In contrast, the rocking curve for another LED structure showed a much narrower substrate peak (though still greater than theoretical), shown in Figure 2(a). Topographs taken with the specimen set at the substrate peak and the epilayer peak respectively are shown in Figs. 2(b) and 2(c). Interfacial dislocations are again seen in this sample, but the epilayers are clearly more perfect than the substrate. However, the sample contains two quaternary peak layers (with an average mismatch of 1240 ppm); whilst the asymmetry of the quaternary peak may indicate a slight difference between them (although this can also be caused by grading), it is not possible to separate the images of the two Q layers.

(a)

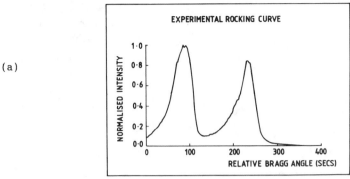

(b)

(c)

Figure 2(a) Rocking curve for another edge emitting laser diode, (b) Topograph with a specimen set on the substrate (left hand) peak, (c) Topograph taken with the specimen set on the quaternary (right hand) peak. First axis: high perfection InP crystal, 400 reflection at 1.5A, slit size 100 x 2000 microns

Application of rocking curve analysis and simulation

A number of workers have developed theoretical models for simulation of rocking curves for crystals in which the strain or lattice parameter varies as a function of depth. This work forms parts of a collaboration with B.K. Tanner's group at the University of Durham and we used the computer program written by M.J. Hill[9]. It should be said that no uniqueness theorem has

been proved for rocking curves, and it is not impossible that two different
structures could be found that would give similar rocking curves. However,
enough is usually known about the structure for the possibility of
confusion to be most unlikely. It is essential to know the substrate
material, the number of epilayers, the element present in the epilayers, an
accurate specimen rocking curve that has been corrected for misorientation,
and the reflection plane and wavelength. For a quaternary layer it is
necessary to have an additional measurement of some sort in order to settle
the composition (because when the lattice parameter is fixed there is still
one degree of freedom in the composition) and a simple measurement is that
of the band gap by means of photoluminescent measurement. Given this
amount of information, although it may take some considerable time to find
the best match between the simulated and experimental rocking curves, the
sensitivity of the rocking curves to minor variations in the structure is
quite remarkable. Hence, it seems unlikely that any ambiguity will remain
at least in the major features of the structure.

An example will clarify this claim. A linearly graded quaternary layer was
grown specifically for testing the methodology. The crystal growers
reported that the layer was approximately 4.2 microns thick, and was
linearly graded in two directions, with first decreasing and then
increasing mismatch, thus a complex rocking curve is to be expected, as
shown in Figure 3. However, despite considerable variation in the amount
and type of grading, no reasonable fit could be obtained between the
experimental rocking curve and that simulated for a doubly graded epilayer.
As seen in Figure 3(a) - (e) a doubly graded epilayer gives a relatively
poor fit which is greatly improved in Figure 3(f) - 3(i) by assuming a
singly graded layer and finally an excellent fit is obtained in Figure 3(j)
by slightly changing the assumed thickness of the layer and introducing a
small (0.7 micron) ungraded region near the interface. Figure 3(j) shows
as good a match as is normally obtained, and in fact the differences
between the experimental and simulated curves in this figure are less than
the differences found between experimental curves from one point to another
on the specimen. We shall return to this evidence of inhomogeneity later.

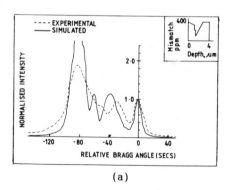

(a)

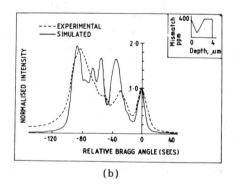

(b)

Figure 3.   Comparison of simulated and experimental rocking curves for a
graded epilayer. The thickness and mismatch assumed is shown in the top
right hand corner inset of each Figure. 400 reflection, 1.5 A.

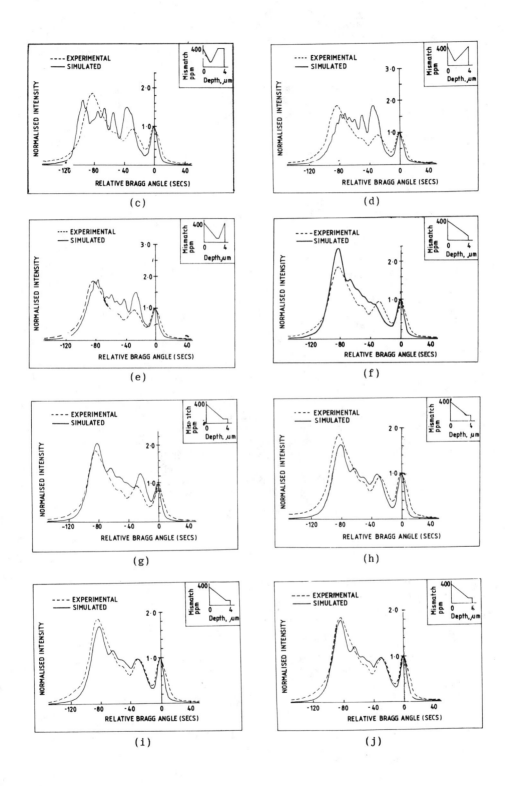

The structure as finally deduced is two-fold: a section adjacent to the substrate having a uniform mismatch of 100 ppm and a thickness of 0.7μm , and a section 3.1μm thick with a linearly graded mismatch varying between a 100 and 395 ppm.

The specimen was then windowed with Lacomit as described above, rocking curves were measured for each window, which now corresponds to different layer thicknesses, then these were compared with simulated rocking curves using the above structure in each case and varying only the thickness. The step changes between adjacent windows were finally compared with Talysurf measurements. The first step could not be measured reliably because of an uneven surface, but the others were as follows:

Position	Talysurf	Simulation
C	Uncertain	0.25 μm
D	0.14 μm	0.15
E	0.15	0.10
F	0.21	0.20

The agreement between the values measured by direct metrology and by non-destructive x-ray data is quite good, bearing in mind the variation from point to point that is undoubedly present (the method does not permit direct measurement of the layer thickness at precisely the same points as those measured by x-rays). It is probably reasonable to claim that thickness measurements at the 0.1 or 0.2 micron level can be made by x-ray rocking curves to an accuracy of about 10%. This is confirmed by another experiment in which the Q cap thickness of a four layer heterostructure laser diode was measured; the predicted thickness of the cap was "at least 0.3 microns", that derived from rocking curve measurements was 1 micron and that determined by Talysurf measurements was 0.92 microns. In this case as well as that of the graded epilayer, it was necessary to determine the grading of the epilayers by means of rocking curve simulation in order to get a reasonable fit with experimental data, which makes the agreement on the thickness determination still more convincing.

REFERENCES

1. J.F. Petroff, M. Sauvage and P. Riglet and H.Hashizume, Phil. Mag., 42A, 319 (1980).
2. B.K. Tanner and M.J. Hill, Advances in X-Ray Analysis, this volume.
3. D.K. Bowen and S.T. Davies, Nucl. Instr. and Methods, 208, 725 (1983).
4. S.B. Phatak and G. Kelner, J. Electrochem. Soc., 126, 287 (1979).
5. S. Komiya and K. Nakajima, J. Cryst. Growth, 48, 403 (1980).
6. R.J. Nelson, P.D. Wright, P.A. Barnes, R.L. Brown, T. Cella and R.G. Sobers, Appl. Phys. Letts., 36, 358 (1980).
7. M.S. Abrahams and S.J. Buiocchi, J. Appl. Phys., 36, 2855 (1965).
8. K. Akita, T. Kusunoki, S. Komiya and T. Kotani, J. Cryst. Growth, 46, 783 (1979).
9. M.J. Hill, B.K. Tanner, M.A.G. Halliwell and M.H. Lyons, J. Appl. Cryst. 18 (1985) in press.

CHARACTERIZATION OF EPITAXIAL FILMS BY X-RAY DIFFRACTION

Armin Segmüller

IBM Thomas J. Watson Research Center

Yorktown Heights, New York 10598

ABSTRACT

In this paper, the application of recently developed x-ray diffraction techniques to the characterization of thin epitaxial films will be discussed. The double-crystal diffractometer, with high resolution in the non-dispersive arrangement, enables the materials scientist to study epitaxial systems having a very small mismatch with high precision. A key part of the characterization of an epitaxial film is the determination of the strain tensor by measuring lattice spacings in various directions. The determination of strain and composition profiles in ion-implanted films, epitaxial layers and superlattices by rocking-curve analysis will also be reviewed. Grazing-incidence diffraction, an emerging new technique, can be used to obtain structural details parallel to the interface on films with thicknesses ranging down to a few atomic layers. The synchroton has now become increasingly available as a powerful source of x radiation which will facilitate the application of conventional and grazing-incidence diffraction to ultra-thin films.

INTRODUCTION

Important processing steps in the manufacturing of integrated circuits are the deposition of epitaxial films onto and the implantation of dopant ions into semiconductor crystals. Structural parameters affecting the band structure of epitaxial and ion-implanted layers, such as lattice parameter, strain and composition, can be obtained in a quantitative, fast and non-destructive way by x-ray diffraction.

X-ray diffraction methods have been widely employed in the past for characterization of epitaxial films. Photographic film methods often give fast information on film/substrate orientation. The Laue back-reflection method, the glancing-angle method, i. e. mounting the sample stationary in the center of a Debye-Scherrer camera with the primary x-ray beam incident at a glancing angle about equal to a low Bragg angle of interest, and the Weissenberg and Buerger moving-film methods are all useful. Diffractometer methods facilitate the quantitative measurement of Bragg angles, intensities and diffraction line profiles. An extensive review of x-ray diffraction methods for thin-film characterization was prepared recently (1). In this paper, emphasis will be given to the rapid development of double-crystal diffractometer methods during the last decade, and to the newly emerging grazing-incidence diffraction technique.

BRAGG DIFFRACTION FROM EPITAXIAL FILMS

Powder diffractometers can be used for epitaxial film characterization if the angle ω between the sample surface and the incident beam can be adjusted so that the single-crystal film is diffracting maximum intensity into the detector set to an angle $2\theta_o$, where θ_o is the Bragg angle for the reflection. If a 2θ-θ-scan is executed with narrow receiving slits in front of the detector, the line width observed in this scan is determined by the crystallite size and strain profile along the diffraction vector. If the detector is kept stationary at the angle $2\theta_o$ and the sample is rotated around the diffractometer axis, the line width observed in this so-called ω-scan for the epitaxial layer, as compared to that of the substrate reflection, is a measure of the orientational spread of the lattice planes in the epitaxial layer.

A curved-crystal monochromator, e. g. pyrolytic graphite, frequently used for powder diffractometry, is not of much advantage, since the single-crystal sample cannot utilize the divergence of the monochromated and focussed beam without broadening the reflection line. A channel-cut, perfect germanium crystal, diffracting the primary beam twice from $(111;\bar{1}\bar{1}\bar{1})$ or $(220;\bar{2}\bar{2}0)$ planes non-dispersively, as shown in Fig. 1, provides a highly monochromatic and well collimated beam of Cu-Kα_1 radiation with a divergence of a few $1/1000°$ (2). Figure 2 shows a 2θ-θ-scan of the (002) reflection of a (Ga,Al)Sb superlattice consisting of 10 double-layers with a period $D=(633\pm1.5)$Å as determined from the spacing of the superlattice reflections around $\theta=14.5°$. At $\theta=15.8°$ the (200) reflection of the GaAs substrate can be observed. The sample was prepared by Molecular Beam Epitaxy (MBE) with the growth sequence: 1000Å GaAs, 3000Å AlSb as a buffer, 10 periods of alternating layers of GaSb and AlSb, 181Å and 452Å thick, repectively, and 200Å GaSb as a protective cover (3).

The angular positions of the superlattice peaks or satellites are given by

$$\sin \theta_n = \frac{\lambda}{2} \left(\frac{1}{d_o} \pm \frac{n}{D} \right) \tag{1}$$

where λ is the x-ray wavelength, d_o the lattice spacing averaged over one period of the superlattice, and n the order of the satellite. The period D is obtained by plotting $\sin\theta_n$ *versus* n, and fitting a straight line to the data points by the least-squares method. The superlattice period D is then obtained from the slope of the straight line. The assignment of the orders n to the satellite peaks is sometimes not unique from one diffraction pattern alone. It is recommended to compare for instance the lattice spacings obtained from the (002), (004) and (006) diffraction patterns with the average spacing d_o calculated from growth data. In Fig. 2 the highest peak at $\theta=14.50°$ is assigned the order $n=0$. From the intensity distribution of the satellite pattern the composition and strain modulation can be obtained (4,5), as discussed later. It should be noted that the superlattice period D, obtained from the diffraction pattern, is rarely precisely the one computed assuming an integer number of molecular layers separated by the strained lattice spacing. This incommensurability is a consequence of the physical reality of the growth process that does not terminate a sublayer with a complete molecular layer.

The diffractometer depicted in Fig. 1 was also used to observe small-angle interferences on thin films and multilayers. With this method the thickness, average electron density and the surface roughness of a thin film can be determined within a certain thickness and absorption range. Periodic or near-periodic multilayers can be modelled by stratified layers with specified thicknesses and refractive indices. More details as well as literature can be found in the recent review (1).

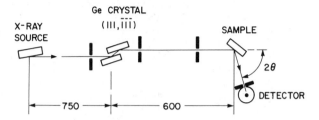

Fig. 1. Schematic diagram of x-ray diffractometer with non-dispersive, channel-cut monochromator of single-crystal germanium. Measures in mm.

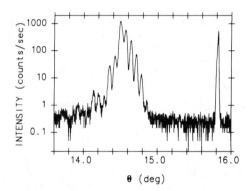

Fig. 2. Diffraction curve of symmetric (002) reflection of a (Ga,Al)Sb super-lattice with a period of $D=633\text{Å}$. At right, GaAs (002) substrate peak.

Using a Double-Crystal Diffractometer (DCD) in the non-dispersive arrangement, the resolution can be improved up to the theoretical limit. The use of the DCD as an analytical tool to measure an extremely small lattice mismatch in epitaxial systems came about 15 years ago with the rising interest in garnet bubble devices (6). Figure 3a shows the DCD for symmetric reflection by monochromator and sample. If the Bragg angles of monochromator and sample are exactly equal the dispersion of the monochromator is cancelled by that of the sample and the twice-diffracted beam is parallel to the primary beam. To obtain a 'rocking curve' the monochromator is kept stationary and the sample is rotated changing the angle ω. The detector is usually also kept stationary with its active area sufficiently large to receive the full width of the rocking curve. The observed rocking curve is the convolution of the diffraction curves of the two crystals. For perfect crystals the diffraction curve is close to the so-called 'Darwin' curve with a width of a few seconds of arc (7). The width of the Darwin curve can be reduced considerably by using asymmetric Bragg reflection from lattice planes that are not parallel to the crystal surface. Figure 3b shows the DCD with an asymmetrically diffracting monochromator. In this arrangement and in exact non-dispersive mode, rocking curves have been observed that are close to the theoretical diffraction curve of the sample crystal (8).

Often an epitaxial film sample is bent due to film stresses. To avoid a broadening of the rocking curve the width of the irradiated sample area has to be reduced, either by slits between monochromator and sample, or by narrow receiving slits in front of the detector and by moving sample and detector in a 2θ-θ-scan.

Figure 4 shows a DCD rocking curve of an epitaxial garnet layer deposited by Liquid Phase Epitaxy (LPE) on a (111) oriented garnet substrate (9). For monochromator and sample the (880) reflection was chosen in the same asymmetric mode, decreasing the rocking curve width. The two high peaks at left and right are from the substrate and film, respectively. Their angular separation is due to a difference of the (880) lattice spacings and a tilt of the planes, as discussed in the next section. On the low-angle side of the epitaxial layer peak secondary interferences with small amplitudes, so-called Pendellösung fringes, are visible. They are caused by the finite thickness of the layer. From their angular spacing $\Delta\theta_p$ the thickness t of the epitaxial layer can be determined by

$$t = \frac{\lambda \, \gamma_h}{\Delta\theta_p \, \sin 2\theta_f}, \qquad (2)$$

where γ_h is the cosine of the angle between the diffracted beam and the interface normal, and θ_f the Bragg angle of the film (9). From the spacing of the secondary maxima of $\Delta\theta_p \sim 20$ arc sec we obtain the thickness of the layer $t=1.6\mu m$. Pendellösung fringes on the high-angle side of the layer peak are very weak due to the higher absorption of the β-branch of the x-ray wavefield; they can be observed more distinctly on thinner layers.

MEASUREMENT OF FILM STRAIN

Strain is an important characteristic of the epitaxial film since it is apt to modify its physical properties. The cause of strain in epitaxial films is primarily the difference of the bulk lattice spacings of substrate and film parallel to the interface, the so-called lattice mismatch. In the literature, the tailoring of magnetic (10,11,12) and electronic (13) properties by means of the lattice mismatch strain has been reported. In the case of perfect epitaxy of two cubic crystals the film strain parallel to the interface is given by

$$e_{epi} = \frac{a_s - a_f}{a_f} \qquad (3)$$

where a_s and a_f are the lattice spacings of the unstrained substrate and film, respectively. Especially in systems with large epitaxial mismatch, the epitaxial-mismatch strain may be partially or fully released by formation of misfit dislocations, but strain may still be caused by

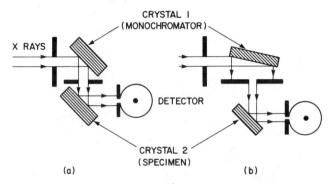

Fig. 3. Schematic diagram of non-dispersive Double-Crystal Diffractometer (DCD) with monochromator diffracting in symmetric (a), and asymmetric (b) Bragg reflection.

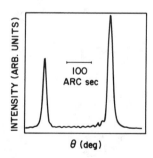

Fig. 4. DCD rocking curve of asymmetric Cu-Kα-(880) reflection of (111) oriented garnet layer, 1.6μm thick. Substrate peak at left, layer peak at right. From Stacy and Janssen (9), © 1974, North-Holland Physics Publishing.

different thermal expansion in substrate and film. The thermal mismatch strain parallel to the interface is given by

$$e_{th} = \int_{T_p}^{T_o}(\alpha_f - \alpha_s)dT \ , \tag{4}$$

where α_f and α_s are the linear thermal expansion coefficients of film and substrate, respectively, and T_p and T_o the temperature during preparation and observation, respectively. If the temperature dependence of α_f and α_s in the range from T_o to T_p can be ignored, Eq. (4) can be written as

$$e_{th} = (\alpha_f - \alpha_s)(T_o - T_p) \ . \tag{4a}$$

In cubic crystals, both, epitaxial and thermal mismatch strain, have rotational symmetry in the interface. We introduce the films orthogonal reference system x'_1, x'_2, x'_3 with the first two axes parallel and the third perpendicular to the interface; these may be different from the principal axes x_1, x_2, x_3 parallel to the $<100>$ directions. Then the components of the films strain tensor in the interface are given by

$$\varepsilon'_{11} = \varepsilon'_{22} = e \ , \quad \varepsilon'_{12} = 0 \ , \tag{5}$$

where we have dropped the subscript of e for convenience. The stress tensor of a thin film also has a simple form. Outside the immediate neighborhood to an edge, no stresses are applied perpendicular to the interface, leading to the expression

$$\sigma'_{33} = \sigma'_{23} = \sigma'_{13} = 0. \tag{6}$$

The relation between the strain tensor ε and the stress tensor σ is given by the elastic equations

$$\varepsilon = S \sigma \quad\quad \sigma = C \varepsilon \ , \tag{7}$$

where S and C are the fourth-rank tensors of the elastic compliances and stiffnesses, respectively. In the cubic crystal system, both tensors have three independent components. Using Eqs. (5) and (7), the three non-zero stress tensor components, $\sigma'_{11}, \sigma'_{22}, \sigma'_{12}$, can be obtained from three linear equations

$$\varepsilon'_{ik} = S'_{iklm}\sigma'_{lm}, \tag{8}$$

where the subscripts i,k,l,m take only the value 1 or 2, and where the 'prime' indicates that the equation is written in the film coordinate system. In order to solve Eq. (8) the tensor of the elastic compliances S has to be transformed from the coordinate system of the crystal axes, x_i, to that of the film, x'_i (14). Once the three independent stress tensor components, $\sigma'_{11}, \sigma'_{22}, \sigma'_{12}$, are obtained the missing three strain tensor components, $\varepsilon'_{33}, \varepsilon'_{23}, \varepsilon'_{13}$, can be computed by Eq. (8), with subscripts $i=3$; $k=1,2,3$; and $l,m=1,2$. An elegant, alternative method avoids the necessity to transform the compliance tensor (15). The strain tensor is presented in the system (x_i) as sum of a hydrostatic strain, equal to the interface strain e, and a strain with zero components in the interface:

$$\varepsilon_{ik} = e \ \delta_{ik} + a_i \ l_k, \tag{9}$$

where $\delta_{ik} = 1$ for $i = k$, and $\delta_{ik} = 0$ for $i \neq k$. The vector l with the components l_i is the unit vector normal to the interface, and the vector a with components a_i is computed using the elastic

Table I. Ratio of film strain values normal and parallel to interface for isotropic films and three interface orientations.

	Isotropy	(001)	(011)	(111)
$-\dfrac{\varepsilon'_{33}}{\varepsilon'_{11}}$	$\dfrac{2\nu}{1-\nu}$	$\dfrac{2C_{12}}{C_{11}}$	$\dfrac{2C_{12}-C_0/2}{C_{11}+C_0/2}$	$\dfrac{2C_{12}-2C_0/3}{C_{11}+2C_0/3}$

Definitions: $C_0 = 2C_{44} - C_{11} + C_{12}$

equations and boundary conditions, Eqs. (5-7). A critical comparison of both methods will be given in a review (16). For the three symmetry cases, i. e. the interface parallel to the (001), (110) or (111) plane, and for elastically isotropic materials, the strain tensor has rotational symmetry with the rotation axis and vector a perpendicular to the interface. For all other orientations of the interface, the strain tensor loses the rotational symmetry, the principal axis and vector a are not perpendicular to the interface anymore, but the interface strain still has rotational symmetry.

Once the strain tensor is determined with either method, the film strain ε_{hkl} parallel to m_{hkl}, the unit vector normal to the (hkl) planes, and the tilt $\Delta\phi_{hkl}$ of the (hkl) planes can be computed. Both quantities can be measured by x-ray diffraction. They are given by the components of the displacement vector $u=\varepsilon m_{hkl}$, parallel and perpendicular to m_{hkl}. In terms of the vector a they assume the simple form:

$$\varepsilon_{hkl} = (a \cdot m_{hkl}) \cos \phi_{hkl} + e, \tag{10a}$$

$$\Delta\phi_{hkl} = (a \cdot m_{hkl}) \sin \phi_{hkl}, \tag{10b}$$

where

$$\cos \phi_{hkl} = (l \cdot m_{hkl}) \tag{10c}$$

defines the angle ϕ_{hkl} between the interface and the (hkl) planes. For the three symmetry cases, mentioned above, and for films with elastic isotropy, a is parallel to the interface normal l and Eqs. (10a,b) take the familiar form

$$\varepsilon_{hkl} = a_o \cos^2 \phi_{hkl} + e, \tag{11a}$$

$$\Delta\phi_{hkl} = a_o \sin \phi_{hkl} \cos \phi_{hkl}, \tag{11b}$$

where the modulus of a, $a_o = \varepsilon'_{33} - e$, is the film strain normal to the interface, relative to the substrate. In Table I values of the strain ratio $\varepsilon'_{33}/e = 1 + a_o/e$ are listed for the symmetry cases.

As shown in Fig. 4, the rocking curves of the (hkl) planes typically show two peaks, from substrate and film, respectively, separated by

$$\Delta\omega_{hkl} = \Delta\theta_{hkl} + \Delta\omega_{\Delta\phi} \tag{12}$$

where

$$\Delta\theta_{hkl} = -(\varepsilon_{hkl}-e) \ \tan\theta_{hkl} \tag{13}$$

is the difference of Bragg angles due to the difference of lattice spacings between substrate and film, and $\Delta\omega_{\Delta\phi}$ is the change of ω due to the tilt of the lattice planes by $\Delta\phi_{hkl}$. If the sample is rotated around the diffraction vector (hkl) ($\parallel \boldsymbol{m}_{hkl}$) the latter contribution changes according to

$$\Delta\omega_{\Delta\phi} = \Delta\phi_{hkl} \ \cos\alpha, \tag{14}$$

where α is the angle between the projections of the incident beam and the interface normal into the (hkl) plane (17). Therefore, $\Delta\theta_{hkl}$ can be determined as the mean value of two measurements $\Delta\omega$ taken at two azimuths α, $180°$ apart. Eqs. (10a,b) and (11a,b) have been checked experimentally with (Ga,Al)As films deposited on (001), (110), ($\bar{1}\bar{1}\bar{1}$), and (113) GaAs surfaces, and the values of ε_{hkl} and $\Delta\phi_{hkl}$ measured for various (hkl) planes were found to be in excellent agreement with theory (8).

If the strain tensor is known, the strainfree lattice parameter of the film can be determined. In substitutional systems, e. g. (Ga,Al)As, the composition of the film can then be obtained from the strainfree lattice parameter by Vegard's law. If the epitaxial strain is partially relieved by formation of misfit dislocations, the value of the interface strain e, determined from strain tensor measurements, will be less than expected from Eq. (3).

If the strain is caused by substitutional or interstitial atoms, introduced into the epitaxial layer for instance as dopant by diffusion or ion implantation, it changes with the distance from the surface or penetration depth due to the concentration profile of the dopant. Strain profiles in ion-implanted garnet films have been determined from DCD rocking curves (18), using a model consisting of thin laminae each with its specific strain perpendicular to the interface. Figure 5 shows at the left experimental rocking curves in comparison with computed ones for the as-implanted sample and for the sample with a surface layer removed by ion-milling in two progressive stages. At the right the strain profile and the damage parameter used for the computed rocking curves is plotted versus the distance from the interface. The damage parameter U is defined as the rms displacement of the atoms. At any depth the uncertainty of the strain value is not more than 2% of the maximum strain.

The modulation of strain and composition in semiconductor superlattices has been determined from DCD rocking curves (5), as a special case of the laminar structure described in the last paragraph. Figure 6 shows the experimental (004) rocking curve of an AlSb/GaSb superlattice, deposited on a (001) GaSb substrate by Metal-Organic Chemical Vapor Deposition (MOCVD), in comparison with the computed one. The model used for the computation is given by a laminar, periodic structure with a superlattice period consisting of two layers, AlSb and GaSb, each with its own structure factor, thickness and strain. The fitting of the experimental data to the theoretical curve was achieved by trial-and-error adjustment of the structural parameters, obtaining a value of 305Å for the thickness of both sublayers, i. e. a value for the superlattice period of 610Å, and a strain, defined relative to the GaSb substrate, of $(1.25\pm0.02)\%$ for the AlSb sublayer, and of $(-0.03\pm0.02)\%$ for the GaSb sublayer. The thickness of the transition layer between the two sublayers was estimated not to exceed 2.5% of the period (610Å). The strain value for the AlSb sublayer is slightly lower than one would expect from the lattice mismatch between the AlSb ($a_o=6.1347\text{Å}$) and GaSb ($a_o=6.095\text{Å}$) of $e=0.65\%$. The perpendicular strain, relative to the substrate, resulting from the mismatch amounts to $\Delta d/d=1.99e=1.29\%$. The lower strain value is indicative of a slight relief of the elastic strain, a conclusion confirmed by measurements of the strain parallel to the interface of $\varepsilon_p=0.03$ (relative to substrate) from DCD rocking curves of an asymmetric (422) reflection. The strain relief occurs by formation of dislocations at the interface, and, consequently, superlattice and substrate are not completely coherent any more.

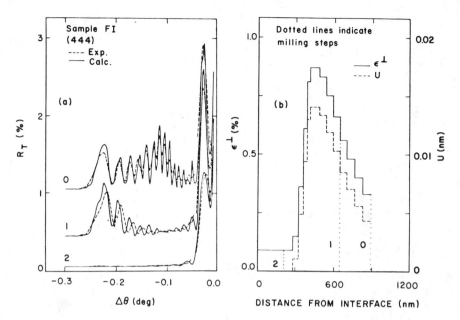

Fig. 5. a) Measured and computed Cu-Kα_1 (444) rocking curves of epitaxial garnet film after ion-implantation. Curve 0: as implanted; curve 1 and 2: top layer of 250 and 680nm removed by ion-milling. b) Distribution of strain ε and damage parameter U used for computation of rocking curves in a). From Speriosu (18), © 1981, American Institute of Physics.

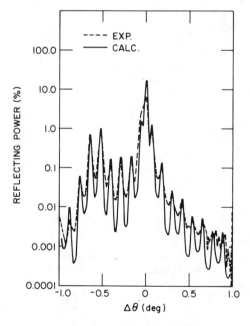

Fig. 6. Measured and computed Fe-Kα_1 (400) rocking curve of (Ga,Al)Sb superlattice with a period of $D=610$Å. From Speriosu and Vreeland (5), © 1984, American Institute of Physics.

GRAZING-INCIDENCE DIFFRACTION

For very thin epitaxial layers ($t \lesssim 100\text{Å}$) conventional Bragg diffraction, as discussed in the preceding sections, does not provide sufficient diffracted intensity due to lack of scattering material. If an x-ray beam irradiates the sample in grazing incidence, either in the total reflection range or close to it, not only an intense specularly reflected beam can be observed, but also strong diffracted beams with the diffraction vector lying in the surface and leaving the surface under an equally grazing angle (19). Figure 7 shows the geometry of Grazing-Incidence Diffraction (GID) or Surface Diffraction. The angle of incidence (and also of exit) ψ is typically chosen between the critical angle for total reflection ($\psi_c \lesssim 0.5°$) and $\sim 1°$. From a GID 2θ-θ-scan the lattice parameter, strain and crystallite size parallel to the interface can be obtained. A GID ω-scan provides information about the orientational spread parallel to the interface of the epitaxial layer relative to the substrate.

Figure 8 shows ω-scan data of the (110) reflection of a 220Å thick Mo film, deposited by MBE on a (001) GaAs substrate (20). By conventional Bragg diffraction a (111) orientation of the film was found. The GID ω-scan data show that two (111) oriented domains are present: A with Mo[110] $\parallel$ GaAs[110], and B with Mo[110] $\parallel$ GaAs[1$\bar{1}$0]. Also, a domain C with Mo[110] $\parallel$ GaAs[100] can be identified which has a (001) orientation and has not been observed by conventional Bragg diffraction. For domains A and B, the lattice mismatch, as defined by Eq. (3), amounts to -10.2% along the Mo[110] direction and to $+3.7\%$ perpendicular to it, whereas for domain C it amounts to $+27\%$. The angular spread of domains A and B is rather large, as evidenced by the half-peak-intensity width of $\Delta\omega \sim 11°$. From the line width observed in a 2θ-θ-scan the domain size parallel to the interface has been estimated to be $\sim 130\text{Å}$. The lattice parameters measured parallel and perpendicular to the surface agreed with the bulk value within the limits of probable error.

An example for epitaxy of two materials with the same structure is the system CdTe/GaAs with a lattice mismatch of -12.8%. Two orientations of the CdTe layer deposited on (001) GaAs substrates have been observed by conventional Bragg diffraction: (A) CdTe(001) $\parallel$ GaAs(001), and (B) CdTe(111) $\parallel$ GaAs(001). Figure 9 shows GID ω-scan data for the (220), (400) and (422) reflections of a 2500Å thick (Cd,Mn)Te film prepared by high-vacuum evaporation (21). For each of the two orientations the in-plane orientation can be identified: A with CdTe[110] $\parallel$ GaAs[110], and B with CdTe[11$\bar{2}$] $\parallel$ GaAs[110.]. For the (111) oriented layer (B) only one of the two possible domains could be observed due probably to substrate preparation or growth conditions. The lattice mismatch for domain A amounts to -12.8%. Domain B has a small mismatch of $+0.7\%$ along CdTe[11$\bar{2}$], but a large one of -12.8% perpendicular to it. The angular spread of the two domains is different: $\Delta\omega \sim 3.3°$ for A, and $\Delta\omega \sim 0.7°$ for B. The domain size, both parallel and perpendicular to the surface, was estimated to be $\sim 150\text{Å}$ and $\sim 250\text{Å}$ for domain A and B, respectively. On a 200Å thick (Cd,Mn)Te film similar well developed GID ω-scan patterns have been observed, reduced in intensity, and with the orientational spread of domain A comparable to that of domain B (21). Often only one of the two orientations could be observed in conventional Bragg diffraction, whereas in GID both were visible, due to the high sensitivity of the method.

SYNCHROTON SOURCE

With the availability of the synchroton as a high-power x-ray source, the study of ultra-thin layers, down into the mono-layer range, by x-ray diffraction methods has become increasingly feasible. The high intensity enabled the application of GID to Pb monolayers on Cu for the study of the liquid-solid phase transition (22), or to films of amorphous germanium selenide for the determination of radial distribution functions (23). It allows the use of perfect-crystal monochromators and analyzers to obtain high resolution in reciprocal space, as demonstrated in a study of Xe monolayers on exfoliated graphite (24). Recently, synchroton x-ray diffraction has been applied to the study of orientational epitaxy of Kr monolayers on single-crystal graphite (25). By use of pinholes the irradiated area on the sample can be made very small to allow diffraction from selected areas. Besides high power, the synchroton source has other unique qualities that can be utilized in diffraction experiments: Extremely small divergence in vertical direction, almost unlimited choice of wavelength, and linear polarization. For instance the small divergence makes the angle of incidence in GID extremely well defined (23).

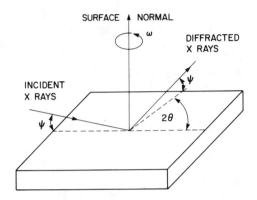

Fig. 7. Schematic diagram of Grazing-Incidence Diffraction (GID). Grazing angle $\psi \lesssim 1°$. Specularly reflected beam not shown.

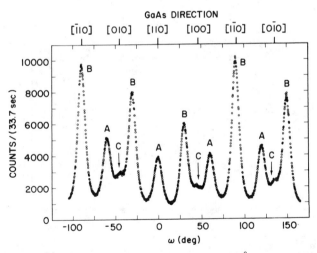

Fig. 8. GID ω-scan data of the (110) reflection of a 220Å thin molybdenum film on GaAs(001) with Cu-Kα radiation, rotating anode, 35kV, 100mA, $1 \times 1mm^2$ focal spot, and vertically bent, pyrolytic graphite monochromator. A, B, C: Domains of different orientation, explained in text. GaAs directions parallel to interface (001) indicated at the top.

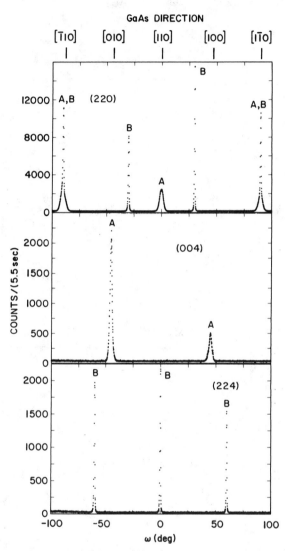

Fig. 9. GID ω-scan data of (220), (004) and (224) reflections of a 2500Å thin (Cd,Mn)Te film on GaAs(001). A, B: Domains of different orientation, explained in text. Same conditions as in Fig. 8.

SUMMARY

The application of x-ray diffraction methods, evolved in the last decade, to the characterization of thin epitaxial films and superlattices has been discussed. Double-crystal diffractometry is a very powerful and precise tool to determine the strain in epitaxial films due to lattice or thermal mismatch. Rocking-curve analysis can provide accurate and detailed information on strain and composition profiles in ion-implanted layers and superlattices. Grazing-incidence diffraction supplements conventional x-ray diffraction by providing information on lattice parameters, domain sizes and orientation in directions parallel to the surface, and it expands the range of detectable structures down into the monolayer region, especially when used with a synchroton x-ray source.

I am indebted to Drs. J. Bloch and M. Heiblum for the Mo/GaAs epitaxy sample, and to Drs. T. Siegrist and V. S. Speriosu for discussion of their work.

REFERENCES

1. A. Segmüller and M. Murakami, "Characterization of thin films by x-ray diffraction," in "Thin Films from Free Atoms and Particles," K. J. Klabunde, Editor, Academic Press, New York, in press.

2. A. Segmüller, "Observation of x-ray interferences on thin films of amorphous silicon," *Thin Solid Films* **18**, 287-294 (1973).

3. P. Voisin, C. Delalande, M. Voos, L. L. Chang, A. Segmüller, C. A. Chang and L. Esaki, "Light and heavy valence subband reversal in GaSb-AlSb superlattices," *Phys. Rev. B* **30**, 2276-2278 (1984).

4. A. Segmüller, P. Krishna and L. Esaki, "X-ray diffraction study of a one-dimensional GaAs-AlAs superlattice," *J. Appl. Cryst.* **10**, 1-6 (1977).

5. V. S. Speriosu and T. Vreeland, Jr., "X-ray rocking curve analysis of superlattices," *J. Appl. Phys.* **56**, 1591-1600 (1984).

6. R. Zeyfang, "Stresses in epitaxially grown single-crystal films: YIG on YAG," *J. Appl. Phys.* **41**, 3718-3721 (1970).

7. B. W. Batterman and H. Cole, "Dynamical diffraction of x rays by perfect crystals," *Rev. Mod. Phys.* **36**, 681-717 (1964).

8. W. J. Bartels and W. Nijman, "X-ray double-crystal diffractometry of $Ga_{1-x}Al_xAs$ epitaxial layers," *J. Cryst. Growth* **44**, 518-525 (1978).

9. W. T. Stacy and M. M. Janssen, "X-ray Pendellösung in garnet epitaxial layers," *J. Cryst. Growth* **27**, 282-286 (1974).

10. P. J. Besser, J. E. Mee, P. E. Elkins and D. M. Heinz, "A stress model for heteroepitaxial magnetic oxide films grown by chemical vapor deposition," *Mat. Res. Bull.* **6**, 1111-1124 (1971).

11. P. J. Besser, J. E. Mee, H. L. Glass, D. M. Heinz, S. B. Austerman, P. E. Elkins, T. N. Hamilton and E. C. Whitcomb, "Film/substrate matching requirements for bubble domain formation in CVD garnet films," *In* "Magnetism and Magnetic Materials 1971," *AIP Conf. Proc.* **5**, Part 1, pp. 125-129, American Institute of Physics, New York. (1972).

12. E. Klokholm, J. W. Matthews, A. F. Mayadas and J. Angilello, "Epitaxial strains and fracture in garnet films," *ibid.* pp. 105-109.

13. G. C. Osbourn, "Strained-layer superlattices from lattice mismatched materials," *J. Appl. Phys.* **53**, 1586-1589 (1982);
"$In_xGa_{1-x}As-In_yGa_{1-y}As$ strained-layer superlattices: A proposal for useful, new electronic materials," *Phys. Rev. B* **27**, 5126-5128 (1983).

14. R. W. Vook and F. Witt, "Thermally induced strains in evaporated films," *J. Appl. Phys.*
     36, 2169-2171 (1965);
     F. Witt and R. W. Vook, "Thermally induced strains in cubic metal films," *ibid.* **39**,
     2773-2776 (1968).

15. J. Hornstra and W. J. Bartels, "Determination of the lattice constant of epitaxial layers
     of III-V compounds," *J. Cryst. Growth* **44**, 513-517 (1978).

16. A. Segmüller and M. Murakami, "X-ray diffraction analysis of strains and stresses in thin
     films," to be published in "Analytical Techniques for Thin Films," Treatise on Materials
     Science and Technology, (K. N. Tu and R. Rosenberg, eds.), Academic Press, New
     York.

17. T. Hattanda and A. Takeda, "Direct measurement of internal strains in liquid phase
     epitaxial garnet films on gadolinium gallium garnet (111) plate," *Japan J. Appl. Phys.* **12**,
     1104-1105 (1973).

18. V. S. Speriosu, "Kinematical x-ray diffraction in nonuniform crystalline films: Strain and
     damage distributions in ion-implanted garnets," *J. Appl. Phys.* **52**, 6094-6103 (1981).

19. W. C. Marra, P. Eisenberger and A. Y. Cho, "X-ray total-external-reflection-Bragg
     diffraction: A structural study of the GaAs-Al interface," *J. Appl. Phys.* **50**, 6927-6933
     (1979).

20. J. Bloch, M. Heiblum and Y. Komem, "Growth of molybdenum and tungsten on GaAs in
     a molecular beam epitaxy system," *Appl. Phys. Lett.* **46**, 1092-1094 (1985).

21. T. Siegrist, A. Segmüller, H. Mariette, and F. Holtzberg, "Epitaxial growth of CdTe and
     (Cd,Mn)Te films on GaAs substrates," to be published.

22. W. C. Marra, P. H. Fuoss and P. E. Eisenberger, "X-ray diffraction studies: Melting of
     Pb monolayers on Cu(110) surfaces," *Phys. Rev. Lett.* **49**, 1169-1172 (1982).

23. A. Fischer-Colbrie and A. I. Bienenstock, personal communication (1985).

24. P. A. Heiny, P. W. Stephens, R. J. Birgeneau, P. M. Horn and D. E. Moncton, "X-ray
     scattering study of the structure and freezing transition of monolayer xenon on gra-
     phite," *Phys. Rev. B* **28**, 6416-6434 (1983).

25. K. L. D'Amico, D. E. Moncton, E. D. Specht, R. J. Birgeneau, S. E. Nagler and P. M.
     Horn, "The rotational transition of incommensurate Kr monolayers on graphite," *Phys.
     Rev. Lett.* **53**, 2250-2253 (1984).

MEASUREMENT OF ELASTIC LATTICE DISTORTION IN PbTe/PbSnTe - STRAINED-LAYER SUPERLATTICES BY ASYMMETRIC HIGH ANGLE X-RAY INTERFERENCES

E.J. Fantner

Institut für Physik, Montanuniversität, A-8700 Leoben, Austria

Present Address: Philips Applications Laboratory,
7602 EA Almelo, Lelyweg 1, The Netherlands

ABSTRACT

Elastic strain significantly affects the electric and optical properties of PbTe/Pb$_{1-x}$Sn$_x$Te - strained-layer superlattices. In the range of 10
- 350K the temperature dependence of the elastic strain present in these
superlattices was measured by double-crystal x-ray diffraction. For superlattice periods smaller than 100nm high-angle x-ray interferences were
observed. Using a novel method, which makes use of the high-angle interferences both for symmetrical as well as for asymmetrical reflections in a
theta-twotheta scan with a narrow detector slit, the relative inclination
of equivalent lattice planes due the elastic strain was measured. The
components of the complete strain tensor of the constituent layers can be
determined seperately even if their unstrained lattice constants are not
known with sufficient accuracy as is the case in ternary and quaternary
compounds. The lattice mismatch of up to 0.4% for Sn-contents smaller than
20% was found to be accommodated almost completely by elastic misfit
strain. The amount of strain is shared between the constituent layers
inversely to their relative thicknesses as long as the superlattice as a
whole is much thicker than the buffer layer. Below room temperature an
additional temperature dependent tensile strain due to differnt thermal
expansion coefficients of the film and the BaF$_2$-substrate is measured
quantitatively.

INTRODUCTION

By the progressive improvement of the thin-film growth techniques
such as molecular beam epitaxy or metal organic chemical vapor deposition
it has been made possible to synthesize high quality multiple thin and
ultrathin epitaxial layers. Establishing a strictly one-dimensional perio-
dicity in the growth direction it has become possible to engeneer the
electrical and optical properties of these epitaxial structures, the
manmade superlattices[1,2]. The interest in a wider spread of the combination of different materials in heteroepitaxy has lead to a quickly
developing investigation of superlattices with considerable lattice mismatch. It was shown that even a lattice mismatch of some percent can be
accommodated completely by elastic strain as long the individual layer
thicknesses do not exceed a critical thickness, which depends on the
amount of lattice mismatch, the elastic constants of the constituent
materials as well as the growth conditions. With increasing film thickness

an accommodation of the lattice misfit shared between misfit dislocations and strain becomes energetically favourable[3]. Being able to avoid the creation of undesired misfit dislocations in lattice mismatched superlattices of sufficiently small periods has offered the opportunity to use the misfit strain even as a tayloring parameter in the so called strained layer superlattices (SLS)[4]. In III-V - SLS, e.g. in GaAs/In$_x$Ga$_{1-x}$As (x = 0.27), which means a lattice mismatch of1.9%, TEM lattice fringe images have shown absolute "coherence" at the heterointerface for layer thicknesses smaller than 18nm[5]. In IV-VI heterostructures it is well known that dislocations increase the interface recombination velocity of electrons and holes[6]. The growth of leadsalt superlattices with low misfit dislocation densities has been reported recently[7]. Partin[8] has demonstrated that it is possible to improve the laser characteristics such as threshold current by growing multi quantum well lasers of sufficiently thin individual layers. Beside the misfit strain an additional strain is expected to be introduced by the substrate due to thermal expansion coefficient misfit. This paper will present elastic strain data at low temperatures obtained by x-ray diffraction on PbTe/PbSnTe - SLS with small periods.

EXPERIMENTAL

The PbTe/PbSnTe - SLS investigated were grown by a modified hot-wall epitaxy[9] on cleaved (111)-oriented BaF$_2$ substrates. The superlattices consist of 60 to 100 alternating layers with a tin content ranging from10 to 20% and superlattice periods of 40 to 200nm being measured by symmetric high-angle x-ray interferences at various wavelengths[10]. A buffer layer of either PbTe or PbSnTe of about 100 - 300nm was grown to overcome the large lattice mismatch of the superlattice constituents and the substrate.

A Siemens D 500 diffractometer with a bent Silicon monochromator (Johansson type) has been used in connection with an optical x-ray cryostat providing temperatures down to 10K by cooling with a two stage closed cycle cryostat (Fig.1). The sample holder is mounted on a standard goniometer head within the insulation vacuum of the cryostat providing a rotation around 2 perpendicular axes as well as translation of the sample even at low temperatures in the whole range of the goniometer head used. The

Fig. 1. Low temperature x-ray diffractometer for single crystal epitaxial
        structures: standard ω-goniometer with optical x-ray cryostat
        and a two-stage closed cycle helium refrigerator.

cooling is provided by a flexible heat wick connecting the sample holder
and the second stage of the cold head, which gives a cooling time of less
than 45 minutes from room temperature to 10K.

Using MoKα_1 radiation theta-twotheta scans have been performed for
various lattice planes of the <211> - crystallographic zone. The detector
slit always was 0.018° or 0.05°, which is smaller than the intrinsic
linewidth of the high-angle Bragg reflections of a thick PbTe- or PbSnTe-
layer or the BaF$_2$. The unstrained lattice constant of the substrate at
room temperature (0.620nm) has been used as internal standard. Absolute
lattice constant measurements at low temperatures have been performed
using Bond's method[11].

RESULTS

The elastic strain in the individual layers, which are of cubic
symmetry in the case of PbTe/PbSnTe - SLS, ought to be manifested both by
different lattice constants for different orientations as well as a strain
induced change of the angle of inclination of lattice planes oblique to
the interface normal direction. This is illustrated in Fig.2. As the
unstrained lattice constant of PbTe (0.646nm at RT) is larger than that of
PbSnTe - according to Vegard's law it decreases linearly with tin content
to the SnTe lattice constant (0.6327nm) - the misfit strain ought to be
compressive for PbTe and tensile for PbSnTe in the filmplane. As a consequence the PbTe exhibits its maximum lattice constant perpendicular to the
interface, its minimum value in the plane. In the PbSnTe-layers the opposite situation occurs. Due to the high crystallographic symmetry in our
samples the strain is isotropic in the film plane. Measuring these actual
lattice spacings for various orientations the complete strain status in
three dimensions can be determined[12-14]. This method fails for superlattices with small periods, where high-angle x-ray interferences occur due

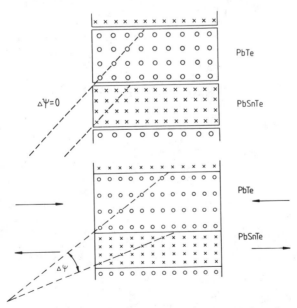

Fig. 2. Schematic illustration of the straininduced lattice distortion of
a misfit dislocation free SLS ($\Delta\psi$ is the straininduced relative
inclination of equivalent oblique lattice planes of the constituents).

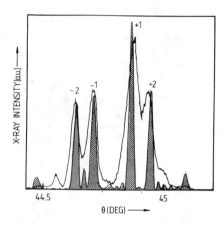

Fig. 3. Measured (unshaded) and calculated (shaded) x-ray interference pattern for the symmetric (444)Mo-Kα_1 reflection at T = 15K. For a PbTe/PbSnTe-SLS on a thick substrate (30 periods of 57nm, equally thick sublayers).

to the one-dimensional structural modulation. Although the angular positions of the maxima of the scattered x-ray intensity are of course related to the lattice constants of the constituent materials, to get the desired information from the high-angle interference data an extensive fit both of the scattered x-ray intensity as well as its angular dependence has to be performed assuming a particular strain distribution in the superlattice. As an example the interference pattern of a PbTe/PbSnTe-SLS for the (444)-Mo Kα_1 reflection is shown in Fig.3. The maxima of the intensity are equidistant, yielding a superlattice period of 57nm. There have been published several methods to calculate the scattered x-ray intensity by a onedimensional periodic structure[15-18]. Nevertheless, to our knowledge all these culculations published up to now have treated symmetric reflections only. This means, that in principle only information on the strain component perpendicular to the surface has been obtained. The strain parallel to the surface has to be calculated using the elastic constants of the constituent layers. In Fig.3 there is shown a calculated interference pattern taking into account coherent interference of the scattered x-ray waves both from the PbTe and the PbSnTe lattice planes. The good agreement with the experimental curve proves the zero order reflection to be extinguished by destructive interference. The fit of course yields the strained lattice constants of the constituent layers in the surface normal direction, too, being 0.6466 and 0.6431nm for PbTe and PbSnTe respectively. To calculate the strain component perpendicular to the surface one has to know accurately the unstrained lattice constant at the temperature of interest. Beside the experimental uncertainty this is not possible to a satisfying extent for ternary and quarternary compounds, where the compositions are determined usually by a lattice constant measurement.

In order to overcome these problems we have performed theta-twotheta scans of asymmetric reflections, too. In contrast to the lattice planes parallel and perpendicular to the interface, for obliquely oriented lattice planes of a SLS the Bragg condition for both SLS-constituents can never be fullfilled simultanously in a theta-twotheta scan with a narrow detector slit. This is illustrated in the insert of Fig.4. Changing the initial $(\theta - \psi)$-offset $-$ ψ represents the angle of inclination between the interface and the plane of interest - the position of the detector, where the interference of the scattered x-rays takes place, is shifted relative to the ideal Bragg configuration of the strained individual

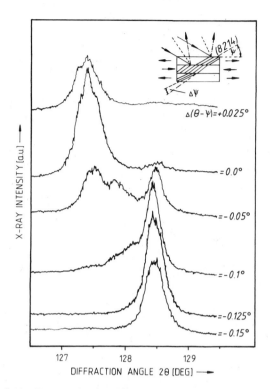

Fig. 4. Measured Mo-Kα₁ theta-twotheta scans of the sample of Fig.3 at
         T = 15K for different offsets of the angle (θ − ψ) of the asym-
         metric (8 2 14) reflection (ψ = 31.43°).

constituent layer for the whole theta-twotheta scan. The consequence is an
envelope of the scattered x-ray intensity integrated by the detector, the
asymmetry of which with respect to the angular position of the main superlattice peak strongly depends on this (θ − ψ)-offset. This is shown in
Fig.4 by high-angle interference pattern of the (8 2 14) MoKα₁ for various
(θ − ψ)-offsets measured at 15K. Presetting the offset-angle to smaller
values, which is closer to the Bragg angle of the strained PbTe, obviously
prefers the constructive interference of the PbTe-part of the scattered x-
ray intensity. The opposite offset prefers the x-rays scattered by the
PbSnTe-layers. For symmetric reflections or in unstrained superlattices a
change of the offset-angle changes the intensities of all superlattice
reflections by the same ratio[19]. In any case, the positions of the interference peaks remain unchanged. This method can be described more formally
in a reciprocal lattice representation as a scanning of the intensity of
the superlattice reflections. The reciprocal lattice points of the crystallographic zone containing the lattice planes of interest are expected
to be accompanied by satellites in the reciprocal growth direction, which
in our case is <111>. A theta-twotheta scan passes through these reciprocal lattice points radially, the change of the (θ − ψ)-offset, which is
equivalent to an ω − scan, corresponds to a scan perpendicular to the
radial direction.
    Plotting the measured x-ray intensities ·vs. the (θ − ψ)-offsets,
two maxima referring to the two SLS-constituents clearly are resolved.
Their angular deviations from the unstrained (θ − ψ)-angle, which is
given by the reflections of the lattice plane parallel to the interface,

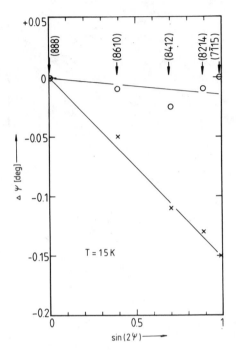

Fig. 5. Measured inclination angles of various lattice planes of the <211>
crystallographic zone at $T = 15K$ for the same SLS as in Fig.3
(x x x refer to PbSnTe, o o o to PbTe).

can be attributed to the strain induced distortion of the individual
layers. The angular separation of the maxima reflects the relative inclination of corresponding oblique lattice planes in the SLS-constituents. In
Fig.5 we have plotted the measured strain induced change of the angles of
inclination of various lattice planes relative to the unstrained values.

DISCUSSION

For the special symmetry in our samples, which is representative for
most of the epitaxial structures of interest - two of the main axes of
strain are parallel to the film plane, the third perpendicular to it - a
linear relation between the components of the strain tensor and the
straininduced changes of the angle of inclination is valid. If we define
in a coordinate system with its z-axis perpendicular to the (111)-inter-
face plane the strain in the individual layers by

$$\varepsilon_i^j = (a_i^j - a_i^o)/a_i^o \qquad (1)$$

where i=1,2 corresponds to the superlattices constituents, j=1,2,3 to the
x-, y- and z-direction and a_i^j and a_i^o denote the strained and unstrained
lattice constants respectively, this relation is given by equ.2[12]:

$$\Delta\psi = - \frac{1}{2} (\varepsilon_i^1 - \varepsilon_i^3) \sin(2\psi) \qquad (2)$$

Due to the high crystallographic symmetry in our samples, ε_i^1 equals ε_i^2
both in the PbTe- (i=1) and the PbSnTe-layers (i=2). The thick substrate
remains unstrained. For the (111)-geometry the relation between ε_i^1 and

$\epsilon_i{}^3$ is given by

$$\epsilon_i{}^1 = - \frac{C_{11} + 2C_{12} + 4C_{44}}{2(C_{11} + 2C_{12} - 2C_{44})} \; \epsilon_i{}^3 \tag{3}$$

For $T \rightarrow 0K$, the ratio of the elastic strain parallel and perpendicular to the interface is -0.93 according to the elastic stiffness coefficients reported by B.Housten et al.[20]

It is evidenced by equ.2 and 3 that the knowledge of one angle of distortion allows the determination of both the parallel and perpendicular component of the strain tensor. Nevertheless, in order to prove the correlation given in equ.2 we have investigated various lattice planes of the <211>-zone. In addition, the determination of the strain tensor components from the slope of the measured angles of inclination vs. $\sin(2\psi)$, which is plotted for various lattice planes in Fig.5, is of course much more accurate. It should be pointed out, in evaluating the strain values the knowledge of the unstrained lattice constants is not neccessary. From the slopes of the two branches the elastic strain of the PbTe-layers (upper branch) and the PbSnTe-layers (lower branch) is found to be 0.5×10^{-3} and $(2.7 + 0.2) \times 10^{-3}$, respectively. The negative slope proves the strain being tensile for both constituents. These data are in quantitative agreement with the calculated distortion assuming the elastic misfit strain to be shared between the two constituents inverse to their relative thicknesses. The solid lines in Fig.5 were calculated using the sample parameters for the fit shown in Fig.3 taking into account the additional tensile strain at low temperatures originating from the different thermal expansion coefficients of the constituents and the BaF_2-substrate.[10] The effect of the large mismatch between the substrate and the superlattice of about 4% is negligibly small due to its accommodation by a bufferlayer of 200-300nm being much thinner than the SLS as a whole.[14]

SUMMARY

In order to determine the complete strain status in three dimensions in the individual constituents of PbTe/PbSnTe-SLS with superlattice periods as small as to produce high-angle x-ray interferences, we have performed theta-twotheta scans with narrow detector slits both for symmetric and asymmetric reflections using a novel low-temperature x-ray diffractometer. The dependence of the high-angle interference intensities for lattice planes inclined to the sample surface is shown to be a direct measure for the relative inclination of equivalent lattice planes in the differently strained layers. This method allows the determination of the complete strain tensor in the individual SLS constituents without having to know their unstrained lattice constants. The quantitative agreement between the experimentally derived and calculated data prove the lattice mismatch in PbTe/PbSnTe-SLS with Sn-contents smaller than 20% and SL-periods below 100nm to be accommodated almost completely by elastic strain.

ACKNOWLEDGEMENTS

The author gratefully acknowledges many helpful discussions with Prof.G.Bauer, B.Ortner and H.Krenn, who has performed the x-ray interference calculations too. The author thanks H.Clemens for growing the samples and G.Rauscher for technical assistance. This work was supported by Forschungsförderungsfonds der gewerblichen Wirtschaft, Steiermärkischer Wissenschafts- und Landesfonds and Fonds zur Förderung der wissenschaftlichen Forschung (Proj. P5321).

REFERENCES

1) L.Esaki, Semiconductor superlattices and quantum wells through development of molecular beam epitaxy, in: "Molecular Beam Epitaxy and Heterostructures", L.L.Chang and K.Ploog, ed., Martinus Nijhoff Publishers, Dordrecht/Boston/Lancaster (1985).

2) J.E.Hilliard, Artificial layer structures and their properties, in: "Modulated Structures", J.M.Cowley, J.B.Cohen, M.B.Salamon, and B.J.Wuensch, ed., American Institute of Physics, New York (1979).

3) C.A.B.Ball and J.H.van der Merwe, The growth of dislocation-free layers, in: "Dislocations in Solids", Vol.6, F.R.N.Nabarro, ed., North-Holland Publishing Company, Amsterdam/New York/Oxford (1983).

4) G.C.Osbourn, Strained-layer superlattices from lattice mismatched materials, J.Appl.Phys. 53:1586 (1982).

5) J.M.Brown, N.Holonyak,Jr., R.W.Kaliski, M.J.Ludowise, W.T.Dietze and C.R.Lewis, Direct observation of lattice distortion in a strained-layer superlattice , Appl.Phys.Lett. 44:1158 (1984).

6) Y.Horikoshi, M.Kawashima, and H.Saito, PbSnSeTe-PbSeTe lattice matched double-heterostructure lasers, Appl.Phys.Lett 21:77 (1982).

7) P.Pongratz, H.Clemens, E.J.Fantner, and G.Bauer, Dislocations and strains in PbTe/PbSnTe superlattices, Electron Microscopy of Semicond. Materials , Inst. of Phys. Conf. Ser., Oxford (1985).

8) D.L.Partin, Lead Salt Quantum Well Diode Lasers,Superlattices and Microstructures 1:131 (1985).

9) H.Clemens, E.J.Fantner, and G.Bauer, Hot wall epitaxy system for growth of multilayer IV-VI - compound heterostructures, Rev.Sci.Instr. 54:685 (1983).

10) E.J.Fantner and G.Bauer, Strained layer IV-VI semiconductor superlattices in: "Two-Dimensional Systems, Heterostructures, and Superlattices", Springer Series in Solid-State Sciences, Vol.53, p.207, G.Bauer, F.Kuchar and H.Heinrich,ed., Springerverlag,Berlin/Heidelberg/New York/Tokyo(1984)

11) W.L.Bond, Precision lattice constant determination, Acta Crystallogr. 13:814 (1960).

12) J.Hornstra and W.J.Bartels, Determination of the lattice constant of epitaxial layers of III-V compounds, J.Crystal Growth 44:513 (1978).

13) B.Ortner, Simultanous Determination of the Lattice Constant and Elastic Strain in Cubic Single Crystals by X-Ray Diffraction; Adv. X-Ray Anal.29 (in press).

14) E.J.Fantner, Elastic strain in PbTe/PbSnTe heterostructures and super-lattices, J. de Physique (in print).

15) D.de Fontane, A theoretical and analogue study of diffraction from one-dimensional modulated structures, in: "Local Atomic Arrangement Studies by X-Ray Diffraction", Metallurgigal Society Conferences, Vol.36, p.51, J.B.Cohen and J.E Hilliard, ed., Gordon and Breach, New York/London/Paris (1966).

16) A.Segmüller and A.E.Blakeslee, X-ray diffraction from one-dimensional superlattices in $GaAs_{1-x}P_x$ crystals, J.Appl.Cryst. 6:19 (1973).

17) R.M.Fleming, D.B.McWhan, A.C.Gossard, W.Wiegmann, and R.A.Logan, X-ray diffraction study of interdiffusion and growth in $(GaAs)_n(AlAs)_m$ multilayers, J.Appl.Phys. 51:357 (1980).

18) M.Quillec, L.Goldstein, G.LeRoux, J.Burgeat, and J.Primot, Growth conditions and characterization of InGaAs/GaAs strained layers super-lattices, J.Appl.Phys. 55:2904 (1984).

19) E.J.Fantner, Direct determination of elastic strain in strained - layer superlattices by high-angle x-ray interferences, Appl.Phys.Lett. (in print).

20) B.Houston, R.E.Strakna, and H.S.Belson, Elastic constants, thermal expansion, and Debye temoerature of lead telluride, J.Appl.Phys. 39:3913 (1968).

CHARACTERIZATION OF TUNGSTEN SILICIDE AND TITANIUM SILICIDE THIN FILMS WITH A FULLY-AUTOMATED SEEMANN-BOHLIN DIFFRACTOMETER

R. J. Matyi

Central Research Laboratories
Texas Instruments, Inc.
Dallas, TX 75265

I.   INTRODUCTION

The continuing drive in the semiconductor industry towards smaller and more complex electronic devices has placed increasing severe demands on the materials systems that are used in semiconductor circuits. The refractory metal silicides that are commonly used for electrodes and interconnects are particularly susceptible to the sometimes conflicting requirements of good electrical properties, ease of processing, and long term reliability. We have found that the characterization of metal silicides with a fully automated diffractometer based on the Seemann- Bohlin (S-B) focusing geometry has proved to be an important aid in the development of several semiconductor metallization systems.

Although a conventional Bragg-Brentano (B-B) diffractometer can provide adequate information for some thin film systems, the use of an S-B diffractometer in which the incident x-ray beam makes a very small angle of incidence with the sample surface results in a substantial improvement in the usable signal from very thin samples. Figure 1 shows two diffraction scans that were obtained from a titanium silicide film that had been deposited on a $<100>$ silicon substrate. The scans were recorded using both Bragg-Brentano and Seemann-Bohlin diffractometers under comparable conditions of beam divergence, receiving slit aperture and x-ray tube power. The most intense peaks in the B-B scan (marked "x" in the figure) arise from spectral impurities in the x-ray tube; the peaks from the silicide are very weak at best. In contrast, the S-B diffraction scan clearly exhibits the desired silicide peaks.

While diffractometers based on the S-B geometry have been employed for thin film characterization for many years, their use has not been widespread. We have learned, however, that a diffractometer of this sort can become a highly productive analytical tool when it is interfaced to one of the many commercially available conventional powder diffraction analysis systems that exist in many advanced x-ray laboratories today. Although our particular hardware configuration is based on the the Diffrac V system manufactured by Siemens, we are not necessarily advocating the use of this particular commercial vendor. Instead, it is the purpose of this paper to demonstrate that commercial x-ray analytical systems that are designed to work strictly with "normal" B-B diffract ometers can be successfully adapted to the somewhat different requirements of the Seemann-Bohlin thin film diffractometer.

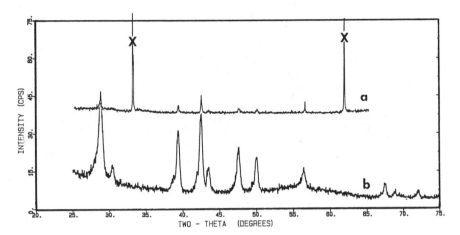

Fig. 1. Typical diffraction scans from a titanium silicide film.
(a) Bragg-Brentano diffractometer
(b) Seemann-Bohlin diffractometer

II.   EXPERIMENTAL

The diffractometer used in this work is the Huber 652 Seemann-Bohlin
diffractometer equipped with a singly-bent germanium incident beam monochromator. A 5 degree angle between the incident x-ray beam and the sample
surface is employed. Soller slits in both the incident beam collimator and in the
receiving slit assembly reduce the vertical divergence to about 3 degrees; the use
of Soller slits results in a major reduction in background scattering at angles less
than $40°\ 2\theta$.   Conventional pulse-height analysis and step counting techniques
are used. As mentioned above, the diffractometer is run using the Diffrac V
analytical system, where a Siemac V controller serves as an interface between the
PDP 11/23 + host computer and the diffractometer.

Previous studies[1-3] of the S-B geometry have demonstrated that aberrations such
as sample displacement from the focusing circle, the use of a flat instead of a
curved sample, and sample transparency all contribute to peak displacements
that are proportional to $1/\sin\alpha$, where α is the incident beam glancing angle. As a
result, the use of a small angle of incidence between the x-ray beam and the
sample surface exacerbates the systematic errors introduced by these aberrations. These factors suggest that accurate diffractometer calibration is perhaps
more important in this system than it is with a B-B diffractometer. The effect of
transparency dictates that a bulk sample (such as a compressed powder briquette)
must not be used to calibrate an instrument that will be used for thin films. The
dependence of peak variance with glancing angle varies as $1/\sin^2\alpha$, so
instrumental broadening is much greater than would normally anticipated from
a focusing geometry. This leads to complications in line broadening analysis for
particle size, since the necessary deconvolution of size and instrumental
broadening becomes much more difficult in this case. Peak shifts and variances
are also inversely proportional to the size of the focusing circle, so the relatively
small radius of 57.3 mm in the Huber diffractometer tends to magnify the impact
of the aberrations.

One peculiarity of the Huber diffractometer arises in the means by which the
incident beam glancing angle is adjusted. In order to maintain focusing with a
continuously variable glancing angle, the diffractometer body and focusing circle
are pivoted through the action of a micrometer screw. The effect of this
mechanical arrangement is that the angular scale readings on the diffractometer
are offset from their "true" angular counterparts by a constant amount equal to

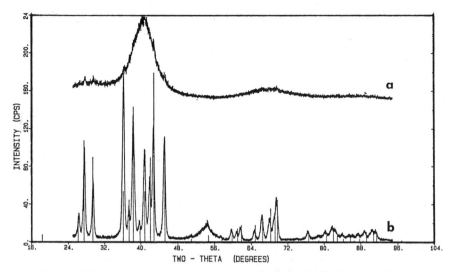

Fig. 2. S-B scans from sputtered tungsten silicide films following
anneals at (a) 850° C and (b) 900° C. Vertical lines indicate
peak positions for W_5Si_3.

the glancing angle setting. We have found that this offset is most easily
compensated through a modification to the instrument control software that
permits the glancing angle to be input along with the conventional angular scan
parameters. This approach has proved to be superior in the technical service
laboratory environment to either large corrections in diffractometer calibration or
the input of apparently "incorrect" angular settings to the instrument control
program.

III.   RESULTS

Figure 2 illustrates a typical analysis that has been performed using the S-B
diffractometer with a powder diffractometer analytical system. The figure shows
two scans from a tungsten silicide thin film that had been deposited on a substrate
by argon sputtering tungsten and silicon sources. The transformation into the
silicide phase by raising the annealing temperature from 850° C to 900° C is
evident in the figure. Following data collection, peak identification and correction
of peak positions, it was possible to perform some of the analyses on the thin film
that are common on conventional powder systems. For instance, the Diffrac V
software package permits an automated search of the JCPDS data base; the
results of this search resulted in an excellent match between the experimental
pattern and the silicide phase W_5Si_3. An examination of the raw experimental
data and the expected JCPDS peak positions shows relatively large discrepancies
between the experimental peaks and the positions indicated in the JCPDS file.
This difference, which would not be acceptable in a conventional diffractometer,
underscores the importance of accurate instrumental calibration when the S-B
diffractometer is employed. After phase identification the lattice constants of the
silicide phase were determined using the unit cell refinement routine in the
Siemens software package. The measured values ($a_0 = 0.9607$ nm and $c_0 =
0.4961$ nm) are in excellent agreement with the literature values [4] for W_5Si_3.

One advantage of the S-B geometry over a B-B diffractometer is that it is
relatively simple to minimize the effects of substrate reflections in this
arrangement. A small rotation of the sample about its surface normal prior to
data collection is sufficient to send a potentially interfering reflection
(particularly the (311) reflection from <100> substrates) out of the plane of the
diffractometer. In contrast, B-B scans from thin films on <100> substrates are

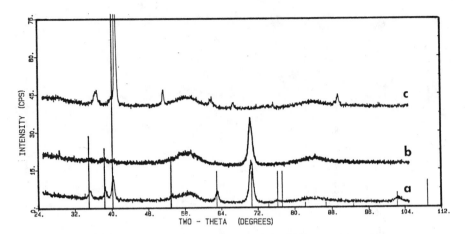

Fig. 3. S-B scans from (a) 100 nm titanium as-deposited; (b) 60 nm
titanium as-deposited; (c) 60 nm titanium (annealed 625° C for
30 min.). Vertical lines indicate peak positions for titanium.

often adversely effected by the unavoidable (400) substrate reflection.

Figure 3 illustrates diffraction scans from titanium films with thick- nesses of 60
and 100 nm that were sputtered onto <100> silicon substrates. Scans (a) and (b)
were recorded from the as-sputtered films and are characterized by intense
Ti(103) reflections at $2\theta \approx 70°$. We attribute this behavior to preferred
orientation in the as-deposited titanium film. If the basal plane of the hexagonal
titanium unit cell is aligned parallel to the silicon substrate surface [5], then the
(103) Ti planes would be inclined to the surface by 31.5°. The divergent incident x-
ray beam, impinging on the sample surface at a nominal angle of 5°, would then
satisfy the Bragg condition for the highly oriented (103) reflection.

The negative effect of preferred orientation on some applications of the S-B
diffractometer is shown in Figure 3(c). This scan, which was obtained from the 60
nm Ti film following an anneal at 625° C, is dominated by a single strong silicide
peak at approximately 41° 2θ. This suggests that the original preferred
orientation was not removed during the silicide phase transformation. Since
many of the titanium silicides exhibit their strongest peaks in this angular region,
the presence of preferred orientation complicates the analysis of phase
transformations in this system.

Figure 4 shows the effects of annealing temperature on the phases that are formed
from the 100 nm titanium film. The upper scan was recorded following a 30
minute anneal at 650° C, while the bottom scan was obtained after an additional
heat treatment at 850° C. A comparison with the JCPDS peak positions
demonstrates that the additional 850° C heat treatment was sufficient to
completely transform the original titanium film to face-centered orthorhombic
$TiSi_2$. In contrast, the film that was only annealed at 650° C shows the presence of
other silicide phases. Table I illustrates that following peak identification and
calibration, the major experimental peaks in the 650° C film could be accounted by
a combination of Ti_5Si_3 and two orthorhombic polymorphs of $TiSi_2$.

The S-B diffractometer has proved to be highly beneficial in the development of
novel thin film deposition technologies. For instance, Figure 5 shows the effect of
ion beam mixing on phase formation in the titanium silicide system. The samples
consisted of 45 nm titanium that had been deposited on 200 nm polysilicon prior to
a 160 KeV ion implant. The unannealed sample (Figure 5a) showed little
evidence of titanium, presumably due to the effect of ion beam mixing. The

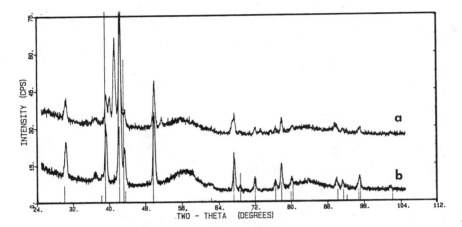

Fig. 4. S-B scans from 100 nm titanium films following anneals at
(a) 625° C for 30 min., and (b) 625° C for 30 min. followed by
850° C for 30 min. Vertical lines indicate peak positions for
face-centered orthorhombic TiSi₂.

Table I.   Titanium silicide phase identification (100 nm Ti on Si, anneal
30 min. at 650° C). All dimensions in nm; numbers beneath chemical
identification indicate JCPDS file number.

Experimental spacings	TiSi₂ 31-1405	TiSi₂ 10-225	Ti5Si₃ 29-1362
0.2970	0.2967		
0.2292	0.2299	0.2288	
0.2253		0.2232	
0.2192		0.2188	0.2203
0.2135	0.2137		0.2151
0.2099	0.2092		0.2114
0.1830	0.1832		

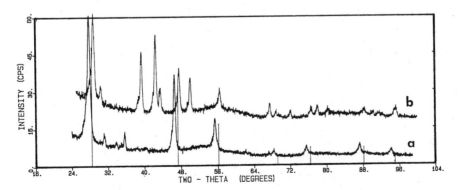

Fig. 5. S-B scans for 45 nm titanium on polysilicon following ion beam
mixing (a) before annealing; and (b) after annealing at 650° C
for 30 min. Vertical lines indicate peak positions for silicon.

increase in the silicon lattice parameter to 0.5632 nm (almost 4% greater than the nominal bulk value) suggests that the ion mixing of the Ti resulted in the formation of a metastable, interstitial solid solution in the polysilicon. Subsequent annealing (Figure 5b) resulted in the formation of the equilibrium silicon and silicide phases.

IV. CONCLUSIONS

We have found that the interfacing of a Seemann-Bohlin thin film diffractometer to a commercial powder diffraction analytical system has greatly extended our capabilities for thin film characterization. Although preferred orientation has been found to complicate analyses, it appears that most routine powder diffraction techniques can be success fully adapted to the S-B diffractometer.

V.   ACKNOWLEDGEMENTS

The author would like to thank C. A. Dearman for technical assistance, and R. Eklund, B. E. Gnade, B. W. Shen and S. K. Tiku for supplying samples used in this work.

VI.   REFERENCES

1.   A Segmuller, Z. Metallkde., 48, 448 (1957).

2.   W. Parrish and M. Mack, Acta Cryst., 23, 687 (1967).

3.   C. J. Gillham, J. Appl. Cryst., 4, 498 (1971).

4.   W. B. Pearson, "Handbook of Lattice Spacings and Structures of Metals and Alloys," Pergamon, Oxford (1967).

5.   H. Kato and Y. Nakamura, Thin Solid Films, 34, 135 (1976).

X-RAY TOPOGRAPHY OF ION-IMPLANTED LASER-ANNEALED Si

R. D. Dragsdorf and C. P. Bhalla

Kansas State University
Department of Physics
Manhattan, Kansas 66506   USA

The introduction of various atoms into crystal surfaces by ion implantation has become common practice in the research laboratory and in many industrial processes. The deceleration of the introduced foreign ions during implantation results in extensive lattice damage. Atom displacement from equilibrium crystal sites and pile-up of the new atoms take place. The use of the pulsed laser for annealing these atomic defects in a single crystal is well documented[1]. The absorption of the total energy in the laser pulse can be rapidly transferred to the lattice, approximately 10^{-9}s, for an elevated temperature anneal of the damaged crystal. The photon energy of the monochromatic laser beam can be tailored so that the absorption of the radiation by a specific crystal can be matched for its penetration into the crystal to that of the depth of the implanted atoms. A highly efficient mechanism is thus utilized to anneal just that portion of the crystal which is damaged by the impinging atoms of the implant.

In this investigation an attempt is made to determine the degree of recrystallization that occurs as a function of the laser energy per unit area that is introduced. Previous experiments have established that annealing energy fluxes from 0.7 - 1. J/cm^2 gives rise to a nearly complete anneal of the crystal surface layer. A determination of the relative x-ray diffracted intensities for various laser anneals in two silicon crystals is made and a comparison to dynamic x-ray diffraction theory is shown.

EXPERIMENT

Two silicon specimens were utilized in this examination. One of these was a silicon on sapphire substrate, SOS, (100) film 1.1 μm thick. This sample was implanted with Si^+ at 1.x 10^{15} ions/cm^2 at 180 KeV for a projected range of 0.27 μm. The other sample was a bulk silicon crystal with (100) face. This was implanted with As^+ at 1.x 10^{15} ions/cm^2 at 200 KeV for a projected range of 0.11 μm. Figure 1 shows the distribution of the implanted ions with the range of lattice damage that occurred in the stopping of the ions.

A Nd:YAG laser was used to anneal the damaged samples. Various anneals in spots up to 0.8 mm in diameter (area of $5 \times 10^{-3} cm^2$) were made with a range from 0.08 to 1.24 J/cm^2 in the SOS sample and from 0.1 to 4.6 J/cm^2 in the bulk crystal. The photon energy of the Nd:YAG laser as used was 2.33 eV (λ = 532. nm). The linear absorption coefficient of this

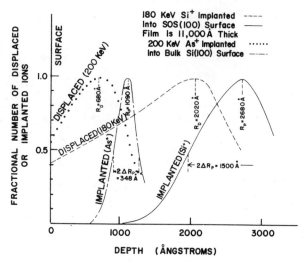

Fig. 1. Fraction of displaced atoms resulting from the implants of 180 KeV
Si$^+$ ions and 200 KeV As$^+$ ions as well as the respective regions of
deposit in a silicon on sapphire film and in a bulk silicon crystal.

radiation in Si at room temperature is 8.9 x 10^3 cm^{-1} giving a penetration
depth of 1.1 μm. The annealing energy then is absorbed in the region of
all the damage and implant.

    The appearance of the two samples with anneals looks much as shown in
the two sketches, figs. 2 and 3. Figure 2 depicts the implanted region of
the SOS sample with various anneals. Several anneals of 1.24 and 0.62
J/cm^2 were made to examine reproducibility. Figure 3 shows the bulk crystal
and the location of various anneals after the As$^+$ implant. The major dot
field was another experiment utilizing this crystal with smaller anneal
spot sizes with numerous anneals sometimes overlapping.[3]

    The samples are then placed on the translation portion of an adapted
Unicam Weissenberg camera to run a Lang type reflection topograph. A
limited beam spread of 3.0 mins of arc is obtained from two slits a small
distance apart (20. cm) to allow as much Kα x-radiation as possible from
a conventional source to be incident on the sample. The crystal is oriented
for diffraction close to the angle for (400) diffraction from the base
silicon. CuKα x-ray penetration is approximately 20. μm perpendicular to
the (100) surface at the (400) Bragg angle. Several different x-ray film
types were tried with type AX Eastman film giving the best compromise
between speed and resolution.

RESULTS

    In the developed x-ray topographs, figs. 4, the contrast appears to
be maximized for the annealed regions in the implant when the crystal is
positioned at about half maximum intensity on the small angle side of the

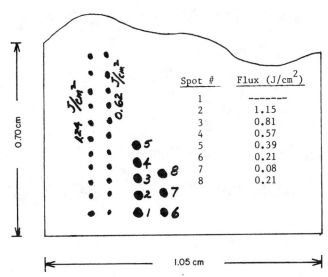

Fig. 2. SOS (silicon on sapphire) film, 1.1 μm thick, showing location of
various laser spot anneals and annealing flux following Si[+] implant.

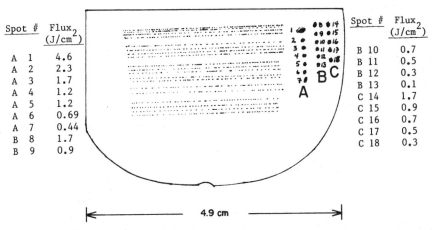

Fig. 3. Bulk silicon showing location of various laser spot anneals and
annealing flux following As[+] implant.

(400) Bragg peak. The laser annealed region gives sharp contrast with the
implanted but unannealed region.

In most cases where the laser beam has been enlarged to $5. \times 10^{-3}$ cm^2,
various maxima and minima are observed within the area of the spot. This
indicates variations in lattice strain in the crystal. This is probably

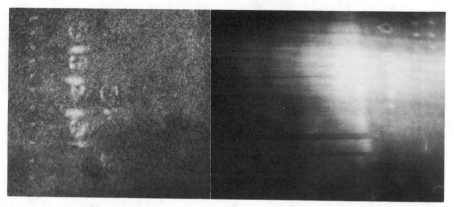

Fig. 4. X-Ray topographs, CuKα radiation, from (400) planes of implanted
and annealed SOS (left) and bulk (right) silicon samples. Annealed
regions show as white spots. Streaking in the bulk results from
smaller overlapping spot anneals.

due to the laser vibrational mode which appears when the laser has been
defocused at the crystal surface giving rise to non-uniform laser intensity
and thus a non-uniform anneal within the confines of the apparent spot.

For only the uniform small topographic images blackening as a function
of laser energy flux is plotted in fig. 5. The blackening from these spots
which gave repeatedly reproducible and recognizable densities above the
background are shown for only the bulk silicon where enough uniform spots
of varying anneal were present to allow a comparison. A normalization is
used for all spots from the given photograph. A limited but uniform
increase in intensity appears for anneal energy fluxes up to about 0.9
J/cm^2 and then a slower fall-off in intensity occurs above this energy
density value.

The only result to be reported for the SOS sample is the uniform
blackening observed for the many spots at both the 0.62 and the 1.24 J/cm^2
anneals. The other, larger spot anneals were non-uniform.

DISCUSSION

The sizes of the implanted ions, arsenic and silicon, R = 1.18 Å and
1.17 Å respectively, are not greatly different from the silicon crystal ions. Of
course, the silicon ions have the identical size. With implant doses of
1. x 10^{15} ions/cm^2 the average mass density increase of the crystals
depends on the depth of the implant. As^+ ions at 200 KeV have a projected
range of 0.11 μm. Assuming uniformity of implant to such depth would add
9 x 10^{19} ions/cm^3 within the boundaries of the implant. With the density
of silicon atoms in the perfect crystal at 5 x 10^{22} ions/cm^3, there would
be a residual lattice strain for the annealed implant of Δd/d = 1.84 x 10^{-3}
coming from both the atom size and increased number of atoms in the
implanted volume. In the SOS sample with Si^+ implant at 180 KeV only the
increased number of ions in the implanted volume would contribute to the
lattice strain. The projected range for the Si^+ is 0.27 μm and with a
dose of 1 x 10^{15} ions/cm^2, assuming uniform deposition, the increase in
ion density would be by 4.07 x 10^{19} ions/cm^3 giving rise to a strain after
anneal of 7.37 x 10^{-4}.

An x-ray diffraction curve was generated for an unstrained single
crystal of silicon using CuKα radiation from the (400) planes after the
theory of Burgeat and Taupin using a Runge-Kutta solution.[4] The

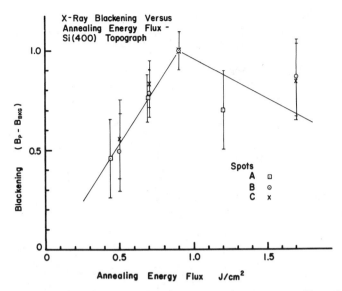

Fig. 5. Measured X-Ray Blackening versus annealing energy flux for (400)
spots indicated in diffraction topograph from bulk silicon sample.

size of the strains mentioned above for the implanted and annealed
samples slightly decrease the peak height and slightly increases the wings
of this curve. This is readily shown from the Burgeat and Taupin equations
by modifying the strain term in the limited region of the thin implanted
layer near the surface of the crystal. There also appears a small and much
broadened peak on the small angle side of the major peak associated with
this modified larger spacing in the thin implant region. Thus, for our
topographs, maximum contrast appears between the implanted annealed portions and the unimplanted portion of the silicon crystals when one sets
the crystal on the small angle side of the Bragg maximum at about half
the peak value.

The topographic spot size should be the same as the size of the laser
beam incident on the crystal for the anneal. The recording film is set
parallel to the diffracting crystal planes for the translation. However,
the x-rays see the entire depth of the annealed implants. Also, some of
the films used are double coated so that spot registration on the two
surfaces is displaced slightly. For a Bragg angle of 34.6° and thin
emulsion separation, this displacement is not large. In all cases the
small single diffraction spots show up as circles on the film and nearly
symmetric peaks on the microdensitometer trace. The grain size of the AX
emulsion is 0.4 μm as compared to the 150 μm diameter of our smallest
anneal spots.

The determined increase in film blackening of the diffraction spots
up to an anneal of 0.9 J/cm^2 indicates that a direct relation of the
amount of anneal or strain relief exists to the amount of energy introduced
into a given volume of sample. At small (less than 0.3 J/cm^2) laser pulse
energy densities, there is not sufficient ion mobility to relieve much of
the strain. However, as the energy flux increases more strain is relieved

by atoms moving into crystal sites of the base crystal. The result is the observation of blackening above the background for the recrystallized system with its excess number of atoms/unit volume.[5]

As the laser energy input becomes greater than 0.9 J/cm^2 for this system, two possible processes occur giving rise to a reduced blackening. First, with larger energy input the annealing process goes beyond what is known as the ideally imperfect crystal toward a condition of the ideally perfect crystal.[6] Strain is further relieved by atom segregation. The diffraction blackening would then decrease. Second, with the higher energy density input the surface of the crystal is modified. Sputtering appears to have occurred as seen in the reflecting optical microscope. There is a region near the center of each laser pulse which is roughened significantly as the laser energy flux is increased beyond 1. J/cm^2. This surface change would give rise to a decrease in the observed x-ray diffracted intensity.

CONCLUSION

Reflection x-ray topographs from implanted and then laser annealed silicon samples have shown that strain is reduced and that the damaged single crystal is restored for laser energy densities up to 0.9 J/cm^2. Energy densities greater than this might cause segregation of any impurity ions and definitely causes spalation or surface roughening of the portion laser irradiated. The perfection of the crystal has been shown by the increased blackening that results for increased annealing fluxes below 0.9 J/cm^2. The choice of the degree of perfection desired for the implanted or damaged surface layer of any crystal would depend on the actual use to be made of the crystal following an implant-anneal sequence. X-ray diffraction topography can show the degree of recovery from the damage.

ACKNOWLEDGMENTS

It is a pleasure to thank Professor Alvin D. Compaan for providing the samples. This work is supported in part by the Department of Energy, Division of Chemical Sciences.

REFERENCES

1. Lo, H. W. and A. Compaan, Phys. Rev. Lett. 44, 1604 (1980); Lee, M. C., H. W. Lo, A. Aydinli and A. Compaan, Appl. Phys. Lett. 38, 499 (1981); Narayan, J., J. Fletcher, C. W. White, and W. H. Christie, J. Appl. Phys. 52, 7121 (1981); Wood, R. F., D. H. Lowndes, G. E. Jellison, Jr. and F. A. Modine, Appl. Phys. Lett. 41, 287 (1982); Larson, B. C., C. W. White, T. S. Noggle, J. F. Barhorst and D. M. Mills, Appl. Phys. Lett. 42, 282 (1983).

2. Johnson, W. S. and J. Gibbons, Projected Range Statistics in Semi-conductors, distributed by the Stanford University Bookstore 1969.

3. Aydinli, A., H. W. Lo, M. C. Lee and A. Compaan, Phys. Rev. Lett. 46, 1640 (1981).

4. Burgeat, J. and D. Taupin, Acta Crystallogr. A24, 99 (1968); Taupin, D., Bull. Soc. Franc. Miner. Crist., 87, 469 (1964).

5. Narayan, J., J. Fletcher, C. W. White and W. H. Christie, J. Appl. Phys. 52, 7121 (1981).

6. Warren, B. W., X-Ray Diffraction, p. 327, Addison-Wesley Publ. Co. 1969.

SIMULTANEOUS DETERMINATION OF THE LATTICE CONSTANT

AND ELASTIC STRAIN IN CUBIC SINGLE CRYSTAL

Balder Ortner

Erich-Schmid-Institut für Festkörperphysik
Österreichische Akademie der Wissenschaften
Leoben, Austria

INTRODUCTION

In epitaxial single crystal growing it happens very often that the composition of the grown film, say $A_x B_{1-x} C$, is not known exactly. Usually it would be possible to find out that composition simply by determining the lattice constants of the film. The problem of such a measurement lies in the fact of internal and unknown stresses in the film. On the other side the knowledge of the state of stress σ and strain ε can also be of great importance. To calculate the stresses from X-ray measurements we need the lattice constants in the unstrained condition.

Hornstra and Bartels[1] have shown that the lattice constant can be determined by X-ray measurements if the strain is isotropic in the film plane. In the following we show that the lattice constant and the full strain tensor can be calculated from at least four X-ray measurements also in the general case.

The information and assumptions needed are: We must know relationships between the six values determining the lattice (a, b, c, α, β, γ) and their dependence on the concentration (x). If, for instance, we know that the ratio c/a in a hexagonal crystal is independent of the composition (x) we would be able to determine a and σ_{ij}. But this kind of problem is not dealt with here, we only considered the case where a cubic lattice is retained for all compositions, i.e. $b = c = a(x)$, $\alpha = \beta = \gamma = 90°$.

Since the film usually is very thin, we can assume it to be stressed homogeneously and that the equilibrium conditions for a free surface hold. We further need the knowledge of the elastic properties of the film (S_{ij} or C_{ij}). If these constants strongly depend on the concentration it is sufficient to know this dependence ($S_{ij}(x)$ or $C_{ij}(x)$) since then we can calculate $a \rightarrow x \rightarrow S_{ij} \rightarrow a \rightarrow \ldots$ in a reiterative procedure. Therefore in the following we assume S_{ij} to be known.

Because of the equilibrium condition we have only three unknown stress components and the unknown lattice constant a_o. The question is: can we find an equation system with which these four unknowns can be calculated

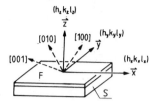

Fig. 1.   Specimen coordinate system $(\vec{x}, \vec{y}, \vec{z})$ and orientation of strained film (F). S.....substrate

from four or more lattice plane distance measurements. The answere is yes,
although this is not a matter of course. (Without using the equilibrium
condition, we would have 7 unknowns, but it is not possible to calculate
these unknowns from 7 or more measurements.)

BASIC EQUATIONS

If σ'_{ij} are stress components in the specimen coordinate system (Fig. 1) the
equilibrium condition reads:

$$\sigma'_{33} = \sigma'_{23} = \sigma'_{13} = 0 \tag{1}$$

The lattice plane distance of a plane (hkl) in the strained crystal
is given by equ. 2.

$$d_{(hkl)} = d_{o,(hkl)} \, (1 + \varepsilon_{(hkl)}) \tag{2}$$

$$d_{o,(hkl)} = a_o/N_{(hkl)} = a_o/\sqrt{h^2 + k^2 + l^2} \tag{2a}$$

$$\varepsilon_{(hkl)} = \varepsilon_i \, \alpha_i \qquad i = 1.....6 \tag{2b}$$

$$\varepsilon_i = (\varepsilon_{11} \quad \varepsilon_{22} \quad \varepsilon_{33} \quad 2\varepsilon_{23} \quad 2\varepsilon_{13} \quad 2\varepsilon_{12}) \tag{2c}$$

$$\alpha_i = (H^2 \quad K^2 \quad L^2 \quad KL \quad HL \quad HK) \tag{2d}$$

$$H = h/N_{(hkl)} \; ; \; K = k/N_{(hkl)} \; ; \; L = l/N_{(hkl)} \tag{2e}$$

For the relation between ε and σ we have equ. 3, where S_{ij} are the compliances of the crystal.

$$\varepsilon_i = S_{ij} \, \sigma_j \tag{3}$$

$$\sigma_j = (\sigma_{11} \quad \sigma_{22} \quad \sigma_{33} \quad \sigma_{23} \quad \sigma_{13} \quad \sigma_{12}) \tag{3a}$$

ε_i, σ_i, S_{ij} are components of the respective tensors in the crystal coordinate system. Since the equilibrium condition (equ. 1) is given in the spe-
cimen coordinate system we need $d_{(hkl)}$ as a function of σ'_i. Using again
the matrix notation the transformation of σ'_i to σ_i reads[2]:

$$\sigma_j = q_{jK} \, \sigma'_K \tag{4}$$

Inserting equs. 2a, 2b, 3 and 4 to equ. 2 we get equs. 5 and 6:

$$d_{(hkl)} = a_o/\sqrt{h^2 + k^2 + l^2} \, (1 + S_{ij} \, q_{jK} \, \sigma'_K \, \alpha_i) \tag{5}$$

$$d_{(hkl)} \sqrt{h^2 + k^2 + l^2} \equiv a_{(hkl)} = a_o + S_{ij} \, q_{jK} \, \alpha_i \, a_o \, \sigma'_K \tag{6}$$

$a(hkl) = a_m = d\sqrt{h_m^2 + k_m^2 + l_m^2}$ is a fictitious lattice constant measured by X-rays in the direction normal to the plane (hkl). m means the number of the plane. Using equ. 1 we come out with:

$$a_m = \bar{A}_{m1} \, a_o \, \sigma_1' + \bar{A}_{m2} \, a_o \, \sigma_2' + \bar{A}_{m3} \, a_o \, \sigma_6' + \bar{A}_{m4} \, a_o \tag{7a}$$

$$\bar{A}_{m1} = S_{ij} q_{j1} \alpha_{mi}; \quad \bar{A}_{m2} = S_{ij} q_{j2} \alpha_{mi} \tag{7b}$$

$$\bar{A}_{m3} = S_{ij} q_{j6} \alpha_{mi}; \quad \bar{A}_{m4} = 1$$

In equ. 7a the unknowns are $a_o \sigma_1'$, $a_o \sigma_2'$, $a_o \sigma_6'$ and a_o.
For practical purposes it would not be very convenient to have different dimensions and very different numbers in the $\bar{A}$'s and in the unknowns. Therefore, we redefine the unknowns in equ. 7a and also their coefficients:

$$a_o S_{11} \sigma_1' = s_1; \quad a_o S_{11} \sigma_2' = s_2; \quad a_o S_{11} \sigma_6' = s_3; \quad a_o = s_4 \tag{7c}$$

$$A_{m1} = \bar{A}_{m1}/S_{11}, \quad A_{m2} = \bar{A}_{m2}/S_{11}, \quad A_{m3} = \bar{A}_{m3}/S_{11}, \quad A_{m4} = \bar{A}_{m4} = 1. \tag{7d}$$

$$a_m = A_{m1} s_1 + A_{m2} s_2 + A_{m3} s_3 + A_{m4} s_4 \tag{7e}$$

To shorten the following formulae we write $S_{12}/S_{11} = \bar{S}_{12}$; $S_{44}/S_{11} = \bar{S}_{44}$.

Using the fact that there are only three independent compliances (S_{11}, S_{12}, S_{44}) in cubic crystals (see e.g. Voigt[3] or Nye[4]) we can reduce equs. 7b,d: (For equs. 8a,b see also[11,12].)

$$A_{m1} = (1 - \bar{S}_{12} - \bar{S}_{44}/2) (\mathbf{V}_m, \mathbf{V}_x) + \bar{S}_{12} + \bar{S}_{44} (\vec{H}_m, \vec{H}_x)^2/2 \tag{8a}$$

$$A_{m2} = (1 - \bar{S}_{12} - \bar{S}_{44}/2) (\mathbf{V}_m, \mathbf{V}_y) + \bar{S}_{12} + \bar{S}_{44} (\vec{H}_m, \vec{H}_y)^2/2 \tag{8b}$$

$$A_{m3} = (2 - 2\bar{S}_{12} - \bar{S}_{44}) (\mathbf{V}_m, \mathbf{V}_6) + \bar{S}_{44} (\vec{H}_m, \vec{H}_x) (\vec{H}_m, \vec{H}_y) \tag{8c}$$

$\mathbf{V}_m, \mathbf{V}_x$ and so on are (3×1) matrices (not vectors, as $\vec{H}_m, \vec{H}_x$ are), $(\mathbf{V}_m, \mathbf{V}_x)$ is defined like the scalar product of two vectors $(\vec{H}_m, \vec{H}_x)$:

$$(\mathbf{V}_m, \mathbf{V}_x) = V_{m1} V_{x1} + V_{m2} V_{x2} + V_{m3} V_{x3} \tag{9a}$$

$$\mathbf{V}_m = \begin{vmatrix} H_m^2 \\ K_m^2 \\ L_m^2 \end{vmatrix} \quad \mathbf{V}_x = \mathbf{V}(1) = \begin{vmatrix} H_x^2 \\ K_x^2 \\ L_x^2 \end{vmatrix} \quad \mathbf{V}_y = \mathbf{V}(2) = \begin{vmatrix} H_y^2 \\ K_y^2 \\ L_y^2 \end{vmatrix} \quad \mathbf{V}(3) = \begin{vmatrix} H_z^2 \\ K_z^2 \\ L_z^2 \end{vmatrix}$$

$$\mathbf{V}(4) = \begin{vmatrix} H_z H_y \\ K_z K_y \\ L_z L_y \end{vmatrix} \quad \mathbf{V}(5) = \begin{vmatrix} H_x H_z \\ K_x K_z \\ L_x L_z \end{vmatrix} \quad \mathbf{V}(6) = \begin{vmatrix} H_x H_y \\ K_x K_y \\ L_x L_y \end{vmatrix} \tag{9b}$$

$\vec{H}_m$; $\vec{H}_x$; ... means normed vectors, see equ. 2e.

$$\vec{H}_m = \begin{vmatrix} H_m \\ K_m \\ L_m \end{vmatrix}; \quad \vec{H}_x = \begin{vmatrix} H_x \\ K_x \\ L_x \end{vmatrix}; \quad \ldots$$

To calculate the unknown values σ_1 to a_o one must measure the lattice plane distances d_m of at least four planes $(h_m\ k_m\ l_m)$. The planes $(h_1\ k_1\ l_1)$ to $(h_N\ k_N\ l_N)$ must be chosen so that the rank of the matrix A_{mi} (m = 1 to N, i = 1 to 4) is 4. The solution of the equation system is then given by:

$$s_1 = a_o\ S_{11}\ \sigma_1' = Q_{1j}\ D_j \qquad j = 1\ \text{to}\ 4$$

$$s_2 = a_o\ S_{11}\ \sigma_2' = Q_{2j}\ D_j$$

$$s_3 = a_o\ S_{11}\ \sigma_6' = Q_{3j}\ D_j \tag{10a}$$

$$s_4 = a_o = Q_{4j}\ D_j$$

$$Q_{ij} = (P_{ij})^{-1} \tag{10b}$$

$$P_{ij} = \sum_{m=1}^{N} A_{mi}\ A_{mj}\ ;\ D_j = \sum_{N=1}^{N} a_m\ A_{mj} \quad \text{(*)} \tag{10c}$$

Equs. 8a to 10c are the only ones needed to calculate a_o, σ_1', σ_2', σ_6'. These equations are brought to a shape so that they can easily be transferred to computer programs, if standard matrix manipulation procedures are used. (The same holds for the following equs. 11 to 13)

If we want to know the strain we can use equation 11

$$\varepsilon_i' = S_{ij}'\ \sigma_j' = S_{i1}'\ \sigma_1' + S_{i2}'\ \sigma_2' + S_{i6}'\ \sigma_6' \tag{11}$$

For the compliances S_{ij}' in the specimen coordinate system we have:

$$S_{ij}' = (S_{11} - S_{12} - S_{44}/2)\,(\mathbf{V}(i),\mathbf{V}(j)) + S_{12} + \delta_{ij}\ S_{44}/2$$

$$= (S_{11} - S_{12} - S_{44}/2)\,(\mathbf{V}(i),\mathbf{V}(j)) - \delta_{ij}) + \delta_{ij}\ S_{11} + (1 - \delta_{ij})\ S_{12}$$

$$i,j \le 3 \tag{12}$$

$$S_{ij}' = 2(S_{11} - S_{12} - S_{44}/2)\,(\mathbf{V}(i),\mathbf{V}(j)) \quad i \le 3,\ j > 3\ \text{or}$$
$$i > 3,\ j \le 3$$

$$S_{ij}' = 4(S_{11} - S_{12} - S_{44}/2)\,(\mathbf{V}(i),\mathbf{V}(j)) + \delta_{ij}\ S_{44} \quad i,j > 3$$

If stresses or strains are needed in the crystal coordinate system, we have to transform the tensors. For σ the transformation equations are[2]:

$$
\begin{vmatrix} \sigma_1 \\ \sigma_2 \\ \sigma_3 \\ \sigma_4 \\ \sigma_5 \\ \sigma_6 \end{vmatrix}
=
\begin{vmatrix}
H_x^2 & H_y^2 & 2H_x H_y \\
K_x^2 & K_y^2 & 2K_x K_y \\
L_x^2 & L_y^2 & 2L_x L_y \\
K_x L_x & K_y L_y & L_x K_y + L_y K_x \\
H_x L_x & H_y L_y & L_x H_y + L_y H_x \\
H_x K_x & H_y K_y & H_x K_y + H_y K_x
\end{vmatrix}
\begin{vmatrix} \sigma_1' \\ \sigma_2' \\ \sigma_6' \end{vmatrix}
\tag{13}
$$

(*) We use the dummy suffix notation no matter whether the suffix runs from 1 to 3, 4 or 6, but not for 1 to N

Having σ_i in the crystal system it is easy to calculate also ε_i with the aid of equ. 3.

It is of course possible to derivate formulae similar to equs. 7, 8, 10, but with three components of ε_i' as the unknown variables. Yet in the general case (any orientation of the crystal) such a procedure would be much more complicated.

A very special state of strain often found in single crystal epitaxial films is isotropic strain in the film plane. Then we have $\varepsilon_1' = \varepsilon_2' = \varepsilon_r'$, $\varepsilon_6' = 2\varepsilon_{12}' = 0$. And σ_1', σ_2', σ_6' can be expressed by ε_1' so that equ. 7 can be reduced to one with only two unknown variables, namely ε_1' and a_0. (But it must be kept in mind that isotropic strain in the film plane does not mean $\sigma_1' = \sigma_2'$, $\sigma_6' = 0$, except if the crystal has a (100) or a (111)-orientation i.e. the free surface is parallel to a (100) or a (111) plane.) This special condition was already treated by Hornstra et al.[1] and measurements on specimens with that type of strain have been done by different authors[1,5-8].

ERROR CALCULUS

If we assume all measurements to have the same mean statistical errors then the errors of s_1 to s_4 are given by:

$$\Delta s_i = \sqrt{Q_{ii}} \ \Delta a_M \tag{14}$$

where Δa_M is the probable error of the single measurement according to equ. 6.

Equ. 15 is already contained in 14, the equs. 16a, b, c to estimate the errors of the stress components can be derived from equ. 14 using equ. 10a.

$$\Delta a_0^2 = Q_{44} \ \Delta a_M \tag{15}$$

$$\Delta \sigma_1'^2 = (\Delta a_M/a_0)^2 \, (Q_{11}/s_{11}^2 + Q_{44} \ \sigma_1'^2 - 2Q_{14} \sigma_1'/s_{11}) \tag{16a}$$

$$\Delta \sigma_2'^2 = (\Delta a_M/a_0)^2 \, (Q_{22}/s_{11}^2 + Q_{44} \ \sigma_2'^2 - 2Q_{24} \sigma_2'/s_{11}) \tag{16b}$$

$$\Delta \sigma_6'^2 = (\Delta a_M/a_0)^2 \, (Q_{33}/s_{11}^2 + Q_{44} \ \sigma_6'^2 - 2Q_{34} \sigma_6'/s_{11}) \tag{16c}$$

Since the Q_{ij}'s are in the same order of magnitude and σ_i' is much smaller than $1/s_{11}$ we can neglect all terms containing σ_i' in the equs. 16. Therefore we get:

$$\Delta \sigma_1' \cong \sqrt{Q_{11}} \ \Delta a_M/a_0/s_{11} \quad ; \quad \Delta \sigma_2' \cong \sqrt{Q_{22}} \ \Delta a_M/a_0/s_{11} \quad ;$$

$$\Delta \sigma_6' \cong \sqrt{Q_{33}} \ \Delta a_M/a_0/s_{11} \tag{17}$$

From equs. 17 we see that the relative errors of σ_1', σ_2', σ_6' are of course much greater than that of a_0.

If the number of measurements is greater than 4 we can estimate the measurement errors by equ. 18 or 19[9]:

$$\Delta a_M^2 = \sum_{i=1}^{N} (a_i - A_{i1} s_1 - A_{i2} s_2 - A_{i3} s_3 - A_{i4} s_4)^2 / (N-4) \tag{18}$$

$$\Delta a_M^2 = (\sum_{i=1}^{N} a_i^2 - D_1 s_1 - D_2 s_2 - D_3 s_3 - D_4 s_4) / (N - 4) \tag{19}$$

SELECTION OF LATTICE PLANES

As already mentioned the lattice planes must be chosen so that the matrix A_{ij} has rank 4. (Although it would be possible to find configurations of planes so that rank $(A_{ij}) < 4$ but a_o and σ'_1, σ'_2 (not σ'_6) can still be calculated.) Yet this is only a necessary condition. From equs. 15 and 17 we see that the probable errors of our unknowns do not only depend on the accuracy of the measurements themselves but also on the Q_{ij} which depend on the number of measured planes and on their orientations relative to each other. Therefore the experimenter should try to choose the lattice planes so that Q_{11} to Q_{44} are as small as possible i.e. he should try to minimize the ratios between the errors of σ'_1, σ'_2, σ'_6, a_o and the measurement errors.

We can say a configuration of lattice planes is the better the smaller Q_{11} to Q_{44} are. But Q_{11} to Q_{44} are not very convenient to evaluate the quality of a configuration because these are four figures to be minimized and Q_{11} to Q_{33} are changed if the coordinate system is rotated around the z-axis. Fortunately we can combine the errors of σ' in an expression which is invariant.

$$\overline{\Delta\sigma^2} = \Delta\sigma'^2_1 + \Delta\sigma'^2_2 + 2\,\Delta\sigma'^2_6 \tag{20}$$

$$\overline{\Delta\sigma^2} \cong (\Delta a_M/a_o)^2/S_{11}^2 \,(Q_{11} + Q_{22} + 2\,Q_{33}) \tag{21}$$

The factor 2 for $\Delta\sigma'^2_6$ seems to be arbitrary, but it is necessary for the invariance of $\Delta\sigma^2$. The proof of this assertions will not be given here, it is similar to the proof of a similar problem published elsewhere[10].

A general rule for the choice of the lattice planes in order to minimize the errors can not be given, since Q_{11} to Q_{44} depend not only on the configuration of the planes but also on the elastic properties of the crystal. But we can deduce general rules for the case of an isotopic crystal. And there is few doubt that applying these rules also in the general case, the result will also be rather good.

To find out the best configurations, a computer program was designed, which generates different configurations by a random procedure and calculates $\Delta a_o/\Delta a_M$ and $S_{11} \overline{\Delta\sigma}/(\Delta a_M/a_o)$. With that program a very great number of configurations was tested. The result of this test can be condensed in the following rule of thumb for the choice of lattice planes: In order to get a small ratio between the errors of the calculated quantities and the inevitable errors of the lattice plane distance measurement one should use a configuration which is similar to a 3-, 5-, 6-,....fold symmetric configuration as shown in the inserted stereograms of Fig. 2. (A configuration with a four-fold symmetry would give rank $(A_{ij}) = 3$.) A configuration can be regarded to be similar to one of Fig. 2 also if the plane(s) in the centre of the pole figure is (are) not exactly in the centre, and/or the other planes do not lie exactly on the small circle, and/or the configuration is not exactly symmetrical. Which configuration one takes, depends on the number of planes to be measured, on Poisson's ratio ν $(= -S_{12}/S_{11}$ for isotropic crystals), and whether the lattice constant a_o or the stress tensor is the most wanted information.

Fig. 2 shows the dependence of the ratios $\Delta a_o/\Delta a_M$, $\overline{\Delta\sigma}/\Delta a_M$ on the angle ψ and on Poisson's ratio ν for some of these symmetrical configurations. From this figure we conclude: If a_o is the data the experimenter is primarily interested in, then he should use a configuration with only one plane parallel to the surface, the other planes inclined to the surface and approximately equally distributed on a small circle. (Only the configuration with N = 5 and an exact four-fold symmetry must be avoided.) The inclination angle ψ should be taken to get the minimum value of $\Delta a_O / \Delta a_M$ (Figs. 2b,c). The angle ψ also depends on ν. If the crystal is not isotropic, we propose to simply use $-S_{12}/S_{11}$ instead of ν, since the dependence of $\Delta a_O / \Delta a_M$ on ν is not too strong, and it is not too important to get the very best configuration. The difference between the configurations b, c and e, f lies not in the smaller values of the error of a_O, but in the fact that the small values of Δa_O can be get with smaller angles of inclination ψ. A small ψ usually is of some advantage when doing the measurement. If σ_1, σ_2, σ_6 is the most needed information then the most suited configuration is one where the number of measured planes parallel or nearly parallel to the surface is about 1/3 of the number of planes on the small circle (Figs. 2e,f). (With e.g. 3 planes parallel to the surface we simply mean to measure 3 times that one plane.) According to Fig. 2 the angle ψ should be taken as large as possible.

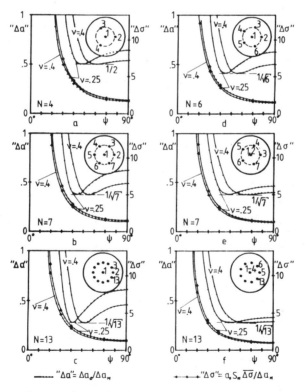

Fig. 2. Dependence of statistical error of a_O and σ on the angle ψ and the Poisson's ratio ν for symmetrical configurations of lattice planes. a_M is the probable error for all the measurements. N is the total number of measured planes. (b,c) configurations better suited for a_O measurement. (e,f) configurations better suited for the stress measurement.

It is noteworthy that the highest possible accuracy for the a_0 - measurement is always $\Delta a_0/\Delta a_M = 1/\sqrt{N}$. That means: If the N measured lattice planes are chosen properly then the accuracy of a_0 is the same as it would be in an unstrained crystal - nevertheless the state of stress can be calculated from the same N measurements. From that we further see: It would be not of any advantage to adopt a measurement strategy where only a_0 can be calculated but not the state of strain. Such an approach can make the evaluation of the measurements easier - if the strain is isotropic in the surface plane, yet in the general case there is no gain in simplicity or in accuracy.

CONCLUSION

It is possible to calculate the unstrained lattice constant and the state of stress (and strain) from at least four measured lattice plane distances, if the elastic properties of the material are known.

The equations given for that purpose are rather easy to handle with a computer if standard matrix manipulation procedures are used.

When choosing the lattice planes to be measured, one can easily get a maximum of accuracy with a minimum of measurements when using simple rules of thumb given in this paper.

ACKNOWLEDGEMENTS

The author is indebted to Prof.H.P. Stüwe for many helpful discussions and to Gesellschaft zur Förderung der Metallforschung and Steiermärkische Landesregierung for financial support.

LITERATURE

1. J. Hornstra, W.J. Bartels, Determination of the lattice constant of epitaxial layers of III-IV compounds, J. Crystal Growth 44: 513 (1978)
2. A. Mälmeisters, V. Tamuzs, G. Teters, "Mechanik der Polymerwerkstoffe", Akademie-Verlag, Berlin (1977)
3. W. Voigt, "Lehrbuch der Kristallphysik", Teubner, Leipzig (1910)
4. J.F. Nye, "Physical Properties of crystals", Clarendon Press, Oxford (1969)
5. W.J. Bartels, W. Nijman, X-ray double crystal diffractometry of $Ga_{1-x}Al_x$ As epitaxial layers, J. Crystal Growth, 44: 518 (1978)
6. E.J. Fantner, B. Ortner, W. Ruhs, A. Lopez-Otero, Misfit strain in epitaxial IV-VI semiconductor films, in: "Physics of Narrow Gap Semiconductors", E. Gornik, H. Heinrich, L. Palmetshofer, Eds., Springer, Berlin, Heidelberg, N.Y. (1982)
7. E.J. Fantner, Elastic strain in PbTe/PbSnTe heterostructures and super-lattices, J. de Physique, in print
8. E.J. Fantner, Direct determination of elastic strain in strained-layer-superlattices by high angle X-ray interferences, Appl.Phys. Lett., in print
9. E. Hardtwig, "Fehler und Ausgleichsrechnung", Bibliographisches Institut, Mannheim (1968)
10. B. Ortner, The choice of lattice planes in X-ray strain measurement of single crystal, 34th Annual Denver X-ray Conference, Denver (1985)
11. V. Hauck, G. Vaessen, Eigenspannungen in Kristallitgruppen texturierter Stähle, Z. Metallkde. 76: 102 (1985)
12. H. Möller, G. Martin, Elastische Anisotropie und röntgenographische Spannungsmessung, Mitt. K.W. Inst. Eisenforschung, 21: 261 (1939)

X-RAY FLUORESCENCE ANALYSIS OF MULTI-LAYER THIN FILMS

T. C. Huang and W. Parrish

IBM Almaden Research Center, K31/80
650 Harry Road
San Jose, CA 95120-6099

ABSTRACT

The characterization of multi-layer thin films by X-ray fluorescence using the fundamental parameter method and the LAMA-III program is described. Analyses of a double-layer FeMn/NiFe and two triple-layer NiFe/Cu/Cr and Cr/Cu/NiFe specimens show that the complex inter-layer absorption and secondary fluorescence effects were properly corrected. The compositions and thicknesses of all layers agreed to ±2% with corresponding single-layer films, a precision comparable with bulk and single-layer thin film analyses.

I. INTRODUCTION

Thin film materials are commonly used in the electronic, magnetic and other high-tech industries. Physical structure and chemistry are two of the important parameters which control their properties. Characterization of these materials is thus essential for the research and development of thin films.

X-ray and electron beam methods are two commonly used techniques for the characterization of thin films. The X-ray technique is used for large area analysis (mm to cm in diameter) and gives better areal average than the electron beam method which sample smaller areas (μm in diameter) The X-ray diffraction method (XRD) is used for the structure characterization of thin films. XRD analysis includes phase identification, preferred orientation, crystalline size, strains and stacking faults of polycrystalline thin films; and crystal orientation, lattice mismatch of epitaxial films on substrate; and the amorphous structure of noncrystalline thin films (1). The X-ray fluorescence method (XRF) is used for the determination of chemical composition and areal density of thin films (2). Therefore, a fairly complete characterization of the physical and chemical structure of thin films can be obtained by the XRD and XRF analyses.

Recently there has been increasing interest in the application of multi-layer thin film materials. The purpose of this paper is to describe the XRF characterization of these materials.

II. XRF ANALYSIS OF BULK AND SINGLE-LAYER THIN FILM MATERIALS

XRF has been used successfully for the analysis of bulk materials since early 1950s. Two commonly used mathematical approaches for quantitative XRF analysis are the empirical coefficient method and the fundamental parameter method. The average precision was found to be about 4% for the empirical coefficient and 2% for the fundamental parameter methods (3).

The empirical coefficient method requires a set of accurate calibration curves of intensity versus thickness and composition which are made using standards prepared to closely approximate the range of the unknowns. This is a difficult and sometimes impractical task, particularly for thin films. However, the fundamental parameters method requires only pure element bulk standards (although compounds or thin film standards can be used if available) which are easily obtainable, thereby greatly simplifying the analysis. The development of the XRF/LAMA method (4,5) made it possible to determine the composition and areal density of single-layer thin films using the fundamental parameter method. The precision is comparable to that of the bulk specimen analysis. For example, elemental concentrations of binary alloy thin films determined by XRF agreed with those obtained by the atomic absorption spectroscopy and the electron microprobe analysis to about 2% (2,5). Similarly, the XRF thicknesses also showed only small average deviations ($\Delta t/t \approx 2\%$) from interferometry and rate monitor values.

III. XRF ANALYSIS OF MULTI-LAYER THIN FILMS

The absorption and fluorescence processes in a multi-layer thin film are very complex because of the inter-layer effects. The main XRF processes that occur include:

- absorption of the incident polychromatic X-rays by upper layers;
- back scattering of the incident polychromatic X-rays by the substrate;
- primary X-ray fluorescence of elements;
- secondary X-ray fluorescence from elements:
  - within the same layer,
  - of other layers,
  - in the substrate;
- absorption of the exit fluorescent X-rays by upper layers.

Mantler developed the mathematical methods and computer programs, LAMA-III, using the fundamental parameter method to compute these interactions in multi-layer thin films (6).

1. Method

The success of XRF analysis with the fundamental parameter method depends on the effectiveness of two main processes: the absorption and secondary fluorescence corrections with the intensity versus composition-thickness equation, and the iteration of compositions and thicknesses so that the results are consistent with the experimental XRF intensities. Mantler recently extended the algorithms and improved the LAMA-III program for the analysis of multi-layer thin films (7). These new programs were used in this study to test the precision in determining the composition and thickness of individual thin film layers. The programs operate on an IBM 3083 computer and require a large amount of memory and long execution time. Extensive modification of the method would be necessary before the feasibility of running them on a mini- or micro-computer is reached.

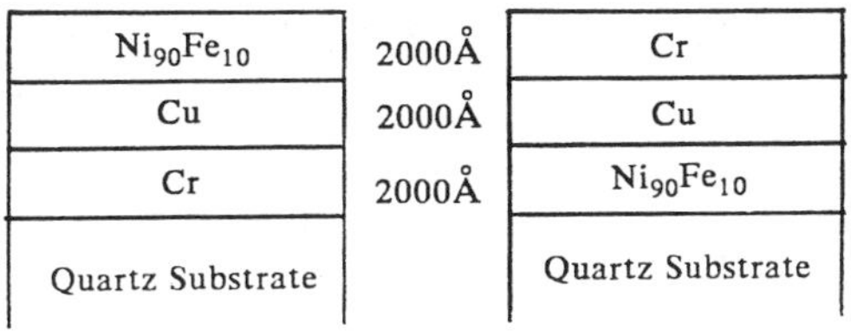

Ni₉₀Fe₁₀	2000Å	Cr
Cu	2000Å	Cu
Cr	2000Å	Ni₉₀Fe₁₀
Quartz Substrate		Quartz Substrate

Figure 1. Schematic picture of the triple-layer thin films:
left, T1 (NiFe/Cu/Cr); right, T2 (Cr/Cu/NiFe).

2. Experimental

The multi-layer thin films used in this study were carefully selected to
evaluate the capability of the method for difficult analyses. Single-layer
films were prepared under identical conditions. Pure element bulk standards
were used. Experimental parameters included Mo X-ray tube operated at 45
kV, LiF(002) analyzing crystal and scintillation counter. Step scanning
method was used to locate the XRF peaks. All peak intensities were collected
with at least 10,000 counts to ensure good counting statistics. The thicknesses of individual layers were calculated from the areal densities and
bulk volume densities.

3. Results and Discussion

Triple-Layer Thin Films of NiFe, Cu and Cr. A schematic of the two triple-layer films used in this study is shown in Figure 1. Both specimens T1
and T2 have identical layers of NiFe, Cu and Cr , but with the top and bottom
layers reversed. The presence of Cu, Ni, Fe and Cr simultaneously in the
specimens provides a very complex matrix for XRF analysis. The absorption
and secondary fluorescence effects of these elements are illustrated in Figure 2. Because of the proximity of the absorption edges and the
characteristic lines of the transition elements, strong absorption and secondary fluorescence effects occur within the same NiFe layer and among all
three layers.

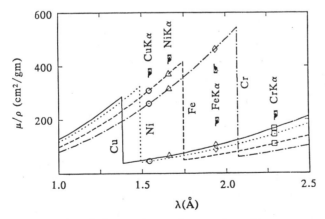

Figure 2. Origin of absorption and secondary fluorescence effects.
The Kβ radiations are omitted.

Table I. Experimental Data of NiFe, Cu and Cr Thin Films

SPECIMEN	Ni	Fe	Cu	Cr
RELATIVE INTENSITY, I(spc)/I(std)				
Triple-Layer Films				
T1(NiFe/Cu/Cr)	0.0404407	0.0055008	0.0442915	0.0356241
T2(Cr/Cu/NiFe)	0.0340010	0.0044069	0.0433732	0.0428958
Single-Layer Films				
S1(NiFe)	0.0400766	0.0054791	---------	---------
S2(Cu)	---------	---------	0.0457358	---------
S3(Cr)	---------	---------	---------	0.0414369
RELATIVE DEVIATIONS				
(T1 - S)/S	0.9%	0.4%	-3.2%	-14.0%
(T2 - S)/S	-15.0	-19.6	-5.2	3.5

The relative intensities of the experimental Ni, Fe, Cu and Cr Kα X-rays
of the triple-layer films compared to those of the pure element bulk standards and the corresponding single-layer NiFe, Cu and Cr films are listed in
Table I. A comparison of the values of these two sets of films shows the severe inter-layer effects of absorption and secondary fluorescence. Ni and
Fe have much lower intensities in T2 than the single-layer films because of
the upper-layer Cr and Cu absorption, but they have slightly higher intensities in T1 due to the secondary fluorescence by Cu in the bottom-layer.
However the Cr intensity is 14% lower in T1 than the single-layer Cr film
but 3.5% higher in T2. The Cu X-rays in both T1 and T2 are lower than the
single-layer Cu film because of the Ni, Fe and Cr absorption in the upper
layers.

The compositions and thicknesses of all layers of a specimen were simultaneously determined from the relative Kα peak intensities and the results
are listed in Table II. The results are consistent for the two triple-layer
films even though they have large differences of XRF intensities (Table I).
Agreement between the triple- and the identical single-layer films are also
good. As shown in the lower part of Table II, the discrepancies are small
and vary from 0.04-0.23% absolute and 0.04-2.2% relative for composition,
and are within 8-88 angstroms or 0.3-3.3% for thickness.

These results show that the analysis of the two triple-layer films has
been successful. The concentrations and thicknesses of the triple-layer
films are consistent with those of the corresponding single-layer films.
The precisions obtained in this analysis are comparable with those of the
bulk and single-layer films (see section II).

The effect of the substrate was also determined with LAMA-III. The quartz
(SiO_2) of course caused no secondary fluorescence and the effect of back
scattering was negligible, causing only ≤ 0.01% difference in composition
and ≤1 angstrom in thickness.

Table II. Analysis Results of NiFe, Cu and Cr Films

SPECIMEN	Ni (wt%)	Fe (wt%)	NiFe t(Å)	Cu t(Å)	Cr t(Å)
XRF/LAMA RESULT					
Triple-Layer Films					
T1(NiFe/Cu/Cr)	89.76	10.24	2123	2470	1653
T2(Cr/Cu/NiFe)	89.72	10.28	2055	2462	1698
Single-Layer Films					
S1(NiFe)	89.53	10.47	2113	----	----
S2(Cu)	-----	-----	----	2418	----
S3(Cr)	-----	-----	----	----	1675
DEVIATIONS					
T1 - T2	0.04	-0.04	68	8	-45
T1 - S	0.23	-0.23	10	52	-22
T2 - S	0.19	-0.19	-58	88	23
RELATIVE DEVIATIONS					
(T1 - T2)/Tavg	0.04%	-0.39%	3.3%	0.3%	-2.7%
(T1 - S)/S	0.26	-2.20	0.5	2.2	-1.3
(T2 - S)/S	0.21	-1.81	-2.7	1.8	1.3

Double-Layer Thin Films with a Common Element. The analysis of multi-layer thin films can be relatively complicated when some layers have one or more common elements. The LAMA-III analysis of these specimens can not be made solely from the experimental XRF intensities. Additional information on the composition and/or thickness of these layers is required. The amount of additional data needed depends on the complexity of the specimen.

A sputtered double-layer FeMn/NiFe thin film was selected for this study. The combination of a common element Fe in both layers and the strong absorption and secondary fluorescence between elements within the same layer and of different layers provide a particular useful test. The Kα intensity ratios of the elements in the double-layer film D1 are listed in Table III. The observed Fe intensity ratio was the sum of the top FeMn and the bottom NiFe layers. The relative intensities of the corresponding single-layer FeMn and NiFe films S4 and S5 are also included. If there were no inter-layer effects, the intensities of D1 should have agreed with those of S4+S5 (S45). A comparison between D1 and S45 shows a 13% intensity reduction of Ni Kα because of the Fe and Mn absorption in the top layer. The Mn enhancement (2.1%) was stronger than that of Fe (0.4%) owing to the secondary fluorescence of Mn by both Ni and Fe K X-rays from the bottom layer.

Table III. Experimental Data of FeMn/NiFe Thin Films

SPECIMEN	Mn	Fe	Ni
RELATIVE INTENSITY, I(spc)/I(std)			
Double-Layer Film			
D1(FeMn/NiFe)	0.0283372	0.0293892	0.0280607
Single-Layer Films			
S4(FeMn)	0.0277469	0.0205245	---------
S5(NiFe)	---------	0.0087504	0.0324742
S45(S4+S5)	0.0277469	0.0292749	0.0324742
RELATIVE DEVIATION			
(D1 - S45)/S45	2.1%	0.4%	-13.0%

The unknown composition (w) and thickness (t) of the double-layer film D1 were determined from the observed Ni, Fe and Mn intensities as well as the known values of w and/or t of the layer. The values of w and/or t determined from the single-layer film S4 or S5 were used and were kept constant during the LAMA-III iteration. Results of the double-layer film are listed in Table IV and the fixed parameters enclosed by <>. Consistent values of XRF composition and thickness were obtained for all six different fixed parameter options: fixed w, t and w & t of the top- and the bottom-layer. Agreement between the results of the double- and single-layer films are good. Deviations in composition are small and vary from 0.12-1.49% absolute and 0.21-7.98% relative. In all analyses the thickness values are within 2-37 angstroms or 0.1-1.7%. As in the results of the triple-layer films, the precision of Δw and $\Delta t/t$ are also within the range of $\pm 2\%$, which is comparable to bulk and single-layer thin films.

The deviations in composition Δw are consistently larger when the parameters w and/or t of the top-layer were fixed than of the bottom-layer (i.e. Δw=0.52, 1.49 and 0.94% compared to 0.29, 0.60 and 0.12% respectively). In addition, the deviations are maximum when t was fixed (i.e. Δw=1.49 and 0.60%).

These dependencies were further investigated using the LAMA-III computed theoretical intensity ratios for various simulated FeMn/NiFe double-layer thin films. Preliminary results showed that Δw generally depended on the mass ratio Mf/Mu, where Mf is the total mass of the common element in the layer with fixed w and/or t, and Mu in the undetermined layer; Mf/Mu=2.6 when the top-layer parameter is fixed and 0.4 for the bottom-layer. The smaller the Mf/Mu, the smaller the Δw. Analysis usually failed when Mf/Mu reached 50 and higher. Among the three options, the fixed t method had the largest Δw and was also most sensitive to the initial thicknesses used in the analysis. The entries of the estimated thicknesses are required by LAMA-III. In principle, the reverse computer calculation using theoretical intensity ratios should give exactly the same results (i.e. Δw=0), which can only be achieved by prefect iteration.

Table IV. Results of Double-Layer FeMn/NiFe Film Analysis

SPECIMEN or PARAMETER	------ Top-Layer ------			---- Bottom-Layer ----		
	Fe (wt%)	Mn (wt%)	t (Å)	Ni (wt%)	Fe (wt%)	t (Å)
XRF/LAMA RESULT						
Single-Layer Films						
S4(FeMn)	44.11	55.89	2150	-----	-----	----
S5(NiFe)	-----	-----	----	81.32	18.68	1905
Double-Layer Film						
Top-Layer						
Fixed w	<44.11>	<55.89>	2175	80.80	19.20	1884
t	43.46	56.54	<2150>	79.83	20.17	1907
w & t	<44.11>	<55.89>	<2150>	80.38	19.62	1892
Bottom-Layer						
Fixed w	44.40	55.60	2187	<81.32>	<18.68>	1872
t	43.51	56.49	2153	79.91	20.09	<1905>
w & t	44.23	55.77	2180	<81.32>	<18.68>	<1905>
DEVIATIONS (D-S)						
Top-Layer						
Fixed w	-----	-----	25	-0.52	0.52	-21
t	-0.65	0.65	----	-1.49	1.49	2
w & t	-----	-----	----	-0.94	0.94	-13
Bottom-Layer						
Fixed w	0.29	-0.29	37	-----	-----	-33
t	-0.60	0.60	3	-1.41	1.41	----
w & t	0.12	-0.12	30	-----	-----	----
RELATIVE DEVIATIONS (D-S)/S						
Top-Layer						
Fixed w	-----	-----	1.2%	0.64%	2.80%	-1.1%
t	-1.47%	1.16%	----	-1.83	7.98	0.1
w & t	-----	-----	----	-1.17	5.03	-0.7
Bottom-Layer						
Fixed w	0.66	-0.52	1.7	-----	-----	-1.5
t	-1.36	1.07	0.1	-1.73	7.55	----
w & t	0.27	-0.21	1.4	-----	-----	----

IV. CONCLUSIONS

The application of XRF method for the determination of composition and thicknesses of multi-layer metal thin films has been successfully tested. The complex inter-layer absorption and secondary fluorescence effects were corrected using the fundamental parameter method and the LAMA-III program. There are no thickness and composition restrictions on the standards and specimens required for the analysis. Results of selected double- and triple-layer specimens were compared to those of single-layer thin films prepared under identical conditions using pure element bulk standards. The agreement of the results with those of the corresponding single-layer films was ≤2% in composition and thickness of layers with a nominal thickness of about 2000 angstroms. This precision is comparable to those of bulk and single-layer thin film analysis. These results demonstrate that the complex inter-layer absorption and secondary fluorescence effects have been properly corrected.

ACKNOWLEGEMENT

The authors are grateful to M. Mantler for the development of the LAMA-III program, and to C. Corpuz, J. K. Howard, W. Lee and F. Sequeda for the preparation of the thin films.

REFERENCES

1. T. C. Huang and W. Parrish, Adv. X-ray Anal. 22:43 (1979).

2. D. Laguitton and W. Parrish, Anal. Chem. 49:1152 (1977).

3. J. W. Criss and L. S. Birks, Anal. Chem. 40:1080 (1968).

4. D. Laguitton and M. Mantler, Adv. X-ray Anal. 20:515 (1977).

5. T. C. Huang, X-Ray Spectrom. 10:28 (1981).

6. M. Mantler, Adv. X-ray Anal. 27:433 (1984).

7. M. Mantler, in preparation.

EDXRF ANALYSIS OF THIN FILMS AND COATINGS

USING A HYBRID ALPHAS APPROACH

Dennis J. Kalnicky

Princeton Gamma-Tech
1200 State Road
Princeton, NJ 08540

INTRODUCTION

In the semiconductor and electronics industries the processes involved in producing integrated circuits often require
thin films such as $P2O5/SiO2$, Si/Al, Ni/Cr, Fe/Ni, or others
to be deposited on a substrate wafer. It is critical to the
performance and lifetime of the device that the composition and
thickness of these films be precisely controlled. The ability
to quickly analyze wafers provides data which can be used to
adjust fabrication parameters before large quantities of
devices are lost.

Energy-Dispersive X-ray Fluorescence (EDXRF) is well suited
for thin specimen analyses and the XRF technique has been
successfully applied to particulate (1-6) as well as thin film
(7-16) systems. This technique is rapid, non-destructive, and
accurate. The composition and thickness of a film on a substrate may be determined by comparison of the characteristic
X-ray lines of the film elements to those of suitable standards
with corrections for interelement, film thickness, and substrate
intensity effects. A number of theoretical and non-theoretical
approaches have been employed to various degrees of success to
achieve this end. The Hybrid Alphas approach recently described
by LaChance (17) combined with Fundamental Parameter computations
(18) provides unique capabilities to accurately apply these
corrections while requiring minimal calibration standards.

THIN FLIM QUANTIFICATION

When a thin film is being analyzed using tube X-rays as
incident radiation, the emitted X-ray intensity depends on the
film thickness as well as concentration within the film (Figure
1). Therefore, with the use of appropriate standards and mathematical models, both thickness and composition can be determined in a single analysis. Quantification of film thickness and
composition may be performed using a Hybrid Alphas/Fundamental
Parameters approach which recognizes X-ray intensities due to
the elements in the film and substrate, and properly calculates
the influence of film thickness and interelement effects on

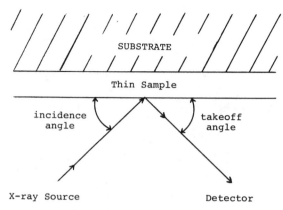

Fig. 1. Thin sample schematic

those intensities (13-18). The measured intensity of an X-ray
line of an element in a film of uniform thickness is given by
(7,8,10,13-15):

$$R(i) = ((W(i)/BCORR(i)) * TCORR(i)) + SCORR(i) \qquad (1)$$

where,

    W(i)   = weight fraction of element i

    R(i)   = ratio of measured to pure intensity
              for element i

    BCORR   = bulk intensity correction factor
              using hybrid alphas

    TCORR   = film thickness intensity correction
              factor

    SCORR   = substrate intensity contribution to
              measured intensity

The details of this algorithm are described elsewhere in these
proceedings (19). An expression of the form given in equation
1 is calibrated for each element of interest (10 maximum) in
the film using suitable standards (2 minimum). The thickness
and composition cannot be obtained directly from these equations
but, rather, they are solved iteratively by the method of
successive approximations. The iterations are continued until
a self-consistent set of concentration and thickness values are
obtained between iteration steps. Generally, values that agree
to within +/- 0.5% relative may be considered self-consistent.

    An advantageous feature of this approach is that an un-
measured element may be determined by stoichiometry and its
effect is combined into the Hybrid Alphas for all measured
elements. This speeds the iteration process and provides a
converging computation for typical thin film applications
where total film composition is normalized to 100%.

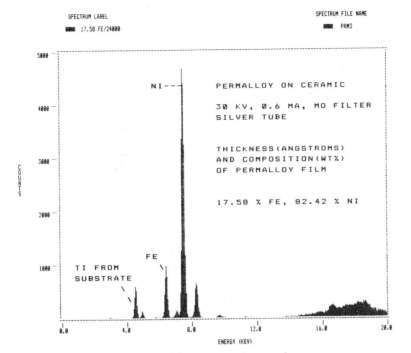

SPECTRUM LABEL

▰▰▰ 17.58 FE/24000

SPECTRUM FILE NAME

▰▰▰ PRM3

NI---| PERMALLOY ON CERAMIC

30 KV, 0.6 MA, MO FILTER
SILVER TUBE

THICKNESS(ANGSTROMS)
AND COMPOSITION(WT%)
OF PERMALLOY FILM

17.58 % FE, 82.42 % NI

FE

TI FROM
SUBSTRATE

ENERGY (KEV)

Fig. 2. Typical Permalloy (Fe/Ni) spectrum

THIN SAMPLE ANALYSIS

Figure 2 shows a typical spectrum of a permalloy (Fe/Ni)
film on a ceramic "AlSiMag" substrate. Note the high peak-
to-background for both the Fe and Ni X-ray lines. The titanium
(Ti) lines are from the substrate and are attenuated in proportion to the film thickness. Typical composition are: 15-25%
Fe and 75-85% Ni for 1.0 to 4.0 micrometer films (10000 to 40000
Angstroms). Figure 3 shows a screen display of typical thin
sample analysis results for permalloy films on glass substrates.
These may be (optionally) listed to a hard copy device along
with the current values of intensity drift correction factors
if that option was included at calibration. The typical performance of this thin film method for permalloy analyses is
shown in Table 1. Note the excellant comparison between given
and fit values; typically agreement within +/- 0.4 wt% absolute
for sample composition (15 to 25% Fe, balance Ni) and +/- 500
Angstroms for thickness (10000 to 40000 Angstroms). This ex-

```
RESULTS FOR SAMPLE NO.   1 1371C
CALIBRATION FILE:   THICK
SPECTRA:   1371C
MONITOR CORRECTION FACTORS:
   FE =   1.0059
   NI =   0.9996
ELEMENT   K-RATIO    WT %   +/-   SIGMA
 FE/KA    0.1327    19.3        0.140
 NI/KA    0.3563    80.7        0.365
 THICKNESS(ANGSTROMS): =   26471. +/-      8.
  COMPUTED
```

Fig. 3. Permalloy results display

Table 1:   Analysis of Permalloy Film Samples
           on Ceramic or Glass Substrates   a)

		Analysis Agreement b)	
Parameter	Range	Range	Typical
Wt.% Fe	15 - 25	0.1 - 0.7	+/- 0.4
Wt.% Ni	75 - 85	0.1 - 0.7	+/- 0.4
Thickness (Angstroms)	10000 - 40000	100 - 1500	+/- 500

a) Conditions: Silver Tube, 30kV, 0.60mA,
   Molybdenum filter, 100 seconds

b) Agreement between given and fit values when
   standards were run as unknowns

cellant performance may be achieved in 100 seconds or less with
the typical EDXRF measurement conditions listed in Table 1.

Figure 4 shows a typical spectrum for very thin (approximately 150 Angstroms) Nickel/Chrome (Ni/Cr) films on silicon
substrates. The excellant peak-to-background characteristics
for the Cr K-alpha and Ni K-alpha lines were achieved by using
a Molybdenum (Mo) filter between the Mo X-ray tube and the film
sample. This eliminated scattered background as well as the
first order Si <100> diffraction (n=1) peak which interferes

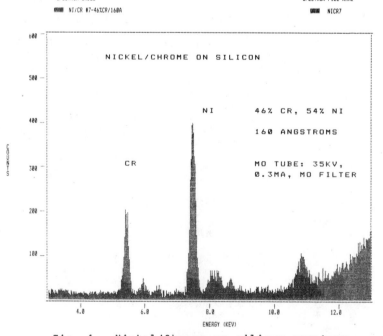

Fig. 4.   Nickel/Chrome on silicon spectrum

Table 2:   Analysis of Nickel/Chrome Film Samples
           on Silicon Substrates   a)

		Analysis Agreement	b)
Parameter	Range	Range	Typical
Wt.% Ni	20 - 60	0.1 - 3.0	+/- 2.0
Wt.% Cr	40 - 80	0.1 - 3.0	+/- 2.0
Thickness (Angstroms)	100 - 200	1 - 20	+/- 7

a) Conditions: Molybdenum tube, 35kV, 0.30mA,
    Molybdenum filter, 600 seconds

b) Agreement between given and fit values when
    standards were run as unknowns

with Cr K-alpha. The geometry of the EDXRF system used here
is such that the first order Si <100> peak occurs at approximately 5300eV and thus, represents a serious overlap at the
5400eV Cr K-alpha X-ray line unless it is eliminated as was done
here. Nickel/Chrome spectra of the type shown in Figure 4 may
be obtained in 600 seconds or less using typical EDXRF measurement conditions. Table 2 shows typical analysis capabilities
for the Ni/Cr thin film system. Again, the analysis method
gives excellent agreement with given values: typically +/- 2.0
wt% absolute for Cr and Ni composition (20-80% range) and
+/- 7 Angstroms for thickness from 100 to 200 Angstroms. The
density of these films is approximately 8.0 grams/cm^3 so that
the total mass per unit area is 8 to 16 micrograms/cm^2. The
viewing field is approximately 5 cm^2 so that total mass being
measured is 40 to 80 micrograms. Thus, an agreement to within
2.0 wt% Ni and Cr is equivalent to measuring mass differences
of 0.8 to 1.6 micrograms for typical films. The detection
limit is less than 2 wt% for these elements which corresponds
to mass detectability on the order of 0.5 micrograms absolute.

The above illustrates the capabilities of EDXRF techniques
to reach very low levels of detectability with proper measurement conditions for thin film systems. There are, of course,
limitations, particularily, as the film thickness approaches
semi-thin regions, ie, not infinitely thick but thick enough
such that significant thickness/interelement corrections are
required and detection limits may suffer. The thin film algorithm given in equation 1 properly accounts for these corrections (19).

A low concentration thin sample spectrum is shown in Figure
5 for Aluminum/Silicon (Al/Si) films on Carbon substrates. The
Si content typically ranges from 0.5 to 1.5 wt% for films from
10000 to 20000 Angstroms. There is an enormous overlap of the
Al K-alpha X-ray line into the Si K-alpha region with typical
EDXRF SiLi detector resolutions for these low levels of Si where
the Al/Si ratio may be as high as 200 to one (see Figure 5).
Proper peak stripping algorithms are important to resolve this
overlap. The method used here employs the digital filter/least
squares fit technique commonly used to resolve EDXRF spectral

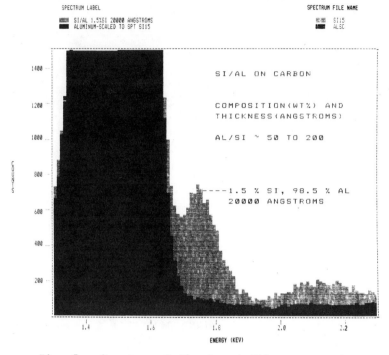

Fig. 5. Spectra of Aluminum/silicon on carbon

overlaps (20). Table 3 shows thin film analysis results for low concentrations of Si in typical Al/Si films. Agreement is very good: +/- 0.07 wt% for 0.3 to 2.0 % Si and +/- 150 Angstroms for 16000 to 18000 Angstrom films. This good agreement indicates that the peak stripping algorithms are properly accounting for the various overlap and background factors to produce good net intensities for both Si K-alpha and Al K-alpha.

Table 3: Analysis of Aluminum/Silicon Film Samples
          on Carbon Substrates   a)

		Analysis	Agreement	b)
Parameter	Range	Range	Typical	
Wt. % Si	0.3 - 2.0	.02 - .10	+/- .07	
Wt. % Al	98 - 99.7	0 - .10	+/- .10	
Thickness (Angstroms)	16000 - 18000	70 - 300	+/- 150	

a) Conditions: Silver tube, 5kV, 0.70mA, no filter,
    600 seconds

b) Agreement between given and fit values when
    standards were run as unknowns

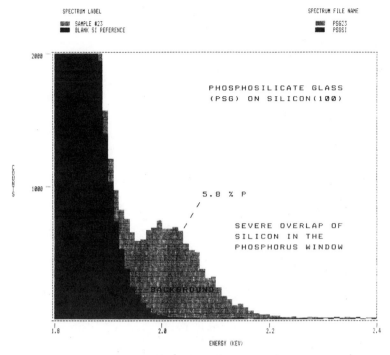

Fig. 6.   Spectra of Phosphosilicate glass on silicon

      A different type of thin film system is shown in Figure 6
which shows spectra of Phosphosilicate (PSG) glass on a Si <100>
substrate.   There is no diffraction peak problem, however, the
vast majority of measured Si K-alpha counts are from the Si
<100> substrate and, hence, the net Si K-alpha counts from the
film must be extracted for quantification.   The substrate intensity correction term in this thin film algorithm (see
equation 1) serves this purpose.   Note also that the Si K-alpha
line severely overlaps P K-alpha as was the case for the Al/Si
films.   Table 4 shows typical results for the determination of
PSG composition where the thickness was entered, NOT computed.
Typical agreements of +/- 0.3 wt% may be obtained for both wt%
P (4-11%) and wt% Si (35-43%).   In this example the oxygen in
the films (approximately 50 to 55 wt%) was included in the
hybrid alphas for P and Si K-alpha assumming the following
stoichiometry: P_2O_5 and SiO_2.   Thickness computations were
not acceptable and, therefore, externally measured values (for
example, by optical interferometry) were entered to fix the
sample thickness.

CONCLUSIONS

      This paper has described a Hybrid Alphas/Fundamental
Parameters approach which is designed to enhance typical XRF
thin sample analysis capabilities.   It is particularily useful for multielement thin sample applications where calibration
may be achieved with as few as two standards while still pre-
serving a high degree of accuracy.   Unmeasured elements may be
determined by stoichiometry and thickness may be computed or
entered (fixed) from an external measurement.   While the method
may be somewhat less accurate than full Fundamental Parameters

Table 4: Analysis of Phosphosilicate Glass Film
Samples on Silicon Substrates   a)

		Analysis	Agreement b)
Parameter	Range	Range	Typical
Wt. % P	4 - 11	.03 - .7	+/- 0.3
Wt. % Si	35 - 43	0 - .8	+/- 0.3

a) Conditions: Silver tube, 5kV, 0.7mA, no filter,
    300 seconds

b) Agreement between given and fit compositions when
    standards were run as unknowns. Thickness entered;
    not computed

(FP) approaches, it retains many basic FP features and computation time is much faster than full theory. The method may
also incorporate corrections for day-to-day instrument drifts
and lends itself to automation for routine use by non-technical
operators on a day-to-day basis with results/accuracy comparible
to the original calibration. This combination of state-of-the-
art user friendly software and EDXRF hardware provides a unique
capability to accurately analyze various thin sample products in
a short time with minimum operator intervention.

REFERENCES

1. Rhodes, J.R., et al.; Air Qual. Instrum.(1974), 2,
    1.; Environ. Science Tech. (1972), 6, 922; ISA Trans.
    (1972), 11, 337.

2. Van Espen, P., Adams, F.; Anal. Chim. Acta. (1974), 75
    61.

3. Vanderstappen, M., Van Grieken, R.; Z. Anal. Chem. (1976),
    282, 25.

4. Kalnicky, D.J.; "Analysis of Air, Stack Gas, and Solution
    Particulates by Secondary Target Energy-Dispersive X-ray
    Fluorescence", paper no. 224, Pittsburgh Conference on
    Analytical Chemistry and Applied Spectroscopy, Cleveland,
    Ohio, 1979.

5. Birks, L.S.; Anal. Chem. (1977), 49, 1505.

6. Davis, D.W., et al.; Anal. Chem. (1977), 49, 1990.

7. Bertin, E.P.; "Principles and Practices of X-ray
    Spectrometic Analysis", 2nd ed.; Plenum: New York, 1975.

8. Bergal, L., Cadieu, F.J.; X-Ray Spectrom. (1980), 9, 19.

9. Birks, L.S.; "X-Ray Spectrochemical Analysis", 2nd. ed.;
    Interscience: New York, 1969.

10. Rhodes, J.R., et al.; Anal. Instrum. (1972), 10, 143.

11. Rhodes, J.R.; "Recommended Procedures for use of C.S.I.
     Thin Standards for X-Ray Fluorescence Spectrometry",
     ARD Internal Report no. 206; Columbia Sciencetific
     Industries: Austin, Texas,1975.

12. Laguitton, D., Parrish, W.; Anal.Chem. (1977), 49, 1152.

13. Kalnicky, D.J., Moustakes, T.D.; Anal.Chem. (1981),53,
     1792.

14. Kalnicky, D.J., Barbi, N.C., Schnerr, G.; "A New Quality
     Control Instrument to Determine Concentration and Thick-
     ness in Films", presented at the International Forum for
     the Production of Electronic Components, Wiesbaden,
     West Germany, 1982.

15. Kalnicky, D.J.; "Mathematical Treatment of X-ray Data
     From Films on Substrates", presented at the SUNYA X-ray
     Clinic, State University of New York at Albany, 1983.

16. Kalnicky, D.J.; "Application of Energy-Dispersive
     Techniques for Materials Characterization", paper
     no. 86; Eastern Analytical Symposium, New York, 1984.

17. LaChance, G.R.; "Introduction to Alpha Coefficients", 1984.
     Available from Corporation Scientifique Claisse, Inc. 2522,
     Chemin Sainte-Foy, SAINTE-FOY (Quebec), G1V 1T5, Canada:
     LaChance, G.R.; "The Family of Alpha Coefficients in X-ray
     Fluorescence Analysis": X-ray Spectrometry, (1979),8, 190.

18. Criss, J.W.; "Fundamental Parameters Calculations on a
     Laboratory Computer", Advances in X-Ray Analysis, (1980),
     23, 93; X-Ray Software Review, No. 2, (1982).

19. Kalnicky, D.J.; "A Combined Fundamental Alphas/Curve
     Fitting Algorithm for Routine XRF Sample Analysis",
     Denver X-ray Conference, Snow Mass, Colorado, 1985.

20. Statham, Peter J.; Anal. Chem. (1977), 49, 2149.

MEASUREMENT OF THE SPECTRAL DISTRIBUTION

EMITTED FROM X-RAY SPECTROGRAPHIC TUBES

Tomoya Arai, Takashi Shoji and Kazuhiko Omote

Rigaku Industrial Corporation
Takatsuki
Osaka, Japan

INTRODUCTION

The fundamental parameter method for quantitative analysis of composite elements has been a powerful technique for x-ray spectrochemical analysis in which the x-ray intensity and spectral distribution from x-ray spectrographic tubes are the most essential factors in the calculating process based on x-ray physics.

In previous papers on this study, there were experimental reports which were presented by Gilfrich and co-workers, using a single crystal monochromator (1,2,3,4) and by Loomis and Keith, using an energy dispersive system equipped with a lithium-drifted silicon detector (5). Pella and co-workers studied spectral distributions using the same detecting system as Loomis and Keith in combination with an electron probe microanalyzer, varying accelerating voltage and target elements (6). Furthermore, in parallel to the use of a side-window x-ray tube, an end-window tube as an x-ray excitation source has become popular because of its better performance for the analysis of light elements.

It is the purpose of this report to investigate the relationship among the construction of x-ray tubes, emitted intensity and spectral distribution, compare side-window tubes with end-window tubes and discuss some considerations.

EXPERIMENTAL I

Structures and geometrical conditions of the x-ray tubes are shown in Fig. 1. The differences between side and end-window tubes are in incident angle of the electron beam and take-off angle of emitted x-rays. For the emission of soft x-rays, a take-off angle of 90 degrees is the most effective configuration because of the smaller absorption in the target metal.

There are two factors which influence the shape of the spectral intensity distribution in the region of the shortest wavelength x-rays. The first is the ripple of the applied voltage (about ± 0.5% at 60Kv and 10mA) and the second is the emission of polarized x-rays. Because, in the case of a side-window x-ray tube, the angle between the electron beam and emitting x-rays is 90 degrees, the emitting probability of polarized

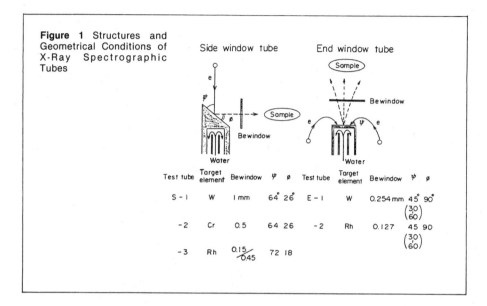

Figure 1 Structures and Geometrical Conditions of X-Ray Spectrographic Tubes

Test tube	Target element	Be window	ψ	ø	Test tube	Target element	Be window	ψ	ø
S - 1	W	1 mm	64°	26°	E - 1	W	0.254 mm	45°	90° $\binom{30}{60}$
-2	Cr	0.5	64	26	-2	Rh	0.127	45	90 $\binom{30}{60}$
-3	Rh	0.15/0.45	72	18					

Table 1 Related Factors between Measured X-Ray Intensity and Emitted X-Ray Intensity from X-Ray Tube

Factors	Note
High voltage supply	KV, mA, Constant potentical including ripple
X-ray tube	Incident angle of electrons Take off angle of X-rays Target element Be-Window Emission of polarized X-rays
Main collimator	Holizontal and vertical width X-ray transmission of edge slits
Single crystal monochromator	Dependency of reflection upon various X-rays Degree of single crystal perfection Polarization of reflected X-rays
Receiving slit	Elimination of stray X-rays
Mylar window	Transmission of Mylar and Nylon mesh
Thin air path	
Detector	Detection efficiency (SC and F-PC)
PHA condition	Transmission and elimination of second and third order X-ray effects
Counting loss	Linearlity check
Background	Measuring technics and correction

Table 2-1. Correcting Numerical Values of Each Factor for Short Wavelength X-Rays

Correcton Factors		X-Rays (Å)					
		0.5	1.0	1.5	2.0	2.5	3.0
Reflectivity of LiF	Integrated reflection (in units of Radian)	3.5×10^{-5}	3.5×10^{-5}	4.09×10^{-5}	4.37×10^{-5}	4.43×10^{-5}	5.20×10^{-5}
Mylarfilm & Mesh	Transmission	0.998	0.990	0.969	0.931	0.876	0.807
Air Path	Transmission	0.998	0.991	0.970	0.932	0.873	0.800
Detector efficiency	Efficiency	0.988	0.976	0.891	0.855	0.747	0.609
PHA Transmission For 1st Order For 2nd Order For 3rd Order		0.965 0.066 0.0	0.979 0.011 0.0005	0.980 0.019 0.0012	0.980 0.0265 0.0023	0.980 0.034 0.0033	0.981 0.046 0.0041
Conversion factor $\Delta\theta$ to $\Delta\lambda$	$1/\cos\theta$	1.008	1.032	1.068	1.149	1.272	1.499

Table 2-2. Correcting Numerical Values of Each Factor for Long Wavelength X-Rays

Correcton Factors		X-Rays (Å)						
		2.0	3.0	4.0	5.0	6.0	7.0	8.0
Reflectivity of PET	Integrated reflection (in units of Radian)	7.4×10^{-5}	6.8×10^{-5}	8.0×10^{-5}	8.7×10^{-5}	9.0×10^{-5}	14.6×10^{-5}	31.8×10^{-5}
Detector efficiency	Efficiency	0.653	0.949	0.565	0.743	0.869	0.951	0.987
PHA Transmission for 1st order		1.00	0.979	1.00	1.00	1.00	1.00	1.00
Conversion factor $\Delta\theta$ to $\Delta\lambda$	$1/\cos\theta$	1.007	1.065	1.125	1.219	1.375	1.670	2.483

continuous x-rays is increased compared with the case of an end-window tube having an angle of about 45 degrees.

In Table 1, the related factors are listed converting from measured x-ray intensity to spectral intensity and in Table 2 the numerical data of each factor are also shown. For these calculations, mass absorption coefficients reported by Victreen (7) and McMaster (8) were used for short and long wavelength x-rays, respectively. In order to estimate detector efficiencies, reports by Parrish and Taylor (9) and Loomis and Keith (10) were consulted. The counting linearity is ± 1% or less at an x-ray intensity of 500,000 c.p.s. Gilfrich's (1) method was used to make background intensity measurements and corrections.

A single crystal spectrometer was used for the measurement of spectral distribution of short wavelength x-rays. The narrow x-ray beam, restricted to a divergence of 0.172 degrees in the horizontal direction and 3.43 degrees in the vertical direction, strikes the single crystal of LiF and monochromated x-rays are detected by a scintillation counter with the pulse height discriminator in the differential mode.

For long wavelength x-ray measurements, a PET single crystal and a gas flow proportional counter equipped with a thin mylar window were used, and the influence of short wavelength higher order x-rays was eliminated by means of pulse height discrimination.

EXPERIMENTAL II

For the determination of the reflectivity of the LiF crystal for short wavelength x-rays, a double crystal spectrometer was used in an air path arrangement. As the first crystal of the double crystal spectrometer, a silicon single crystal (220) was adopted in order to

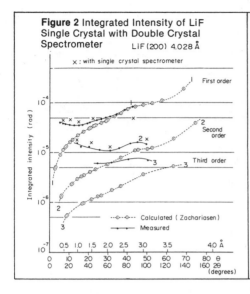

Figure 2 Integrated Intensity of LiF Single Crystal with Double Crystal Spectrometer   LiF (200) 4.028 Å

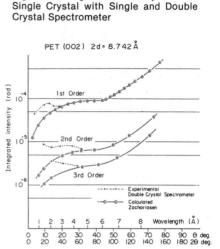

Figure 3 Integrated Intensity of PET Single Crystal with Single and Double Crystal Spectrometer

obtain a parallel beam. The LiF crystal was prepared by means of cleaving and etching because of the necessity of an optical system of high resolution. The divergence of the reflected x-rays was approximately one minute of θ. Because the measured intensity of a second crystal contains some amount of polarized x-rays, the polarization correction was made for the purpose of providing the reflectivity for unpolarized incident x-rays. In Fig. 2, the integrated reflectivity of the LiF single crystal is shown. Dotted lines show the calculated intensities after Zachariasen's formula (11) and solid lines are the intensities measured by this method. There are discrepancies between calculated and measured intensities in the region of short wavelength x-rays of first order reflection. In order to check the perfection of the deeper portion of a single crystal, a topographic camera technique was applied, employing Ag-Kα x-rays.

Structural defects consisting of a conglomeration of subgrains could be found on photographic film, giving the single crystal increased reflected intensity. On the other hand, the calculated intensity of a perfect crystal is reflected from a very thin surface layer (about 2 microns) because of the strong extinction effect and would therefore be lower. For second and third order x-ray reflections, the discrepancies are more clearly exhibited since the measured x-rays are reflected from a deeper portion in the LiF crystal. The measured reflected x-rays contain some. amount of polarization at the angular range from 30 to 60 degrees and the polarization correction should be applied. The small x are the intensities measured by a single crystal spectrometer. This measuring method is very similar to that of Gilfrich and co-workers (12). In the case of the measured intensities of Ti-Kα x-rays (2.75Å) of the first order reflection, there is a discrepancy between the values from a single and a double crystal spectrometer, which is caused by the difference in polarization.

For the determination of crystal reflectivity of a PET crystal, the double crystal spectrometer was used for the shorter wavelength x-rays (one to three Å) adopting a PET crystal as the first crystal, and the single crystal spectrometer was used for the long wavelength x-rays under vacuum. The PET crystal was prepared by means of cleaving and etching. The divergence of the relfected x-rays, when using the double crystal spectrometer, was approximately one minute of θ.

In Fig. 3, the measured and calculated intensities of a PET crystal
are shown. Since the reflected intensities as measured by the single
crystal spectrometer are almost the same as the calculated values which
were derived by Zachariasen using unpolarized x-rays, the calculated
reflectivity was used in the calculation for the spectral distribution
of the x-ray tube. For the calculations of integrated intensities,
scattering factors reported by Cromer and co-workers (13,14) and the
crystal structure of PET according to Shiono (15) and Hvoslef (16) were
used. Numerical results were referenced to reports by Brown and co-
workers (17,18) for LiF and by Evans and co-workers (19) for PET.

EXPERIMENTAL RESULTS AND DISCUSSION

Fig. 4 is a plot of 2θ degrees of LiF versus measured intensity as
emitted from the side-window W-target tube (S-1) at 45 kV c.p. It clearly
shows completely the relation among characteristic x-rays of the W
L-series, absorption edges of the L-series and the continuum. Fig. 5 is
a similar plot for the end-window Rh-target tube (E-2) at 45 kV c.p.

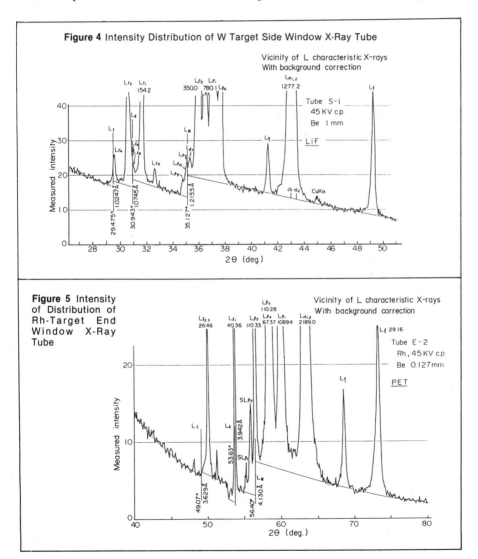

Figure 4 Intensity Distribution of W Target Side Window X-Ray Tube

Figure 5 Intensity of Distribution of Rh-Target End Window X-Ray Tube

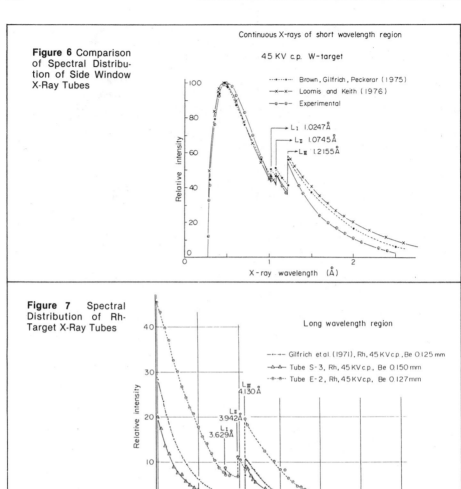

Figure 6 Comparison of Spectral Distribution of Side Window X-Ray Tubes

Figure 7 Spectral Distribution of Rh-Target X-Ray Tubes

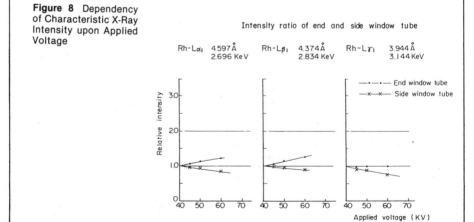

Figure 8 Dependency of Characteristic X-Ray Intensity upon Applied Voltage

There is an intensity gap at 52.57 degrees 2θ which arises from the gap
of the Argon K absorption edge of the proportional counter gas.

    Correction factors to be applied for the short wavelength x-ray
region are effective slit width of primary slits; mylar window with nylon
mesh; air path between mylar window and the scintillation counter window;
the detector efficiency consisting of beryllium foil, aluminum reflector
and thickness of NaI(Tl); the transmission of pulse height discrimination
for the elimination of second and third order x-rays especially for scintillation counter; the conversion factor from Δθ to Δλ and the measured
integral reflectivity of the LiF crystal for unpolarized x-rays,
listed for several wavelengths in Tables 1 and 2.

    Fig. 6 shows the spectral distribution of side-window x-ray tubes
with W-target at 45 kV c.p. in the short wavelength region. Comparison
with reported results by Gilfrich and co-workers and Loomis and Keith is
also shown. Below the L absorption edges of the W-target, all of Gilfrich
(1975) Loomis and Keith and this experiment are in fairly good agreement.
Beyond the L absorption edges of the W-target, Loomis and Keith's data
are highest. This may be because of a thinner Beryllium window of the
x-ray tube.

    Correction factors for the long wavelength x-ray region are the de-
tector efficiency, consisting of the aluminized mylar window and density
of the detecting gas; the transmission of pulse height discrimination
containing the elimination of second and third order x-rays; the conversion factor from Δθ to Δλ and the integral reflectivity of the PET
crystal for unpolarized x-rays, which is listed for several wavelengths
in Tables 1 and 2.

    Fig. 7 shows the spectral distribution of Rh-target x-ray tubes at
45 kV c.p. (in the long wavelength region) and compares it with reported
results by Gilfrich and co-workers. End-window x-ray tube data are
included.

    Even though an accurate comparison could not be made because of dif-
ferences between side and end-window x-ray tubes in shape and size of
focal spot as well as distance from focal spot to primary slits of the
spectrometer, it can be stated that the intensity from an end-window
x-ray tube is two to three times higher than that from a side-window
tube in the long wavelength region. The spectral distributions of the
x-ray tubes used are tabulated in Table 3. All the intensities were
normalized to the maximum intensity of the continuum at 45 kV c.p.

    The dependency of the relative intensity upon applied voltage for
the short wavelength region of continuous x-rays and for the characteristic x-rays Rh-Kα and W-Lα can be derived from values listed in Table 3.
For continuous x-rays, the relation between measured intensity and applied voltage is given by the equations derived by Kulenkampff (19) or
Kramers (20). For characteristic x-rays, the relationship between
measured intensity and applied voltage is given by the equations derived
by Webster and Clark (21) or Wooten (22). Because of the smaller take-
off angle of x-rays in the case of the side-window x-ray tube, smaller
dependency upon applied voltage can be derived. In the case of Rh-Kα
x-rays, there is no difference between side and end-window.

    The dependency of the relative intensity upon applied voltage for
characteristic x-rays of Rh-Lα$_1$, Lβ$_1$ and Lγ$_1$ shows the take-off angle
effect, mentioned above, more clearly (Fig.8). It is clearly shown that,
in the case of a side-window x-ray tube, an intensity decrease takes
place with an increase in applied voltage, because of the high penetra-

Table 3-1. Spectral Distribution of X-Ray Tube
Tube S-1, W, Be 1mm, Side Window

Applied Voltage	40kV	45kV	50kV	60kV
λ(Å)		I •Δλ*		
0.20	—	—	—	8.1
0.24	—	—	6.0	87.0
0.28	—	17.9	61.2	129.2
0.32	23.2	61.5	94.1	152.0
0.36	49.3	78.9	105.0	162.7
0.40	65.5	92.9	117.1	166.3
0.44	74.3	98.9	120.9	161.9
0.48	78.5	100.0	120.1	154.9
0.52	79.5	98.5	118.5	147.8
0.56	80.0	97.1	114.0	139.0
0.60	79.3	93.3	110.6	129.0
0.64	78.0	89.5	102.0	120.0
0.7	74.1	83.9	94.2	108.6
1.0	42.8	45.8	49.0	52.6
1.025 –**	40.5	43.0	46.1	48.0
1.025 +	43.4	46.2	49.5	51.7
1.075 –	39.5	42.0	44.0	45.9
1.075 +	45.3	48.2	50.6	54.0
1.1	43.0	46.0	48.0	51.2
1.2	33.5	36.5	39.7	41.2
1.216 –	32.7	35.3	38.0	40.1
1.216 +	48.8	54.0	57.6	68.4
1.3	41.8	46.5	50.0	56.8
1.4	35.1	37.6	43.2	46.1
1.5	29.0	30.1	35.1	37.0
2.0	11.2	11.8	12.0	12.7
2.5	4.0	4.0	4.1	3.1

Characteristic Lines				
λ(Å)		I •Δλ*		
WLα_1	1.4764	803.0	936.0	1067.0	1274.0
Lα_2	1.4874	82.4	107.0	115.0	143.0
Lβ_1	1.2818	474.0	570.0	659.0	823.0
Lβ_2	1.2466	208.0	249.0	283.0	354.0
Lβ_3	1.2627	73.2	90.9	106.0	136.0
Lγ_1	1.0986	85.9	101.2	115.9	142.0

The values listed are the intensity ratios to the peak intensity of the continuum at an applied voltage of 45kV.

*Δλ = 0.02Å; natural line breadth for characteristic lines.

**The (–) and (+) signs denote that the intensity at the bottom and top of the absorption edge jump is listed respectively.

Table 3-2. Spectral Distribution of X-Ray Tube
Tube S-3, Rh, Be 0.15mm, Side Window

Applied Voltage	40kV	45kV	50kV	60kV
λ(Å)		I •Δλ*		
0.20	—	—	—	—
0.24	—	—	4.2	70.0
0.28	—	13.6	47.6	105.5
0.32	17.1	50.5	75.4	129.1
0.36	38.5	66.8	85.4	135.0
0.40	54.7	78.7	96.6	138.8
0.44	61.6	84.0	98.3	134.0
0.48	64.8	82.5	95.9	124.7
0.52	65.8	81.0	92.8	117.4
0.534 –	66.0	79.3	91.1	116.1
0.534 +	77.6	100.0	116.6	159.9
0.56	77.6	98.8	115.6	156.4
0.60	77.7	97.0	114.6	148.7
0.70	76.6	92.2	105.1	132.9
0.80	72.7	84.4	95.4	112.8
0.90	65.7	75.5	84.6	101.3
1.00	58.2	66.2	73.4	86.9
1.50	29.7	31.2	31.6	33.8
2.00	13.5	13.5	13.7	12.8
2.50	6.6	6.4	5.5	5.1
3.00	3.38	3.21	2.94	2.45
3.50	1.72	1.59	1.50	1.08
3.60	1.52	1.43	1.32	0.98
3.629 –	1.57	1.37	1.22	0.93
3.629 +	1.91	1.74	1.60	1.31
3.70	1.65	1.52	1.43	1.01
3.80	1.43	1.27	1.18	0.83
3.90	1.15	1.06	1.01	0.66
3.9425 –	1.05	0.98	0.89	0.54
3.9425 +	1.86	1.77	1.69	1.06
4.00	1.77	1.65	1.54	0.93
4.10	1.52	1.43	1.27	0.69
4.1299 –	1.40	1.35	1.22	0.61
4.1299 +	5.91	5.59	5.40	3.88
4.20	5.40	5.15	5.00	3.53
4.50	3.80	3.61	3.38	2.26
5.00	2.13	1.96	1.77	0.89
5.50	1.22	1.06	0.89	0.37
6.00	0.56	0.55	0.47	0.13
6.50	0.25	0.24	0.21	0.04
7.00	0.12	0.11	0.10	0.04
7.50	0.04	0.04	0.03	0.01
8.00	0.01	0.01	0.01	—

Characteristic Lines				
λ(Å)		I •Δλ*		
Rh K$\alpha_{1,2}$	0.6147	580.0	869.0	1203.0	1965.0
K$\beta_{1,3}$	0.546	110.0	165.0	230.0	370.0
L$\alpha_{1,2}$	4.5984	909.0	878.0	848.0	783.0
Lβ_1	4.3741	474.0	466.0	448.0	426.0
Lβ_2	4.1310	38.1	36.2	32.9	32.1
Lβ_3	4.2522	46.2	45.6	43.9	41.9
Lγ_1	3.9437	11.7	10.5	10.5	9.3

The values listed are the intensity ratios to the peak intensity of the continuum at an applied voltage of 45kV.

*Δλ = 0.02Å; natural line breadth for characteristic lines.

tion of incident electrons. The mass absorption coefficient of Rh-Lα and RL-Lβ_1 for Rh metal is relatively small but that of Rh-Lγ_1 is larger. Therefore, smaller dependency of Rh-Lγ_1 x-rays upon applied voltage can be expected. For the end-window tube, an intensity increase of Rh-Lα_1, and Rh-Lβ_1 and no intensity change of Rh-Lγ_1 are found with an increase of applied voltage.

CONCLUSION

Intensity measurements of x-rays emitted from x-ray spectrographic tubes were carried out using a single crystal spectrometer and spectral distributions were estimated after corrections were made. The experimental results are almost the same as previously reported results; however, small differences could be found.

Table 3-3. Spectral Distribution of X-Ray Tube
Tube E-1, W. Be 0.254mm. End Window

Applied Voltage	40kV	45kV	50kV	60kV
λ (Å)		I •Δλ*		
0.20	—	—	—	2.5
0.24	—	—	0.9	63.5
0.28	—	5.5	40.0	105.5
0.32	7.8	43.9	75.1	136.1
0.36	34.0	64.0	91.2	147.8
0.40	53.0	81.0	108.0	161.9
0.44	65.6	92.5	116.0	165.0
0.48	72.3	97.0	120.0	164.1
0.52	77.0	99.0	119.8	159.1
0.56	79.8	99.7	117.2	154.3
0.60	80.5	98.9	114.8	151.2
0.64	80.0	96.5	111.4	145.9
0.70	77.8	92.1	106.9	134.5
1.00	54.2	62.0	69.5	81.5
1.025 –	52.2	59.0	65.5	77.2
1.025 +	53.8	62.1	68.8	82.3
1.075 –	50.7	57.5	63.2	75.1
1.075 +	54.5	63.2	68.9	81.4
1.10	52.7	60.8	66.4	78.2
1.20	44.9	50.9	55.6	66.7
1.216 –	43.8	49.0	53.5	65.0
1.216 +	59.2	64.8	70.5	87.0
1.30	51.5	57.2	62.9	75.5
1.40	43.1	50.5	54.7	66.5
1.50	38.0	45.0	48.6	59.1
1.60	34.0	39.1	44.1	52.5
2.00	23.0	25.3	27.2	30.2
2.50	14.4	15.1	15.9	17.1
3.00	8.9	8.8	9.5	9.8
3.50	4.4	4.4	4.5	4.3
4.00	2.2	2.2	2.2	2.0
4.50	1.2	1.3	1.3	1.1
5.00	0.7	0.7	0.7	0.7
5.50	0.4	0.4	0.4	0.4
6.00	0.2	0.2	0.2	0.2
6.50	0.1	0.1	0.1	0.1
Characteristic Lines				
λ (Å)		I •Δλ*		
W Lα₁ 1.4764	1099.0	1451.0	1551.0	2022.0
Lα₂ 1.4874	118.0	154.0	179.0	238.0
Lβ₁ 1.2818	586.0	725.0	922.0	1143.0
Lβ₂ 1.2466	251.0	306.0	365.0	484.0
Lβ₃ 1.2627	87.2	113.0	141.0	189.0
Lγ₁ 1.0986	114.0	141.0	166.0	217.0

The values listed are the intensity ratios to the peak intensity of the continuum at an applied voltage of 45kV.

*Δλ = 0.02Å; natural line breadth for characteristic lines.

Table 3-4. Spectral Distribution of X-Ray Tube
Tube E-2, Rh 0.127mm, End Window

Applied Voltage	40kV	45kV	50kV	60kV
λ (Å)		I • Δλ*		
0.20	—	—	0.51	2.0
0.24	—	—	0.81	51.0
0.28	—	3.2	13.3	94.6
0.32	5.1	32.6	64.7	123.7
0.36	27.0	52.4	81.5	133.7
0.40	44.6	69.3	97.0	154.3
0.44	57.5	80.7	106.4	154.8
0.48	64.2	84.4	110.3	151.8
0.52	70.8	86.4	111.0	149.2
0.534 –	72.3	86.7	110.5	148.5
0.534 +	76.5	98.9	127.6	175.2
0.56	77.5	100.0	125.5	168.0
0.60	78.7	99.8	123.2	166.1
0.70	80.3	98.3	114.0	151.6
0.80	76.4	92.0	103.9	141.5
0.90	71.8	85.0	96.2	124.2
1.00	65.5	76.9	87.6	110.8
1.50	40.7	45.9	51.7	60.2
2.00	24.4	26.8	28.7	31.8
2.50	15.6	15.8	16.4	17.7
3.00	9.7	9.8	9.8	9.0
3.50	4.8	4.8	4.6	4.3
3.60	4.2	4.2	3.9	3.7
3.629 –	4.1	3.9	3.7	3.6
3.629 +	4.7	4.4	4.4	4.5
3.70	4.3	4.0	4.0	3.9
3.80	3.7	3.6	3.4	3.1
3.90	3.1	2.9	2.9	2.6
3.9425 –	3.0	2.7	2.4	2.2
3.9425 +	4.7	4.1	4.3	4.2
4.00	4.2	3.9	3.9	3.9
4.10	3.5	3.3	3.3	3.2
4.1299 –	3.4	3.2	2.9	3.0
4.1299 +	8.6	9.2	9.2	10.0
4.20	7.8	8.5	8.7	9.0
4.50	5.9	6.1	6.4	6.4
5.00	3.7	3.7	3.7	3.7
5.50	2.2	2.2	2.2	2.1
6.00	1.35	1.4	1.2	1.2
6.50	0.7	0.75	0.71	0.65
7.00	0.4	0.4	0.40	0.35
7.50	0.2	0.2	0.2	0.15
8.00	0.06	0.07	0.06	0.05
Characteristic Lines				
λ (Å)		I • Δλ*		
Rh Kα₁,₂ 0.6147	759.0	1202.0	1604.0	2578.0
Kβ₁,₃ 0.546	120.0	187.0	262.0	440.0
Lα₁,₂ 4.5984	1401.0	1482.0	1584.0	1720.0
Lβ₁,₂ 4.3741	677.0	724.0	784.0	849.0
Lβ₂ 4.1310	69.2	71.3	74.6	76.2
Lβ₃ 4.2522	66.2	70.9	78.1	85.6
Lγ₁ 3.9437	29.2	28.5	29.1	28.6

The values listed are the intensity ratios to the peak intensity of the continuum at an applied voltage of 45kV.

*Δλ = 0.02Å; natural line breadth for characteristic lines.

The spectral distribution from end-window tubes is unique compared with the results from side-window tubes.

It can be conjectured that, in the long wavelength region, intensity from an end-window x-ray tube is two or three times higher than that from side-window x-ray tubes.

The intensity dependence of characteristic x-rays of long wavelength upon applied voltage of an x-ray tube should be effective in the study of the generation of x-rays in the target metal, which is governed by the complex relation among accelerating voltage of electron, incident angle of the electron beam and the take-off angle of the x-rays.

REFERENCES

1. Gilfrich and Birks: Anal. Chem. Vol. 40 No. 7 (1968), 1077-1080.

2. Gilfrich, Burkhalter, Whitlock, Norden and Birks: Anal. Chem.
    Vol. 43 No. 7 (1971), 934-936.

3. Brown and Gilfrich: Journal of Applied Physics, Vol. 42 No. 10
    (1971), 4044-4046.

4. Brown, Gilfrich, Peckerar: Journal of Applied Physics, Vol. 46
    No. 10 (1975), 4537-4540.

5. Keith and Loomis: X-Ray Spectrometry, Vol. 5 (1976), 104-114.

6. Pella, Feng and Small: X-Ray Spectrometry, Vol. 14 No. 3 (1985),
    125-135.

7. Victreen: Journal of Applied Physics, Vol. 14 (1943), Vol. 19 (1948)
    855, Vol. 20 (1949), 1141.

8. McMaster, Kerr, Del Grand, Hallett and Hubbell: Compilation of
    X-Ray Cross Sections by Lawrence Radiation Laboratory, University
    of California, Livermore.

9. Taylar and Parrish, R.S.I. Vol. 26 (1955), 367-373.

10. Loomis and Keith: Applied Spectroscopy, Vol. 29 No. 4 (1975),
    316-322.

11. Zachariasen: "Theory of X-ray Diffraction in Crystals." Dover
    Publications, Inc. New York (1967), 120-123.

12. Gilfrich, Brown and Burkhalter: Applied Spectroscopy, Vol. 29
    No. 4 (1975), 322-326.

13. Cromer and Waber: Acta Cryst. Vol. 18 (1965), 104-109.

14. Cromer and Liberman: Journal of Chemical Physics, Vol. 53, No. 4
    (1970), 1891-1898.

15. Shiono: Acta Cryst, Vol. 11 (1958), 389.

16. Hvoslef: Acta Cryst, Vol. 11 (1958), 383.

17. Brown and Fatemi: Journal of Applied Physics, Vol. 45, No. 4
    (1974), 1544-1554.

18. Brown, Fatemi and Birks: Journal of Applied Physics, Vol. 45,
    No. 4 (1974), 1555-1561.

19. Kulenkampff: Ann. Phys., Vol. 69 (1922), 548.

20. Kramers: Philos. Mag., Vol. 46 (1923), 836.

21. Webster and Clark: Phy. Rev., Vol. 9 (1917), 571.

22. Wooten: Phy. Rev., Vol. 13 (1919), 27.

AN X-RAY MICRO-FLUORESCENCE ANALYSIS SYSTEM WITH DIFFRACTION CAPABILITIES

M. C. Nichols

Materials Science Dept.
Sandia Natl. Laboratory
Livermore, California

R. W. Ryon

Non-Destructive Evaluation Section
Lawrence Livermore Natl. Laboratory
Livermore, California

Abstract

A prototype X-ray fluorescence system for chemical and phase microanalysis of materials has been developed and tested. Preliminary work with this system has indicated X-ray fluorescence detection limits on the order of 40 picograms for heavier elements such as gold when using a 100 micron collimator, 400 second counting time and a silver anode operating at 12 Kw. Phase identification by X-ray diffraction can be obtained from the same spot. A proposed design for an improved system providing greater elemental sensitivities and capable of semi-automated operation has been completed.

Introduction

There have been a number of instruments designed in the past to allow chemical analysis of small areas on a sample using X-ray fluorescence techniques (1-4). Some of these instruments made use of a wide target X-ray source and flat or curved crystal optics and used collimation of the secondary beam to limit the area of the sample that was being observed. Most of these instruments were attachments that were meant to fit onto conventional broad-band spectrometers of their day and required long counting times in order to provide reasonable counting statistics. More recently, electron optical instrumentation in the form of the electron microprobe, scanning electron microscope and the scanning transmission electron microscope have pretty much taken over the job of analysis of small areas.

It was the intent of this investigation to determine whether an X-ray fluorescence instrument could be assembled that could a) provide an analysis to a greater depth within the sample than the near surface analysis obtainable from the electron optical instrumentation, b) do so for small areas on large samples, and c) be capable of collecting data in air or helium rather than in vacuum. It was important that sample damage be minimized, that the sample not be subjected to any coating procedure, and that we be able to examine materials that were both inorganic and organic in nature. It was also desirable to be able to collect this data in both a transmission and a reflection mode and to be able to provide signals of sufficient strength that data collection would require only a short time so that an X-ray map of the sample could be made if desired. It was also of interest to make the instrument as sensitive as possible so that the

423

minimum detection limit for transition and high Z elements would be as low
as possible. In addition, we needed an instrument that could perform mass
thickness measurements of small areas on "thin" samples using X-ray
absorption methods in which the attenuation of a direct beam of X-rays by a
sample is used to measure the mass thickness of the sample.

Approach

The initial approach was to design a prototype that would establish
the feasibility of such a system and would use as much equipment in an
"off-the-shelf" form as was possible in order to minimize the risks
involved and to cut down on the initial design and fabrication time. To
accomplish this, a Rigaku micro-diffractometer, normally used to collect
diffraction data from small sample areas as discussed by Goldsmith (5), was
mounted on a Rigaku 18Kw rotating anode X-ray source. A Kevex Si(Li)
detector and analysis system was used to detect and process the X-ray
fluorescence signal generated by the very small beam of X-ray from the
micro-diffractometer striking the sample. The micro-diffractometer provided
nominal beam diameters of 10, 30 and 100 microns at the sample surface. The
micro-diffractometer provided manual sample translation capability for step
sizes as small as 10 microns and was also the means by which diffraction
information was obtained. The energy dispersive detector was mounted on a
moveable support to facilitate the alignment and the positioning of the
detector close to the sample.
        The use of the micro-diffractometer as a collimation system had a
number of advantages over past systems. First, such a system provided a
small spot size of reasonably high intensity without the need for any
additional attachments. Second, the micro-diffractometer contained a
microscope as an integral component. After the microscope was aligned with
the beam, it was possible to directly observe the area from which the
fluorescence signal would be produced. Third, the geometry of the
micro-diffractometer insured that the X-ray beam could be normal to the
surface of the sample which minimized the divergence observed in previous
systems in which the sample was not normal to the beam. Finally, the
micro-diffractometer is itself an instrument rather than an attachment, and
as such was capable of obtaining X-ray diffraction information from the
same small area used to perform the fluorescence analysis. Although the
collimation system for the micro-diffractometer was really an aperture
rather than a true collimator, this would prove to be a problem only when
dealing with thick samples in the transmission mode. In the present
collimation system for the micro-diffractometer, the beam grows in diameter
by approximately 2% as it passes through a matrix with a thickness of 1mm.
        The use of a rotating anode source of X-radiation provided us with
a much more intense beam than could have been obtained using normal broad
spectrographic sources. Using such a source allowed us to use an incident
collimator and to accept the reduction in intensity that accompanied it. In
its normal mode of operation for the collection of diffraction data, the
micro-diffractometer works best using a copper anode as the source of
radiation. In order to more efficiently excite the K and L fluorescent
lines of the transition and higher elements, we chose to use a silver anode
for our initial experiments. This resulted in a diminished ability to
perform diffraction work due to the very short wavelength of the silver
radiation. While the detection limits would not be as favorable and the
counting times for many elements would be longer if a copper anode were
to be used, it would still be possible to use the system described here to
obtain very useful qualitative and quantitative information.
        The Si(Li) energy dispersive system used was also an off-the-shelf
item, although we did use a detector that was inclined at an angle of
approximately 37 deg. to the horizontal and was at the end of a "snout" 12
in. long. The base on which the detector was mounted was designed and built
especially for this application and the plans for such a base can be

obtained from the authors. The base contained rollers to facilitate
positioning the detector system so it could be used to either collect data
a) in reflection mode by looking down over the tube tower from the same
direction as the collimator for the micro-diffractometer, or b) by being
positioned on the side of the sample opposite the incoming X-ray beam. In
either case the use of a motorized slider was very useful in that it
allowed the detector to be brought into a reproducible position close to
the sample. The sample-to-detector distances used to date vary from a mean
distance of 1 to 1.5 cm.

Results

        The system has been used to perform a number of analyses. Figure 1
shows the results of a scan across a pair of lines printed onto a filter
media using inks containing gold and platinum. We wanted to find out
whether there was any mixing of the two inks in the region where they most
closely approached each other and we also wanted to see how the
concentration of each element fell off toward the edges of the inked lines.
It can be seen from Figure 1 that there was essentially no interaction
between the elements in each line and that the concentration of Pt fell off
very rapidly at the edge of the line. Other scans along the length of each
of the lines served to show how the concentration of each element varied
along the line. The missing data points for the gold line point out that
one of the disadvantages of using the off-the-shelf micro-diffractometer is
its lack of automation for scanning the sample. This is a difficulty
because, in order to acquire data from a number of points on a sample, a
rather elaborate series of steps needs to be performed; viz. a) shut beam
shutter, b) press override to prevent X-rays off when safety door is
opened, c) open safety door, d) manually move sample by desired amount,

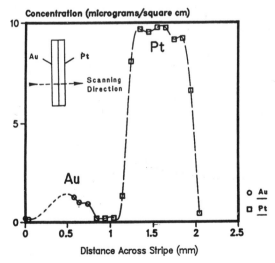

Fig. 1   Data from the micro-fluorescence system showing the variation
         in concentration of inked lines of gold and platinum made by
         stepping a 140 micron beam across them in the direction shown.

e) shut safety door, f) open X-ray shutter, g) start data collection,
h) wait for data to be collected, i) check data, etc. This time consuming
set of steps has many times been the limiting factor in the collection of
the data and certainly has precluded making any sort of detailed X-ray map
of a reasonably sized region on a sample.

We have also collected data from a number of standard thin film
samples in order to better understand and calibrate our system. In this
process we have obtained X-ray fluorescence detection limits on the order
of 40 picograms for heavier elements such as gold when using a 100 micron
collimator, 400 second counting time and a silver anode operating at 12 Kw.

Conclusions

We have demonstrated that by using off-the-shelf components it is
possible to build a micro-fluorescence system with diffraction capability.
We have also found that the use of this system to collect data sufficient
to provide a detailed elemental map of a portion of a sample is impractical
due to the lack of any automated sample translation and data collection
capability. To solve this problem we have completed the design and
fabrication of a system that allows automated sample translation and also
positions the detector even closer to the sample than was possible using
the micro-diffractometer.

References

1. I. Adler, and J.M. Axelrod, The Curved-Crystal X-ray Spectrometer, A
      Mineralogical Tool, American Mineralogist, 41:524 (1956).
2. I. Adler, J. Axelrod, and J.J.R. Branco, Further Application of the
      Intermediate X-ray Probe, in "Adv. in X-ray Anal.", W.M. Mueller,
      ed., Plenum Press, NY, Vol. 2, (1960).
3. K.F.J. Heinrich, X-ray Probe With Collimation of the Secondary Beam,
      in "Adv. in X-ray Anal.", W.M. Mueller, ed., Plunum Press, NY,
      Vol. 5, (1962).
4. H.J. Rose Jr., R.P. Christian, J.R. Lindsay and R.R. Larson,
      Microanalysis With the X-ray Milliprobe, U.S. Geol. Survey Prof.
      Paper 650-B, pp B128-B135, (1969).
5. C.C. Goldsmith and G.A. Walker, Small Area X-ray Diffraction
      Techniques; Applications of the Micro-Diffractometer to Phase
      Identification and Strain Determination, in "Adv. in X-ray Anal.",
      J.B. Cohen et al., ed., Plenum Press, NY, Vol. 27, (1984).

This work was supported by the Department of Energy under contracts
with Sandia National Laboratory, AT-(29-1)-789 and Lawrence Livermore
National Laboratory, W-7405-ENG-48.

APPLICATION OF SYNCHROTRON RADIATION EXCITED X-RAY FLUORESCENCE ANALYSIS TO MICRO AND TRACE ELEMENT DETERMINATION

Atsuo Iida and Yohichi Gohshi

Department of Industrial Chemistry, Faculty of Engineering,
University of Tokyo, Hongo, Bunkyo-ku, Tokyo 113, Japan

Hideki Maezawa

Photon Factory, National Laboratory for High Energy Physics,
Oho-machi, Tsukubagun, Ibaraki 305, Japan

ABSTRACT

Micro and trace element analysis by X-ray fluorescence was carried out using synchrotron radiation from a bending magnet and an undulator for hard and soft X-ray excitation respectively. The minimum detection limits obtained in the hard X-ray region were less than pg, which corresponds to a spatial resolution of less than a hundred micronmeters, with a detection limit of a few ppm. Light elements such as oxygen, nitrogen and carbon in silicon compounds were analyzed by soft X-ray emission spectroscopy using undulator radiation. The minimum detectable amount of the light elements was greatly improved, since undulator radiation is very strong in intensity, and is highly collimated.

INTRODUCTION

Recent research has shown that a synchrotron radiation (SR) source is a powerful analytical tool for trace element analysis by X-ray fluorescence (XRF). The minimum detection limits (MDL) obtained have been less than ppm or pg[1-7]. This high sensitivity with SR excited XRF (SRXRF) is due to the high brightness, polarization, natural collimation and energy tunability of SR. One of the most interesting applications of SRXRF is the micro analysis of trace element[1,8]. SR microanalyzers are being planned at various SR facilities for this purpose[9,10,11]. By using SR continuum or a wide bandpass monochromator, element mapping has already been carried out with a spatial resolution of less than a hundred micronmeters.

The SRXRF experiments performed to date have been concerned mainly with elements whose atomic numbers were higher than 11. In recent years, there has been increasing interest in light element analysis for material science. For instance, oxygen and carbon in semiconductors play an important role during the device fabrication process. However there have been few methods available to analyze light elements at a trace level

427

nondestructively. Although light element analysis by soft X-ray emission spectroscopy is appropriate for this purpose[12], there are problems in using it for trace element analysis because of the low fluorescence yield of the light element and the strong absorption of emitted X-rays in the material. The undulator is a promising light source because the spectral brightness of the radiation is much higher than that obtained using bending magnets. Since the undulator at the Photon Factory (PF) provides a spectrum in the soft X-ray region ranging from about 400 eV to 1 keV and is, in addition, well collimated, it would seem suitable for micro analysis of light elements at the trace level.

In the present paper, the results of trace element analysis using SR are given, with emphasis on the micro analysis. First, the results obtained using hard X-rays from the bending magnet are summarized. Then the light element analysis using undulator radiation is described. A comparison with the electron excitation was also made.

MICRO ANALYSIS IN THE HARD X-RAY REGION

XRF experiments in the hard X-ray part of the spectrum ranging from 4 keV to 25 keV have been carried out at the PF on beam line 4A. Table 1 shows the excitation mode we used for energy dispersive XRF analysis[13]. With monochromatic excitation using a crystal monochromator, the spatial resolution is in the order of one or two mm with a detection limit of less than ppm. Longitudinal element analysis of a single strand of human hair was carried out using this excitation mode.

The most efficient excitation mode for micro analysis is the wide bandpass monochromator using a mirror system, which consists of an X-ray reflection mirror and an absorber, and provides a high photon flux density. The spatial resolution is less than a hundred µm with a detection limit of less than ppm. An analysis of the calcium distribution in a corn root is now under way using this excitation mode.

To achieve high spatial resolution of less than a few tens of micronmeters, it is necessary to use focusing X-ray optics. This SR microanalyzer will open up new applications for SRXRF.

Table 1. SR excitation modes for energy dispersive X-ray fluorescence at the Photon Factory in the hard X-ray region. The typical values for various parameters are also shown.

	Energy Resolution $(\Delta E/E)$	Spatial Resolution 2 (mm^2)	MDL Rel. (ppm)	MDL Abs. (pg)
(1) SR Continuum	1	$\sim 1 \times 10^{-2}$	~ 0.5	~ 0.1
(2) Wide Bandpass Monochromator				
a) Mirror System	> 0.1	$\sim 1 \times 10^{-2}$	> 0.1	< 0.1
b) Layered Synthetic Microstructure	~ 0.01	$\sim 5 \times 10^{-2}$	~ 0.1	~ 0.1
(3) Monochromatic Mode	$10^{-4} \sim 10^{-5}$	~ 5	~ 0.05	~ 5
Si with Various Surface Treatments				
Graphite				

Table 2. Parameters of the undulator at the Photon Factory.

Total length	3.8 m
Length of a period	6 cm
Number of periods	60
Pole width	9 cm
Range of magnet gap	2.7 - 8 cm
Range of peak magnetic field	3180 - 190 gauss
Energy of electrons	2.5 GeV

LIGHT ELEMENT ANALYSIS USING UNDULATOR RADIATION

Features of Undulator Radiation

An undulator is a periodic electromagnetic structure forcing relativistic electrons to wiggle many times. Interference effects in radiation result in narrow bands at a few wavelengths with strong intensity. The principal parameters of the undulator at the PF are summarized in Table 2 [14]. The main features of undulator radiation are as follows [15].

(1). The spectrum of radiation consists of strong first harmonic and weaker higher order bands.

(2). The peak brightness of undulator radiation is stronger than that from a bending magnet by a factor of a few hundred.

(3). The energy of the first harmonic of the undulator radiation is continuously varied from 400 eV to about 900 eV by changing the magnitude of the magnetic field.

(4). Since undulator radiation is highly collimated, the beam size at the experimental station is less than 2 mm^2.

Experimental

Fig.1 shows the experimental arrangement for light element analysis. We used beam line 2B. Undulator radiation is deflected by the Pt coated SiC mirror to avoid the γ-ray background from the source [16], and then directly impinged on a sample without a monochromator. The samples used

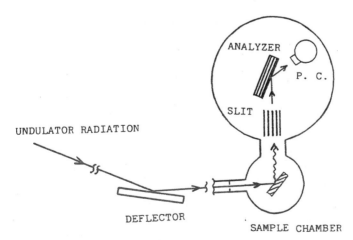

Fig. 1. Experimental arrangement for light element analysis using undulator radiation.

were thin films of SiO_2 and Si_3O_4 on Si wafers, and SiC. The irradiation area was about 1 x $2.4 mm^2$.

The experiment was carried out with a wavelength dispersive mode. The spectrometer used was a modification of the two-axis goniometer. For trace element analysis by wavelength dispersive spectrometer, the key factor in achieving high sensitivity is the dispersing element itself. The dispersing elements used were a layered synthetic microstructure (LSM, W/C 200 layer pairs, 2d=59.1 Å) [17] , and a Langmuir-Blodgett film of a lead stearate for oxygen, nitrogen, and carbon K lines. A TAP was also used for the oxygen K line All elements were flat type. The X-ray detector was a gas flow proportional counter using atmospheric PR gas.

A differential pumping system was used between the beam line and the sample chamber.   The sample chamber and the spectrometer were evacuated by a turbo molecular pump to 10^{-5} Torr, while the beam line pressure was maintained at 10^{-7} by ion pumps.   The rather poor vacuum in the sample chamber resulted in some hydrocarbon contamination during the experiment.

For purpose of comparison, electron excitation analysis was also carried out. A JEOL JCXA-733 electron probe micro analyzer with a wave length spectrometer was used. The dispersing elements used were a lead stearate and a RAP. The acceleration voltage was 4 keV and the beam current was less than μA.

Results and Disccussion

Since one of the main features of undulator radiation at the PF is its energy tunability, the optimum excitation conditions for high signal to background (S/B) ratio were investigated by changing the excitation energy. Fig. 2 shows the calculated dependence of the energy of the first harmonic on the gap width of the undulator magnet. The energy of the first harmonic is almost linearly proportional to the gap width. Fig.3   shows the peak and the background intensities of the oxygen K emission line as a function of the gap width (the energy of the first harmonic). Since the absorption energy of the oxygen K is 530 eV, the S/B ratio increases at around this energy, and reaches a peak at over 600 eV. At a gap width of less than 32 mm, since the energy of the 1st harmonic radiation is lower than the absorption edge of O K , the sample was excited by higher harmonics. The S/B ratio is not high in this region. Fig. 3 indicates

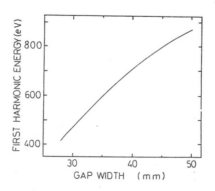

Fig. 2. Dependence of the 1st harmonic energy of undulator radiation on the gap width of the undulator magnet.

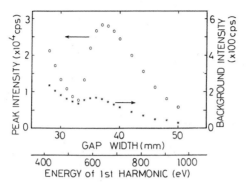

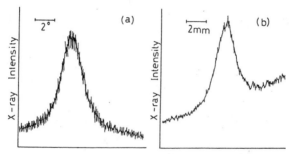

Fig. 3. Intensity of the peak (o) and the background (*) of O K   as a
function of the gap width of the undulator magnet. The energy of 1st
harminic is also shown for reference.

Fig. 4. Comparison of N K spectra from Si_3N_4 film (518 Å thick) on Si.
(a) Undulator radiation excitation with an LSM analyzer. (b) Electron
excitation with a lead stearate analyzer. A Rowland circle radius is   140
mm.

that the optimum conditions for high S/B ratio are obtained by adjusting
the magnitude of the magnetic field, even without a monochroamtor.   From
nitrogen to fluorine, the optimum conditions can be obtained following the
same procedure.

The   intensities of the fluorescence signal and the background from
the thin film sample were studied in detail. Figs.4 (a) and (b) show
typical spectra of the N K emission line obtained from a Si_3N_4 film, (a)
using undulator radiation with an LSM analyzer, and (b) by electron
excitation with a lead stearate analyzer. The background in the undulator
radiation excitation is constant, and lower than that in the electron
excitation.

Figs.5 (a) and (b) show the peak and the background intensities of
the O K emission as a function of the SiO_2 film thickness for undulator
radiation and   electron excitation respectively.   The peak intensities
increase with increasing film thickness for both excitation methods.   The
main difference between the two excitation methods is the background
dependence on the film thickness. For the electron excitation the
background increases as the film thickness decreases. The apparent
increase in the background intensity in undulator excitation is due to the
low angular resolution of the LSM. These results suggest that a higher S/B

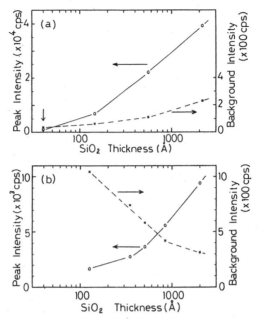

Fig. 5. The intensities of the peak (o) and the background (*) of O K
as a function of SiO$_2$ thickness. (a) Undulator excitation with an LSM
analyzer.   The natural oxide of 40 Å thick was assumed for the   reference
Si wafer (arrow).(b) Electron excitation with a lead stearate.

ratio can be obtained using undulator excitation for an   extremely   thin
sample .

        The intensity obtained using the LSM was stronger than that using the
lead stearate by a factor of about 5. The LSM analyzer is   suitable   for
trace   element analysis. The characterization of the   dispersing   elements
used here will be reported on elsewhere [18].

        Table 3 summarizes the minimum detection thicknesses (MDT) for oxygen
and nitrogen.   The minimum detection thickness is defined as the thickness
whose signal intensity is equal to three times the square   root of   the
background.   With an LSM or a lead stearate analyzer,   MDTs for oxygen   in
SiO$_2$ are   less   than   an angstrom. For nitrogen in   Si$_3$N$_4$, the   MDT   is
degraded   by a factor of 40, because of the stronger   absorption   in   the
window material of the proportional counter, the lower reflectivity of the
analyzer, and the lower fluorescence yield.

Table 3. Comparisons of the minimum detection thickness and the absolute
detectability. Counting time was 20 seconds for the peak and the background.
The electron excitation measurement was made with a lead stearate analyzer.

	L S M		L S	TAP	EPMA
O (SiO$_2$ on Si)	0.1 Å	(7 x 10^{11} atoms)	0.2 Å	2.7 Å	4.4 Å
N (SiN$_x$ on Si)	4.2 Å	(4 x 10^{13} atoms)	8.5 Å	——	11 Å
C (SiC)	——	(3 x 10^{13} atoms)	——	——	——

If we assume that the composition of the thin film is stoichiometric, the MDL in absolute amounts is in the order of 10^{11} atoms for oxygen, and 10^{13} for nitrogen. Since the irradiation area using the conventional soft X-ray spectrometry with the X-ray tube is in the order of a few cm^2, the improvement in the MDL is mainly due to the highly collimated X-rays from the undulator. During the C K measurement, however, the intensity increased as a function of time due to the contamination of the sample. The thick sample was assumed in the calculation, and the value for carbon cited in the table is just for reference.

The MDT obtained is better by a factor of a few tens than that obtained using electron excitation. Though the absolute detectability in electron excitation is superior to that using undulator radiation excitation at present, if the microbeam for use with undulator radiation can be made using a zone plate or a focusing X-ray mirror, micro and trace element analysis of light elements will be possible.

Since the present experimental conditions were far from optimal, higher sensitivity will be obtained by improving the experimental arrangements. Energy dispersive experiments with a windowless Si(Li) detector are now under way.

ACKNOWLEDGMENT

The authors would like to thank Prof. T. Sasaki for his continuous encouragement. They also thank Dr. T. Matsushita and other PF staff for their help during the experiments. They are also indebted to Dr. K. Usami and K. Honda for providing them with samples, and Dr. K. Kawabe for providing them with a lead stearate.

REFERENCES

1. C. J. Sparks, Jr., X-Ray Fluorescence Microprobe for Chemical Analysis, in: "Synchrotron Radiation Research," H. Winick, and S. Doniach, eds., Plenum Press, New York (1980)
2. M. Prins, W. Dries, W. Lenglet, S. T. Davies and K. Bowen, Trace Element Analysis with Synchrotron Radiation at SRS Daresbury, Nucl. Instr. and Methods, B10/11:299 (1985)
3. A. Knoechel, W. Petersen, and G. Tolkiehn, X-Ray Fluorescence Analysis with Synchrotron Radiation, Nucl. Instr. and Methods, 208:659 (1983)
4. J. M. Jaklevic, R. D. Giauque and A. C. Thompson, Quantitative X-Ray Fluorescence Analysis using Monochromatic Synchrotron Radiation, B10/11:303 (1985)
5. J. V. Gilfrich, E. F. Skelton, S. B. Odari, J. P. Kirkland, and D. J. Nagel, Synchrotron Radiation X-ray Fluorescence Analysis, Anal. Chem., 55:187 (1983)
6. K. W. Jones, B. M. Gordon, A. L. Hanson, J. B. Hastings, M. R. Howell, H. W. Kraner, and J. R. Chen, Application of Synchrotron Radiation to Elemental Analysis, Nucl. Instr. and Methods, B3:225 (1984)
7. A. Iida, K. Sakurai, T. Matsushita, and Y. Gohshi, Energy Dispersive X-ray Fluorescence Analysis with Synchrotron Radiation, Nucl. Instr. and Methods, 228:556 (1985)
8. P. Horowitz and J. A. Howell, A scanning X-Ray Microscope using Synchrotron Radiation, Science, 178:608 (1972)
9. B. M. Gordon and K. W. Jones, Design Criteria and Sensitivity Calculation for Multielemental Trace Analysis at the NSLS X-Ray Microprobe, Nucl. Instr. and Methods, B10/11:293 (1985)

10. G. E. Ice and C. J. Sparks, Jr., Nucl. Instr. and Methods,
     Focusing Optics for a Synchrotron X-Radiation Microprobe, 222:121 (1984)
11. M. Prins, S. T. Davies, and D. K. Bowen, Trace Element Analysis and
     Element Mapping by Scanning X-Ray Fluorescence at Daresbury SRS, Nucl.
     Instr. and Methods, 222:324 (1984)
12. B. L. Henke, R. C. C. Perera, E. M. Gullikson and M. L. Schattenberg,
     High Efficiency Low-Energy X-Ray Spectroscopy in the 100 - 500 eV
     Region, J. Appl. Phys. 49:480 (1978)
13. A. Iida, T. Matsushita, and Y. Gohshi, Energy Dispersive X-Ray
     Fluorescence Analysis using Synchrotron Radiation, Adv. in X-Ray Anal.
     28:61 (1985)
14. Photon Factory Activity Report 1982/1983 IV-48
15. H. Maezawa, Y. Suzuki, H. Kitamura, and T. Sasaki, Characterization of
     Undulator Radiation at Photon Factory, to be published in Nucl. Instr.
     and Methods.
16. Photon Factory Activity Report 1983/1984 V-1
17. J. V. Gilfrich, D. J. Nagel, and T. W. Barbee, Jr., Layered Synthetic
     Microstructure as Dispersing Devices in X-Ray Spectrometers, Appl.
     Spectrosc., 36:58 (1982)
18. T. Ishikawa, A. Iida, and T. Matsushita, to be submitted to Japan. J.
     Appl. Phys.

MULTILAYER SCATTERERS FOR USE IN POLARIZED X-RAY FLOURESCENCE

John D. Zahrt
Los Alamos National Laboratory[1]

Richard W. Ryon
Lawrence Livermore National Laboratory

INTRODUCTION

In the EDXRF analysis of minor and trace elements in a variety of matrices, the use of a polarized x-ray source incident on a sample will provide minimum detection limits three to five times lower than the use of non-polarized sources (1,2). There are various methods of producing mono-chromatic polarized x-rays for specimen excitation (3,4,5,6). Such x-ray sources may produce the lowest detection limits for a single element or a narrow range of elements. However, if one is interested in simultaneously analyzing a broad range of elements, a polychromatic source is desired (7,8,9). We present here a new method for producing broad-band polarized x-rays.

Polarized x-rays are conveniently produced in the laboratory by scat-tering the x-rays through 90° (Bragg or Thomson scattering). The x-rays may be characteristic and/or Bremsstrahlung. For broad band analyses the use of the Bremsstrahlung is desirable. For low energy x-rays low Z mate-rials are the better Thomson scatterers but they are more transparent to high energy x-rays. If high Z scatterers are used the low energy x-rays are absorbed and lost to the polarized beam. It is suggested then to use a layered scatterer with a low Z material on the front side and a higher Z scattering material on the back side. The low energy x-rays scatter off of the low Z material while the high energy x-rays transmitted by the low Z material are then scattered by the high Z material. In principle more layers could be added until, in the limit, the scattering cross sections of the scatterer would vary continuously with respect to depth of penetra-tion. This suggestion is born out and the cases of the bilayer and con-tinuous scatterer are treated theoretically.

EXPERIMENTAL

The experiments were performed on a rectilinear polarized x-ray spec-trometer as schematically shown in Fig. 1. It is more common to have the analyte and scatterer interchanged. However, we used the arrangement as shown to keep a constant analyte signal and for ease of manipulation.

---

[1]Formerly Northern Arizona University

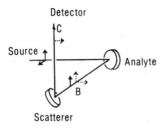

Detector

Source

Analyte

Scatterer

A) Unpolarized Source X-Rays

B) Polarized Source Plus
   Unpolarized Analyte X-Rays

C) Polarized Analyte X-Rays

Fig. 1.   Experimental design and
          radiation polarization.
          Both scattering angles
          are 90°.

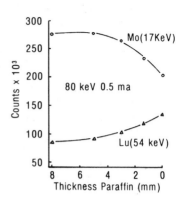

Fig. 2.   Intensities of Mo and
          Lu K_α x-rays off of
          various bilayer scat-
          terers.

We chose as our scatterers paraffin and teflon. Paraffin being $(CH_2)_n$ acts effectively as boron as a scatterer while teflon being $(CF_2)_n$ acts essentially like oxygen as a scatterer.

As the collimators being used had a diameter of 9.5 mm, no single scattered x-rays would enter the exit collimator if they scattered more than 6.7 mm below the surface of the scatterer. Thus the total thickness of the bilayer was always kept at least 7 mm thick. Any material farther than 6.7 mm from the surface will contribute only multiple scattered x-rays to the intensity. We have ignored this but the effect is usually small (8).

For each analyte a series of at least 5 measurements were made beginning with 8 mm of paraffin and ending with 7 mm of teflon. Measurements were occasionally repeated to test the counting statistics and/or the reproducibility.

The voltage and current on the x-ray tube were adjusted to give reasonable values of count rates and dead time. Thus not all analytes were run under the same conditions. Figures 2 and 3 show two pairs of analytes that were run under identical conditions.

THEORY

σ As A Continuous Function of Depth

In this model we assume the scattering cross sections to be a function of energy and depth, that is $\sigma = \sigma(E,z)$. For convenience we assume the x-rays are delivered to the scatterer and depart from it via collimators of square cross section.[2]

Now the probability of an x-ray photon entering the scatterer in a dxdy neighborhood of x,y, penetrating to a depth within a dz neighborhood

---

[2]Previous studies have indicated only small quantitative differences and no qualitative differences between the use of square and circular collimators.

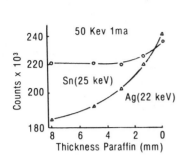

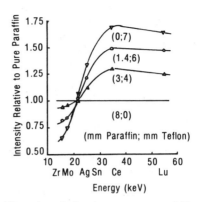

Fig. 3. Intensities of Ag and
         Sn K$_\alpha$ x-rays off of
         various bilayers.

Fig. 4. Relative scatter effi-
         ciencies for various
         bilayers.

of z where a certain fraction f, is Thomson scattered through an angle θ
(θ = 90° here) is given by

$$dP = \frac{dxdy}{\sqrt{2}d^2} \cdot \sigma_T e^{-\sigma_T\sqrt{2}z} \, dz \cdot f \cdot \frac{3}{16\pi} (1+\cos^2\theta) \cdot e^{-\sigma_T\sqrt{2}z} \tag{1}$$

For square collimators of cross section d we have

$$-d/2 < x < d/2$$
$$-\sqrt{2}d/2 + z < y < \sqrt{2}d/2 - z$$
$$0 < z < d/\sqrt{2}$$

We also are assuming a perfectly parallel source of x-rays which scatter
through 90° and form a parallel beam in the exit collimator.[3]

The x and y integrations are easily accomplished to obtain

$$dP = \frac{3}{16\pi\sqrt{2}d} \cdot 2(\frac{d}{\sqrt{2}} - z) \, \sigma_s \, e^{-2\sqrt{2}\sigma_T z} \, dz \tag{2}$$

We now need to specify the forms of $\sigma_s(z,E)$ and $\sigma_T(z,E)$. They are to be
monotonic increasing functions of z. For reasons of tractability we take

$$\sigma_s = (az^n + b)/E^{cz^n+d} \tag{3a}$$

$$\sigma_T = (Az^n + B)/E^{Cz^{n+D}} \tag{3b}$$

where n may be 1 or 2. With a, c, A and C all taken to be zero we fit b,
d, B and D to the surface scattering material. Here this is taken to be
boron which approximates to paraffin. The constants a, c, A and C were
determined so that at z = d/√2 the cross sections were those of oxygen
which in turn approximates to teflon in its scattering qualities.

----
[3]While not physically realizable this assumption likewise has been tested
elsewhere in different context. No significant differences in polarization
or scattering probabilities were observed for a divergent beam.

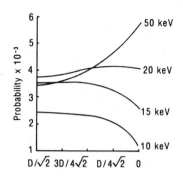

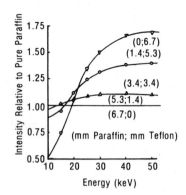

Fig. 5. Scattering probabilities of various energy x-rays as a function of paraffin thickness. The sum of paraffin and teflon thicknesses add to $D/\sqrt{2}$. D = 9.5 mm.

Fig. 6. Calculated relative scatter efficiencies for various bilayers.

The integrations were done numerically using Simpson's rule and the interval was made successively smaller until two runs were identical to better than 1%.

σ as a Step Function of Depth

In this model σ is a function of energy only and we make a discontinuous jump in the σ's at a boundary of the two materials in the bilayer. The expression for the probability of single scatter is then

$$
P = \frac{3}{8\sqrt{2}\pi d} \{\sigma_1 S \int_0^{T_1} \left(\frac{d}{\sqrt{2}} - z\right) e^{-2\sqrt{2}\sigma_{1T} z} \, dz
$$

$$
+ e^{-2\sqrt{2}\sigma_{1T} T_1} \sigma_{2s} \int_{T_1}^{d/\sqrt{2}} \left(\frac{d}{\sqrt{2}} - z\right) e^{-2\sqrt{2}\sigma_{2T}\left(z - T_1\right)} \, dz\} \tag{4}
$$

where the extra exponential factor in front of the second integral allows for attenuation by the first layer. The remaining integrations are simple.

RESULTS AND CONCLUSIONS

Figure 2 shows typical data for molybdenum and lutecium both run at 80 keV and 0.5 ma. For pure paraffin as the scatterer Mo gives the higher count rate as paraffin is more transparent to LuKα x-rays. As the amount of teflon (behind the paraffin) is increased - keeping the total thickness about 8 mm - the intensity of Mo falls off while that of Lu increases. The teflon absorbs the MoKα while interacting with the LuKα to cause scattering.

Under different conditions Fig. 3 shows the results of the tin and silver measurements. Paraffin is transparent to both SnKα and AgKα and both intensities increase as the amount of teflon backing increases. The intensity from Sn increases more rapidly. The intensity from Ag is so nearly constant that we may interpret this to mean that AgKα is nearly equally penetrating to both paraffin and teflon.

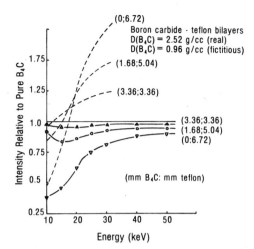

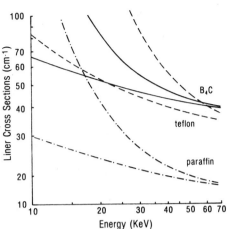

Fig. 7.   Scatter efficiencies
          relative to pure B₄C
          (——— experimental,
          ------- fictitious).

Fig. 8.   Scattering and total cross
          sections for paraffin, B_4C
          and teflon cross sections
          at 22 keV and 60 keV.

Figure 4 shows data for all analytes for pure paraffin and pure teflon and for two bilayers. For each analyte pure paraffin was chosen as
the reference (relative intensity = 1). Below Ag (or about 21 keV)
paraffin is the major scatterer while above Ag teflon is the more efficient scatterer. For broad band analyses, however, a bilayer may prove
more efficient. For example, a scatterer of 3 mm paraffin backed by 4 mm
teflon would reduce the intensity of Zr by only about 6% (over that of
pure paraffin) while increasing the signal of Ce and Lu by about 30%
(although this is about 23% less than what could be gained by pure teflon).

The experimental data presented in Figs. 2, 3, and 4 are quite well
reproduced by the use of eqn. (4). Figure 5 shows the calculated probabilities of scatter as a function of layer thickness for four energies.
The curves for 15 keV and 50 keV reproduce reasonably well the curves for
Mo and Lu in Fig. 2. Figure 6 shows the theoretical intensity (probability) versus energy curves for pure paraffin, pure teflon and three bilayers. These curves agree quite well with the experimental curves of
Fig. 4.

The results of a second series of experiments on B_4C and teflon indicated that B_4C was the superior scatterer (at least to 70 keV) and thus
although teflon has the higher effective Z bilayer scatterers do not improve the intensities over those of pure B_4C. the calculations on this
system are shown in Fig. 7. As is indicated by the solid lines every
bilayer computed and pure teflon are inferior scatterers compared to pure
B_4C. By arbitrarily changing the density of B_4C to 0.96 g/cc and repeating the computations the dashed lines of Fig. 7 were obtained. Now the
behavior of this system is similar to the paraffin-teflon system.

These B_4C results were quite unexpected. A graph of the linear cross
sections versus energy are shown in Fig. 8 for paraffin, teflon and B_4C.
The paraffin-teflon system works because both σ_s and σ_T for paraffin lie
below those of teflon. However, because of B_4C's high density (2.52 g/cc)
the σ_s curve for B_4C crosses that of teflon at about 22 keV and crosses

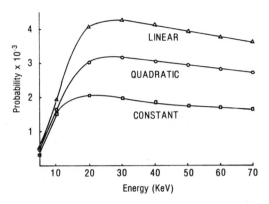

Fig. 9. Probabilities of single scatter as function
         of energy for a scatterer which has its scat-
         tering and absorption cross sections constant
         with depth, varying linearly with depth, and
         varying quadratically with depth; collimator
         cross section is 0.95 cm, density of material
         is 1 g/cc.

the σ_T curve of teflon at about 60 keV. The σ_T curve of B_4C crosses the
σ_T curve of teflon at about 57 keV. It is these crossings, rather than
the high density per se, that are responsible for the great success of B_4C
as a scatterer in polarized EDXRF.

    In conclusion, it appears that B_4C is the best material to use for
polarized EDXRF at least from say 10 - 70 keV. If such material is una-
vailable, bilayers such as paraffin-teflon appear to be better over the
whole energy range rather than pure teflon or paraffin.

    Calculations performed on ficticious materials which would vary their
cross sections continuously with depth were also accomplished using eqns.
(3a) and (3b) in eqn. (2). The results are shown in Fig. 9. Here we see
the linear variation of σ's with depth is superior to a quadratic variation or constant σ's (e.g., pure paraffin). With this limited information
and approximating the linear variation by one step (a bilayer) we might
expect the best results to be obtained by equal thicknesses of each layer.

This is essentially born out by our earlier suggestion that 3 mm paraffin,
4 mm teflon give perhaps the best broad band results. If one progressed
to trilayers or higher multilayers we hypothesize that the thickness of
each layer be such that the resulting many step function be the best approximation to the linear curve.

ACKNOWLEDGMENTS

    One of the authors (J.D.Z.) would like to thank Northern Arizona
University for an Institutional Research Grant under which much of this
work was performed.

REFERENCES

(1) Ryon, Richard W., Adv. in X-Ray Anal., 20, 575-590 (1977).

(2) Wobrauschek, P., and H. Aiginger, X-Ray Spect. 8, 57-62 (1979).

(3) Sparks, Jr., Cully, in "Synchrotron Radiation Research," H. Winick
     and S. Doniach, eds., Plenum Pub. Corp., 1980.

(4) Howell, R. H. and W. L . Pickles, Nuc. Inst. Meth., $\underline{120}$, 187-8
     (1974).

(5) Wobrauschek, P., and H. Aiginger, X-Ray Spect., $\underline{12}$, 72-78 (1983).

(6) Zahrt, John D., Adv. in X-Ray Anal. $\underline{26}$, 331-336 (1983).

(7) Ryon, Richard and J. Zahrt, Adv. in X-Ray Anal., $\underline{22}$, 453-60 (1979).

(8) Zahrt, John and Richard Ryon, Adv. in X-Ray Anal., $\underline{24}$, 345-50
     (1981).

(9) Zahrt, John D., Adv. in X-Ray Anal., $\underline{27}$, I. (with Richard W. Ryon)
     505-511; II. 513-517 (1984).

A NEW METHOD FOR ORIENTATION DISTRIBUTION FUNCTION ANALYSIS

Munetsugu Matsuo, Koichi Kawasaki,
and Tetsuya Sugai

Nippon Steel Corporation
R & D Laboratories I
Ida, Kawasaki, 211 Japan

Abstract

As a means for quantitative texture analysis, the crystallite orientation distribution function analysis has an important drawback: to bring ghosts as a consequence of the presence of a non-trivial kernel which consists of the spherical harmonics of odd order terms. In the spherical hamonic analysis, ghosts occur in the particular orientations by symmetry operation from the real orientation in accordance with the symmetry of the harmonics of even orders. For recovery of the odd order harmonics, the 9th-order generalized spherical harmonics are linearly combined and added to the orientation distribution function reconstructed from pole figures to a composite function. The coefficients of the linear combination are optimized to minimize the sum of negative values in the composite function. Reproducibility was simulated by using artificial pole figures of single or multiple component textures. Elimination of the ghosts is accompanied by increase in the height of real peak in the composite function of a single preferred orientation. Relative fractions of both major and minor textural components are reproduced with satisfactory fidelity in the simulation for analysis of multi-component textures.

Introduction

The crystallite orientation distribution function (ODF) analysis[1] has come to be widely used as a means for quantitative texture analysis. However, there remain the problems associated with its reliability[2]. Two methods, the spherical harmonic analysis[1] and the vector method[3], are now in practical use. The analysis can be thought as a three dimensional image reconstruction procedure from its two dimensional projections which are measured as pole figures (PF). Without regard to the procedures of the analysis, the occurrence of ghosts in the reconstructed image is inevitable as a consequence of a non-trivial kernel present in the mapping of ODF space into PF space which is not isomorphic[4].

Following the implementation of the harmonic method by Morris and Heckler[5], the authors have demonstrated the appearence of ghosts in the reconstructed ODF's[6]. The orientation of every ghost has been shown to be correlated by a symmetry operation with the real orientation. The symmetrical characteristics of ghosts is a consequence of the symmetry of the spherical harmonics of even-order terms. Furthermore, the authors have indicated that the spherical harmonics of odd-order terms in ODF's send a null space into the pole figures[7]. It means that the kernel consists of the sperical harmonics of odd-order terms which are missed in the measurement of pole figures by x-ray diffraction.

The vector method is a discrete approach to approximating an ODF in terms of vectors of high dimensions that are coupled with the pole densities on PF's by a corresponding matrix.   The transformation matrix is degenerated to provide solutions with the dimensions of the nullity, therefore a unique solution cannot be obtained by the vector method; selection of the algorithms for stablility of convergence to a solution has been shown to result in a variety of the orientation vectors[8].   If the non-negative condition is removed together with the routine of peak-selection procedure in the algorithm, the ghosts occur at exactly the same orientations as in the harmonic analysis. This proves that the vector method is not free from the ghost phenomena in contrast to the claim[9] that it was not vulnerable to the appearence of ghosts.   Moreover, the process of ghost elimination is shown to smear the real peaks of weak textural components. Therefore, the vector method is not reliable as a means for quantitative texture analysis.

As a consequence of the previous results on the characterization of the ghosts in the spherical harmonic analysis, it is thought that, if the harmonics of odd terms lost in the pole-figure measurement can be recovered, the ghosts will be eliminated by compounding an odd-term harmonic function, $f_{odd}(g)$, with an ODF, $f_{even}(g)$, reconstructed from PF's which consists only of the harmonics of even terms. The harmonics of odd orders, $f_{odd}(g)$, having opposite phases to each of the ghosts and the negative regions in the reconstructed ODF, $f_{even}(g)$, is most likely the kernel in the projection of ODF space into PF space.

Characterization of ghosts

A simulation was made to determine the orientations and heights of ghosts for a variety of preferred orientations found in the rolling and annealing textures of iron by executing the ODF analysis by inputting the artificial pole figures having a single textural component. The pole distribution around each ideal orientation was assumed to have a Gaussian distribution with a half-breadth value of 17.6 degrees.   Computation was performed by series expansion of a set of pole figures $\{001\}$, $\{111\}$, and $\{110\}$ up to 22nd order.

Figure 1 shows the $\phi = 45^o$ section of the ODF reconstructed from the artificial pole figures only of $\{110\}\langle001\rangle$ orientation. In the figure, the real peak of $\{110\}\langle001\rangle$ is accompanied by five types of ghosts with an identical peak height as high as about 12.5% of the true peak.   Each orientation of the ghosts is correlated with that of the real peak by a rotation about a common axis of high symmetry, as shown in Fig.2.   The ghost peaks A and B are generated from the real peak by a rotation of 45° about a common $\langle001\rangle$ axis.   The peak C is in the orientation by a rotation of 45° about a common $\langle110\rangle$ axis.   The peak D is a mirror image of the real orientation by a $\{111\}$ plane which can be also represented by a rotation of 70.53^o about a common $\langle111\rangle$ axis.   The peak E is correlated with the real peak with a rotation of 35.25^o about a common $\langle111\rangle$ axis. This symmetrical characteristics of the ghost orientations, which is found without regard to the orientations of the real peak, is a reflecton of the symmetry of spherical harmonics of even order.

Selection of an optimum odd-order harmonic function for elimination of ghosts

The 9th-order generalized spherical-harmonics, P_9^{mn}, are of the lowest odd-order terms, of which four types of the harmonics are of significance: n=2, 4, 6, and 8, with m=0. Plot of each harmonics on the $\phi = 45^o$ section is shown in Fig.3. Linear combinations of these harmonics,

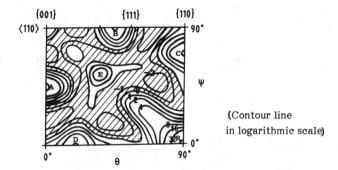

Fig.1   Reconstructed ODF from artificial pole figures singly oriented in {110}⟨001⟩ with ghosts A, B, C, D, and E.

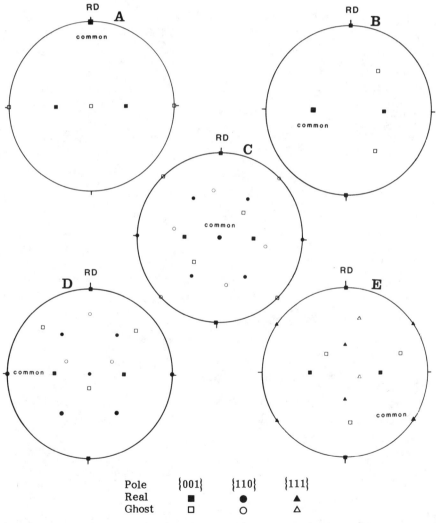

Pole	{001}	{110}	{111}
Real	■	●	▲
Ghost	□	○	△

Fig.2   Orientations of the ghosts associated with the {110}⟨001⟩ real orientation: A, B, C, D, and E in Fig.1.

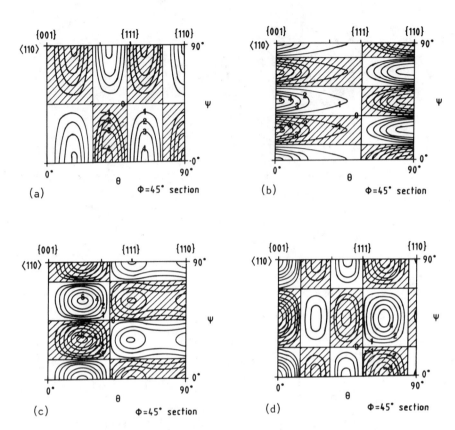

Fig.3   Generalized spherical harmonics of the 9th order: (a) m=0, n=2,
(b) m=0, n=4, (c) m=0, n=6, and (d) m=0, n=8.

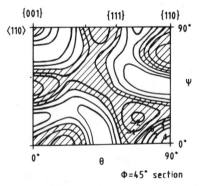

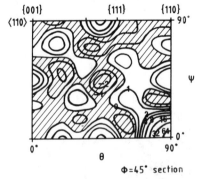

Fig.4   Optimum combination of the
9th-order harmonics for
elimination of ghosts.

(Contour line in logarithmic scale)

Fig.5   Composite ODF of
$\{110\}\langle 001\rangle$ preferred
orientation.

(Contour line in logarithmic scale)

$$f_{9th}(g) = a_1 P_9^{02} + a_2 P_9^{04} + a_3 P_9^{06} + a_4 P_9^{06},$$

are added systematically to the reconstructed ODF, $f_{even}(g)$, to find an optimum combination of the coefficients, a_i, giving the minimum of the sum of negative values in the composite function.

For the $\{110\}\langle001\rangle$ orientation, the combination of $a_1 = -6$, $a_2 = 2$, $a_3 = -3$, and $a_4 = -6$ was shown to reduce most effectively the negative values in the reconstructed function consisting of the harmonics only of the even orders from 4th to 22nd order.   The optimum combination of the 9th-order harmonics, $f_{9th}(\{110\}\langle001\rangle)$, is shown in Fig.4 on the $\phi = 45^o$ section. The odd-term function is superposed on the reconstructed ODF, $f_{even}(\{110\}\langle001\rangle)$, shown in Fig.1 and the composite ODF is shown in Fig.5.   As evident from the figure, both the ghosts and the negative regions are eliminated as a result of the superposition.

For the $\{001\}\langle100\rangle$ orientation, because of its higher symmetry, only the terms of n= 2nd and 4th degree is significant: $a_2 = -6$ and $a_4 = 5$.   The result of the superposition is shown in Figs. 6a and 6b in terms of the profiles of the reconstructed and the composite ODF on the sections of $\phi = 45^o$, $\theta = 0^o$, and $\phi = 45^o$, $\theta = 70^o$.   On these sections, an increase in the real peak height is noticed together with marked decreases in the heights of ghost peaks and the negative values; the ratio of the highest ghost peak to the real peak height is reduced by half in the composite function.   A summary of the improvements is given in Table 1.

Table 1   Ghosts and negative values in the reconstructed and the composite ODF.

Orientation	ODF	Peak height			Negative values	
		Real	Ghost	Ratio*	Minimum	Sum
$\{110\}\langle001\rangle$	Reconstructed	66.8	8.4	12.6%	-3.6	-281.8
	Composite	75.9	4.7	6.2%	-3.0	-168.2
$\{001\}\langle100\rangle$	Reconstructed	65.1	8.4	12.9%	-3.7	-272.2
	Composite	72.8	4.8	6.6%	-2.3	-153.0

* The ratio of the highest ghost peak to the real peak height.

Extension to textures of multiple components

In the preceding results, the superposition of odd term harmonics is proved to be applicable to more accurate reconstruction of the orientation distribution function of a single textural component.   For wider practical applications, the procedure should be verified for the textures of multi-components.   In the present study, a set of artificial pole figures was prepared to have a combination of five types of preferred orientations with different volume fractions: 30% of $\{001\}\langle110\rangle$, 30% of $\{112\}\langle110\rangle$, 20% of $\{111\}\langle112\rangle$, 15% of $\{111\}\langle110\rangle$, and 5% of $\{110\}\langle001\rangle$.   Optimization of the coefficients, a_i, of the 9th-order harmonic function was made to give the minimum value of the sum of negative values in the composite function, $f_{even}(g) + f_{9th}(g)$.   The minimum is found to be obtained when the coefficients, a_i, take the values of $a_1 = -0.3$, $a_2 = 2.2$, $a_3 = -1.2$, and $a_4 = 1.8$.   Figure 7 shows the profiles of the reconstructed and the optimized composite ODF on the sections $\phi = 45^o$, $\psi = 0^o$, and $\phi = 45^o$, $\psi = 90^o$. The peaks of the real orientations are increased in their heights and separated clearly from the other peaks.   Table 2 shows the relative fraction of each textural component. The initial input ratio of the prefered orientations is substantially recovered in the

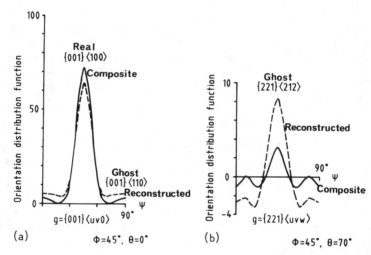

Fig.6   Reconstructed and composite ODF with preferred orientation
         in {001}⟨100⟩ : profile on the section $\phi = 45^\circ$, $\theta = 0^\circ$ (a)
         and $\phi = 45^\circ$, $\theta = 70^\circ$ (b).

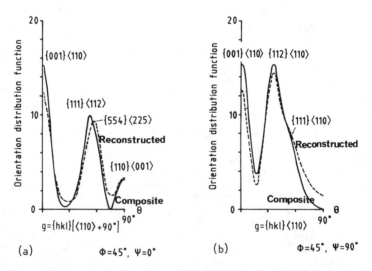

Fig.7   Reconstructed and composite ODF having multiple textural
         components: profile on the section of $\phi = 45^\circ$, $\psi = 0^\circ$ (a)
         and $\phi = 45^\circ$, $\psi = 90^\circ$ (b).

Table 2.   Relative fractions* of textural components

ODF	{001}⟨110⟩	{112}⟨110⟩	{111}⟨112⟩	{111}⟨110⟩	{110}⟨001⟩
Initial	30	30	20	15	5
Reconstructed	26.5	30.3	19.2	16.9	7.1
Composite	30.3	29.7	19.1	14.8	6.1

* Percentage of the respective peak-height divided by the sum of the heights of the five peaks.   The sum in the reconstructed ODF is 94% of that in the composite.

composite ODF. Thus the composite of the reconstructed ODF and the optimum odd-order function is proved to ensure the accuracy of quantitative texture measurement.

Further improvements can reasonably be expected by composition of the harmonics of the higher odd orders. The present procedure is also applicable to the vector method to eliminate the ghosts without recourse to the non-negativity condition which necessarily accompanies the smearing of real peaks of weak textural component.

References

(1)   H.J. Bunge: "Texture Analysis in Materials Science" (P.R. Morris, trans.), Butterworths, London, 1982.
(2)   S. Matthies: "Aktuelle Probleme der Texturanalyse", ZfK-480, Akademie der Wissenschaften der D.D.R., Dresden, 1982.
(3)   D. Ruer and R. Baro: Adv. X-ray Anal., vol.20, 1977, p.187.
(4)   S. Tani, M. Matsuo, T. Sugai, and S. Sekino: Proc. 6th Int. Conf. on Textures of Materials, Iron and Steel Institute of Japan, Tokyo, 1981, p.1345.
(5)   P.R. Morris and A.J. Heckler: Adv. X-ray Anal., vol.11, 1968, p.454.
(6)   M. Matsuo, S. Tani, and S. Hayami: Acta Cryst., vol.A28, 1972, S233.
(7)   M. Matsuo, S. Tani, and K. Kawasaki: Proc. 6th Int. Conf. on Textures of Materials, Iron and Steel Institute of Japan, Tokyo, 1981, p.1250.
(8)   M. Matsuo, K. Kawasaki, and T. Sugai: Tetsu to Haganè, vol.71, 1985, S1351.
(9)   J. Pospiech, D. Ruer, and R. Baro: J. Appl. Cryst., vol.14, 1981, 230.

A COMBINED FUNDAMENTAL ALPHAS/CURVE FITTING
ALGORITHM FOR ROUTINE XRF SAMPLE ANALYSIS

Dennis J. Kalnicky

Princeton Gamma-Tech
1200 State Road
Princeton, NJ 08540

INTRODUCTION

Analysis of sample composition and/or thickness in a routine, process-control or monitoring environment generally requires rapid turn-around time with minimal sample handling. X-ray Fluorescence (XRF) is well-suited for these kinds of analyses and has been applied to various bulk and thin sample applications (1-7). This technique is rapid, precise, non-destructive, and requires minimal sample handling.

X-ray Fluorescence is generally considered a secondary analysis technique, that is, instrumentation must be calibrated using known standards before unknown samples may be analyzed (quantitative analysis). Continuing work on standardless approaches, particularly for Energy-Dispersive XRF (EDXRF) systems, relax this requirement for some samples but the large share of XRF analyses still require calibration with standards. A number of mathematical and non-mathematical techniques have been devised to calibrate XRF measurements for many different sample types (1,2,8,9). The "influence coefficient" models fall into the mathematical category and are commonly applied to correct for interelement effects (8,9), ie, the effects of all other elements in the sample on the X-ray line of the element being measured. The "alpha coefficient" methods (9) are the most frequently applied approaches within the influence coefficient category and have their basis in the full theoretical, ie, Fundamental Parameters, description of X-ray generation.

Fundamental Parameters XRF approaches (10) can be extremely useful for accomplishing reliable and accurate calibrations for many types of samples. These techniques require very few standards (as few as one per element) and have the added advantage of being able to predict the interelement effects to be expected for the various elements in a sample for different measurement conditions and compositions. The analyst can conduct a theoretical study of the order and magnitude of these effects without making a

single measurement!   This information can then be used to
help select appropriate standards and the best measurement
conditions for the sample types to be analyzed.   These methods
are time-consuming and may require several minutes to compute
the composition of an unknown sample and, hence, may not be
practical for some routine sample applications where quick
turn-around is of prime importance.

    Because Fundamental Parameters can be used to accurately
predict interelement effects, the analyst need not measure
special standards to empirically determine these effects. A
few standard samples are measured in order to compute
empirical constants related to sample geometry and X-ray
collection efficiency.   The empirical fitting algorithm may
include various tests for inconsistent data which are due to
faulty or deteriorated standards, improper net intensity
measurements, or instrument instability.   This paper will
discuss an XRF method combining Fundamental Parameters and
Curve Fitting approaches for various bulk and thin sample
systems.   The basic equations will be reviewed where
Fundamental Parameters will be presented from the viewpoint
of Modified or Hybrid Alphas described by LaChance (9).   This
provides a unique capability to accurately apply interelement
and sample thickness corrections while requiring as few as two
calibration standards.   The composition is determined directly
from the measured characteristic X-ray intensities of the
elements of interest with corrections for interelement effects.
For thin samples, thickness is also computed using additional
corrections for film thickness and substrate intensity effects
(11).

INTERELEMENT CORRECTIONS

    In general, measured intensity is not directly propor-
tional to the concentration of the element of interest but,
rather, is a complex integral function of concentrations of
all sample elements, instrument geometry, mass absorption
coefficients, fluorescence yields, etc. (2,8,10):

$$I_i = F\ (C_i,\ C_j,\ \omega_i,\ \mu_i,\ \mu_j,...) \qquad\qquad (1)$$

where,

    F = multiple-integral function
    I_i = intensity of X-ray line for element of interest
    j = subscript denoting other elements
    C = concentration (usually weight fraction)
    ω_i = fluorescence yield for the X-ray line
    μ = mass absorption coefficient

and, other terms have been omitted for simplicity (see
references 2,8,10).

    An interelement coefficient correction model is a math-
ematical technique (approximation) used to determine elemental
concentrations by correcting measured intensities for the
effects of the other elements present in the sample without
solving the full complex integral function.   With this
definition, equation 1 may be rewritten as

$$C_i = R_i * F'(C_j) \qquad\qquad (2)$$

where, the measured intensity is adjusted by a factor, F',
which is a function of the concentrations of the other elements
present in the sample. The R_i term in equation 2 is generally
the ratio of the measured to pure-element intensity for the
X-ray line of interest, ie, $(I_i/I_i, \text{pure})$. Equation 2 represents the general concentration-based interelement coefficient correction model where $F'(C_j)$ may take one of several
forms.

ALPHA COEFFICIENT MODELS

Alpha Coefficient correction models (9) generally assume
that the total correction factor is a linear sum of binary
element effects. The LaChance-Traill model uses a single
alpha coefficient, α_{ij}, to compensate for interelement effects
(9);

$$W_i = R_i \left[1 + \sum_{j \neq i} \alpha_{ij} W_j \right] \qquad (3)$$

where,

W_i = weight fraction of measured element, i.
W_j = weight fraction of interfering element, j.
R_i = ratio of measured to pure intensity for element i.
    = $(I_i/I_i, \text{pure})$
α_{ij} = alpha coefficient for the effect of j on i.
    (Fundamental Alpha - LaChance form)
$\sum W_j$ = 1.0, ie, all elements present add to 100%

This method has been successfully applied to many types
of bulk sample analyses. This form of interelement correction
may be derived from full theory and, hence, may be termed
Fundamental Alphas (9,12). Recent work (12) has extended this
approach to three-term alpha coefficients and cross-coeffi-
cients to more accurately describe the variable nature of these
correction factors. For the purposes of this paper we will
restrict ourselves to the single-term alpha model given in
equation 3.

It may be inconvenient to consider all elements as required by equation 3, for example, when there is a significant unmeasured (low atomic number) element present in the
sample. Under these circumstances equation 3 may be rewritten in the modified alphas form (7,9)

$$W_i = R_i (A_i) \left[1 + \sum_{j \neq m} Z_{ij} W_j \right] \qquad (4)$$

where,

A_i = intensity scaler for element i,
    = $(1 + \alpha_{im})$
Z_{ij} = modified alpha for j on i,
    = $(\alpha_{ij} - \alpha_{im}) / (1 + \alpha_{im})$
α_{im} = Fundamental Alpha (LaChance form) for the modifying
    element or compound on the measured element, i,

and other symbols as previously defined. The above approach
has been successfully applied to routine bulk analyses (7) and
has the advantages that: the modifying element/compound is not

measured, but, rather its effect is combined into the modified alphas and, the concentration computation is self-converging, that is, the $\Sigma W_j=1.0$ constraint is not needed.

For samples where stoichiometry is important, eg, oxides, equation 3 may be written in the hybrid alphas form (9)

$$W_i = R_i \left[1 + \sum_{j \neq se} Z_{ij} S_j W_j \right] \tag{5}$$

where,

$$
\begin{aligned}
Z_{ij} &= \text{hybrid alpha for } j \text{ on } i, \\
&= W_j^! \, \alpha_{ij} + W_{se}^! \, \alpha_{i,se} \\
se &= \text{stoichiometric element}, \\
W_j^!, W_{se}^! &= \text{weight fraction of } j, se \text{ in the stoichiometric} \\
&\quad \text{compound}, \\
S_j &= \text{stoichiometric factor}, \\
&= (\text{compound weight})/(\text{atomic weight of } j)
\end{aligned}
$$

and other symbols as previously defined. This approach has been used for analysis of cements (13) in combination with a delta method using known standards (14). It has the advantage that the unmeasured element is treated in its compound form, eg, oxide, and the approach may be self-converging. In addition, the delta method relaxes the need for the interelement coefficients to be correct from zero to 100% wt%.

BULK SAMPLES - CURVE FITTING TECHNIQUES

This paper describes a bulk sample analysis method combining hybrid or modified alphas computed by Fundamental Parameters (10) with curve fitting techniques using as few as two calibration standards. The LaChance alphas expression, equation 3, may be re-arranged into the linear form

$$Y_i = S_i^! W_m^! + C_i^! \tag{6}$$

where,

$$
\begin{aligned}
C_i^! &= \text{intercept} = (1/I_{i,\text{pure}}); \quad 0 < C_i^! < 10 \\
S_i^! &= \text{slope} = \alpha_{im} C_i^! \quad ; -1 < \alpha_{im} < 10 \\
W_m^! &= \text{wt. fraction of modifying element/compound} \\
Y_i &= (W_i/I_i) - C_i \sum_{j \neq m} \alpha_{ij} W_j
\end{aligned}
$$

The "best fit" values of slope and intercept $(S_i^!, C_i^!)$ are obtained using two or more calibration standards. Note that the intercept, $C_i^!$, occures on both sides of equation 6 so that $S_i^!$, $C_i^!$ are obtained iteratively starting with a theoretical value for α_{im} and an average value for $C_i^!$ based on the standards used for calibration. Only $C_i^!$ is fit when there are no unmeasured modifying elements.

The general fitting procedure is as follows:

1. Form Y values for each non-zero standard. (minimum two, flag and quit if less than two).

2. Solve for starting values of S_i' and C_i' for standards using theoretical α_{im} when there are unmeasured elements.

3. If all elements are measured then go to step 8.

4. Solve for "best fit" values of S_i', C_i' for standards.

5. Test α_{im}, C_i values at each iteration; note, new Y_i values are generated for the standards for each C_i during the iterations.

6. Error flag if exceed maximum number of iterations (20 max).

7. If best fit α_{im} out of bounds use the theoretical α_{im} and compute corresponding C_i ($-1 < \alpha_{im} < 10$).

8. Show given vs. fit values for all standards (whether included or not, flag standards not used).

9. User selected fit options: store current fit, hard copy of all relevant fit parameters, delete some standards (minimum two for fit), and run normal or run using theoretical modifying alpha or run a non-weighted fit.

10. If user does not store this fit then go to step 1.

11. If user stores this fit then go to next element and then to step 1. If last element go to step 12.

12. Create a calibration file (using modified alphas as in equation 4) and query user for other options to store with the fit.

In this approach the user is not required to interpret the "goodness" of interelement corrections since they are computed theoretically but, rather, the agreement between given and fit wt.% values for the element of interest in the non-zero calibration standards included for a given fit. These are presented in terms of a "chi-square of fit" for each standard defined by

$$\text{ChiSqr} = (\text{delta squared})/(\text{error squared}) \qquad (7)$$

where

delta = (given wt%) - (fit wt%)

error = given error

and the error term is specified by the user during the initial calibration steps. Thus, a chi-squared of 1.0 means the fit agrees to within the error estimate, 4.0 means the fit is within a factor-of-two of the error estimate, and 9.0 means the fit agreement is three times the error estimate. This approach provides a reliable goodness-of-fit criterion when realistic errors are specified for calibration standards and elemental count data are collected sufficiently long to guarantee good statistical presision.

Additional options that may be included with the calibration curve (equation 4) are:

1. Delta calculation using a Quantitative Match Standard (QMS) file

2. Intensity drift correction file

The QMS file stores calibration standards with known intensities and concentrations for use in a delta computation. The standards stored are user-specified and should be those with acceptable agreement between given and fit values for all elements included in the calibration (see equation 7). The procedure to apply QMS computations during sample analysis is as follows:

1. Compare net intensities in unknown samples to those of stored QMS standards.

2. Select standard with "closest" intensity match.

3. Use delta (difference) computation with full interelement correction based on known standard.

4. Valid "error" estimate for each element based on statistics, calibration uncertainties, and error for known standard.

When the delta model is applied with respect to the midpoint between the unknown and selected reference sample composition, it may be written as

$$W_i(unk) = W_i(std)$$
$$+ A_i(R_i(unk)-R_i(std))\{1+0.5 \sum_{j \neq m} Z_{ij}(W_j(unk)+W_j(std))\}$$
$$+ 0.5(A_i)(R_i(unk)+R_i(std)) \{\sum_{j \neq m} Z_{ij}(W_j(unk)-W_j(std))\} \qquad (8)$$

where, (unk) and (std) refer to the unknown and standard respectively and the other symbols are as defined previously. This is a different delta model than that given in reference 14 but accomplishes essentially the same result, ie, it relaxes the need for absolutely correct interelement correction factors from 0 to 100 percent and allows error estimates based on the error for the QMS reference standard used in the computation.

The intensity drift correction option allows the user to correct for day-to-day instrument drifts after calibration files have been created. This method uses a stable reference sample which is measured along with quantitative standards prior to calibration. The steps are summarized below:

1. Store average net intensities for all elements of reference sample at calibration.

2. Compare current intensities for the reference sample to stored values.

3. Compute correction factor, IDF_i, for each element.

4. Apply correction factors to subsequent unknown intensities BEFORE quantification.

$$IDF_i = I_i(REF)/I_i(NOW) \tag{9}$$

where,

IDF_i = intensity drift correction factor for i

I_i (REF) = average reference intensity at calibration

I_i (NOW) = reference intensity just measured

The user periodically checks the intensity monitor sample to establish current intensity drift correction factors which normalize measured intensities (for unknowns) to the scale existing at calibration (for standards).

THIN FILM SAMPLES - CURVE FITTING TECHNIQUES

Combined alphas/curve fitting methods may be applied to thin samples in a fashion similar to those used for bulk samples above. The method described here combines hybrid alphas and effective mass absorption coefficients computed by Fundamental Parameters (10) with linear fits using two or more quantitative standards. This method provides analysis of sample composition and thickness for a single thin sample on a fixed substrate. Appropriate corrections are applied for sample elements that are also in the substrate. Additional terms could be added to extend its usefullness to multi-layer thin film samples.

The general form of the thin sample expression for the measured intensity of an element in a film of uniform thickness is given by (2,5,11)

$$R_i = \frac{W_i}{\left[1+ \sum_{j \neq se} Z_{ij}S_jW_j\right]}\{1 - \exp(-\bar{\mu}_i\rho\tau)\} \tag{10}$$
$$+SI_i \exp(-\bar{\mu}_i\rho\tau)$$

where,

Z_{ij} = hybrid alpha for j on i,

$\bar{\mu}_i$ = total mass absorption coefficient of the film for the measured x-ray line of element i (cm^2/g) over all incident exciting radiation,

ρ = thin film density (g/cm^3),

τ = film thickness (cm)

$SI_i = (I_i,sub/I_i,pure)$

I_i,sub = substrate intensity for element i without overlying film,

and, other symbols as previously defined. In this formulation, unmeasured elements are determined by stoichiometry where their interelement effects are included in the hybrid alpha and mass absorption coefficient terms. The advantages of the approach

are: semi-thick samples may be considered as well as thin
samples, substrate intensity contributions are included in the
fit/analysis, and only two standards are needed for the fit.
In practice, 5-10 standards may provide a better fit particu-
lairly when substrate intensity is significant.

As for the bulk sample case discussed above, the thin
sample expression (equation 10) may be rewritten in a linear
form

$$Y_i = S_i' \, X_i + C_i' \tag{11}$$

where,

$$C_i' = \text{intercept} = SI_i; \quad 0 < C_i' \le 1$$

$$S_i' = \text{slope} \quad = (1/I_i,\text{pure}); \quad 0 < S_i' < 10$$

$$X_i = \exp(+\bar{\mu}_i \rho \tau)$$

$$Y_i = \frac{W_i}{\left[1 + \sum_{j \ne se} Z_{ij} S_j W_j\right]} \left[\frac{1 - \exp(-\bar{\mu}_i \rho \tau)}{\exp(-\bar{\mu}_i \rho \tau)}\right]$$

In this form the coefficients of fit are the reciprocal pure
element intensity (slope = S_i') and the substrate intensity
(intercept = $C_i' = SI_i$).

The general procedure to obtain best fit coefficients for
the slope (S_i') and intercept (C_i') in equation 11 is as follows:

1. Compute $\bar{\mu}$, Z_{ij}, S_j from theory.

2. Form X, Y for each standard (2 minimum).

3. Solve for "best fit" S_i', C_i'

4. Test S_i', C_i': warning for bad values
   ($0 < S_i' < 10; 0 < C_i' < 1$)

5. Show Given Wt% vs. Fit Wt% for standards used.

6. User selected fit options: store current fit,
   hard copy of all relevant fit parameters, delete
   some standards, and run next fit.

7. If user does not store this fit then to to step 2.

8. If user stores this fit then go to next element
   and then go to step 2. If last element then
   create calibration file.

As for the bulk sample fit previously described the user
interprets "best fit" based on the chi-square criterion
(equation 7) for goodness-of-fit between given and fit
elemental weight precent. The thickness is considered a
known quantity for this chi-square test which is based solely
on composition.

When a thin sample is analyzed thickness may be computed
along with sample composition or fixed if it is entered prior
to quantitative computations (11). Intensity drift corrections
may also be applied to thin sample measurements using the same
procedure as described above for bulk samples (equation 9).

ANALYSIS EXAMPLE

Table 1 shows results for the application of the modified
alphas method for the analysis of synthetic uranium ore samples
which were analyzed as glass disks after dilution (5-fold) with
lithium tetraborate. This is a complex 17-element example
where lithium, boron and oxygen were the modifying elements in
the ratio 1:4:9 for boron:lithium:oxygen. Note the very good
analysis agreement for the 14 measured elements even though the
analysis was done without a Quantitative Match Standard (QMS)
file. The agreement was significantly improved when a QMS file
was included (4 of 12 standards). This excellant performance
is typical for complex systems with low atomic number unmeasured
elements which are included in the modified alphas. Generally,
calibrations of this type may be reliabley achieved with two to
five quantitative standards.

Table 1. Analysis of Synthetic Uranium Ore Samples a)
with a Modified Alphas/Curve Fitting Approach

Element	Concentration Range (wt%)	Analysis Agreement b) Range	Typical
Mg	.04 - .80	.01 - .40	+/- .10
Al	.20 - 8.0	.02 - 1.2	+/- .20
Si	.40 - 9.2	.01 - 2.0	+/- .30
P	.005 - .015	.001 - .004	+/- .002
S	.02 - .55	.007 - .17	+/- .03
K	.002 - .50	.001 - .09	+/- .01
Ca	.004 - 11.7	.006 - 2.0	+/- .20
Ti	.009 - .45	.001 - .08	+/- .01
Cr	.001 - .01	.001 - .005	+/- .002
Mn	.001 - .05	.001 - .030	+/- .008
Fe	.07 - 5.0	.01 - 1.2	+/- .05
Sr	.001 - .04	.001 - .08	+/- .005
Zr	.01 - .03	.005 - .02	+/- .007
U	.008 - .35	.001 - .01	+/- .002

a) Elemental composition after samples diluted (5-fold) with
   Lithium tetraborate and cast into glass beads.

b) Analysis after calibration; agreement between given and
   analyzed concentrations, no quantitative match standard
   file.

CONCLUSIONS

The Fundamental Alphas/Curve Fitting method described here
has been successfully applied to various bulk (7) and thin (11)
sample EDXRF analyses. It has a number of advantages when
compared to totally empirical or theoretical methods: 1) Few
standards are required (two minimum), 2) Interelement and mass
absorption coefficients obtained from theory not calibration
standards, 3) The method retains much of the accuracy of full
Fundamental Parameters approaches while utilizing a fast

coefficient based algorithm, 4) Tests may be applied for inconsistent data, 5) Error estimates may be carried through the
quantification process, and 6) The method lends itself to intensity drift correction and delta calculation techniques.

The advantages and ease-of-use of this method generally
outweigh any limitations imposed by calibration standard requirements for typical routine sample analyses. Furthermore,
the method lends itself to automated on-line instrument
analysis schemes for specific applications (15).

REFERENCES

1. Jenkins, R.; "An Introduction to X-Ray Spectrometry",
   Heyden and Son Ltd., London, 1976.
2. Bertin, E. P.; "Principals and Practices of X-Ray
   Spectrometry", Plenum Press, New York, 1975.
3. Analytical Chemistry Reviews, published April of each even
   year.
4. Advances in X-ray Analysis, Vols. 1-27, Plenum,
   New York, 1957-1983.
5. Dzubay, T.G.; "X-ray Fluorescence Analysis of Environmental
   Samples", Ann Arbor Science Publishers, Inc., Ann Arbor,
   Michigan, 1977.
6. Kalnicky, D. J.; "Application of Energy-Dispersive Techniques
   for Materials Characterization", paper No. 86, the Eastern
   Analytical Symposium, New York, 1984.
7. Kalnicky, D. J.; "EDXRF Analysis of Petroleum Products Using
   a Modified Alphas Approach", to be published in American
   Laboratory.
8. Jenkins, R., Gould R.W., and Gedeke, D.; "Quantitative X-ray
   Spectrometry", Marcel Dekker, Inc., New York, 1981.
9. LaChance, G. R.; "Introduction to Alpha Coefficients", 1984.
   Available from Corporation Scientifique Claisse, Inc., 2522,
   Chemin Sainte-Foy, SAINTE-FOY (Quebec), G1V 1T5, Canada:
   LaChance, G.R.; "The Family of Alpha Coefficients in X-ray
   Fluorescence Analysis", X-ray Spectrometry, (1979), 8, 190.
10. Criss, J. W.; "Fundamental Parameters Calculations on a
    Laboratory Computer", Advances in X-Ray Analysis, (1980),
    23, 93: X-Ray Software Review, No. 2, (1982).
11. Kalnicky, D. J.; "EDXRF Analysis of Thin Films and Coatings
    Using a Hybrid Alphas Approach", Denver X-ray Conference,
    Snow Mass, Colorado, 1985.
12. Tao, G. C., Pella, P. A. and Rousseau, R. M.; "NBSGSC - A
    FORTRAN Program for Quantitative X-ray Fluorescence
    Analysis", National Bureau of Standards Technical Note 1213,
    U.S. Dept. of Commerce, April, 1985.
13. Frechette, G., et al.; Analytical Chemistry, (1979), 51,
    957.
14. Claisse, F., Thinh, T. P.; Analytical Chemistry, (1979),
    51, 954.
15. Miller, R. A., Kalnicky, D. J., and Rizzo, T. V.; "Highly
    Automated Techniques for On-Line Liquid Analysis with an
    XRF System", Denver X-ray Conference, Snow Mass, Colorado,
    1985.

RESOLUTION OF OVERLAPPING X-RAY FLUORESCENCE PEAKS

WITH THE PSEUDO-VOIGT FUNCTION

T. C. Huang and G. Lim

IBM Almaden Research Center
650 Harry Road
San Jose, CA 95120-6099

ABSTRACT

A method for resolving overlapping X-ray fluorescence spectra by curve fitting is described. The profile shape of an experimental fluorescence line obtained by wavelength dispersive method is represented by a simple pseudo-Voigt function, i.e. a sum of an asymmetric Gaussian and Lorentzian, each of equal width. Results showed that the pseudo-Voigt function matched the experimental profiles with high reliability. The relative Gaussian and Lorentzian contents and the asymmetry of the profiles depended upon the analyzing crystal, collimating system and the 2θ peak position. For fixed crystal and collimator the smaller the 2θ, the larger the Gaussian content and the lower the asymmetry. The original Gaussian and Loretzian components of the exact Voigt function calculated from the parameters of the fitted pseudo-Voigt function explain the broadening effects of the X-ray emission lines and the instrumental aberrations on observed spectra. Curve fitting method with the psuedo-Voigt function has been used successfully to analyze overlapping fluorescence spectra. Examples and applications include a thin film sample where the $K\alpha$ and the $K\beta$ lines of adjacent transition elements overlap, and a strontium zirconium oxide specimen where the Zr $K\alpha$ and the Sr $K\beta$ lines strongly interfere. Concentrations obtained from the resolved individual peak intensities of Zr and Sr $K\alpha$ lines are within $\pm 1\%$ of the true values.

INTRODUCTION

In X-ray fluorescence (XRF) spectrometry, there are a number of instances where spectral lines lie close together and interfere. Proper use of analyzing crystal, collimator and pulse height analyzer will usually permit the discrimination of higher order peaks, unless both are the first order. One of the common interferences between the first order peaks is the overlap of the $K\alpha$ line of an element with the $K\beta$ peak of a lighter element such as Cr $K\alpha$/V $K\beta$ and Zr $K\alpha$/Sr $K\beta$ etc. (Examples of these two cases will be used in this paper). Interferences of the $K\alpha$ may also come from another $L\alpha$ or $L\beta$ peak (e.g., As $K\alpha$/Pb $L\alpha$ and Cu $K\alpha$/Tu $L\beta$ etc.). Similarly, overlaps also occur between the $L\alpha$ line of an element and the $K\beta$ or $L\beta$ peak of another, e.g. Tl $L\alpha$/Ga $K\beta$ and W $L\alpha$/Yb $L\beta$ etc. The overlapped peaks are usually avoided in XRF and quantitative analysis is made with the much weaker intensity $K\beta$, $L\beta$ or $L\gamma$ peaks. Therefore, the development of an effective method to resolve overlapping XRF spectra would be useful for XRF analysis. Curve fitting of complex spectra is commonly used in Mossbauer, MNR and other spectroscopies. In addition to high precision in resolving overlapping peaks, this method also has the advantages of minimizing the possible errors due to poor counting statistics and peak shifts arising from specimen displacement and spectrometer positioning.

In this paper, a new curve fitting method for resolving overlapping spectra is described. Its applications to selected examples are also given.

SHAPE OF EXPERMENTAL XRF PROFILES

The success of spectral analysis by curve fitting relies heavily on the analytical expression used to represent the shape of an experimental profile. The line shape of an observed XRF peak is known to lie between a Lorentzian and a Gaussian. Mathematically, it can be represented by the convolution of a Gaussian (G) and a Lorentzian (L), namely, the Voigt function.[1] It is well known that the X-ray emission line is a Lorentzian[2] with a narrow width (Γ_l) and the instrumental aberration function can be approximated by a broadened Gaussian[3] with width (Γ_g). As a result, the Voigt function is dominated by the G component. In other words, the shape of the experimental profile is strongly dependent upon the instrument (namely the collimator and the analyzing crystal) used. The instrumental effects on the experimental XRF profiles is illustrated in Figure 1. The Cu Kα and Gd Lα peaks obtained with a fine (4 x 0.005 inch spacing) or a course (4 x 0.020) source collimator and a LiF or a Graphite (GP) crystal have different profile shapes. The fine collimator with the LiF crystal gives narrower widths, and the GP crystal causes slower rates of decay at the tails.

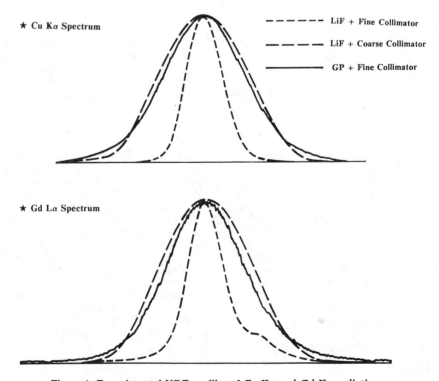

Figure 1. Experimental XRF profiles of Cu Kα and Gd Kα radiations.

CURVE FITTING METHOD WITH THE PSEUDO-VOIGT FUNCTION

In principle, an XRF line $y(2\theta)$ can be calculated from the Voigt function as follows:

$$y(2\theta) = G * L = I \times \int_{-\infty}^{\infty} \{ e^{-ln2 (\frac{2\theta'}{\Gamma_g/2})^2} \times \frac{1}{1 + [\frac{(2\theta-2\theta_o)-2\theta'}{\Gamma_l/2}]^2} \} \, d(2\theta') \qquad (1)$$

where I and $2\theta_o$ are the intensity and position of the XRF line. Direct calculation of Equation (1) is complicated and time consuming, thus making it impractical for routine use in curve fitting analysis. Instead, the analytical or empirical approximations to the Voigt function have been commonly used.[4,5] The linear combination of a Gaussian (G') and a Lorentzian (L'), also called the pseudo-Voigt function, has been shown to be a good representation of experimental spectral profiles.[6,7] Mathematically, a single experimental XRF peak $y_i(2\theta)$ can be approximated by the psuedo-Voigt function as follows:

$$y_i(2\theta) = C_g \times G' + C_l \times L' = I_i [C_g \, e^{-ln2 (\frac{2\theta-2\theta_i}{W})^2} + \frac{C_l}{1 + (\frac{2\theta-2\theta_i}{W})^2}] \qquad (2)$$

where C_g ($0 \leq C_g \leq 1$) and C_l ($=1-C_g$) are the G' and the L' contents, respectively; I_i the peak intensity, $2\theta_i$ the angle of component i, and W the half width at half maximum for both G' and L'. The replacement the convolution integral in Equation (1) by a sum of G' and L' in Equation (2) has greatly simplified the curve fitting calculation. To account for the asymmetry of the XRF peak, two values of W are used i.e., $W=W_r$ if $2\theta \geq 2\theta_i$ and $W=W_l$ if $2\theta < 2\theta_i$. The degree of asymmetry is given in terms of the asymmetric index AI ($=W_l/W_r$). Note that both G' and L' have the same parameters I_i, $2\theta_i$, W_r and W_r but different rates of decay.

The actual analytical equation used for fitting an experimental spectrum containing n components is given as follows:

$$Y(2\theta) = \sum_{i=1}^{n} y_i(2\theta) + BG(2\theta) \qquad (3)$$

where $BG(2\theta)$ is the background intensity. Generally speaking, there will be 2n+5 independent parameters where 2n+3 occur in the $\Sigma y(2\theta)$ term and 2 in the linear $BG(2\theta)$. However due to component overlaps, values of some of these parameters may be fixed and the number of variables to be determined in Equation (3) by curve fitting is reduced. For example, both the peak intensity and 2θ angle of the $K\alpha_2$ component may be fixed to those of the $K\alpha_1$ by setting $I_2 = 0.5 \times I_1$ and $2\theta_2 = 2\theta_1 + (\lambda_2 - \lambda_1)/d\cos\theta_1$ (where d is the d-spacing of the analyzing crystal). Therefore, the independent variables required by the curve fitting a $K\alpha$ spectrum is reduced to 7.

The goodness of match between the experimental $Y_e(2\theta)$ and the calculated data $Y_c(2\theta)$ is expressed by the reliability R as follows:

$$R(\%) = \sqrt{\frac{\sum_{2\theta} [Y_e(2\theta) - Y_c(2\theta)]^2}{\sum_{2\theta} Y_e(2\theta)^2}} \times 100\% \qquad (4)$$

The smaller the value of R, the better the match between the pseudo-Voigt function and the experimental XRF profile.

RESULTS AND DISCUSSION

The method described above has been used successfully to analyze various XRF spectra of single- and multi-element specimens.

Single-Element Specimens

Typical examples are the curve fitting of two Cu Kα spectra obtained with a fine source collimator and LIF and GP crystals. As shown in Figure 2, the fitted Kα_{1+2} profiles (solid curves) match the experimental data (x's) well with R~1%. Notice the difference in the G' content where C_g reached almost 95% for LiF but only about 90% for GP. The lower value of C_g (or the higher C_l) and the larger width Γ for GP indicate that the distribution of the GP (002) grains was relatively less Gaussian (or more Lorentzian) and had a larger "mosaic" spread than those of the LiF single crystal. The individual Kα_1 and Kα_2 components resolved by curve fitting have also been plotted (short dash profiles). Either the peak intensity of the single Kα_1 or the combined Kα_{1+2} can be used in the quantitative XRF analysis. Profile obtained with the LiF crystal showed slight asymmetry with AI=W_l/W_r=1.04.

Analysis of XRF profiles obtained with other analyzing crystal, specimens, data collection and reduction conditions were also conducted and similar reliabilities (R≤2%) have been obtained.[8] For fixed analyzing crystal and collimator, profile shapes varied systematically with the peak position 2θ. For example, the variations of C_g and AI obtained with a LiF crystal, fine collimator and ten single-element standards (namely Cs, Gd, Zr, Se, Ga, Cu, Co, Fe, Mn and Cr) are plotted in Figure 3. Profiles at 20° or less are pure Gaussian but gradually the Lorentzian content C_l becomes greater than zero with increasing 2θ and reaches almost 10% at 70°. At low 2θ the profiles are symmetric, but AI increases slowly to 1.06 at 70°. These dependencies can be explained by studying the parameters of the actual XRF emission profile L and instrumental function G as described below.

The widths Γ of the experimental profile or the Voigt function, Γ_l of the original Lorentzian and Γ_g of the Gaussian components can be calculated either numerically[9] or graphically[6] from the values of W_l, W_r and C_g. As shown in Figure 4, Γ increases slowly, but Γ_g remains constant with 2θ. Since the crystal and the collimator were fixed in the measurements, the geometrical aberration or the G function should remain unchanged. However, Γ_l is

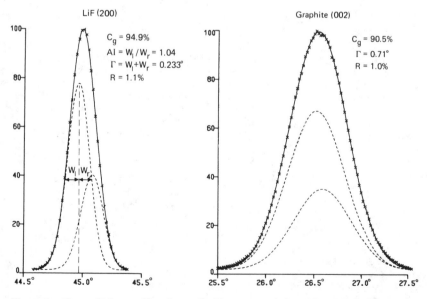

Figure 2.   Curve fitting results of two Cu Kα spectra: left, with a LiF (200) crystal and a fine collimator; right, a GP (002) crystal and a fine collimator.

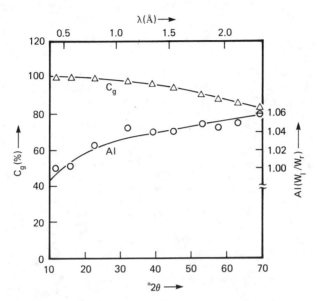

Figure 3. Plot of Gaussian content C_g and asymmetric index AI vs. 2θ.

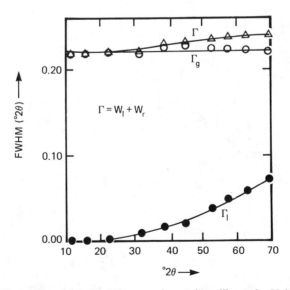

Figure 4. Plot of line widths Γ of the experimental profile or the Voigt function,
Γ_g of the original Guassian and Γ_l of the Lorentzian components vs. 2θ.

seen to increase with 2θ, which is explained by the increasing wavelength dispersion of the emission line with 2θ causing broadened experimental profiles. The asymmetry of the emission line[10] and its increasing contribution to the observed profiles at higher 2θ were the reasons for increased AI.

Multi-Element Specimens

In actual analyses, specimens generally contain two or more elements. The application of this curve fitting method to multi-element analysis has also been successful. One of the examples is the resolution of the XRF spectra of a 5350Å thick Ti-V-Cr film. As shown in Figure 5, the separation of Cr $K\alpha$ from V $K\beta$ and V $K\alpha$ from Ti $K\beta$ as well as W $L\beta_{2,3}$ was successful with $R \leq 1.4\%$. (W $L\beta_{2,3}$ peaks are from the X-ray tube.)

Another application was the analysis of a $SrZrO_3$ specimen. Experimental spectra showed that Sr $K\alpha$ interferes severely with the Zr $K\beta$ line. The correct intensities of the Sr $K\alpha$ peak can best been obtained by curve fitting. As shown in Figure 6, the match between the fitted (solid curves) and experimental data (x's) was good, $R.=0.8-1.3\%$. The Sr $K\alpha$ peak intensity obtained by curve fitting was almost 15% below the unresolved intensity. The curve fitted (resolved) intensities of the Sr and Zr $K\alpha$ have been used to correctly determine the relative concentration of the two elements. Data and the LAMA results[11] obtained with both resolved and unresolved intensities are listed in Table I. It can be note that reliable concentration has been obtained with the former. The normalized numbers of Zr and Sr atoms in the specimen was found to be within ±0.8% of the expected value. Without curve fitting, the discrepancy was ten times higher (i.e., ±8.7%).

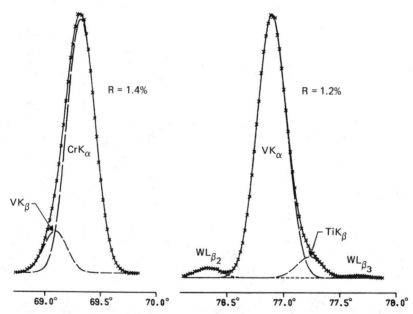

Figure 5. Resolution of overlapped spetra of a Ti-V-Cr thin film.
(W $L\beta_{2,3}$ peaks are from the X-ray tube.)

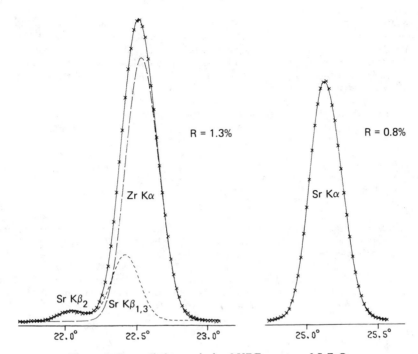

Figure 6. Curve fitting analysis of XRF spectra of SrZrO₃.

Table I. Quantitative XRF Analysis of ZrSrO₃

	Zr Kα	Sr Kα
• X-RAY LINE		
• STANDARD	Zr	Sr(NO₃)₂
• INTENSITY RATIO		
Curve Fitting	0.5122	0.5032
w/o Curve Fitting	0.5931	0.4995
• LAMA RESULTS		
Expected Number of Atom	1.0	1.0
Curve Fitting	1.008	0.992
w/o Curve Fitting	1.087	0.913

CONCLUSION

An effective curve fitting method for the resolution of overlapping XRF spectra has been presented. A simple psuedo-Voigt function is used to represent an experimental fluorescence line. Its Gaussian and Lorentzian components (G' and L') have the same peak intensity, width and peak location. To account for the asymmetry of the emission spectrum, two different values of half width, W_l and W_r on the low and high angle sides of the peak respectively, have been used.

Analysis of XRF spectra of single-element specimens showed that profile shapes were dependent upon the colimator, analyzing crystal and the peak position 2θ. For fixed instrumental parameters, the Gaussian content C_g or Lorentzian content C_l and the asymmetric index AI varied systematically with 2θ. The parameters of the fitted psuedo-Voigt function, i.e., C_g, W_l and W_r were used to calculate the actual widths Γ_l and Γ_g of the original Gaussian G and Lorentzian L components of the exact Voigt function. This explained the broadening effects of the XRF emission line and the instrumental aberration on the observed spectra.

The method has also been applied to resolve the overlaps of multi-element specimens. Examples were the separation of Cr $K\alpha$ from V $K\beta$, and of V $K\alpha$ from Ti $K\beta$ and W $L\beta_{2,3}$. Another example was the the analysis of a $ZrSrO_3$ specimen where the Zr $K\alpha$ line had almost 15% error in intensity due to gross interference from the Sr $K\beta$ line. The normalized numbers of Zr and Sr atoms obtained with the curve fitted intensities and the LAMA program were found to be within ±0.8% of the expected values. On the other hand, the discrepancy was ten times higher when the unresolved intensities were used.

REFERENCES

1. W. Voigt, Munch. Ber., 603 (1912).

2. A. Hoyt, Phys. Rev., 40:477 (1932).

3. B. L. Henke and K. Taniguchi, J. Appl. Phys., 47:1027 (1976).

4. E. E. Whiting, J. Quant. Spectrosc. Radiat. Tranfer., 8:1379 (1968)

5. J. E. Kielkopf, J. Opt. Soc. Am., 63:987 (1973).

6. G. K. Wertheim, M. A. Butler, K. W. West and D. N. E. Buchanan, Rev. Sci. Instrum., 45:1369 (1974).

7. S. Enzo and W. Parrish, Adv. X-Ray Anal., 27:37 (1984).

8. T. C. Huang and G. Lim, submitted to X-Ray Spectrom.

9. J. B. Hasting, W. Thomlinson and D. E. Cox, J. Appl. Cryst., 17:85 (1984).

10. M. A. Blokhin, The Physics of X-Rays (2nd Edition), State Publishing House of Technical-Theoretical Literature, Moscow 1957.

11. D. Laguitton and M. Mantler, Adv. X-Ray Anal., 20:515 (1977).

COMPARISON OF STRATEGIES FOR DEALING WITH UNANALYZED LIGHT ELEMENTS IN

FUNDAMENTAL PARAMETERS X-RAY FLUORESCENCE ANALYSIS

Peter B. DeGroot

Celanese Technical Center
P. O. Box 9077
Corpus Christi, TX   78469

ABSTRACT

The total elemental composition of a sample must be measured or specified for successful application of fundamental parameters type matrix corrections in x-ray fluorescence analyses. Unanalyzed light elements can be dealt with in three ways. They can be related to the concentration of another element by stoichiometry, calculated by difference, or their effects minimized by addition of a diluent of known composition. The relative success of these approaches depends on the composition of the sample. Limiting cases tested are high or low concentrations of analytes having long or short wavelength emission lines, in the presence of additional heavy or light elements. Molybdenum oxide is used as the analyte, with the oxygen serving as a typical unanalyzed light element to be treated by stoichiometry, difference, or dilution. The accuracy and precision of the analysis for molybdenum, using either the K_α or L_α line, is the criterion for judging the success of the strategy. Experimental results and theoretical calculations using the XRF11* matrix correction program are employed. Generally, the difference strategy is inferior to stoichiometry or dilution. However, difference methods can give acceptable results except in the case of a high concentration of short wavelength analyte in a light element matrix. Where stoichiometry is not known, an assumed stoichiometry that is correct within one oxidation number gives results comparable to the difference method.

INTRODUCTION

Light elements which are difficult, inconvenient, or impossible to analyze by x-ray fluorescence spectroscopy (XRF) are often present in XRF samples. There are three principal means of dealing with such elements. The first strategy is to add a relatively large amount of diluent of known composition to the sample. The inter-element effects on XRF intensities are then determined primarily by the diluent, and the effects of the unanalyzed light element can be ignored. If the identity of the light element present is not known, this is the only practical method.

---

*Criss Software, Inc., 12204 Blaketon St., Largo, MD   20870

The dilution method can still, of course, be applied if the identity
of the light element is known. Alternatively, the unanalyzed light element
can be dealt with within the computation processes of the fundamental parameters (FP) method of calculating concentrations from raw intensities. The
light element can either be related to the concentration of unanalyzed
elements by stoichiometry, or calculated by difference.

The purpose of this investigation is to determine the relative success
of each of the three strategies in analyzing samples of different compositions. Analytes with long or short wavelength XRF emissions, present in
high or low concentrations in samples with additional heavy or light elements
are used as test samples. Both theoretical predictions of the XRF11 FP
matrix correction program and actual analysis results are used to determine
the precision and accuracy of the analysis for the analyte's metallic element
obtained using each of the three strategies on the test samples.

EXPERIMENTAL

X-Ray intensities were measured with a Philips Universal Spectrograph,
using LiF or EDDT crystals and a Cr-target x-ray tube. Analysis time was
adjusted to give approximately ±1% (2σ) counting statistics for each intensity
measurement.

Samples were ground to -325 mesh, thoroughly blended, and pressed into
planchets. Samples for testing the dilution strategy were prepared in two
different ways. Either solid mixtures were made by blending the samples
with starch, or solutions were prepared by dissolving the samples in 5% HF.
In either case, sufficient diluent was added to lower the analyte concentration to less than 1%.

Fundamental parameter matrix correction calculations were performed
using XRF11 (version of July, 1981). Molybdenum and oxygen concentrations
were calculated. Concentrations of the other matrix components were
considered known, and fixed in the calculation.

Table 1. Composition of test samples

SAMPLE	ANALYTE CONC.	MATRIX	COMPOSITION, WT%		
			MoO_3	Sb_2O_3	Na_2CO_3
A	HIGH	HEAVY	69.96	30.04	
A-std	HIGH	HEAVY	75.01	24.99	
B	HIGH	LIGHT	70.10		29.90
B-std	HIGH	LIGHT	74.99		25.01
C	LOW	HEAVY	2.20	97.80	
C-std	LOW	HEAVY	3.29	96.71	
D	LOW	LIGHT	2.20		97.80
D-std	LOW	LIGHT	3.37		96.63

RESULTS AND DISCUSSION

Test Sample Compositions

    Molybdenum (VI) oxide was used as the analyte, with its oxygen serving
as a typical unanalyzed light element to be treated by stoichiometry,
difference, or dilution. To simulate a "heavy" or "light" matrix containing
the Mo and oxygen, either Sb_2O_3 or Na_2CO_3 was added.

    Four test samples were prepared representing "high" (~70% MoO_3) and
"low" (~2% MoO_3) concentrations of the analyte with either Sb_2O_3 or Na_2CO_3
as the balance of the sample. For each test sample, a corresponding
"standard" of slightly higher MoO_3 concentration was also prepared. Table
1 lists the actual compositions of the samples and standards. Short or long
wavelength analyte radiation was selected by measuring either the Mo K_α or
L_α intensity. There were thus eight possible combinations of the variables
"concentration", "matrix", and "wavelength" to be tested using each of the
three strategies for dealing with the unanalyzed oxygen in the MoO_3.

Theoretical Predictions of Precision

    The capability of XRF11 to predict theoretical intensities given sample
composition allows the calculation of the relative change in intensity for
a given change in composition, or, more usefully, the reciprocal of this
ratio. Consider a simple binary system of a metal plus a light element,
oxygen, for instance. The effects of varying the proportions of the two
components on the relative uncertainty in concentrations calculated from
the K_α intensities of several metals is shown in Figure 1. As can be seen,
for long wavelength radiation (low atomic number elements), the relative
uncertainty in concentration for a given relative precision of intensity

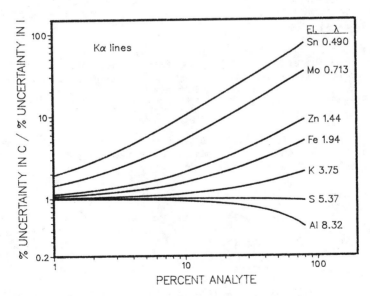

Fig. 1. Ratio of % uncertainty in concentration to %
         uncertainty in intensity for various elements
         combined with oxygen as a function of concen-
         tration and wavelength.

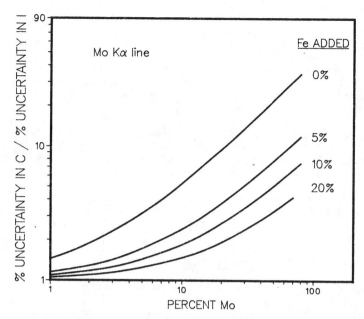

Fig. 2. Improvement in uncertainty in Mo concentration
         in Mo-O system by adding heavy element to the
         matrix.

measurement is nearly constant or even decreases as the metal content
increases. However, as the analyte wavelength becomes shorter and shorter,
the relative uncertainty in the calculated concentration rises more and
more rapidly with increasing metal content. The L and M emission lines
show similar behavior.

    The only matrix effects in these cases are due to self-absorption by
the metal and the effects of the very light element oxygen. Addition of
an even moderately heavy additional matrix component improves the predicted
analysis precision considerably. Figure 2 shows the effect on the theoretical relative precision curve for the Mo K_α line of adding various amounts
of Fe to the matrix.

    Qualitatively, then, analysis precision is expected to become poorer
as the analyte concentration becomes higher, the analyte wavelength becomes
shorter, and the matrix becomes lighter.

Observed Precision Using Different Light Element Strategies

    The precision of analysis of five, or in some cases ten, replicate
analyses of the test samples was determined using each of the three
strategies for dealing with the light element oxygen. Results are given
in Table 2, along with a theoretical prediction of uncertainty based on the
actual composition of the sample.

    Qualitatively, the observed results for the difference strategy match
the theoretical predictions fairly well. The worst case for this strategy
is clearly the high concentration of short wavelength analyte in a light
matrix. For all of the other combinations of matrix, wavelength, and concentration, the difference strategy would probably give results of sufficient
precision for routine analyses.

Table 2. Observed and theoretical precision of test analyses. Precision
         = T x σ, in %, of individual observation, where T is the value
         for the 95% confidence interval

					STRATEGY			
	ANALYTE						DILUTION	
SAMP.	WAVEL.	CONC.	MATRIX	THEOR.(a)	DIFF.	STOICH.	MIXT.	SOLN.
A	SHORT	HIGH	HEAVY	±4.6	2.3	0.8	0.2	1.6
B	SHORT	HIGH	LIGHT	25.5	13.6	1.4	1.7	1.1
C	SHORT	LOW	HEAVY	1.4	5.1	2.9	2.0	0.6
D	SHORT	LOW	LIGHT	2.0	1.5	0.8	3.3	0.9
A	LONG	HIGH	HEAVY	0.7	1.1	0.4	1.6	1.5
B	LONG	HIGH	LIGHT	1.0	2.9	1.0	3.0	1.6
C	LONG	LOW	HEAVY	1.4	5.4	2.7	10.7	1.5
D	LONG	LOW	LIGHT	1.4	3.8	1.7	3.2	1.7

(a) Percent relative error in analysis for ±1% relative variation in
    measured intensity of both standard and test sample.

The stoichiometry strategy yields precision superior to the difference
method. In the worst case by the difference method, the improvement is
enormous; about a factor of ten. In the other cases, an improvement by
about a factor of two is seen.

The "mixture" variation of the dilution strategy, in which dry powders
were blended was less successful. When grinding and blending dry components,
there is always the potential for problems caused by inhomogeneity, differences in particle size, and differences in density. This is apparently the
case for the dry mixture dilution strategy. While the precision of analysis
for all the samples using short wavelength radiation is fairly good, the
precision deteriorates considerably when measuring long wavelength radiation.
These effects have an even greater influence on the accuracy of the analysis,
as opposed to the precision, and will be discussed further in the next
section.

The dilution strategy employing a homogeneous solution gives good
precision for most combinations of analyte wavelength, original matrix,
and concentration. Solid solution methods such as the borate flux method
would be expected to give similar good results. There is, of course, a
trade-off. An extra sample preparation step is introduced, with the potential for dilution or preparation errors. The reduction in concentration of
the analyte can also mean much longer counting times to reach the desired
intensity precision. In the two cases where original concentration was
low, and the less intense Mo L_α line was used, very long counting times
were required after dilution. This may have contributed to the somewhat
poorer precision in these instances.

The stoichiometry strategy, which yields precision comparable to
homogeneous dilution, looks attractive because it involves no extra sample
preparation step. However, one must know both the identity and stoichiometry of the element in question. What are the consequences of an incorrect
assumption of stoichiometry? Table 3 shows the effect of errors in stoichiometry for the set of test samples. In almost all cases, the precision

Table 3.   Percent relative error in calculated concentration for
           incorrect assumptions of stoichiometry

	ANALYTE			ERROR IN STOICHIOMETRY			
SAMP.	WAVEL.	CONC.	MATRIX	±1	±2	±3	±4
A	SHORT	HIGH	HEAVY	±0.9	1.7	2.6	3.5
B	SHORT	HIGH	LIGHT	0.2	0.3	0.5	0.6
C	SHORT	LOW	HEAVY	2.7	5.5	8.1	11
D	SHORT	LOW	LIGHT	1.8	4.1	6.0	8.0
A	LONG	HIGH	HEAVY	2.1	4.2	6.3	8.5
B	LONG	HIGH	LIGHT	2.0	4.0	5.9	7.9
C	LONG	LOW	HEAVY	2.7	5.5	8.3	11
D	LONG	LOW	LIGHT	2.7	5.5.	8.3	11

obtained with an error ±1 oxidation number gives results no worse than
those obtained with the difference method. In the high concentration,
short wavelength, light matrix case, even an assumed oxidation state that
is off by four oxidation numbers produces less than a 1% difference in the
calculated concentration of the test samples. This is especially fortunate,
because this is just the type of sample where results from the difference
strategy are the worst.

Sample C (low concentration, heavy matrix) when analyzed in the solid
state gave precisions significantly worse than the other samples with either
short or long wavelength radiation. Several replicate preparations of this
sample were made with the same results. As discussed above in connection
with the dry mixture dilution strategy, this may be due to particle size
effects. Again, the effect on analysis accuracy was even greater, and will
be discussed in the next section.

Observed Accuracy Using Different Light Element Strategies

The precision tests discussed above reflected matrix absorption effects
on analysis strategies. If no other variables affect the analyses, then

Table 4.   Observed accuracy of test analyses. Accuracy = % deviation
           of mean of analyses of 5 samples from true value

	ANALYTE			STRATEGY		DILUTION	
SAMP.	WAVEL.	CONC.	MATRIX	DIFF.	STOICH.	MIXT.	SOLN.
A	SHORT	HIGH	HEAVY	-2.4	-0.3	+0.4	-0.8
B	SHORT	HIGH	LIGHT	+3.5	+0.2	-1.6	-1.7
C	SHORT	LOW	HEAVY	+6.8	+3.4	+2.7	+1.4
D	SHORT	LOW	LIGHT	-4.4	-1.9	+4.5	+1.4
A	LONG	HIGH	HEAVY	+1.1	+0.4	-8.0	+0.5
B	LONG	HIGH	LIGHT	+2.8	+1.0	-8.0	-0.9
C	LONG	LOW	HEAVY	+23.8	+11.6	+6.1	+2.4
D	LONG	LOW	LIGHT	+0.6	+0.6	+10.8	+2.8

analysis accuracy for a single analysis will be determined statistically
by the precision. A determination of the accuracy of analysis obtained by
averaging several analyses should minimize the effects of differences in
precision and expose any other variables affecting the analyses. Results
of such a determination are given in Table 4.

As with precision, accuracy by the difference strategy is inferior to
that obtained with the stoichiometry strategy by about a factor of two in
most cases. It appears that the difference strategy magnifies errors from
any source compared to the stoichiometry strategy.

Sample C again stands out as being significantly less accurate than the
other samples in all cases except dilution in a homogeneous solution. The
effect is fairly small with short wavelength radiation, but large inaccuracies
are produced when long wavelength radiation is measured. It appears that
differences in particle sizes between the MoO_3 and Sb_2O_3 are responsible.
Scanning electron microscopic examination of the sample shows the MoO_3 to
have a significantly larger average particle size than the Sb_2O_3 (although
both are smaller than 325 mesh). Such differences would be expected to have
a more pronounced effect on samples with higher concentrations of antimony
and when measuring more highly absorbed long wavelength radiation[1].

Similar effects may account for the poor accuracy of the dry mixture
dilution strategy when using long wavelength radiation. The subject of
particle size differences and their effects in connection with analysis
strategies is under further investigation.

Assuring Proper Convergence of FP Methods

A brief digression into the criteria for proper convergence of the FP
method applied to the analyses of the test samples is worthwhile. For
accurate comparison of results by the various strategies employed, one
must be certain that the mathematical procedure involved is actually
approaching the correct result. In the case of light matrices, and especially with the difference strategy, convergence can be very slow, and many
iterations are required. In these cases, convergence proceeds in very
small steps (very small changes in concentration at each iteration). Iteration is stopped when the relative change in calculated concentration from
one iteration to the next falls below a preset convergence test tolerance.
If this tolerance is too large, iteration can terminate permaturely. For
example, in the present high concentration, short wavelength, light matrix
case by the difference strategy, about 140 iterations and a convergence
tolerance of 5×10^{-4} were required.

How does one determine the required convergence tolerance and number
of iterations? A convenient method is the "self replication" test. In this
test, the composition of a material similar to the unknown is entered as a
standard. An arbitrary intensity for the analyte line is also entered. The
same intensity is then entered for an unknown, and the iteration limits and
convergence tolerance adjusted until the FP procedure reproduces the input
concentration to within, say, 0.1% relative accuracy. With the XRF11 program,
it is worthwhile to observe the results of each iteration as the calculation
progresses. With light element matrices especially, setting the iteration
limit too high for the initial stage of iteration using only alpha factors
can allow the calculation to overshoot the correct concentration. Subsequent slower iterations with the full FP theory matrix correction must then
work their way back to the correct answer. Setting the initial iteration
limit near the point where it overshoots the correct concentration will
speed up the data reduction considerably.

Table 5. Guidelines for selecting light element strategy

WAVEL.	CONC.	MATRIX	NOTE	STRATEGIES IN ORDER OF PREFERENCE
SHORT	HIGH	HEAVY		DIL(SOLN) ~ STOICH ~ DIL(MIXT) > DIFF
SHORT	HIGH	LIGHT		DIL(SOLN) ~ STOICH ~ DIL(MIXT) >> DIFF
SHORT	LOW	HEAVY	(a)	DIL(SOLN) > STOICH ~ DIL(MIXT) > DIFF
SHORT	LOW	LIGHT		DIL(SOLN) ~ STOICH > DIL(MIXT) ~ DIFF
LONG	HIGH	HEAVY		DIL(SOLN) ~ STOICH ~ DIFF > DIL(MIXT)
LONG	HIGH	LIGHT		DIL(SOLN) ~ STOICH > DIFF > DIL(MIXT)
LONG	LOW	HEAVY	(a)	DIL(SOLN) > STOICH >> DIL(MIXT) ~ DIFF
LONG	LOW	LIGHT		STOICH > DIL(SOLN) ~ DIFF > DIL(MIXT)

(a) Differences in particle size in matrix components probably
      contribute significantly to decreased precision and accuracy
      when analyzing solids.

CONCLUSIONS

The results of this study allow one to develop some guidelines for
choosing strategies for dealing with unanalyzed light elements. Table 5
shows a rank-ordering of the strategies for different analytical situations.
The conclusions as to the effects of differences in particle size are ten-
tative. There may be other situations in which particle size has a significant effect which will be revealed by further investigation.

Dilution in a homogeneous solution or stoichiometry are the methods of
choice in all cases. However, there may be complications in applying the
dilution method such as inability to dissolve the sample or an unacceptable
decrease in the emission intensity of the analyte(s). The stoichiometry
strategy is both convenient and generally as good as homogeneous dilution
if the identity of the light element and its stoichiometry are known. Where
stoichiometry is not known with certainty, the effects on accuracy of
incorrect assumptions of stoichiometry should be assessed. Also, particle
size effects may be severe for low concentrations of analyte in heavy
matrices when using long wavelength radiation.

The difference strategy can give acceptable results for routine analyses
in most cases, except with high concentrations of short wavelength analytes
in light matrices. Again, particle size effects may be a problem, as noted
above.

Dilution by mixing dry powders can give acceptable results for routine
analyses when measuring short wavelength analyte lines.

REFERENCE

1. P. F. Berry, Particle size heterogeneity phenomena in x-ray
       analysis, in: "Applications of Low Energy X- and Gamma
       Rays," C. A. Ziegler, ed., Gordon and Breach, New York (1971).

A UNIVERSAL DATA BASE APPROACH FOR QUANTITATIVE WDXRF ANALYSIS WITH GENERAL PURPOSE DATA ACQUISITION ALGORITHMS

M. J. Rokosz and B. E. Artz

Ford Motor Company
Research Staff
P. O. Box 2053
Dearborn, MI 48121

INTRODUCTION

Acquisition of new x-ray fluorescence (XRF) hardware or a data reduction computer can be a particularly frustrating experience for analysts who depend upon programs not supplied by the XRF spectrometer manufacturer. Computerized data collection and reduction programs generated for a particular spectrometer/computer system can become virtually useless when a significant part of the spectrometer/computer system is replaced. The problem is compounded in a research environment where many different kinds of samples are encountered and many unique analysis programs developed.

The potential magnitude of this problem was recently demonstrated at the Ford Motor Company Scientific Research Laboratory when the thirteen year old computer-contolled XRF analysis system was replaced by a new state-of-the-art spectrometer/computer system. The old spectrometer had been computerized using third party interface hardware with extensive modifications to the spectrometer hardware. All the control and analysis software had been internally generated(1). Even though the new spectrometer and computer were manufactured by the same companies as before, the degree of software compatibility was small.

Several months were required to convert only the most important analysis programs. However, there was a further complication in that, as delivered, the spectrometer microprocessor program could not support the sequential data collection method employed by our software. The microprocessor program required that the entire data collection information, for all the samples being processed, be present in its memory before the job was started. In addition, once started the data collection could not be

altered. Our existing programs generated data collection
commands as required and sometimes even modified the data
collection from sample to sample depending upon measurments
made on the particular sample itself. The effort required to
change from this approach would have been enormous as well as
totally undesirable. Fortunately, a revised microprocessor
program was supplied by the manufacturer several months later
that was compatible with both modes of operation.

MODULAR APPROACH

In order to avoid the aforementioned problems associated
with the change in XRF analysis hardware or data reduction
procedures a modular approach is now being adopted. Figure 1
shows the planned configuration along with the interaction
between the indivdual modules. The advantage of this
approach is the stand-alone nature of the individual modules.
As long as the interaction between modules remains constant
any module can be modified without necessitating any change
in another module. This approach minimizes the amount of
work required to change any aspect of the XRF analysis
procedure.

The approach consists of a minimum of three software
modules that interact with a universal data base, the XRF
spectrometer hardware, and the specific application program
for data-reduction. The nature of, and requirements for, the
various aspects of the approach will now be discussed.

X-ray spectrometer and computer interface

This is the XRF hardware that performs the actual x-ray
measurement. Its purpose is to recieve a request for a
certain operation, perform that operation and return the
required information.

An important aspect of the modular approach is that it
be applicable to any computer-controllable XRF spectrometer.
This has significant impact on what is required of the
spectrometer/interface system. XRF hardware has evolved from
manually operated equipment, to host computer-controllable
equipment, to host/microprocessor-controlled equipment.
Often, the implementation of the last step has caused the
most difficulty for the user wishing to implement other than
manufacturer-supplied software. The apparent desire to
offload the data collection and even some of the data
reduction from the host computer to the microprocessor has
resulted in a reduction of the overall flexibility of the XRF
hardware. It is our recommendation that the responsibility
for data collection and reduction be left to the host
computer and that the microprocessor be limited to
controlling and monitoring the spectrometer.

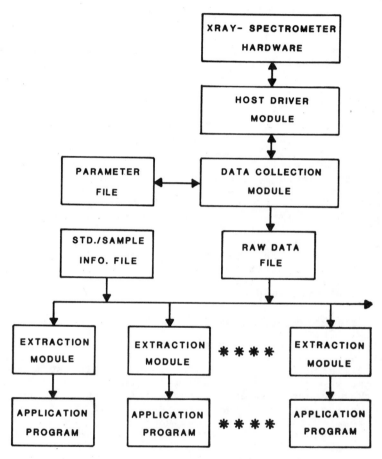

Figure 1.   Overview of modular approach showing the
interactions between the software modules and data base
files.

The x-ray spectrometer/computer interface should perform the following tasks with the last item being desirable but not absolutely necessary.

* Respond to a request for a parameter change indicating busy, done, or error condition.

* Respond to a request for a measurement, returning the time and counts or an error condition.

* Respond to a request for current hardware parameter configuration and status.

* Monitor hardware for error conditions and hardware malfunctions.

The method used to interface the spectrometer to the computer should be as simple as possible. Since XRF data rates are relatively slow, terminal emulation with x-on/x-off protocol is adequate. This usually results in a command set consising of relatively simple ASCII character strings. A good example of this type of interface is the HP7470A plotter(2). However, regardless of the interface method employed, it is essential that complete and accurate documentation be available during equipment evaluation and be supplied when the equipment is purchased. It is usually possible with the more recently developed computer systems to easily develop interface software providing the hardware requirements are completely known. Experience has shown, however, that this information is not always easy to get.

Host Driver Software Module

This module consists of a set of subroutines, accessible through high level languages such as Fortran or Basic, which perform all possible equipment functions. This software is spectrometer and computer specific and should be the only module that requires extensive modification when either the host computer or spectrometer is changed. It should be written in as high a level language as possible and while assembly language may be necessary in some cases, it should be avoided if possible.

This module should perform the following services.

* Generate the appropriate, spectrometer specific, ASCII command string corresponding to a particular subroutine call.

* Wait for and interpret spectrometer response, returning appropriate information to calling program.

* Keep track of current instrument parameters if these are not simply available from the spectrometer.

The independence of this module is the key to easily updating the spectrometer and computer hardware. Whenever spectrometer and computer dependent code is imbedded in data collection and application specific programs, these programs will all require major modifications as a result of any hardware change.

Data Acquisition Module

The data acquisition module is a general purpose program designed to allow the operator to develop and perhaps store a set of data collection instructions specific to a given type of sample. This program should be in a high level language that is easily transportable from one computer to another. The program allows the user to access a parameter data base which contains the most current information on parameter settings for the measurement of a given x-ray line. Using this information and supplying any required new information the operator can develop and store the proper series of measurements for any analytical problem. The output from this module is all of the information in the raw data file.

The data acquisition software can be of varying degrees of sophistication. It is essential that this software communicate with the spectrometer only through the driver subroutines and that it generate all the information required by the raw data file. A simple data collection program might require the operator to specifically input each task to be done while a more complex program might assemble the task from previously stored information and even perform parameter optimization during the data collection. However, it is important to note that the relationship with the driver module, parameter data base, and raw data base be invariant. This allows the data collection program to be modified without changing any other part of the system.

Data Extraction/ Reduction Module

The ultimate goal of quantitative XRF is the conversion of measured x-ray intensities to corresponding elemental concentrations. The modular approach stores the measured x-ray intensities, along with all other pertinent information, in the raw data file. All that is required to make use of any data reduction program is an interface to this file. Since the content and format of the raw data file is known, the user can write a data extraction program that extracts the required data from the raw data file and puts it in the format required by the data reduction program. In some cases, for user-developed programs the raw data file may be used directly. This approach allows relatively easy implementation of public domain programs such as NRLXRF(3), NBSGSC(4), and CIROU(5). In addition, the analyst can run a variety of programs on the same raw data and compare the results.

UNIVERSAL DATA BASE

The data base associated with the modular approach consists of three data files. These are the parameter file, raw data file, and standard/sample information file. All of these files should be in easily readable ASCII format and their contents completely documented. A brief description of the contents of these files follows.

Parameter File

This is a dynamic data file which contains the most recently determined equipment parameters required to measure a specific x-ray line. A typical parameter file might contain the following information.

* measurement identification code
* element name   * crystal
* x-ray line   * wavelength
* peak position   * order
* low background position   * detector
* high background position   * soller slit value
* detector PHA settings   * mask diameter
* generator voltage   * generator current
* chamber gas/ vacuum   * rotation on/off
* detector gas   * x-ray tube target
* x-ray tube window thickness
* tube and sample takeoff angles
* sample mask size and material
* space for information to be defined later

The most important aspect of this file is that it contain all the equipment information necessary to measure a particular x-ray line as well as the equipment parameters needed to analyse the resulting intensity. This data file is dynamic in the sense that it is constantly being updated and expanded. Only the latest information is retained and old information is deleted.

Raw Data File

The raw data file contains, for each sample, the raw measured counts and times as well as all the values of the instrument parameters when the measurements were made. This is a read only file. Once the data has been generated it is permanent and cannot be modified or changed. If it is necessary to remeasure a sample or add additional measurements to a sample an entirely new raw data file is created taking data from the instrument and old data file. This approach guarantees the integrity of the raw data. Raw data files may be deleted or archived as necessary.

The raw data file may be quite large and may contain considerable extraneous and duplicate information. This is allowed in order to have a data file that is easy to use when developing data extraction programs. A complex, but more efficient, data file would complicate the development of the data extraction programs and consequently limit the usefulness of the approach. In general, computer mass storage is relatively inexpensive compared to the cost of program development.

Standard/Sample Information Data File

This file contains all of the currently known information about a given standard or sample. Typically this file contains the following type of information.

* Sample name
* Sample type (standard, sample, etc.)
* Source (NBS, Analysis Request, etc.)
* Sample form (casting, fusion, pressed powder, etc.)
* Diluent (type and amount)
* Concentration units (wt.fr., %, ppm, etc)
* Additional information space
Followed by compositional information
   Component 1, Composition 1
   Component 2, Composition 2
      ....
   Component n, Composition n

This file along with the raw data file is accessed by the data extraction program to produce the input file for the data reduction program.

CONCLUSION

Completion of this plan is well underway in our laboratory. The driver module is complete and the remaining modules nearly complete. The requirements for the data base have been defined and the programs are being developed accordingly. We sincerely expect to avoid any future transitionary problems by using this approach.

We would also like to recommend that XRF equipment manufacturers adopt a similar approach in their own software development. Currently, it is extremely difficult, if not impossible, to utilize any part of the supplied software unless it is used exactly as intended. Most manufacturer supplied software seems to be in the form of a single large program that is supposed to cover all XRF analysis possibilities. Trying to extract subsets of this code for specific user applications is difficult. Indeed, it seems as though the manufacturers would benefit most from the modular approach since they frequently introduce new spectrometers and associated computers. Adoption of this approach would benefit users also in minimizing the effort necessary to develop their own data reduction programs and also in implementing third party programs.

Lastly, we cannot emphasize enough the requirement for complete documentation of all aspects of this approach. It can only work if the intermodule requirements and data base format are completely known to all users.

REFERENCES

(1) B. E. Artz, E. C. Kao, M. A. Short, Advances in X-ray Analysis, Vol. 22, p. 425, (1979)

(2) HP7470A Graphics Plotter; Interfacing and Programming Manual, Hewlett-Packard Co., San Diego, CA (1982)

(3) J. W. Criss, NRLXRF, Cosmic Program No. DOD-0065, Univ. of Ga. Computer Center, Athens, GA. (1977)

(4) P. A. Pella, NBSGSC, Nat. Bureau of Stds., Gaithersberg, MD. (1985)

(5) R. M. Rousseau, Geological Survey of Canada, Ottawa, Ont, Canada. (1985)

SAMPLING AND DATA-TAKING STRATEGIES IN X-RAY FLUORESCENCE ASSAY OF LOW S/N SOLUTIONS

Claude R. Hudgens

Monsanto Research Corporation
Mound*
Miamisburg, Ohio 45342

INTRODUCTION

This project was initiated for the purpose of demonstrating the feasibility of on-line x-ray fluorescence (XRF) analysis for the nondestructive assay of fissile elements (SNM) in reactor fuel reprocessing (dissolver) solutions, using wavelength dispersive x-ray fluorescence analysis because of its high immunity to the intense gamma emissions of the solutions.[1] A prime objective of this project was the identification and dimensioning of the parameters critical to XRF assays of high accuracy. The concepts presented herein, though directed primarily to assay of solutions with emphasis on low signal-to-noise conditions and low count rates, are applicable to all assays of solids, slurries, and gases.

MASS ASSAY BY XRF

If a weighed mass (M_{std}) of internal standard is added to a sample, then the mass (M_a) of analyt is immediately calculable if the mass ratio of analyt to standard is known.

For an absolute mass assay a gravimetric factor will have been predetermined by measuring the x-ray intensity ratios of analyts and standards in samples of known composition. The gravimetric factor (K) is

$$K = (M_a/M_s)*(I_s/I_a)*F(),$$ Eq. 1

*Mound is operated by Monsanto Research Corporation for the U. S. Department of Energy under Contract No. DE-AC04-76DP00053.

where M_s, M_a, I_s, and I_a are the masses and background-corrected x-ray intensities of analyt and internal standard in the standardizing sample, and F() is the usual correction factor. The gravimetric factor may be considered as constant for less demanding assays, thus simplifying Eq. 1 to

$$K = (M_a/M_s)*(I_s/I_a). \qquad\qquad\qquad\qquad\qquad\qquad\qquad\text{Eq. 1a}$$

The range of application of this simplified equation depends, of course, on the targeted standard deviation. For thorium and uranium with strontium as internal standard, deviations of less than 0.5% from linearity over compositions ranges covering over an order of magnitude are usual.

With the gravimetric factor (K) defined, the mass (M_a) of the analyt contained in the solution batch becomes

$$M_a = K * M_{std} * I_a/I_s. \qquad\qquad\qquad\qquad\qquad\qquad\text{Eq. 2}$$

TOTAL SAMPLING

Samples presented for XRF assays are usually small aliquots extracted from a bulk of material. By adding weighed internal standard to an aliquot, the absolute analyt mass content of the aliquot can be determined, and if the mass relationship of batch to aliquot is accurately known, a valid mass assay of the batch is obtained. With aliquotting the only rigorous check on the uniformity of batch composition is by the assay of replicate samples; also, extrapolating from the mass contents of a small aliquot to that of a large batch is very error-prone. Aliquotting errors may be decreased by increasing the aliquot size, a process which if continued leads ultimately to the total sample for which the aliquot becomes identical with the batch (the sampling system is now on-line). Total sampling requires that internal standard be added to the entire batch, after which the entire contents of the batch tank are circulated through the sample cell. The actual pumping of an entire tank content through a small cell is, of course, feasible only with small tanks; with large tanks a stream-splitting stratagem would divert a representative sample through the cell. Solution homogeneity is verified by continuously monitoring the x-ray intensities.[2,3]

OPTIMIZATION OF THE DATA-TAKING PROCESS

The conventional "constant time method" of taking XRF data tends to weight data in proportion to their intensities. There are two shortcomings

to this method: (1) the error of ratios, with which we are concerned, is
dominated by the lower concentration element; (2) background is given
excessive weight: only in the limiting case in which the signal approaches
background level does the importance of the background measurement approach
that of the analyt. In the limiting case of zero background intensity,
it obviously would be a waste of time to measure it. When measurement
times become long, as in the accurate assay of a minor component under
high noise conditions, a more efficient data-taking strategy is required.

The error propagation equation of the ration I_a/I_s of Eq. 2 is
approximately

$$\sigma^2 = \sigma_r^2/R^2 = (\sigma_s/I_s)^2 + (\sigma_a/I_a)^2,$$

where σ_r, σ_s, and σ_a are the standard deviations of R, I_s, and I_a; σ is
thus the relative standard deviation of R.

The available counting time is used most efficiently for attaining a
targeted standard deviation if the count accumulations on each element are
such that no one measurement dominates the total error, viz:

$\sigma_s/I_s \cong \sigma_a/I_a$, which leads to

$$\sigma^2 \cong 2*\sigma_s^2/I_s^2. \qquad\qquad\text{Eq. 3}$$

For the characteristic x-ray intensity (either I_s or I_a),

$$I_s = I_g - I_b,$$

where I_g is the measured gross intensity, and I_b is the corresponding
background intensity. The error propagation equation of I_s is

$$\sigma_s^2 = \sigma_g^2 + \sigma_b^2,$$

with the σ's in intensity units. If neither error term is to dominate σ_s,
then

$$\sigma_g = \sigma_b, \text{ and}$$

$$\sigma_s^2 = 2 * \sigma_g^2.$$

σ_g, which has intensity as its dimension, is related to the relative standard deviation (σ_1) by

$$\sigma_1 = \sigma_g I_g, \text{ from which}$$

$$\sigma_s^2 = 2 * I_g^2 * \sigma_1^2$$

Substituting for σ_s^2 in Eq. 3, and rearranging terms gives

$$1./\sigma_1^2 = [(2 * I_g)/(\sigma * I_s)]^2.$$

Since counts (C) is related to relative standard deviation (σ_{rel}) by

$$C = (\sigma_{rel})^{-2},$$

C_g, the optimal required gross count accumulation of an analyt x-ray intensity as measured by an absolutely stable XRF system, becomes

$$C_g = [2*I_g/(I_s*\sigma)]^2. \hspace{3cm} \text{Eq. 3a}$$

Substitution of subscripts gives for the optimum background count associated with C_g

$$C_b = [2*I_b/(I_s*\sigma)]^2. \hspace{3cm} \text{Eq. 3b}$$

For background intensity determinations by interpolation, error propagation considerations lead to the target counts (C_x) at each of the two background positions

$$C_x = C_b*[(\phi_y - \phi_{bkg})/\Delta\phi]^2*[I_x/I_{bkg}]^2, \hspace{2cm} \text{Eq. 4}$$

where: $\phi_{x,y}$ denotes the background angles;
         ϕ_{bkg} = analyt angle;
         $\Delta\phi$ = interval between ϕ_x and ϕ_y;
         I_x = gross intensity at ϕ_{bkg};
         I_{bkg} = interpolated background intensity.

Finally, additional counts must be taken in order to compensate for instrumental variation (σ_n^2). Incorporating σ_n^2 into the count demand equation gives the adjusted target count

$$C = 1./(1./Ci - \sigma_n^2),$$ Eq. 5

where Ci is any of the counts calculated by Eqs. 3-4.

Blokhin has derived equations qualitatively similar to Eqs. 3, based
on a constant time regime by which a "limit of detectability" is defined.[4]
The above equations recognize only one limit -- that defined by the
instrumental variation which sets the lowest standard deviation attainable
with even an infinite number of counts. For all other cases, in which the
targeted standard deviation is greater than the instrumental deviation,
the assay times are calculable.

COUNTING STRATEGY

Errors of long term drift are significantly reduced by using a short
dwell time at each spectrometer setting. Simultaneous completion times
are assured by allocating the dwell time (T) at each setting such that

$$T = C*I_{max}*D/(C_{max}*I_i),$$ Eq. 6

where D is the operator-assigned dwell time for the element having the
    longest data accumulation time;

I_{max} = intensity at maximum dwell 2θ;

C_{max} = target count at maximum dwell 2θ;

I_i = intensity at ith (any member) 2θ;

C = target count at ith 2θ,

The conventional data accumulation procedure employed early in this
project, which used one uninterrupted count for each analyt or background
measurement, gave at best standard deviations of 2%. These were improved
by an order of magnitude by the short dwell strategy.

If the analyts targeted counts were accumulated as only one counting
set, the standard deviations of the gravimetric factors could be calculated only from counting statistics theory with assumed values for external
perturbations. Measured standard deviations require the accumulation of
several sets of data; thus for a minimum set N (e.g., N=10) we divide C
by N - 1, giving the targeted counts per set. For a completely unknown
sample, initial low precision intensity measurements can be used for
determination of approximate target counts. As each subsequent counting
set member is completed, the accumulated data are reentered into Eqs. 3-5,
yielding continually improving approximations to the ultimate count target.

On completing the required number of count sets, the means and standard
deviations are calculated. The assay or standardization is terminated
only when the required minimum number of counting sets, total counts, and
targeted standard deviations for the intensity ratios are all satisfied.

Times for assays with the low-power XRF system used in this study
ranged from two to thirty-two hours, which imposed severe tests on the
stability of the XRF system and on the sampling and data-taking strategies.
However, in spite of the signal-to-noise ratios of 0.1-0.2 which are inherent in very dilute SNM solutions, the system has demonstrated capability of assaying, in eight hours, low-density uranium solutions at 2 g/L
concentrations with a mass content accuracy of 0.2%.

During standardizing runs with calibrating samples, the gravimetric
factors progressively decreased as zero concentrations were approached.
After the more obvious possible error sources were eliminated, calculation
showed that biased background intensities would strongly affect the gravimetric factors at low S/N. In particular, analyt adsorbed on the boron
carbide sample cell window[5] would elevate both background and analyt readings. Background readings, being the lowest, naturally would be the most
affected. For the U vs Sr assay, the uranium was relatively more affected
than the internal standard strontium, which accounts for the observed
depression of the gravimetric factor. Examination of the cell window by
electron microprobe x-ray analysis showed scattered regions of weakly
adsorbed analyts. The results emphasize that accurate assays of very
dilute analyt demand accurate -- not merely precisely approximate --
background intensities.

When S/N became small -- less than about 0.5 -- targeted standard
deviations were reached more slowly than predicted. The most probable
source for this error is random drift in the x-ray tube voltage which,
because its effects are nonlinear and accumulate with time, is aggravated
by low S/N. This error is only partially averaged out by the short dwell
strategy, and is best reduced by improved voltage regulation of the x-ray
generator.

EFFECT OF CRYSTAL DIFFRACTING HALF-WIDTH ON ASSAY TIME

Table 1, calculated by Eqs. 4-6 using data accumulated on model systems,[6] illustrates the effect on assay time of crystal reflectivity and
diffracting half-width. The effect of reflectivity is predictably linear,
but the effect of half-width at these very low signal-to-noise ratios is

remarkable; e.g., halving the diffracting half-width (exploited by appropriate reduction of the slit widths) almost halves assay time with a
simultaneous count-rate reduction of about 25%. The attendant count rate
reduction benefits accuracy, since deadtime loss is a difficult measurement to make. Deadtime loss also adversely affects the standard deviation
since precision is determined by counts actually observed.

Table 1. Effect of Diffraction Halfwidth and Reflecting Power
          on Pu Assay Time of Modeled System.
          Targeted Standard Deviation = 0.5%
          Instrumental Standard Deviation = 0.1%

Analyzing Crystal	LiF(220)	LiF(220)	LiF(200)	LiF(200)
Halfwidth	0.27°	0.14°	0.14°	0.076°
Assay Time (hr:min)	3:40	1:55	0:36	0:20
Max Cnts/Sec	23333	17235	55151	45543
S/N	0.1	0.2	0.2	0.4

The use of monochromatic excitation could, in principle, increase
S/N by approximately an order of magnitude. Monochromatic excitation has
been demonstrated with energy dispersive XRF,[7] and with a fixed crystal
spectrometer using a position sensitive detector,[8] but its utility with a
conventional wavelength dispersive spectrometer has not been established.

CONCLUSIONS

For solution analysis "total sampling" gives total mass assays with
no need for solution density or tank volume measurements. Time savings
and standard deviations are both benefited by systematically predetermin-
ing the count requirements of analyts, standards, and backgrounds by the
use of equations based on propagation of error considerations. This
becomes quite important when assaying dilute solutions, in which the
signal-to-noise ratios of the x-ray intensities are very low. When
counting times are long, short dwell times at each spectrometer setting
significantly counteract error accumulation arising from long-term instrumental drift.

REFERENCES

1. A. H. F. von Baeckmann, D. Ertel, and I. Nueber, Determination of
    Actinide Elements in Nuclear Fuels by X-Ray Analysis, in "Advances
    in X-Ray Analysis," Vol. 18, W. L. Pickles, ed., Plenum Press, New
    York (1975).

2. C. R. Hudgens and B. D. Craft, "Feasibility Study of the Proposed Use
    of Automated X-Ray Fluorescence Analysis for Measurement of U and
    Pu in Dissolver Tanks," MLM-2533, Monsanto Research Corporation,
    Miamisburg, Ohio (September 1978).

3. C. R. Hudgens and B. D. Craft, "Demonstration of Totally Sampled
    Wavelength Dispersive XRF for Use in the Assay of the SNM Content
    of Dissolver Solutions," Proceedings of the American Nuclear
    Society Topical Meeting, Kiawah Island, South Carolina,
    November 26-30, 1979; NBS Special Publication 582.

4. M. A. Blokhin, "Methods of X-Ray Spectroscopic Research, Pergamon
    Press, New York (1965).

5. C. R. Hudgens, "Selection and Testing of Materials for Use in a Flow-
    Through Liquid Sampling Cell for X-Ray Fluorescence Analysis of
    Special Nuclear Materials in Dissolver Solution," MLM-3175,
    Monsanto Research Corporation, Miamisburg, Ohio (August 1984).

6. C. R. Hudgens, "In-Line X-Ray Fluorescence Analysis of Special
    Nuclear Materials in Dissolver Solution: Laboratory Development
    and Simulation Studies," in press.

7. T. C. Furnas and R. L. Towns, "High Intensity, Monochromatic X-Ray
    Excitation for Clinical Analysis," presented at the Pittsburgh
    Conference on Analytical Chemistry and Applied Spectroscopy,
    Cleveland, Ohio, March 3-7, 1975.

8. J. M. Keller and C. J. Sparks, Jr., "Evaluation of a Wavelength
    Dispersive X-Ray System for the Determination of Uranium and
    Plutonium in Highly Radioactive Solutions," ORNL-5971, Oak Ridge
    National Laboratory, Oak Ridge, Tennessee (September 1983).

INCIDENT AND TAKE-OFF ANGLES FOR COMMERCIAL X-RAY SPECTROMETERS

FOR USE WITH FUNDAMENTAL PARAMETER SOFTWARE

D.E. Leyden and D.B. Bilbrey

Department of Chemistry
Colorado State University
Fort Collins, CO 80523

In recent years there has been substantial development of computer programs which permit the computation of elemental concentration in a variety of samples from basic principles of X-ray absorption and emission. These programs are generally called "fundamental parameter" programs. For many years these programs required large main frame computers for execution. An example is the well known NRLXRF program.[1] For the past few years, programs have been available for minicomputers such as the PDP/11 OR LSI/11 systems. The XRF-11 program from Criss Software,[2] the SAP3 program from Batelle Pacific Northwest Laboratory,[3] and a variety of programs from the X-ray instrumentation vendors are examples. Some fundamental parameter programs have been available for several years that can be executed on microcomputers or the so-called personal computers. Some programs use an "effective wavelength" approximation of the tube output to simplify computations. Examples are a program from Scientific Microprograms [4] which runs on APPLE computers and CORSET [5] which was recently made available for use with the IBM-PC family as well as other CP/M based computers.[6] A more accurate but computationally demanding approach is to use modeling of the tube intensity distribution. Recently Tracor Xray, Inc. introduced a new program known as PC-XRF which runs on the IBM-PC family.[7]

All of these programs require as parameters the incidence and take-off angles for the instrumentation used to acquire X-ray intensities. Because these angles are not the same for all instruments, the spectroscopist must have acess to this information for the instrument in use. In most cases, to obtain the data requires contact with the vendor. Because fundamental parameter programs are clearly becoming routinely available to a variety of users, a tabulation of these data is useful. This report makes available incident and take-off angle information for instrumentation from all vendors who chose to respond to a request to provide the data. All major vendors of X-ray spectrometers in the United States for the past fifteen years were contacted at least twice with a request for this information. If data for a vendor does not appear, it is because they failed to respond to the requests. All angles are measured from the plane in which the irradiated surface of the sample is located.

Table I. INCIDENCE AND TAKE-OFF ANGLES FOR COMMERCIAL X-RAY SPECTROMETERS

VENDOR	MODEL	INCIDENT ANGLE (deg)		TAKE-OFF ANGLE
Applied Research Laboratories Inc.	8400	Goniometer	70	38
		Monochromators	70	30/40/45
	8600	Goniometer	90	35
		Monochromators	90	35 or 45
		Goniometer	90	35
		Monochromators	90	30 or 40
	72000	Scanner	90	35
		Monochromators	90	30 or 40
EG&G ORTEC	6100-6143	50		50
EDAX	EXAM II	67.5		67.5
	EXAM III & MAX	45		45
	EXAM 6(Primary)	55.5		41
	EXAM 6(Secondary)	30		41
	PV 9500	45		45
Link Systems Inc.	XR100	44		50
	XR500	41		45
	MECA 10-40	41		45
Philips	PW1400,1404 1410,1450	64		40
	PW 1600,1606	Upper Level	90	44
		Lower Level	90	29
Princeton Gamma-Tech	PGT 500	57		63
	PGT 800	57		63
	PGT 100	45		30
	PGT 400/X-pert	60		53
	PGT QXA1	57		63
Siemens-Allis	SRS 1	60		45
	SRS 200	60		45
	SRS 300	67		45
	MRS 400	90		45 or 30
Rigaku	S/MAX	63		40
	Simultix	90	Single Channel	35
			Dual Channel { Lower	30
			Upper	40
Tracor X-ray Inc.	Spectrace 420,430,431 440,450	41		45
	4020,4050,5000	45		45

ACKNOWLEDGEMENT

The authors appreciate the cooperation of the vendors listed in Table I.

REFERENCES

1. J.W. Criss, L.S. Birks, and J.V. Gilfrich, Anal. Chem., 50: 33 (1978).
2. J.W. Criss, Adv. X-ray Anal., 23: 93 (1980).
3. K.K. Nielson and R.W. Sanders, THE SAP3 COMPUTER PROGRAM FOR QUANTITATIVE MULTIELEMENT ANALYSIS BY ENERGY DISPERSIVE X-RAY FLUORESCENCE, Report PNL-4173, Pacific Northwest Laboratories, Richland, WA 99352, 1982.
4. J.C. Russ, Scientific MicroPrograms, Raleigh, N.C.
5. Tracor Xray Inc., Mountain View, CA.
6. D.A. Stephenson, Anal. Chem., 43: 1761 (1971).
7. This and other X-ray programs adapted by R.L. Martin are available from The Research Corporation/Research Software, 6840 E. Broadway Ave. Tucson, AZ 85710.

X-RAY FLUORESCENCE ANALYSIS OF WEAR METALS IN USED LUBRICATING OILS*

William E. Maddox[1] and Warren C. Kelliher[2]

[1]Murray State University
Murray, Kentucky
[2]NASA Langley Research Center
Hampton, Virginia

INTRODUCTION

Analyses of lubricating oils from aircraft engines, gear boxes and other lubricated mechanisms have been routinely performed by the military since the 1960's. The monitoring of the wear metal concentrations in the oil can lead to an early detection of abnormal wear and, consequently, the prevention of a malfunction or a complete failure of the aircraft. At the present time, almost all the analysis programs use atomic emission (AES) and/or atomic absorption (AAS) spectroscopy to determine elemental concentrations in the oils (1). These types of analysis require the close support of a laboratory to minimize the delays in obtaining the results of the measurements. The AES and AAS methods are very inefficient for particles 3 - 6 µm in size and are essentially blind to particles larger than 6 - 10 µm (2,3). Since the quantity and size of the primary wear particles increase significantly at the onset of abnormal wear (4), an engine may fail with no prior indication from the AAS and AES oil analysis.

Recently, the military has expressed an interest in developing an oil monitoring device that would be man portable and easy to operate (5,6). The instrumentation used to conduct AES and AAS are usually bulky, costly and too delicate for easy transportation. To meet the needs of a portable device, the use of x-ray fluorescence spectrometry (XRF) has been investigated by several laboratories (7,8,9). Their findings indicate this technique can be miniaturized and made into a rugged, reliable and transportable instrument. The XRF technique would also not have the particle size limitations associated with AAS and AES.

In this work, we have used the XRF technique at two different labs to monitor the oil concentrations of the principal alloying metals used in lubricated parts of aircraft engines. Oil samples are routinely taken from all aircraft operated at the NASA, Langley Research Center and are analyzed for wear metals by a laboratory using the AES technique. Over a period of three years, we have analyzed these same samples using XRF and the results are presented here with comparisons with the AES measurements.

* Work supported by NASA Contract NCC1 - 92.

EQUIPMENT AND PROCEDURE

Three different x-ray spectrometers were used in this work, two at the NASA - Langley Research Center (NASA) and the other at Murray State University (MSU). The first (NASA) was a system using ^{55}Fe, ^{109}Cd and ^{241}Am radioactive sources for excitation, the second (MSU) a polarized x-ray spectrometer (10), and the third (NASA) a direct excitation spectrometer using a molybdenum anode x-ray tube. All three spectrometers used cryogenically cooled Si(Li) detectors with resolutions approximately 160 eV.

A sample was prepared for analysis by first filtering 1-5 ml of the used lubricating oil through a 0.45 µm filtering membrane. The membrane was then analyzed by XRF to determine the amounts of various wear metals on it. Calibrations of the spectrometers were performed using the 0.45 µm filter membranes on which known amounts of finely ground metal oxides were deposited by filtration. Measurements by XRF were made on these oxide/oil mixtures before and after filtration to insure the complete removal of the suspended oxide particles by the membranes. A typical spectrum of the residue on a membrane is shown in Figure 1. This spectrum was obtained using excitation radiation from a ^{109}Cd radioactive source.

RESULTS

The NASA Langley test fleet of aircraft range from single internal combustion engine aircraft such as a Cessna 172 to multi-engine jet aircraft such as a Boeing 737. Due to mission orientation and differing maintenance schedules, the aircraft supplying the data used in this comparison study are those designated in Figures 2-8. Each sample was analyzed by AES for 12 metals and by XRF for all metals from Ti through Pb. Except for iron, which was usually present in 1-2 ppm, negligible concentrations of all metals were found in the samples from the jet aircraft. The oils from the internal combustion engines, on the other hand, showed large concentrations of a number of elements. Br and Pb were usually present in concentrations of 1000-6000 ppm, presumably as a result of the oil coming in contact with the combustion products of the leaded gasolines. Fe and Cu were the only two wear metals consistently

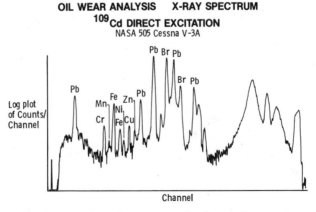

Fig. 1. Typical spectrum of wear metals in an internal combustion engine oil.

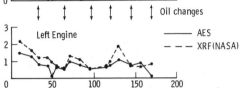

Fig. 2   Comparison of AES and XRF data for Fe in oil from Cessna
          engines. The XRF data is from the NASA Lab.

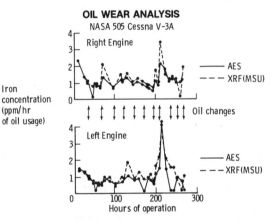

Fig. 3   Comparison of AES and XRF data for Fe in oil from Cessna
          engines. The XRF data is from the MSU Lab.

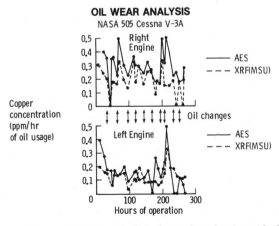

Fig. 4   Comparison of AES and XRF data for Cu in oil from Cessna
          engines.

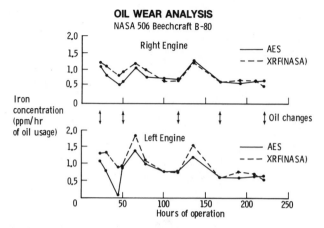

Fig. 5   Comparison of AES and XRF data for Fe in oil from Beechcraft
          engines. The XRF data is from the NASA Lab.

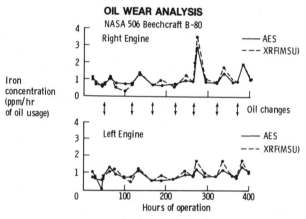

Fig. 6   Comparison of AES and XRF data for Fe in oil from Beechcraft
          engines. The XRF data is from the MSU Lab.

present in relatively large amounts in the samples. For Fe, the
concentration was 1-4 and for Cu, 0.1 - 0.5 ppm per hour of oil usage.
Fe and Cu data for oil from four internal combustion engines are shown
in Figures 2-7. For comparison, the results of atomic emission
spectroscopy (AES) measurements on the same oil samples are shown along
with the XRF data. The data labelled as MSU were taken with the
polarized x-ray spectrometer and the NASA data used direct excitation
from radioactive sources. As can be seen from the figures, the AES and
XRF measurements from both MSU and NASA are in excellent agreement. The
data from MSU repeats the NASA data and extends the analysis data for
engine hours of operation to usage as of July 1985.

    Figure 8 demonstrates a case where there was a difference between
the AES and XRF data. These data were for oil from a jet aircraft
engine. XRF direct analysis of the liquid oil itself gave a 13-ppm
concentration of iron in both the used and unused oils, indicating the
iron was in solution in the oil and was not due to frictional wear in

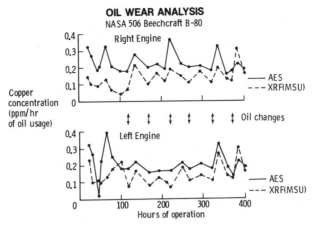

Fig. 7   Comparison of AES and XRF data for Cu in oil from Beechcraft engines.

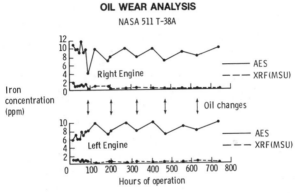

Fig. 8   Comparison of AES and XRF data for Fe in oil from T-38A engines.

the engine. The AES method gives values for total iron concentration in the oil and does not distinguish between suspended particles and oil additives. This limitation is not present in the filtration process used in this XRF work.

DISCUSSION

The analysis of used oils from several aircraft at NASA, Langley Research Center performed over a period of three years indicates XRF is as reliable and accurate as AES for determining concentrations of wear metals in the oils. XRF does not have the particle-size limitation associated with AES and AAS, and since XRF makes simultaneous measurements on all elements of interest, it can be a faster process than AES and AAS. These observations along with the fact that XRF can be easily miniaturized and made into a rugged device clearly indicate this should be the preferred method to be used for constructing a man-portable oil wear analysis instrument.

ACKNOWLEDGEMENTS

    The authors wish to thank Mr. Steave Blivin and Ms. Annette Skaggs,
undergraduate students at Murray State University, Mr. Robert Herrmann,
undergraduate student at the University of Virginia, and Mr. Marty
Baxter and the maintenance personnel of the Research Aircraft Support
Section at Langley Research Center for their assistance in this work.

REFERENCES

1)   Proceedings of the Joint Oil Analysis Program Symposium, JOAP
     Technical Support Center, Pennacola, Florida (1983).

2   Lukas, M., "Development of a Particle Size Independent Method for
     Wear Metal Analysis", JOAP International Symposium Proceedings, P.
     356, May, 1983.

3.   Rhine, W. E., Saba, C. S., and Kauffman, R. E., "Wear Metal
     Particle Detection Capabilities of Rotating Disk Emission
     Spectrometers", JOAP International Symposium Proceedings, p. 379,
     May, 1983.

4)   Peterson, M. B., "Mechanism of Wear", Boundary Lubrication, F. F.
     Ling, et al., Editors, ASME, NY (1969).

5)   Newman, R. W., Niu, W. H., and O'Connor, J. J., "Development of a
     Portable Wear Metal Analyzer", JOAP International Symposium
     Proceedings, p. 337, May, 1983.

6)   Clark, B. C., Woerdeman, V. P., Thornton, M. G., Cook, B. J.,
     Centers, P. W., and Kelliher, W. C., "A Portable X-Ray Analyzer for
     Wear Metal Particles in Lubricants", Proceedings of the 36th
     Meeting of the Mechanical Failures Prevention Group (Scotsdale, AZ
     1982), Cambridge University Press.

7)   Parker, L. L., and Golden, G. S., "X-Ray Wear Metal Monitor", JOAP
     International Symposium Proceedings, p. 118, May, 1983.

8)   Vienot, D. E., "X-Ray Fluorescence Spectrometric Analysis of Wear
     Metals in Used Lubricating Oils", JOAP International Symposium
     Proceedings, p. 142, May, 1983.

9)   Clark, B. C., et al., "An Instrument for Oil Wear Metal Analysis by
     X-Rays (OWAX)" JOAP International Symposium Proceedings, p. 162,
     May, 1983.

10) Maddox, W. E., "Application of a Polarized X-Ray Spectrometer for
     Analysis of Ash From a Refuse-Fired Steam Generating Facility",
     Advances in X-Ray Analysis, Vol. 27, pp. 519-526, 1984.

DETERMINATION OF WEAR METALS IN OIL BY X-RAY SPECTROMETRY

Ya-Wen Liu, A.R. Harding and D.E. Leyden

Department of Chemistry
Colorado State University
Fort Collins, CO 80523

ABSTRACT

Trace elements in oil may be determined by adsorption of the oil sample onto magnesium oxide followed by thermal degradation of the organic material. The resulting powder is easily pressed into a pellet suitable for X-ray spectrometric analysis. The lower limit of detection depends upon the trace impurities in the MgO and is a few parts per million for most elements determined.

INTRODUCTION

Determination of metals in oil is an analytical problem known world wide. A number of investigations of trace element determinations have been reported.[1-4] There are a variety of reasons for performing these determinations. One need is to monitor engine performance by the determination of wear metals in lubricating oils taken from engines in use or under test. Elements such as As, Se and Pb may poison catalytic converters and these and other elements are potentially harmfull to the enviroment. These determinations rely heavily on atomic absorption and inductively coupled plasma optical emission spectrometry.[5] Atomic absorption methods require tedious digestion procedures and inductively coupled plasma emission methods are hampered by problems with viscosity of the oil samples.

X-ray spectrometry should be a useful method for the determination of metals in oil. For the most part, the elements of interest lie in the practical range of atomic number for determinations by X-ray spectrometry. Use of energy dispersive X-ray spectrometry provides the capability for simultaneous determination of several elements. However, like other methods X-ray spectrometry has limitations. In the case of the determination of metals in oil, the limitations are primarily lack of adequate detection limit and matrix effects due to the composition of the oil including metal content as well as the content of carbon, hydrogen and sulfur. As a working figure, most transition elements may be determined in the range of a few parts per million or greater in a matrix such as a hydrocarbon. Further improvement will require some type of preconcentration technique. The matrix effect problem may be approached in two ways. The sample may be prepared as an "infinitely thin" specimen in which matrix effects are minimized, or some type of data treatment using computer software may be employed. Kubo, Bernthal and Wildeman[6] prepared specimens by pipeting 0.05 to 5.0 mg of oil sample spiked with a sulfuric acid solution of Rh, or Rh and Cr as an internal standard onto Formvar film. The two element spike was required for viscous samples which did not flow into films on the Formvar sufficiently thin to ignore absorption by the sample matrix. The technique was tested on only two types of samples only one of which was a standard reference material. The results that were obtained were good, however, the major deficiency in the approach is the long counting times of several hours required for acceptable X-ray statistics.

Giauque, Garrett and Goda[7] used the incoherent scatter of the source X-rays from a Mo X-ray tube as an "internal standard" following standardization of the instrumentation with solutions of dissolved metals. Specimen preparation probably resulted in a thickness intermediate between infinitely thin and infinitely thick specimens. Results reported for NBS SRM 1634 fuel oil were not acceptable for most applications.

Christensen and Agerbo[8] employed a combination of secondary target excitation and quantification calculations based on the Fundamental Parameter approach[9] for the determination of sulfur and heavy metals in crude oil and petroleum products. Samples were analyzed as liquids placed in commercial specimen cups with liquid depths of approximately 2 cm. X-ray counting times ranged from 100 sec. to 1000 sec., and a variety of secondary targets were required. This work presented a methodical investigation of the effect of the carbon to hydrogen ratio in the matrix. However, results were presented for only S, V and Ni. The V and Ni results were compared to another X-ray spectrometric determination and instrumental neutron activation analysis.

Sanders et. al[10] employed a combination of the measurement of the scatter of source radiation and fundamental parameter approach[11] for the determination of heavy elements in unweighed oil samples. Specimen preparation consisted of pipeting sufficient sample into a commercial liquid specimen cup to assure that the bottom of the cup was covered. Ti and Zr secondary target sources were used sequentially. The Ti excited spectrum is collected in a helium atmosphere and is used for elements in the Al-Ca range. The Zr source is used for elements in the Ti-Pb range. The two spectra are stripped of the unused portions and added to obtain a composite spectrum. The Zr radiation scatter is used in conjunction with first approximations of the heavy element content obtained from fluorescencent intensities to obtain the mass of the specimen. A computer program SAP3 was used to perform the computations.[12] Results for sulfur in NBS oils were reported and were good. However, few analysts report significant problems in the determination of sulfur in oil by X-ray spectrometry. The results of trace element determination were reported for only one NBS oil standard reference material. Although the results were good, the test did not prove the method applicable to a variety of oil types. The SAP3 program is not readily available to users in a well documented form and requires a specific and relatively expensive minicomputer.

An alternative approach to those given above is to prepare a sample which has an essentially constant matrix to minimize matrix effects, and to make the sample infinitely thick so that sample mass is not a issue. Adsorption of oil onto an inert substrate with subsequent thermal degredation of the organic material is the approach reported in this paper.

EXPERIMENTAL

Apparatus

All X-ray measurements were performed using a Tracor X-ray Inc. (Mountain View, CA) Spectrace 440 equipped with a Tracor Northern (Madison WI) TN 2000 analyzer. A Mo X-ray tube operated at 25kV and 0.30 mA with a 0.05 mm Mo primary filter was used as the X-ray source. The detector employed is a 10 mm^2 Si(Li) semiconductor detector with a resolution of 155 ev FWHM at 5.9 keV. X-ray emission intensities were extracted from the spectra using the Tracor 'XML' spectrum fitting software. Counting times were 100 s. (livetime) for elemental concentrations above 100 ppm, and 500-1000 s. for concentrations below 100 ppm.

Materials

Magnesium oxide used as the substrate was either J.T. Baker Analyzed, or Aldrich Chemical Co. Gold Lable (99.999%). Standards were CONOSTAN (Conoco Oil) organometallic standards. Test samples for the results shown in Table 2 were prepared analytically by addition of weighed amounts of CONOSTAN standards to a base oil stabilizer also obtained from Conoco. These samples were prepared by a colleague and the values remained unknown to the analyst until after the test runs were completed.

Specimen Preparation

Specimens were prepared by thoroughly mixing 1.1 g of MgO powder with 1.3g of oil in a crucible. If necessary, the oil was preheated to 60-70 °C to lower the viscosity. The mixture was heated to 260-270 °C for 20 min. with occasional stirring on a hot plate in a fume hood. The powder was removed and pressed into a 2.5 cm diameter pellet using 28,000 psi pressure. The pellet was placed in a sample cup (SPEX Industries) and supported using Kapton film (SPEX Industries).

RESULTS AND DISCUSSION

An investigation of the loading of the MgO substrate showed that the amount of oil that could be adsorbed was approximately equal in weight to the amount of substrate. Because the lower limit of detection was an issue, the maximum reliable ratio of oil to substrate was sought. An additional factor to consider is that the final specimen was to be "infinitely thick" for the most energetic X-ray measured. A convenient test was Zn which showed that this criterion was met if greater than 0.5 g of MgO were used. A compromise sample preparation procdure was to add 1.3 g of oil sample to 1.1 g of MgO. If a greater ratio of oil to MgO was used, the oil separated from the substrate and adhered to the wall of the crucible.

It is possible to add oil to the substrate repeatedly after each heating. As many as five additions were performed in this way to improve the detection limit by preconcentration of the trace elements from the oil onto the substrate. The practical upper limit of repeated addtions was not determined.

Heating time was found not to be a critical parameter. If the oil-substrate mixture is occasionally stirred, after approximately 15 minutes the mixture becomes an almost characteristic gray-beige color. After 20 minutes, an investigation showed that the mean percent weight

loss of the organic material is 97.6±0.2%. Further heating increased
the weight loss to near 100% of the weight of the organic material with
no measurable affect on the results. Furthermore, it was found that if
the specimen was not heated beyond 20 minutes, the stability of the
resulting pellet was improved. This may be a result of a small amount
of residual organic material acting as a binding agent for the MgO.

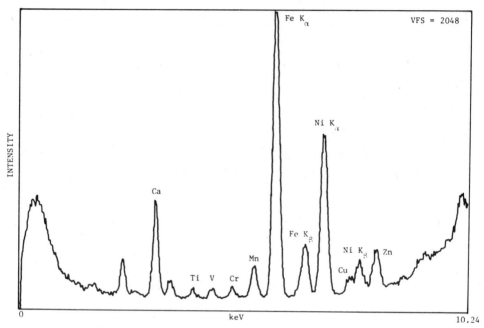

Figure 1. X-ray emission spectrum of an oil sample.

Although many variations in the trace element contents of oils are
encountered, Figure 1 shows as typical a X-ray spectrum of an oil sample
as one can select. In this sample, qualitative identification of Ca,
Ti, V, Cr, Mn, Fe, Ni, Cu, and Zn are obvious. In other samples,
additional elements may be observed. Concentrations standards of 100,
500, 1000, 1500 and 2500 ppm were prepared for V, Cr, Mn, Fe, Ni, Cu,
and Zn using the CONOSTAN standards and the sample preparation method
described. Plots of X-ray intensity versus elemental concentration
resulted in excellent working curves. Figure 2 shows the working curve
for nickel which is representative of all others. Table 1 contains the
statistical parameters for linear regression fits to these data.
Replicates determinations performed on all standards showed that
relative standard deviations of less than 4.5% were obtained for all
elements except V which gave 8.5%.

To estimate the lower limit of detection for the elements
investigated, a set of replicate blank specimens were prepared from the
two types of MgO available. From the standard deviation of the x-ray
intensities for the element peaks in these blank specimens and the slope
of the working curves given in Table 1, the lower limit of detection was
estimated as that concentration which would give a signal three times
the standard deviation of the intensities of the blank. The results are
shown in Table 2. The difference is due to the impurities in the MgO
from J.T. Baker which give rise to a high background signal, especially
for iron.

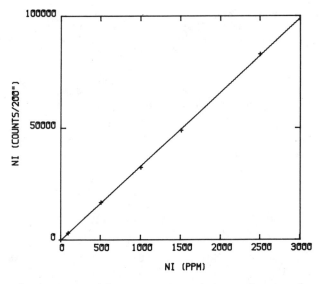

Figure 2. Working curve for Ni determination in oil.

TABLE 1 Statistical parameters for the linear regression fits of
standards data

Element	Slope+SD (Cts./200S./ppm)	Intercept+SD	Correlation Coefficient
V	2.75+0.02	72.6+34.2	0.99988
Cr	5.45+0.02	26.5+21.5	0.99999
Mn	8.27+0.05	98.3+63.3	0.99995
Fe	14.32+0.20	-59.5+283	0.9997
Ni	33.15+0.34	-366 +474	0.99984
Cu	41.93+0.03	-599 +421	0.99992
Zn	65.91+0.68	-1009 +958	0.99984

Table 2. Comparison of the lower limit of detection for two sources
of MgO

	LLD (ppm)						
	V	Cr	Mn	Fe	Ni	Cu	Zn
J.T. Baker	3.6	4.0	7.6	11.5	1.7	1.0	0.7
Aldrich Gold	2.8	2.0	1.5	3.8	0.6	0.9	0.7

To be certain that the recovery was high for the trace elements, several samples were analyzed by standard additions and compared with the results from the working curve method. Four additions were used in the standard additions method and plots of the intensity versus concentration added were excellent. The largest difference between the results from the standard additions method and the working curve method was 6% relative error. This was taken as evidence that the sample preparation method gave consistent recovery.

Table 3. Results of analysis of test samples prepared from standards (ppm)

No.	V	Cr	Mn	Fe	Ni	Cu	Zn
1	300+15 (304)	310+12 (309)	398+9 (408)	620+5 (612)	107+3 (108)	1220+12 (1220)	1048+3 (1048)
2	400+17 (393)	200+15 (197)	950+10 (981)	428+11 (411)	1009+3 (1013)	120+6 (120)	515+8 (515)
3	203+15 (210)	1995+14 (1981)	205+12 (205)	200+5 (199)	594+7 (591)	395+2 (398)	120+3 (114)
4	1010+29 (1010)	420+13 (403)	110+4 (125)	1200+5 (1202)	295+7 (294)	202+5 (206)	403+5 (410)
5	705+14 (701)	580+19 (568)	480+9 (504)	105+2 (102)	998+12 (1001)	253+5 (251)	530+4 (529)
6	140+7 (159)	63+8 (64)	14+1 (20)	164+1 (185)	47+2 (46)	30+1 (33)	60+1 (65)
7	28+4 (34)	307+8 (320)	33+2 (33)	23+1 (32)	95+1 (96)	63+1 (64)	16+1 (18)
8	98+6 (109)	83+4 (89)	70+2 (78)	ND (16)	156+1 (154)	36+1 (39)	80+1 (82)

The precision is reported as the standard deviation for five replicate specimens. The values in parentheses are the concentrations (ppm) calculated from the analytical preparation.

The evaluation of the method was to determine V, Cr, Mn, Fe, Ni, Cu and Zn in a suite of test samples which were prepared by dissolving CONOSTAN standards in a base oil. The results of these determinations are shown in Table 3. With the exception of vanadium, the precision of the determinations is better than 10% in all cases. The relative error between the value reported for the preparation and the results is also less than 10% in most cases. These results demonstrate the reliability of the working curve approach, and that the detection limit achieved is suitable for wear metals in oil.

One of the factors which can effect results of trace element determination by X-ray spectrometry performed directly on oil specimens is the carbon to hydrogen ratio. Oil samples of different carbon to hydrogen ratios were identified and analyzed. These were a "gas oil" which is a high viscosity oil, an SAE 10W and an SAE 30W oil. A set of standards were prepared from mixtures of toluene and cyclohexane. A plot of the ratio of the elastic (Rayleigh) to inelastic (Compton) scatter intensity of the Mo K-line from the X-ray tube versus the carbon to hydrogen ratio gives a smoothly increasing curve. The curve may be used to estimate the carbon to hydrogen ratio in oil. The samples chosen ranged in carbon to hydrogen ratio from approximately 9.3 to 8.0. No significant difference was observed between results obtained using the working curve method for samples of these oils which had been doped to 10, 30 and 50 ppm of the elements determined using the CONOSTAN standards.

CONCLUSIONS

Use of MgO as a substrate for the adsorption of oil provides a simple and rapid specimen preparation technique for the determination of trace elements suche as wear metals in oil. Successive adsorption of oil followed by heating provides for preconcentration of the trace elements. Detections limits are a few ppm, and the accuracy and precision on the order of 5-10%. The method appears to be independent of the matrix composition of the oil sample.

This research was supported in part by a gift from Tracor Xray Inc., Mountain View, CA.

REFERENCES

1.   J.H. Runnels, R. Merryfield, and H.B. Fisher, Anal. Chem., 47:1258 (1975).
2.   H.E. Knauer and G.E. Millman, Anal. Chem, 47:1263 (1975).
3.   W.K. Robbins and H.H. Walker, Anal. Chem., 47:1269 (1975).
4.   H.H. Walker, J.H. Runnels and R. Merryfield, Anal. Chem., 48:2056 (1976).
5.   V.A. Fassel, C.A. Peterson, F.N. Abercrombie and R.N. Kniseley, Anal. Chem., 48:516 (1976).
6.   H. Kubo, R. Bernthal and T.R. Wildeman, Anal. Chem., 50:899 (1978).
7.   R.D. Giauque, R.B. Garrett and L.Y. Goda, Anal. Chem., 51:511 (1979).
8.   L.H. Christensen and A. Agerbo, Anal. Chem. 53:1788 (1981).
9.   C.J. Sparks Jr., Adv. X-ray Anal., 19:19 (1976).
10. R.W. Sanders, K.B. Olsen, W.C. Weimer and K.K. Nielson, Anal. Chem., 55:1911 (1983).
11. K.K. Nielson, Anal. Chem., 49:641 (1977).
12. K.K. Nielson and R.W. Sanders, "The SAP3 Computer Program for Quantitative Multielement Analysis by Energy Dispersive X-Ray Fluorescence"; Pacific Northwest Laboratory Report to U.S. Department of Energy, PNL-4173, Richland, WA, 1982.

DETERMINATION OF NICKEL AND VANADIUM IN CRUDE OIL USING WAVELENGTH
DISPERSIVE X-RAY FLUORESCENCE SPECTROMETRY AND A TIGHT STANDARDIZATION
METHOD *

S. Mansour, F. Abu-Dagga and R. Sabri

Kuwait Institute for Scientific Research
P.O. Box 24885, Safat, Kuwait

INTRODUCTION

Metallic impurities present in crude oils lead to the rapid deactivation of catalysts used in refineries and a drop in the yield of oil
products. The determination of these impurities is therefore of primary
importance. The conventional and most widely used methods to determine
trace metals in oils and petroleum products are chemical analysis methods
such as atomic absorption spectrometry (AAS) and inductively coupled plasma-
optical emission spectrometry (ICP-OES). Sample preparation for these
methods includes decomposition of the crude sample by dry ashing and combustion of the oil sample followed by chemical treatment of the dry ash
residue (ASTM, 1983; Fassel et al., 1976). Major drawbacks of chemical
analysis methods are the length of time required to prepare the sample and
the possibility of losing trace metallorganic compounds in the form of
volatile compounds.

X-ray fluorescence spectrometry, on the other hand, is a non-destructive technique and requires minimum sample preparation. Furthermore, in
oil samples, the net impulse rate of the characteristic x-ray radiations
of the analyte is generally high since the matrix is composed mainly of
hydrocarbons, which results in the detection of parts-per-million (ppm)
concentration levels.

Recently, there have been a number of studies on the analysis of oils
and petroleum products for trace elements using energy dispersive x-ray
fluorescence spectrometry (EDS)(Kubo et al., 1978; Sanders et al., 1983;
Christensen and Agerbo, 1981). In 1974, wavelength dispersive x-ray
fluorescence spectrometry (WDS) was approved as an ASTM standard method for
determining Ni and V in a broad range of hydrocarbons (ASTM, 1983). The
ASTM method and the other EDS methods use different standardization
techniques to compensate for absorptions due to the sample matrix. (Kubo
et al., 1978; Sanders et al., 1983; Christensen and Agerbo, 1981). The
ASTM method determines the Ni and V traces by comparing the respective
corrected x-ray line intensities of the analyte with those obtained in
heavy mineral oil synthetic standards (ASTM, 1983).

---

* Kuwait Institute for Scientific Research Publication No. KISR-1835,
Kuwait.

Table-1. Composition Variation among Three Kuwaiti Crudes

Sample	C/H ratio	S(%)	N(%)
PEF—14 (light crude)	6.5	3.3	0.11
PEF—15 (medium crude)	6.9	4.2	0.17
PEF—16 (heavy crude)	7.5	5.2	0.2

The formula used to correct the x-ray line intensities does not account for the primary absorptions by the sample matrix (ASTM, 1983). Thus, if the analyte composition is different from that of the heavy mineral oil, absorption of primary x-rays will be different and comparing line intensities to calculate concentrations leads to inaccurate results. A comparison of the major matrix components of three Kuwaiti crude oils (Table 1) shows wide variations in composition. Thus, primary absorptions are different from one crude to another and, of course, different from that for mineral oil (C/H=6).

In this article, a rapid and simple method to determine Ni and V in crude oils is described. The method is designed to suit compositional differences of crude oils. It uses wavelength dispersive x-ray fluorescence spectrometry in combination with a standardization method that minimizes the interferences from the sample matrix by diluting the sample in a solvent to the extent that its matrix becomes similar or close to the solvent matrix. The x-ray line intensities of the method are then compared with calibration standards. The method was tested by analysing three Kuwaiti crudes and one EPA reference crude oil. The results obtained were in good agreement with quoted results using other analytical methods.

EXPERIMENTAL

Selection of Solvent

Organic solvents such as toluene, MIBK, petroleum spirit, xylene and lubricating oil were tested for their miscibility with crude oil by diluting one weight of heavy crude oil in one weight of the solvent, and transferring

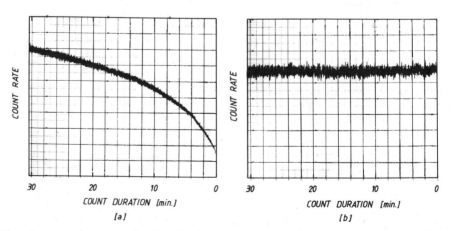

Fig. 1. Counting rate for V Kα line versus duration (a) crude oil diluted by toluene and (b) crude oil diluted by lubricating oil.

an amount of the solution to an x-ray spectro cup. The cup was placed in the x-ray spectrometer, and the pulse rates of Ni and V over a period of time were recorded separately with a chart recorder.

The test of V indicated that crude oil diluted with toluene precipitates at the bottom of the spectro cup with time, resulting in higher counting rates as time increases (Fig.1(a)). Dilution with MIBK or petroleum spirit produced similar profiles, and tests for Ni showed similar behaviour with these solvents. It may be concluded that toluene, MIBK, and petroleum spirit are not miscible with crude oil. On the other hand, the test for V when lubricating oil was used as a solvent shows a constant counting rate over time, indicating that it is miscible with crude oil (see Fig. 1(b); similar behaviour was obtained when the crude was diluted with xylene). Since xylene is volatile, lubricating oil was chosen as a solvent to dilute crude oil samples for this study.

Preparation of Standard Solutions

Synthetic nickel and vanadium calibration standards were prepared using a modified NBS procedure (NBS, 1967): 0.270 g of nickel cyclohexanebutyrate was weighed and transferred to a 100 ml Erlenmeyer flask; 3 ml xylene and 5 ml of 2-ethylhexanoic acid were added, and the flask was heated on a hot plate until a clear solution formed. Enough lubricating oil was added to bring the total weight of the contents of the flask to 75 g which gives a Ni stock solution of 500 ppm. A 100 ppm solution was prepared by dilution. Sub-standards in the concentration range 0-20 ppm were also prepared by dilution.

Similarly, 0.288 g of bis(1-phenyl-1,3-butanediono) oxovanadium (IV) was weighed and transferred to a 100 ml flask. Then, 3 ml each of xylene, MIBK, toluene and 2-ethylhexanoic acid were added, and the flask was heated until a clear solution formed. Enough lubricating oil to bring the total weight of the contents of the flask to 75 g was added. Vanadium sub-standard in the concentration range 0-20 ppm were prepared by dilution.

Calibration standards were prepared for XRF analysis by transferring 5 ml of each sub-standard to x-ray spectro cups*. The cups were covered with 6.3 μ x-ray Mylar film** and sealed by a ring supplied with the spectro cups. An exact control was maintained on standards volume to insure a constant solution thickness.

Sample Dilution

Crude oil sample (PEF-14, 15 and 16 and the EPA reference standard) were prepared in triplicate by mixing a known weight of the sample with a known weight of lubricating oil. The dilution ratios at which the crude oil samples were diluted depend on the density of the crude oil sample. The diluted samples were then prepared for x-ray analysis in a way similar to that described for preparing standard solutions.

Apparatus

A Philips PW 1410 x-ray fluorescence spectrometer was used throughout this study. The spectrometer was operated under He atmosphere for both Ni and V. The spectrometer measuring conditions are given in Table 2.

---

*   Spex Industries Inc., Cat No. 3529
** Chemplex Ind. Inc., Cat. No. 250

Table 2.   Instrumental Conditions for a Philips PW 1410 Spectrometer
           Operated under Helium Atmosphere

Analyte	Line	Tube	Tube power (kV/mA)	Al-filter	Colli-mator (mm)	Crystal	Detector	Pulse Height Selection		Counting time (s)
								LL	W	
Ni	Kα	Cr	40/50	Y	550	LiF 200	GFPC	340	140	40
V	Kα	W	60/30	N	550	LiF 200	GFPC	300	160	20

The objectives followed when selecting conditions were to optimize the
peak to background ratio of the analyte x-ray line and to avoid overlapping
between the x-ray lines with the tube excited lines. To optimize Ni measuring conditions, a Cr tube was used instead of a W tube because the Ni Kα
line (7.472 keV) overlaps with W tube L_ℓ line (7.384 keV). The tube was
operated at 40 kV because higher voltages result in a significant increase
of the background continuum. To further suppress the background an Al filter
was placed between the tube and the sample.

The V Kα line was excited using a W tube at 60 kV excitation voltage.
Unlike the Ni Kα line which was excited only by the background continuum,
the V Kα line was excited by both the continuum and the tube x-ray lines.

RESULTS AND DISCUSSION

Calibration of the Spectrometer

The instrumental conditions quoted in Table 2 were used to measure the
Kα line intensities for both Ni and V. The net intensity of the Ni Kα x-ray
line was measured by recording the line intensities at 2θ angles of 48.67

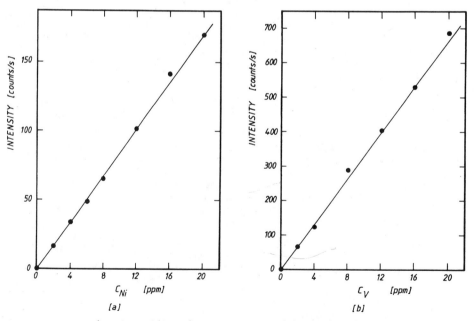

Fig. 2. Calibration curves for (a) Ni and (b) V.

Table 3. Background Intensities, Sensitivities and
Calculated Detection Limits for Ni and V

Analyte	Background (C/S)	Sensitivity (C/S)	Detection Limits (ppm)
Ni	89	9	0.5
V	350	34	0.37

deg, 47.80 deg and 50.00 deg. The average of the intensities measured at
47.80 deg and 50.00 deg (i.e., the background counts under the x-ray peak)
was then subtracted from that measured at 48.67 deg. This was repeated for
each of calibration standards and the blank.

Fig. 2(a) shows the calibration curve obtained by plotting the Ni Kα
line net intensity against the Ni concentration (C_{Ni}). A linear regression
program was used to fit the points and to calculate the spectrometer sensitivity (C/S/ppm).

In a similar way, the V Kα line intensity was measured by recording
the line intensities at 2θ angles of 76.14 deg, 76.00 deg and 78.00 deg.
The average of the background counts measured at 76.00 deg and 78.00 deg
was then subtracted from the line intensity measured at 76.94 deg. Fig.
2(b) shows the net intensity of V Kα line plotted against the V concentration
(C_v). Sensitivity (C/S/ppm) was determined after a linear regression fit.

The minimum detection limit (MDL) for both Ni and V was calculated
using the measured sensitivities. The MDL (in ppm units) equals $3\sqrt{R_b/T_b}/m$,
where R_b is background counts (C/S) under the x-ray line, T_b is the counting
time (S) and m is the sensitivity (C/S/ppm)(See Table 3).

The detection limit for Ni was ultimately improved by using an Al filter
between the tube and the sample, and a Cr anode x-ray tube and by a proper
selection of pulse height and other instrumental conditions. Minimum limits
of 2 ppm and 0.9 ppm were achieved using a Mo tube operated at 50 kV and 40
mA, and a W tube operated at 60 kV and 30 mA and an Al filter, instead of a
Cr tube, respectively.

The Effect of Sulfur

The method described here is a comparative method in which the Ni and
V concentrations in the sample were determined by comparing their net intensities with the corresponding intensities of the calibration standards.

The matrix composition of samples and standards are different. Lubricating oil consists of hydrocarbons only with C/H=6, whereas, the matrix of
the samples consisted of hydrogen, carbon, sulfur, nitrogen and other traces.
As a result, matrix correction for the x-ray line intensities of the analyte
must be applied.

The basic empirical relationship used to correct the measured intensities $I_{meas.}$, in relation to the elemental weight fraction, W_i of the sample
is given by Eq.(1)(ASTM, 1983). This relationship is valid only when
primary absorptions in both samples and standards are similar, which was
achieved in the present study by extensive sample dilution.

$$I_{corr} = I_{meas} \frac{[\mu_C(1-W_S-W_N-W_H)+\mu_S W_S+\mu_N W_N+\mu_H W_H]-I_{blk}(\mu_C W_C+\mu_H W_H)_{std.}}{(\mu_C W_C+\mu_H W_H)_{std.}} \qquad (1)$$

Table 4. Results for Three Kuwaiti Crudes and One EPA Reference Standard
Crude Oil Obtained by XRF in this Work and by Other Methods

Sample	API Gravity (degrees)	Dilution Ratio (D)[a]	Ni conc. (ppm)		V. conc. (ppm)	
			XRF Analysis (this work)	Quoted Values	XRF Analysis (this work)	Quoted Values
PEF—14 (light crude)	31.2	5.0	8.8 ± 0.26	8.4 ± 1.4[b]	28.3 ± 0.73	30.8 ± 1.0[b]
PEF—15 (medium crude)	23.9	7.5	24.2 ± 0.98	19.9 ± 1.1[b]	40.1 ± 0.85	40.4 ± 2.2[b]
PEF—16 (heavy crude)	14	12.0 – 18.0	29.3 ± 0.32	24.0 ± 2.3[b]	105.2 ± 1.02	110.3 ± 2.5[b]
Prudhoe Bay Crude	26.8	6.5	11.3 ± 0.30	11.00[c]	22.2 ± 0.86	21.00[c]

[a]Dilution Ratio=sample wt + solvent wt/sample wt. [b]Measured by ICP-OES. [c]EPA reference values determined by ASTM standard methods.

W_S, W_N, W_H are weight fractions of sulfur, nitrogen and hydrogen, respectively. μ_C, μ_S, μ_N and μ_H are mass absorption coefficients of carbon, sulfur, nitrogen and hydrogen, respectively. I_{blk} is the blank intensity, which is close to zero.

For a diluted sample with dilution ratio D:

$$I_{corr} = I_{meas} \frac{[\mu_C(1-W_S') + \mu_S W_S']}{\mu_C} \tag{2}$$

where $W_S' = W_S/D$

$$I_{corr}]_{Ni} = I_{meas} [(1-W_S') + \frac{112}{5.4} W_S'] \tag{3}$$

$$I_{corr}]_V = I_{meas} [(1-W_S') + \frac{342}{19.2} W_S'] \tag{4}$$

The μ_S/μ_C for the Ni Kα line is 112/5.4 and is 342/19.2 for the V Kα line.

Depending on the density of the sample, samples were diluted, in the range of 5-18 (Table 4). Differences in primary absorptions were thus minimized. The nitrogen effect was also minimized and ignored. The hydrogen effect was further ignored because its mass absorption coefficient (μ_H) is 0.45 and 0.57 (cm²/g) for Ni and V Kα x-ray lines, respectively, whereas corresponding values for carbon are 5.4 and 19.2 [Eq. (2)]. Standards were thus assumed to be primarily composed of carbon (Christensen and Agerbo, 1981) and samples were assumed to be composed of carbon and sulfur (ASTM, 1983). Self-absorptions and effect of other traces in samples were also omitted. Eqs. (3) and (4) represent the approximate matrix correction required for the analysis of Ni and V in crude oil samples, respectively.

The influence of sulfur on Ni and V determination was experimentally investigated using the PEF-16 crude oil samples with 5.2% sulfur. The sample was diluted at different dilution ratios and Ni and V concentrations were determined. Fig. 3 shows the relative concentrations of C/C_0 for Ni and V at different dilution ratios. C_0 was considered to be the corrected concentration obtained at the highest dilution ratio value. The lower part of the curves represent the relationship before the matrix correction was applied (C_{meas}/C_0 Vs. D); the upper parts represent the relationship after the matrix correction (C_{corr}/C_0 Vs. D).

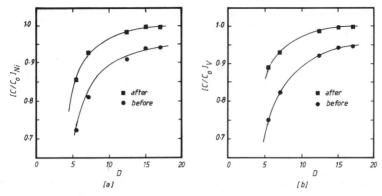

Fig. 3. Relative concentrations (C/C$_0$) before and after correction
versus dilution ratios (D) for (a) Ni and (b) V.

Two major conclusions can be made on the relationship with the sulfur
content of the sample. First, there is a significant drop in the analyte
concentration as a result of the sulfur present in the sample. Second, the
magnitude of the sulfur effect depends on the sulfur concentration in the
sample and the rate of S and C mass absorption coefficients (μ_S/μ_C).
Therefore, it becomes clear that matrix correction for S content is a
necessity.

Effect of Dilution Ratio

The effect of the dilution ratio on Ni and V determination in crude oil
could be demonstrated using Fig. 3. It is shown that the concentration
increases as the dilution ratio increases. Further, this increase is
observed after matrix correction for S content.

That the corrected concentrations vary with dilution ratio indicates
that matrix components other than sulfur are important. However, as the
dilution ratio increases, the variation in C_{corr}/C_0 becomes less and is
expected to disappear as D value increases over a critical D value. Hence,
by sample dilution above this critical value the matrix differences between
samples and calibration standards were minimized.

It was mentioned earlier that the dilution ratio used in this study to
dilute different samples depend on sample density. The higher the API
gravity*, the lower the dilution ratio needed to bring the sample matrix
close or similar to that of calibration standards. This reciprocal relationship was utilized to estimate the proper dilution for any sample. Empirically, this could be presented by:

$$D = \frac{K}{API\ gravity} \tag{5}$$

where K is a constant that could be determined from Fig. 3. To determine K,
it is necessary to determine the dilution ratio at which matrix effects are
minimal.

The plateau curve obtained in Fig. 3 at dilution ratios higher than 12
indicates a constant corrected concentration regardless of D value which

---

* API (American Petroleum Institute) gravity $\equiv \dfrac{141.5}{specific\ gravity} - 131.5$

means that other matrix components such as the C/H ratio and N content have
minimum effects on analysis. D values less than 12, however do not eliminate
matrix effects completely. This D value was thus considered critical for
diluting the PEF-16 crude oil sample (API gravity = 14 degrees) and the K
value was thus estimated using Eq. (5) to be 170 degrees. Accordingly, the
PEF-14, PEF-15 and Prudhoe Bay crude samples were diluted with 5, 7.5 and
6.5 dilution ratios, respectively (Table 4).

Analysis of Crude Oil Samples

Four different crude oil samples, PEF-14 (light crude), PEF-15 (medium
crude), PEF-16 (heavy crude) and Prudhoe Bay reference crude, were prepared
in triplicate by dilution in lubricating oil and were analysed by x-ray
fluorescence (XRF) as was described. The results are presented in Table 4
which also gives the values obtained by other analytical techniques.
Although the results for V obtained by XRF are in good agreement with those
obtained by the ICP-OES method, the results obtained for Ni are slightly
higher, except for the Prudhoe reference standards. One possible explanation
for the higher Ni results is the possible loss of Ni during sample preparation for ICP-OES analysis. This is one major drawback to chemical analysis
methods.

To assess the reproducibility of the sample preparation procedure and
the intensity measurement, samples were prepared in triplicate. The
estimated precision (1σ) obtained is 1-4%. In general, the results obtained
by XRF are in good agreement with values quoted by other analytical techniques and the reproducibility of the sample preparation is better than the
reproducibility of those for chemical analysis methods.

CONCLUSION

The method described here is a rapid, simple, accurate and reproducible
method to determine Ni and V in crude oil samples. The basic principle
applied in this study to minimize the matrix absorption effects was dilution
of the sample in lubricating oil to the degree that the sample has nearly
the same mass absorption coefficient as the calibration standards. For crude
oil samples with high sulfur content, a matrix absorption correction is
necessary for sulfur.

The method described was designed mainly for Ni and V in crude oils.
Other trace elements could also be analysed if their concentrations in the
diluted sample are higher than their MDL. Whether this method could be
used to determine trace metals in other petroleum products depends mainly
on the product matrix composition compared with the lubricating oil matrix
and on the concentration level of traces in the sample.

Acknowledgement

We thank the Petroleum Technology Department for providing the samples
and Dr. Abdul Aziz Inayatullah, QC/QA Unit Head in the Central Analytical
Laboratory for monitoring the results obtained.

REFERENCES

1. ASTM, 1983, Analysis of selected elements in waterborne oils, Water
        and Envir. Tech., Section II, 263.

2. Christensen, L. H., and Agerbo, A., 1981, Determination of sulfur and
    heavy metals in crude oil and petroleum products by energy disper-
    sive x-ray fluorescence spectrometry and fundamental parameter
    approach, Anal. Chem., 53, 1788.
3. Fassel, V. A., Peterson, C. A., Abercrombie, F. N., and Kaiseley, R. N.,
    1976, Simultaneous determination of wear metals in lubricating oils
    by inductively coupled plasma atomic emission spectrometry, Anal.
    Chem., 48, 516.
4. Kubo, H., Bernthal, R., and Wildeman, T., 1978, Energy dispersive x-ray
    fluorescence spectrometric determination in trace elements in oil
    samples, Anal. Chem., 50, 899.
5. NBS (National Bureau of Standards), 1967, Standard reference materials
    1052b and 1065b, U.S. Department of Commerce, Washington, D.C.
    20234.
6. Sanders, R. W., Olson, K. B., Weinier, W. C., and Bielson, K. K., 1983,
    Multielement analysis of unweighed oil samples by x-ray fluores-
    cence spectrometry with two excitation sources, Anal. Chem., 55,
    1911.

HIGHLY AUTOMATED TECHNIQUES FOR ON-LINE LIQUID ANALYSIS
WITH AN XRF SYSTEM

R. A. Miller, D.J. Kalnicky, and T. V. Rizzo

Princeton Gamma-Tech
1200 State Road
Princeton, NJ   08540

INTRODUCTION

Sophisticated matrix correction techniques have been successfully applied to analysis in an interactive laboratory environment (1-3). The present paper describes their application to an on-line industrial environment, where reliable operation is to be maintained with a minimum of operator intervention. Out of bound situations must be detected and corrected automatically to the extent possible. Programming must be substituted for human judgement, within restricted circumstances.

The application described is analysis of gold plating solutions. On-line analysis of gold plating solution is desirable because gold is expensive and monitoring of the plating bath concentration is an important inventory function. In addition, a continuous, automatic record of concentration greatly facilitates traceability and accountability: that is, "when did the concentration change?," and "who changed it?"

On-line analysis is at least as important for quality control. The chemistry of gold plating is complicated and not completely understood. More immediate and more detailed concentration information will lead to better control of the quality of the plated product and, over the long run, to a better understanding of the chemistry.

OVERALL STRUCTURE

The analysis we will describe is embodied in the PGT X-PERT, which consists of a Clean Air Unit (CAU) for display of results and control of up to four Tank Site Units (TSUs) which perform the measurements, (See Figure 1). The Tank Site Units pump solution from plating tanks through a flow cell for analysis by specialized analog and digital electronics. The Clean Air Unit is essentially an industrialized desk top (or "personal") computer with appropriate software and communications ports.

The Clean Air Unit has a 3.5" disk for program and data storage. The Tank Site Unit program is down loaded from the

CAU.   Downloaded code is used instead of ROM code for greater
flexibility.

TSU STRUCTURE

The Tank Site Unit is controlled by a microprocessor which
receives instructions from the Clean Air Unit and sends results
back to it. (See Figure 2.) A sample changer brings a sample
flow cell or one of three standard cells into the analysis
position.   The microprocessor programs the X-ray tube to a
given voltage and current and can read the actual values for
verification.   The microprocessor also programs the
proportional counter bias voltage. Since the system gain or
energy scaling is sensitively dependent on the proportional
counter voltage, it is adjusted automatically until the
spectrum is in the desired position.   This is the method of
obtaining energy calibration.

The preamp and amplifier are of spectroscopic quality.
The analog to digital converter (ADC) uses direct memory access
to build the spectrum in the microprocessor's memory without
diverting the processor from its other tasks.

The microprocessor runs under the control of a state-of-
the-art multiprocessing executive (4) so that a variety of

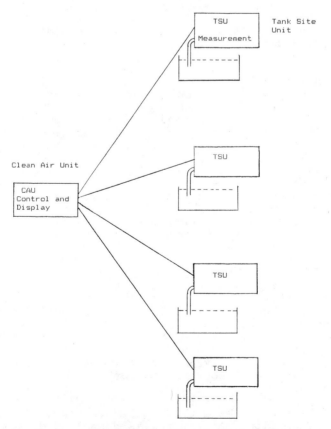

Fig. 1.   The analysis system consists of a Clean Air Unit,
which displays results and allows operator control,
and a Tank Site Unit, which performs the measurements.

processes--such as servo control of the X-ray tube voltage--can
be programmed as if each had complete control of the computer.
This pays great dividends in simplicity of design and ease of
maintenance.

MONITORING AND CONTROL

As mentioned before, the microprocessor exercises
servo control over the X-ray tube voltage and current and
proportional counter bias voltage as well as the sample
position.

1) In normal operation the microprocessor reads the X-ray
   tube voltage and current ten times a second and
   adjusts the power controller as necessary. The
   voltage and current are brought to operating levels in
   a controlled manner to prevent tube arcing and
   maximize tube life.

2) The proportional tube voltage is adjusted by observing
   the spectra obtained and testing them for proper
   energy calibration. If the calibration starts to
   drift, the voltage is adjusted to bring it back to its
   proper value.

3) The sample position is changed by a motorized screw
   jack. There is a positive indication of position.

4) The pump that passes solution through the sample flow
   cell may be turned on or off by the microprocessor.

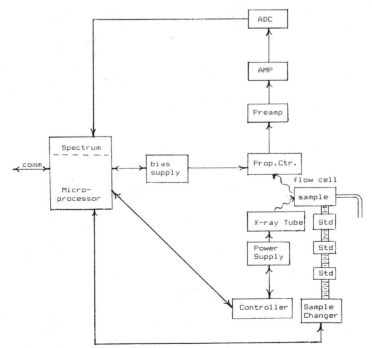

Fig. 2. Tank Site unit structure. The microprocessor controls
the X-ray excitation, proportional counter voltage,
and sample position. It also calculates
concentrations and sends them to the Clean Air Unit.

In addition to active controls, the microprocessor continually checks for problems such as leaks, failure of cell flow, high internal temperature, X-ray tube overvoltage or overcurrent, and power supply voltages out of their specified ranges. If any error condition is detected, the high voltages are shut down and a descriptive error message is sent to the Clean Air Unit.

OPERATOR INTERACTION

The instrument was designed for minimal operator interaction. (See Figure 3.) In normal operation it displays the most recent measurements from up to 4 Tank Site Units continuously with no operator intervention. The operator may choose to see more detail on any one of the TSUs by pressing the corresponding number.

The detailed display presents the last ten values of each measured parameter (solution temperature and pH and concentrations of Au, Co, K) as well as the TSU status, solution fill dates, and customer data. This display updates as new measurements are received from the TSU. There is only one control function that the non-supervisory operator can perform from this display: he can turn off the audio alarm if an error condition has turned it on.

There are a number of other functions that a supervisor can perform by giving a password: He can down load the program to the TSU if this is necessary because the TSU has been turned off. The supervisor can start and stop analysis, set acceptable limits for results, set units for results, change the customer data, and cause the TSU to recalibrate the quantitation. This last function is described in more detail below.

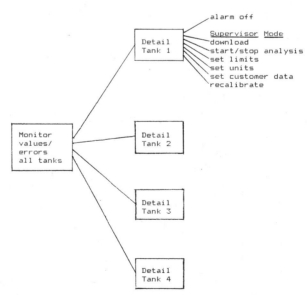

Fig. 3.   Operator interaction.   Beginning with a screen which
          displays results and errors, the operator may request
          detailed displays for each Tank Site Unit.
          Supervisors may also perform certain control
          functions.

QUANTITATION CALIBRATION

This calibration determines free parameters in the quantitation algorithm by fitting standards. (See Figure 4.) The free parameters are an effective pure element intensity and effective unmeasured matrix alpha for each measured element. Remember that there will be no operator intervention. The procedure can only terminate normally or with an error indication. Either termination returns control to a master program described under "INSTRUMENT STRATEGY" below. There is a circumstance that represents only a partial failure. If the measured intensity of one element is too small to be used reliably, then the coefficients are adjusted so that element's concentration will always be reported as zero until a new calibration is done.

The mathematical notation is that of the Modified Alphas technique described by D.J. Kalnicky elsewhere in these Proceedings (5). An overview of the calibration procedure is as follows:

1) Collect spectra from two different standards.

2) Digitally filter the spectra to eliminate background.

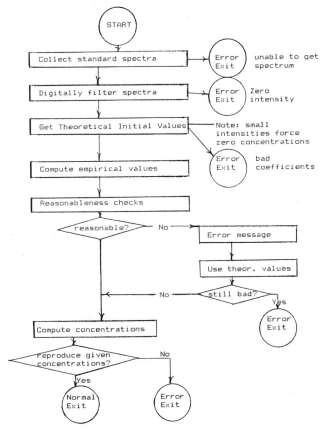

Fig. 4. Logic of the quantitation calibration. Normal and error exits return to the instrument strategy routine shown in Fig. 6.

3) Element intensities are window integrals taken from the filtered spectra.

4) Calculate theoretical approximations to the desired free parameters.

5) Fit empirical values of the free parameters.

6) Check the empirical values for reasonableness. The reasonableness criteria are described by Kalnicky (5).

7) If the empirical values are not reasonable issue an error message and check the theoretical values.

8) If the theoretical values are not reasonable the calibration fails.

9) If either of the theoretical or empirical values pass the reasonableness test, use it to calculate the known concentrations of the standards. If these calculated concentrations are close enough to the known concentrations the calibration succeeds.

Some of the errors detected, such as the inability to collect a spectrum or zero intensities, may be due to transient hardware problems such as a tripped X-ray tube power supply. Errors of this nature may be corrected by retrying. The procedure for retrying is described under "INSTRUMENT STRATEGY". Other errors, such as a divide by zero in calculating theoretical values, are due to invalid data loaded into the machine initially. This should be a rare occurrence as for example when the storage medium fails. Under this circumstance the machine will refuse to provide readings until the situation is corrected.

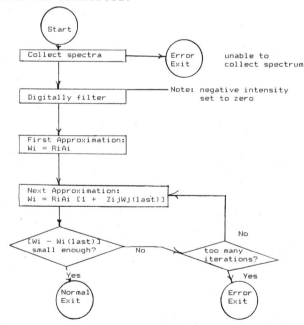

Fig. 5. Logic of the quantitative analysis. Normal and error exits return to the instrument strategy routine shown in Fig. 6.

ANALYSIS

The first steps of analysis are the same as those for
calibration: collect a spectrum and digitally filter it. (See
Figure 5.) Here, as in the case of the quantitative
calibration, the software can assume that the X-ray tube
voltage and current and the proportional counter bias voltage
are correctly set since these tasks are performed independently
under the multitasking executive. Similarly, the sample
changer will be in the proper position for analysis of the
unknown solution.

If a spectrum cannot be collected, this procedure will
take an error exit. The master procedure can then attempt to
restart the X-ray generator if necessary and retry the
analysis.

The first approximation to the concentration is simply the
scaled intensity. Subsequent approximations are obtained by
using the last approximation to concentration as input to the
modified alphas matrix correction algorithm. The procedure is
repeated until the concentrations converge or the predetermined
number of iterations is exceeded.

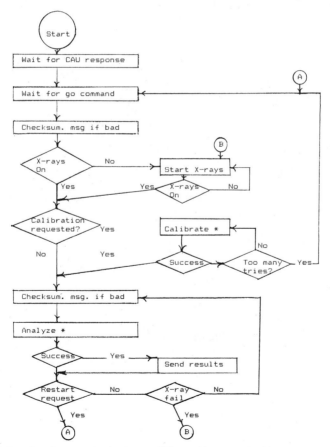

Fig. 6.   Instrument strategy. This routine never exits.
Analysis and calibration subtasks are shown in Fig. 4
and 5.

INSTRUMENT STRATEGY

Figure 6 shows what we call the instrument strategy for
the TSU. Because operator intervention is not normally
available, calibration, analysis and certain other housekeeping
functions have to be woven together in a way which will cover
most operating situations. This is the highest level
procedure. It has no end and no error exits. To the extent
possible retries are performed. If the X-ray generator trips
off, it is restarted if possible. Calibration is retried a
set number of times.

When the program first starts up, it checks for the
presence of the Clean Air Unit by attempting to exchange
messages with it. No further action is taken until this has
been accomplished. At that point the instrument waits for the
command to start analyzing.

When the command to analyze is received, the X-ray
generator is started if necessary and quantitative calibration
is done if requested. Thereafter analyses are run and results
reported indefinitely until the CAU requests an end to analysis
or the X-ray controller trips off. Periodically the program
checksum is computed and any discrepancy reported.

CONCLUSION

An analyzer has been built and used which performs an
analysis within a limited range of samples without operator
intervention except to start it up. The analysis method is
familiar within a laboratory environment where direction by a
skilled operator is normal. In its limited way this points to
the increasing use of sophisticated analysis techniques in
production environments by substituting machine intelligence
for human intelligence.

REFERENCES

1. Advances in X-ray Analysis, Vols. 1-28, Plenum, New York,
    1957-1984.

2. X-ray Spectrometry, Vols. 1-14, Wiley Heyden, Chichester,
    Sussex, England, 1972-1985.

3. R. Jenkins, R. W. Gould, and D. Gedcke, "Quantitative X-ray
    Spectrometry", Marcel Dekker, Inc., New York, 1981.

4. Electronics, Jan. 27, 1983, p. 134.

5. D. J. Kalnicky, "A Combined Fundamental Alphas/Curve
    Fitting Algorithm for Routine XRF Sample Analysis", Denver
    X-ray Conference, Snow Mass, Colorado, 1985.

X-RAY FLUORESCENCE SPECTROSCOPY AS A TECHNIQUE FOR STUDYING

SURFACE CONCENTRATION PROFILES OF HETEROGENEOUS CATALYSTS

Eva M. Kenny and Burton J. Palmer

Halcon Research

1 Philips Parkway, Montvale, New Jersey

ABSTRACT

X-ray fluorescence spectroscopy is usually not considered a surface technique in the same sense as ESCA or Auger spectroscopies. However, it can be a useful tool in studying the "surface" and bulk concentrations of elements in heterogeneous catalysts. The "surface" is defined by the effective penetration depth of the analyte line of the element of interest in a specific matrix. In the example which is presented, silver on alumina, the "surface" using the Ag $K\beta_{1,3}$ is defined by a shell 3000 microns deep while for the Ag $L\alpha_1$ it is 15 microns. Two methods of sample preparation are discussed and data obtained using three different calculation methods is presented. Data from the x-ray fluorescence method is compared to the silver depth profile in the alumina pellets obtained by electron microprobe analysis. The XRFS method allows rapid screening of many catalyst samples for the fraction of the cost of the microprobe technique.

INTRODUCTION

X-ray fluorescence spectroscopy is usually considered only as a bulk analytical method for studying concentrations of elements in catalysts. Actually, it is a near-surface analysis technique since emitted x-rays originate from a near-surface layer or film of a finite thickness (1). The conventional x-ray emission technique can therefore be a useful tool for obtaining concentration profiles of elements in heterogeneous catalysts. The thickness of the surface layer analyzed by XRFS is on the order of a few microns to a few millimeters and is defined by the effective penetration depth of the analyte line in a specific, well defined matrix (2,3). By using a K-line and an L-line of an element, a two-point concentration profile can be obtained. For lighter elements which have only a K-line available for analysis, surface vs. bulk concentration can still be determined by using the proper specimen preparation technique (4).

X-ray fluorescence spectroscopy (XRFS) is complimentary to conventional surface analysis techniques such as x-ray photoelectron spectroscopy (XPS), Auger electron spectroscopy (AES), and secondary ion mass spectrometry (SIMS) (5). With these techniques, the surface depth analyzed is on the order of 0.5 nm to 2 nm. Electron probe microanalysis

(EPMA) is better suited as a referee method for XRFS than these other surface spectroscopies. It provides a more accurate means of determining spatial variation of composition on a microscopic scale with a resolution of 1 μ to 5 μ. With minimal specimen preparation, XRFS provides a coarse but similar picture of the concentration profile. To obtain a concentration profile with XPS, AES, or SIMS requires the use of argon etching. Preferential sputtering of the surface or variations in the sputtering rate caused by surface irregularities can produce erroneous depth profile data.

Qualitative monitoring of samples by XPS, AES, and SIMS is straight forward but quantitation is often difficult. EPMA can be quantitative when fundamental theory and the proper calibration standards are used (6). This can only be achieved however, when instrumentation is available in-house since consulting laboratories are expensive and ill-equipped to deal with specific standard preparation problems. These labs provide semiquantitative data in the form of an intensity ratio of the element in the sample to a pure element standard. An on-site XRFS lab on the other hand can produce highly quantitative data by using type standards or well characterized catalyst samples as standards.

Catalyst characterization represents a thin film problem rather than a true surface problem because most catalysts are non-homogeneous, porous materials where boundaries are poorly defined (7-9). For identical chemical compositions and overall loading, the performance of a supported catalyst depends critically on the distribution of it active ingredient within the support (8-10). Figure 1 illustrates several types of distribution which can occur. XRFS can easily differentiate between a uniform distribution (Fig. 1A) and the egg-shell type (Fig. 1B).

1A. Uniform   1B. Egg-Shell   1C. Egg-White   1D. Egg-Yolk

Figure 1.   Types of catalyst distribution (8).

Recognizing egg-white or egg-yolk distribution types (Figs. 1C and 1D) requires more work since only a two-point concentration profile is obtained. Various other problems in heterogeneous catalysis can be solved using XRFS. Surface contamination by elements (< fluorine) can be measured usually in PPM to % levels (4). Depletion of an active component from the surface of supported or mixed oxide catalysts can also be determined.

XRFS is inexpensive, faster, and requires less stringent experimental conditions (vacuum, specimen preparation, etc.) than any of the surface spectroscopies and EPMA (5). The surface concentration of the active component or surface contaminant of a catalyst may vary from pellet to pellet as well as within the pellet itself. For this reason, many samples of a catalyst must be analyzed by any surface analysis technique including EPMA to insure that the data is representative of the entire catalyst sample. XRFS presents a better overall or average picture of a catalyst system rather than an isolated and shallow view of one or two areas on a few select catalyst pellets. X-ray fluorescence spectroscopy is a practical analytical method which can be used for screening many samples for catalyst engineering and quality control in catalyst production. XPS,

AES, SIMS, and EPMA are highly specialized research techniques for obtaining surface measurements. Expense and time requirements of these techniques limit the number of samples which can be analyzed on a routine basis.

EXPERIMENTAL

Two techniques of sample preparation were used for obtaining the concentration profile of silver on alumina rings by x-ray fluorescence spectroscopy - the single catalyst pellet method and the pressed disk method (4). The first technique required essentially no preparation for the surface analysis except for blowing off any dust with a stream of dry nitrogen. The interior of the catalyst pellet was analyzed after approximately 0.5 MM to 1 MM of the surface was shaved off using a small drill with a cutting wheel. With this method all standards and samples were placed behind a 3 MM gold aperture which was supported on the bottom of a regular gold masked specimen holder. The catalyst pellets were held in place with a plastic ring. The thickness of the gold aperture was critical. The foil used was 0.1 MM thick to eliminate intensity errors caused by AG KB1,3 radiation originating from behind the gold foil.

The pressed disk technique (4) required more sample preparation. For the surface analysis, whole catalyst pellets were placed on their sides in a tared .31 MM aluminum cap which was in a 31 MM die. The catalyst pellets were stacked at least three layers high and all voids were filled with catalyst rings which were split parallel to their vertical axes. The specimen was then briquetted under 20 tons of pressure. The pressed disk was weighed and its thickness was recorded with a micrometer to the nearest 0.01 CM. For the bulk analysis, approximately 10 GM of catalyst was pulverized in a puck and ring grinding system using tungsten carbide containers. The entire ground specimen was briquetted with no binder following the same procedure that was used for the surface specimen. All standards were briquetted as if they were bulk specimens. The same standards were used for the single pellet method and the pressed disk method however, for the single pellet analysis the standard was placed behind another 3 MM gold aperture.

Three types of standards were used for generation of calibration curves. The first type were prepared by mixing silver powder (-325 mesh) and pre-ground α-alumina catalyst rings (-325 mesh) in a plastic vial with mixing balls on a mixer/mill for 5 minutes. The thoroughly mixed powders were transferred to a tungsten carbide grinding dish and ground for an additional 3 minutes in the shatterbox before being briquetted. The second type of standard was prepared in the same manner but silver oxide powder (-325 mesh) was used. Concentrations of type-one and type-two standards varied from 5 WT% to 50 WT%. The third type consisted of a series of well characterized "standard" catalysts. The silver concentration of these catalysts was determined by a potassium thiocyanate titration after the silver was extracted out of the α-alumina rings with 10% nitric acid solution. A minimum of five titrations were preformed on each "standard" catalyst and averaged to obtained the silver concentration. A 10 GM portion of the "standard" catalyst was ground for 3 minutes in a tungsten carbide grinding dish before being briquetted.

The instrumentation used for the analysis was a Siemens' SRS200 wavelength dispersive spectrometer. The data system consisted of a Digital Equipment PDP11/03 computer and two software packages. The first was Siemens' Spectra I which was used to generate calibration curves by multiple regression (11-15) and do ratio standard calculations. The second was Criss's XRF11 fundamental parameters program (16,17) which used x-ray physics and theoretical formulas along with two "standard" catalysts to calculate the silver concentration in the surface and bulk specimens. The Spectra I programs were also responsible for controlling spectrometer.

A chromium target x-ray tube operated at 50 KV and 40 MA was used for excitation. For measuring the AG KB1,3 (14.18°), a LiF 100 crystal was used as a dispersive element with a scintillation counter as the detector. A 0.15 collimator was used between the sample and the crystal and a 10 MM aperture was place in front of the x-ray tube window. The AG LA1 (56.75°) was measured using a PET crystal and a gas flow proportional counter (P-10 gas). The 0.15 collimator and 10 MM x-ray tube aperture were also used. Pressed disk specimens were analyzed behind a 23 MM gold aperture with a counting time of 40 seconds on the peak, low angle background, and high angle background. Single pellet specimens and standards were analyzed behind a 3 MM gold aperture with a counting time of 100 seconds on the peak and backgrounds.

Silver concentrations of the surface and bulk pressed disk specimens were calculated three ways for both the AG KB1,3 and the AG LA1 – by fundamental parameters, by the ratio standard method, and from calibration curves. Fundamental parameters calculations were done using a 15% Ag "standard" catalyst, next using a 7.4% Ag "standard" catalyst, and finally using both standards together. The ratio standard calculations were preformed using the 15% Ag standard and also with the 7.4% Ag standard. Calibration curves were generated using only the pressed powder standards since there were not enough "standard" catalysts available. Surface and bulk silver concentration of the single pellet catalyst samples were obtained using two standards and fundamental parameters calculations only.

CALCULATIONS

Calculation of the surface depth analyzed by x-ray fluorescence is straight forward (4). First the volume of the pressed disk specimen is calculated. Typical volume of a pressed specimen disk is 3.2 CM^3.

$$V = \pi * (D/2)^2 * T$$

where V = volume in CM^3
D = diameter of disk in CM
T = thickness of disk in CM

The volume is used to calculate the density of the pressed disk which is typically 2.5 GM CM for Ag/Al_2O_3 samples regardless if it is the surface or bulk specimen.

$$\rho = \frac{M}{V}$$

where ρ = density in GM CM^{-3}
M = mass of disk in GM
V = volume of disk in CM^3

Next the matrix mass absorption coefficient is determined (2,3,18). This parameter is dependent on each element present, its weight fraction, and the wavelength of the analyte line. For the AG KB1,3 analyzed in a 15% Ag/Al_2O_3 matrix, μ (MATRIX) = 2.7 $CM^2 GM^{-1}$. While for the AG LA1 it is 512 $CM^2 GM^{-1}$. This indicates that the AG LA1 x-rays are being more heavily absorbed and originate from a much shallower depth than the AG KB1,3 x-rays.

$$\mu_M = \sum_i u_i * w_i$$

where μ_M = matrix mass absorption coefficient of analyte line in CM GM
μ_i = individual mass absorption coefficient for each element i in the matrix
w_i = weight fraction of element i

The effective penetration depth or surface thickness is calculated using the approximation below (4). Table 1 illustrates the dependence of surface thickness on the analyte wavelength and the composition of the matrix.

$$d = \frac{4.6 * 10^4}{\mu_M * \rho} * SIN\ \Psi$$

where μ_M = matrix mass absorption coefficient in $CM^2\ GM^{-1}$

ρ = density of disk in $GM\ CM^{-3}$

Ψ = spectrometer take-off angle (45° SRS200)

d = surface thickness in μM

TABLE 1: SURFACE FILM THICKNESS

Analyte Line	%Ag	Depth μM
AG KB1,3 0.497Å	5	4000
	15	2500
	25	1900
AG LA1 4.154Å	5	15
	15	12
	25	14

RESULTS AND DISCUSSION

The accuracy and precision of the results depended on the calculation method and the specimen preparation method. Results obtained using fundamental parameters, ratio standard, and calibration curves are compared in Table 2. The concentration of silver obtained by potassium thiocyanate titration was 7.8% for sample 44-1. Based on this, fundamental parameters calculations using two standard catalysts gave the most accurate results as indicated by the bulk analysis using the Ag LA1 and the Ag KB1,3. Fundamental parameters calculations using one standard and the ratio standard method also gave accurate results but only when the standard's concentration matched that of the sample. Calibration curves generated using mixed powder standards produced unsatisfactory results. The reason is apparent in the curves themselves (Figure 2). The standards prepared with silver powder (-325 mesh) are not consistent with those prepared from silver oxide powder (-325 mesh).

TABLE 2: COMPARISON OF XRFS CALCULATION METHODS FOR WT% AG
IN PRESSED DISK SPECIMEN 44-1

2A. AG LA1

Area Analyzed	Fundamental Parameters			Ratio Standard		M.R. Curve
	2 Stds	15% Std	7.4% Std	15% Std	7.4% Std	
SURF	23.6	25.3	20.4	23.8	18.9	20.6
	20.7	20.7	16.7	20.1	15.8	17.1
	32.4	32.4	25.9	29.6	23.3	25.6
BULK	7.7	9.4	7.7	9.8	7.7	7.8
	7.7	9.4	7.7	9.8	7.7	7.8
	7.7	9.4	7.7	9.7	7.7	7.7

2B. AG KB1,3

Area Analyzed	Fundamental Parameters			Ratio Standard		M.R. Curve
	2 Stds	15% Std	7.4% Std	15% Std	7.4% Std	
SURF	8.7	8.4	8.9	13.0	7.8	11.2
	9.2	8.9	9.4	13.2	7.9	11.8
	9.0	9.1	9.6	13.1	8.0	12.0
BULK	7.7	7.5	7.9	12.5	7.5	9.8
	7.9	7.8	8.2	12.7	7.6	10.3
	8.0	7.8	8.2	12.7	7.6	10.3

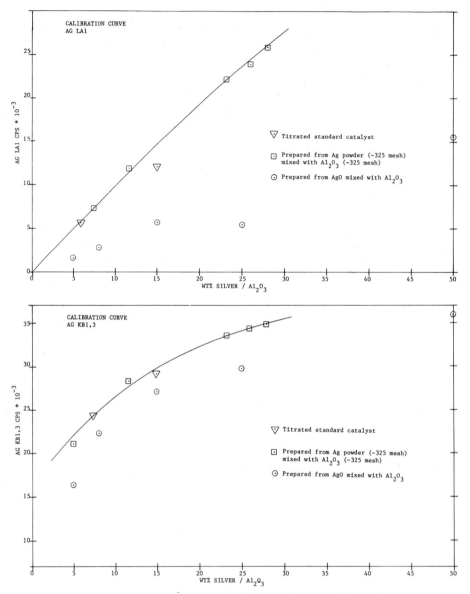

Figure 2. Calibration curves for the Ag $L\alpha_1$ and Ag $K\beta_{1,3}$.

This is more pronounced for the Ag LA1 where particle size or homogeneity problems would manifest themselves. The ideal standards for calibration curves would be a set of standard catalysts which have a similar particle size and density as the samples. To accumulate enough "standards" to generate reliable calibration curves would require much work and fundamental parameters would be the desirable alternative.

The specimen preparation method also influences the accuracy and precision of the data. The single pellet analyses produced more scattered and less accurate results than the pressed disk method. The titrated silver concentration for samples 44-1 and 58-1 were 7.8% and 15.0%

TABLE 3: COMPARISON OF SPECIMEN PREPARATION METHOD FOR SAMPLES
44-1 AND 58-1 USING 2 STDS AND FUNDAMENTAL PARAMETERS

Sample Reference	Area Analyzed	Pressed Disks AG LA1	AG KB1,3	Area Analyzed	Single Pellets AG LA1	AG KB1,3
44-1	SURFACE	23.6	8.7	EXTERIOR	27.5	5.5
		19.8	9.2		24.7	9.7
		29.9	9.0		20.5	9.3
	BULK	7.7	7.7	INTERIOR	2.9	7.2
		7.7	7.9		8.8	10.3
		7.7	8.0		11.5	4.5
58-1	SURFACE	18.5	18.6	EXTERIOR	16.6	14.3
		17.9	17.9		16.9	14.8
		17.5	19.1		17.5	13.7
	BULK	14.9	15.1	INTERIOR	14.8	13.8
		14.8	15.5		15.7	14.9
		14.9	15.2		14.4	12.7

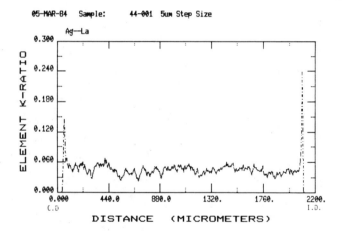

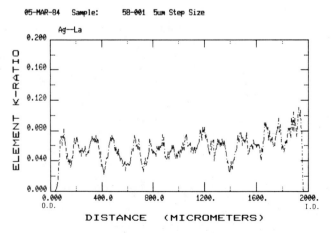

Figure 3.   Electron probe microanalysis data.

respectively. Single pellet analyses are prone to more errors because of the smaller area analyzed (3 MM), pellet non-homogeneity, surface roughness, pellet curvature, and pellet wall thickness (<2 MM in some cases). One internal method check is the agreement between the data obtained using the Ag LA1 and the Ag KB1,3 for the bulk analysis. Regardless of the sample preparation method used, similar trends in the silver concentration profile can be seen. Sample 44-1 has almost three times the silver concentration on the surface as in the bulk while sample 58-1 has only slightly more silver on the surface than in the bulk.

Compared to the electron probe microanalysis data in Figure 3, the XRFS method provides similar results. For sample 44-1, EPMA indicates that there is 3.5 times the concentration of silver in the first 0 – 20 μ's of a catalyst pellet than in the pellet interior. XRFS analysis shows 3.2 times the concentration of silver in a surface depth of 15 μ when compared to the bulk analysis. Sample 58-1 has a uniform distribution of silver throughout the catalyst pellet as indicated by both EPMA and XRFS. The surface silver by XRFS is only 1.2 times that of the bulk silver concentration.   In Figure 3B, the scatter in the EPMA data is due to the large silver particle size in sample 58-1. The silver particle size is 3000 Å for sample 58-1 and 400 Å for sample 44-1.

SUMMARY AND CONCLUSIONS

A method using x-ray fluorescence spectroscopy for analyzing the surface and bulk concentrations of silver on an -alumina support has been developed and compares favorably with data obtained by electron probe microanalysis. The XRFS method has many advantages over EPMA and other surface analysis techniques.   It is highly quantitative provided that "standard" catalysts along with fundamental parameters calculations, and pressed disk specimen preparation are used. It can easily differentiate between a uniform distribution of silver within the catalyst support or a high concentration in a surface film. This same data was obtained by EPMA which only views a very select area of one catalyst pellet and may be biased. Being relatively inexpensive compared to EPMA, x-ray fluorescence can be used to screen many samples for catalyst engineering and for quality control in catalyst production. This makes XRFS a very practical tool for catalyst characterization.

This technique is applicable to other catalyst systems. Even if the analyte has only one line available for measurement, such as in the case of a lighter element, XRFS can be used to obtain surface vs. bulk information.   It can be a very useful tool for determining if surface contamination has occurred from corrosion products generated in a chemical reactor, or from halogens, sulfur, or phosphorous which may be present in feed gas. Another prospect for the technique in general, is that it can be applied to x-ray powder diffraction. The same principles would apply to surface and bulk analysis of crystalline phases by XRD. Surface depletion of a active phase could be monitored as well as the contamination of the surface by a specific compound. This could aid in the elucidation of deactivation mechanisms of a catalyst or identifying specific corrosion product materials.

REFERENCES

1. Josseaux, P., et al, J. Electrochem. Soc., 132, 684 (1985).
2. Jenkins, R., and Devries, J.L., "Practical X-ray Spectrometry", Springer-Verlag; New York, 1967; Chapters 1, 6, and 8.
3. Bertin, E.P., "Principles and Practice of X-ray Spectrometric Analysis"; Plenum; New York, 1970.
4. Cornelius, C., Anal. Chem., 53, 2361 (1981).

5. Russ, J.C., Advan. X-ray Anal., 28, 11 (1984).
6. Wernisch, J., X-ray Spectrom., 14, 109 (1985).
7. Worthy, W., C&EN, April 8, 1985, 28.
8. Lee, S.Y., and Aris, R., Catal. Rev.-Sci. Eng., 27, 207 (1985).
9. Komiyama, M., Catal. Rev.-Sci. Eng., 27, 341 (1985).
10. Harriott, P., J. Catalysis, 14, 43 (1969).
11. Alley, B.J., and Myers, R.H., Anal. Chem., 37, 1685 (1965).
12. Stephenson, D.A., Anal. Chem., 43, 310 (1971).
13. Plesch, R., Siemens Analytical Application Note, No.28.
14. Plesch, R., and Thiele, B., Siemens Analytical Application Note, No.33.
15. Thiele, B., and Plesch, R., Siemens Analytical Application Note, No.41.
16. Criss, J.W., and Birks, L.S., Anal. Chem. 40, 1080 (1968).
17. Criss, J.W., Advan. X-ray Anal., 23, 93 (1980).
18. Jenkins, R., and DeVries, J.L., "Worked Examples in X-ray Analysis";
    Springer-Verlag; New York, 1970; pp 74 - 79.

EVALUATION OF A MOBILE XRF ANALYZER IN AN INDUSTRIAL LABORATORY

Norman F. Johnson

TRIDENT Refit Facility
Code 440
Bremerton, WA   98315

INTRODUCTION

Traditionally x-ray spectrometers have been designed to analyze perfectly prepared specimens that are flat and highly polished such as 32mm discs. Certainly the best analytical results are obtained on such samples, however in an industrial laboratory it is also neccessary to obtain statistically valid results on samples that are less than ideal in size and geometry. The problem at the TRIDENT Refit Facility is to analyze large metal parts such as pump casings, valve bodys, flanges, and finished piping sections that are going into the repair of TRIDENT class submarines. Energy dispersive X-ray techniques have long been recognized as fast and non-destructive and are therefore ideally suited to the analysis of these finished metal parts.

One manufacturer has built a commercially available instrument that does not require that the sample be cut to size to fit the constraints of a small sample chamber. The evaluation of this instrument for the analysis of three common industrial alloys, Inconel, Stainless Steel, and Monel is the subject of this paper.

EXPERIMENTAL

The equipment chosen for this work was the KEVEX Model 750A Mobile XRF Analyzer. It is a free standing x-ray tube and Si-Li detector mounted on a moveable platform raised and lowered on a column that can be rolled around on rubber tires anywhere within a shielded room. The platform has manually driven gear adjustments in the X, Y, and Z directions. The X-ray tube has a rhodium target and can be operated at a maximum of 60 KV and 3.3 Ma. A secondary target material is mounted to the x-ray tube and a choice of collimation can be selected. A dual laser and incandescent lamp source is used to optically position the x-ray tube and detector so as to analyze a particular point of interest on the sample. A system overview is given in figure 1.

After the equipment arrived the first application was to use the analyzer in a qualitative manner for alloy sorting. A data collection time of 10 to 20 seconds was frequently

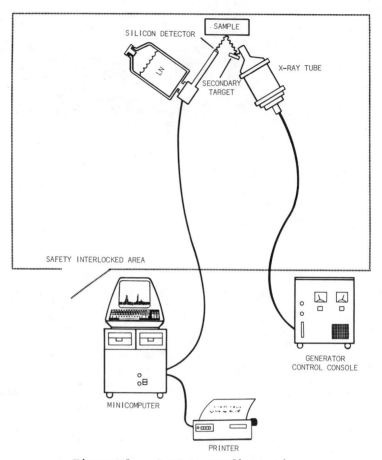

Figure 1.   System configuration.

sufficient to gather enough information to recognize many of
the common alloys by observing the spectrum on the video
screen.   Amoung the easy to recognize alloys were Brass,
Inconel, Copper-Nickel, Steel, Monel, Stainless Steel, and
Bronze.
      The real test of the equipment was to see how accurate and
reproducible it was in producing quantitative data on a variety
of sample matrices.   A number of mathematical schemes for
calibration were tested including; a simple ratio of sample
peak intensities to a standard, a least squares fit with
multiple standards (1), and a fundamental parameters program
called EXACT (2). These initial tests quickly revealed that the
fundamental parameters program offered the most flexibility and
attained the best accuracy for the alloys tested.
      Calibration was done by gathering counts per second on the
Ka peak of the principal constituent elements from well
certified standards.   Brammer Standards BS 825 and BS 400, and
the National Bureau of Standards SRM 1155 were used to
calibrate Inconel, Monel, and Stainless Steel respectively.
During the Monel calibration it was learned that the copper Ka
peak was severely overlapped by the nickel Kb in this alloy
that contains nearly 70% nickel, so the copper Kb was used

instead. The count data for the major elements plus the
concentration data for all elements was entered into the
computer program that calculates calibration constants for the
fundamental parameters calculations. Once the calibration
constants were derived they were tested by comparing the
calculated concentrations with the certified values for a
variety of standards covering the entire range of interest.
This process required repeated tries to find the optimum
calibration constants.

After the operating parameters were settled upon, a
computer file was written that could be recalled later to setup
all of the important conditions needed to obtain quantitative
results. A different file was written for each different alloy
type. The parameters included count time, an element list, and
the calibration constant for each element. Some parameters were
fixed for consistency and to save time during actual analyses.
The KV was always 35, Ma was 0.5, and a silver secondary target
with 6 mm collimators was used for all measurements. A second
computer file for each alloy was used to run the actual samples
and to print the concentrations as calculated by the EXACT
program. The mathematical technique of forcing the total of all
elements to 100% was used to correct for the variations in
sample size and shape. A count time of 100 seconds was used
throughout the statistical study.

RESULTS

Now that the instrumental parameters and calibration
constants for each alloy type could be recalled from disc, a
standard that was not part of the original calibration was
selected for repetitive study. The standard was positioned in a
variety of ways in front of the detector to duplicate the odd
geometries encountered on real samples. The round barstock
piece was run in the center of the flat side, near the outer
edge of the flat side, and on the curved circumference like a
section of pipe. All of these configurations were averaged into
the statistical data to gain a better understanding of the
range of values that the analyst could expect during everyday
operations on difficult samples. This testing scheme was not
designed to demonstrate the most accurate results possible from
the equipment on perfectly prepared specimens. In an industrial
setting time pressures do not normally allow for complete

Table 1. Principal Constituent Elements in Inconel, %

Sample		Cr	Fe	Ni	Mo
BS 825	Calibration Standard	21.85	28.65	43.37	2.82
BS 600B	Certified Values	15.54	7.84	75.23	.075
BS 600B	Average of 42 Runs	16.11	7.97	75.51	.097
BS 600B	Standard Deviation	.15	.09	.18	.006
BS 600B	% Relative Deviation	.93	1.1	.24	6.2
BS 600B	High Reading	16.52	8.20	75.88	.113
BS 600B	Low Reading	15.82	7.79	75.07	.079

optimization of all parameters, yet a quantitative answer
supportable by standards is required.

Table 1 shows the kind of results that can be obtained with
this equipment for a group of nickel-chromium-iron alloys. The
calibration constants were obtained from data collected on BS
825 Inconel composed of the elements shown in the first row. A
different alloy, Inconel 600, was used as if it were an unknown
sample. The third row displays the average of 42 runs on BS
600B made over a period of three weeks. These averages
correlate well with the certified values shown in the second
row. The low standard deviations suggest that a type
calibration with an Inconel 600 standard would yield even
better accuracy.

In Table 2 the calibration was done with an NBS 316
Stainless Steel standard. The "unknown" in this case was a 304
Stainless Steel. Notice that although the calibration for
molybdenum was done on a standard that contained 2.38% of the
analyte, the average of 36 runs on the test standard gave the
correct concentration of .32%. This demonstrates the broad
range of compositions covered by the fundamental parameters
program. However when an Inconel sample was analyzed with
calibration constants obtained from a Stainless Steel standard,
the results were unacceptable, even though they both contain
the same three major elements: nickel, chromium, and iron.

Table 3 shows a true type calibration where the calibration
constants were derived using data from a Monel 400 alloy and
the specimen studied was also Monel 400 with only slightly
different composition. As expected there is an increase in
accuracy while the reproducibility remains about the same as
the previous two alloys. These results exceed the accuracy that
is needed in most routine industrial inspections.

LIMITATIONS

As with all energy dispersive x-ray measurements, this
method lacks the ability to analyze carbon, a most important
element in the classification of carbon and low alloy steels.
Since the instrument operates in air, it is also unable to do
sulfur, phosphorous, aluminum, magnesium, and silicon. Accurate
trace element analysis appeared to be possible with this

Table 2. Principal Constituent Elements in Stainless Steel, %

Sample	Cr	Fe	Ni	Mo
NBS SRM1155 (316) Calibration Standard	18.45	*64.46	12.18	2.38
BS 81D (304) Certified Values	18.33	*70.20	8.46	.32
BS 81D Average of 36 Runs	18.02	71.00	8.67	.32
BS 81D Standard Deviation	.17	.21	.18	.006
BS 81D % Relative Deviation	.94	.29	2.1	1.4
BS 81D High Reading	18.66	71.45	9.00	.34
BS 81D Low Reading	17.58	70.45	8.32	.31

* By difference.

Table 3.   Principal Constituent Elements in Monel, %

Sample		Cr	Mn	Fe	Ni	Cu	Mo
BS 400	(original) Calibrating Standard	.16	1.04	2.17	62.70	33.00	.011
BS 400B	Certified Values	.55	1.04	1.81	64.05	32.09	.011
BS 400B	Average of 55 Runs	.51	.96	1.83	64.33	32.05	.011
BS 400B	Standard Deviation	.02	.02	.03	.25	.26	.0005
BS 400B	% Relative Deviation	3.9	2.1	1.6	.39	.81	4.5
BS 400B	High Reading	.55	1.02	1.89	65.13	32.48	.012
BS 400B	Low Reading	.47	.92	1.76	63.87	31.45	.010

equipment, although it requires significantly more time to collect the data, subtract background, subtract interferences, and calculate concentrations. Each sample becomes a separate study, and it was beyond the capabilities of a single program to quickly measure trace elements in a wide range of routine industrial samples. Interferences can create problems even on elements that exist in relatively large concentrations; manganese at 0.5% could not be quantitated within acceptable limits in the presence of 18% chromium in a Stainless Steel.

CONCLUSIONS

The total analysis time for a full quantitative answer is about 4 minutes from the time that the safety interlocked door is closed until the complete composition is typed out on the printer. Qualitative data to quickly identify alloy types can be obtained in as little as 10 seconds. The Mobile XRF Analyzer has been in operation at the TRIDENT Refit Facility for over half a year now, and it has proven to be an effective tool for the speedy and non-destructive analysis of the critical parts used to keep the TRIDENT submarines in safe operation.

REFERENCES

1. J.C. Russ, "Fundamentals of Energy Dispersive X-ray Analysis," Butterworths & Co., Boston, (1984).

2. J.W. Otvos, G. Wyld and T.C. Yao, "Fundamental Parameter Method for Quantitative Elemental Analysis with Monochromatic X-ray Sources," paper presented at the 25th annual Denver X-ray Conference, (1976).

THE RECESSED SOURCE GEOMETRY FOR SOURCE EXCITED X-RAY FLUORESCENCE

ANALYSIS

C.A.N. Conde and J.M.F. dos Santos

Physics Department
University of Coimbra
3000 Coimbra, Portugal

ABSTRACT

Different geometries are considered for source excited energy-dispersive X-ray fluorescence (EDXRF) analysis systems, including the recessed source geometry introduced in the present work. The calculated physical excitation-detection efficiencies, for the side (or annular), central, receded and recessed source geometries are presented as a function of the target to source distance, for Ca, K, S and Si targets excited with a Fe-55 XBF-3 X-ray source and xenon filled gas proportional scintillation counters. The last two geometries present in general the highest efficiencies. The recessed source geometry present the best performance with peak efficiencies a factor of 3.3 better than those for the standard side or annular source geometries.

INTRODUCTION

There has been a growing interest in portable energy-dispersive X-ray fluorescence (EDXRF) systems, due to their usefulness in field applications. A few commercial models that use a radioactive source to excite the target are already available [1-3]. The performance of these source--excited EDXRF systems depends on the source-target-detector geometry. Recent results [4] for the calculation of the so-called "geometric excitation-detection efficiency", η , favour the then proposed, "receded source geometry" relatively to the central source and the annular or side source geometries. The receded source geometry (Fig. 1) showed geometric efficiencies [4], which were significantly better than those for the other geometries. To make an assessment of the potential of the various geometries, including the "recessed source geometry" introduced in the present work, we carried out calculations for the "physical excitation-detection efficiency", η_0, in situations close to reality, which therefore must also include the target self-absorption and the detector efficiency. We present the results of such calculations for an Fe-55 Isotope Products Laboratories XBF-3 source [5] and 100% pure Ca, K, S and Si targets. The detectors considered were of the gas proportional scintillation counter type.

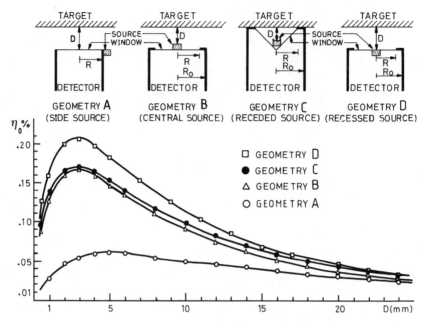

Fig. 1. Physical excitation-detection efficiencies, η_0, as a function of the distance, D, for a Fe-55 XBF-3 source exciting a Si target and various geometries with R=25mm and R_0=30mm detectors filled with Xe at atmospheric pressure.

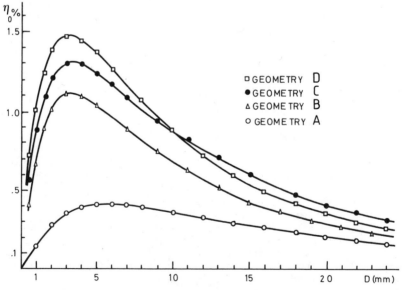

Fig. 2. Physical excitation-detection efficiencies, η_0, as a function of the distance, D, for a Fe-55 XBF-3 source exciting a Ca target and various geometries with R=25mm and R_0=30mm detectors filled with Xe at atmospheric pressure.

THE SOURCE-DETECTOR SYSTEM

The source-target-detector geometries considered in this work are shown in Fig. 1, and were the following: side source (A), central source (B), receded source (C) and recessed source (D). Geometries A, B and C have been described before [4]; geometry D, the recessed source geometry, is similar to the central source one, but with the source recessed in the middle of the window, so that its front surface is in the same plane as the detector window. The gas proportional scintillation counter type detectors considered in the present work, were assumed to be filled with xenon at atmospheric pressure. The windows have a radius R=25mm and, except for geometry A, the body of the detector has a radius R_0=30mm so that radiation entering it, close to the rim of the window, is easily absorbed in the lateral volume corresponding to the shell between R and R_0. For the Si and Ca targets, the efficiency for a smaller detector with R=12.5mm and R_0=17.5mm was also calculated.

The Fe-55 XBF-3 source [5] considered has an active diameter of 4mm, an outside diameter of 8mm and a thickness, h=5mm. For large distances, i.e., D>>h, geometries B and D are equivalent and can be taken as limiting cases for the receded source geometry (C). For D>>R geometries A, B and D are equivalent.

The targets are considered as thick with a large size flat surface perpendicular to the detector axis. D is the distance from the target to the front surface of the XBF-3 source. For geometry C, the rim of the detector is at the fixed distance of 1mm away from the target. Absorption in the detector window is assumed to be negligible. As before [4] whenever necessary a radiation shield for the source was assumed to prevent primary photons entering the detector directly. Only the 5.90 keV K_α radiation emitted by the source was considered in the excitation process of the target; the 6.49 keV K_β radiation, since it is weaker by one order of magnitude, will not affect significantly the relative performances of the different geometries. It is assumed that the secondary radiation emitted by the targets is only K_α. All the data here presented concern calculations for these X-ray lines.

THE CALCULATION METHOD

For all geometries, the calculations were carried out using a Monte Carlo method. The outline of the method is described next. A monoenergetic 5.90 keV K_α photon was assumed to be emitted isotropically (4π) by the source. A photon reaching the target was assumed to be absorbed after a path, ℓ, that was calculated by the Monte Carlo method, assuming an absorption length of 0.0165mm for Ca, 0.0346mm for K, 0.0234mm for S and 0.0284mm for Si; all the targets are assumed to be 100% pure.

From the absorption point, a secondary K_α photon was assumed to be emitted isotropically (4π) with a probability, F, given by

$$F = \omega_k \frac{r_k-1}{r_k} \cdot \frac{\tau}{\mu}$$

where ω_k is the X-ray fluorescence yield, r_k is the K-edge absorption jump factor, μ is the mass total attenuation coefficient and τ the mass photoelectric absorption coefficient; the values used were taken from references 6 and 7. Afterwards, the secondary photon can be either: self absorbed in the target, hit a dead surface (like the source or detector rim), escape through the region between the detector and the target, escape after passing through the detector or be absorbed in the detector.

The absorption lengths used were the following: 0.0436mm for Ca-K_α photons in calcium, 0.0695mm for K-K_α photons in potassium, 0.0192mm for S-K_α photons in sulfur and 0.0120mm for Si-K_α photons in silicon.

The calculated physical excitation-detection efficiency, η_0, is given by the ratio of the number of photons absorbed in the detector to the number of photons initially emitted by the source (4π).

RESULTS AND DISCUSSION

The calculated efficiencies, η_0, for the Fe-55 XBF-3 source with the different geometries are shown in Fig. 1 for a Si target and in Fig. 2 for a Ca target, assuming detectors with R=25mm and whenever applicable R_0=30mm. As shown η_0 is for the case of a Ca target about a factor of 10 smaller than the previously calculated [4] "geometric excitation-detection efficiency" η. For the case of a Si target η_0 is almost another order of magnitude smaller. This is a consequence of the target self-absorption and the low values for the F factor defined before (0.146 for Ca and 0.043 for Si). The ratio η_0/η is smaller for the receded source geometry than for the other geometries since for this geometry photons emitted at near grazing angles can still be detected and for these angles self-absorption is more important. This is also the reason why geometry D is, for small distances, more efficient than geometry C. Actually, the recessed source geometry, D, is the one with the highest efficiencies, η_0, which peak at 1.5% for Ca and 0.21% for Si, and so are a factor of 3.3 better than the peak efficiencies for the standard side (or annular) source geometry, A.

The lateral volume corresponding to the difference between R_0 and R in geometries B, C and D has the effect of increasing the detection efficiency for the harder X-rays. For a typical case with a Ca target and a geometry B detector with R=25 and D=3mm, the efficiency η_0 is equal to 0.98% for R_0=R, 1.03% for R_0=R + 2mm, 1.06% for R_0=R + 5mm and 1.06% for R_0=R + 10mm. For geometries C and D the effect of the lateral volume is less important than for geometry B. Therefore we can conclude that for a Xe filled counter at atmospheric pressure a lateral volume between 2 to 5mm thick is sufficient for Ca-K_α X-rays.

In Figs. 3 and 4 smaller detectors were considered with R=12.5mm and R_0=17.5mm. The efficiency η_0 for the different geometries have now peak values of 0.038%, 0.076%, 0.104% and 0.139% for geometries A, B, C and D, respectively, for a Si target and 0.26%, 0.47%, 0.80% and 1.02% for geometries A, B, C and D, respectively, for a Ca target. As shown, although the area corresponding to R for Figs. 1 and 2 is 4 times larger than the same area for Figs. 3 and 4, the physical excitation-detection efficiency η_0 is less than 4 times larger, which is in agreement with the variation with D/R of the geometric excitation-detection efficiency, η, calculated before [4]. As can be seen by comparison of Figs. 3 and 4 with Figs. 1 and 2 the advantage of geometries C and D over geometry B is enhanced for the smaller radius detectors.

We also have carried out calculations concerning the influence of the lateral volume corresponding to the shell between R_0 and R for the smaller detector (R=12.5mm). They show the same trend as for R=25mm case but with a stronger dependence on the R_0-R difference as can be expected.

Fig. 5 shows the efficiency, η_0, as a function of the source-to-target distance with geometry D for Ca, K, S and Si targets. The decrease of η_0 with the atomic number is as expected.

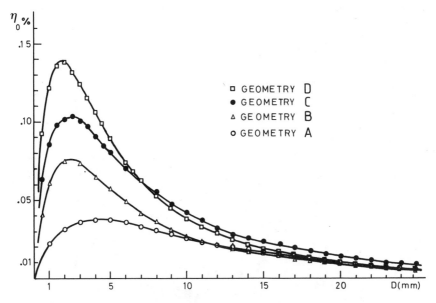

Fig. 3. Physical excitation-detection efficiencies, η_0, as a function of the distance, D, for a Fe-55 XBF-3 source exciting a Si target and various geometries with R=12.5mm and R_0=17.5mm detectors filled with Xe at atmospheric pressure.

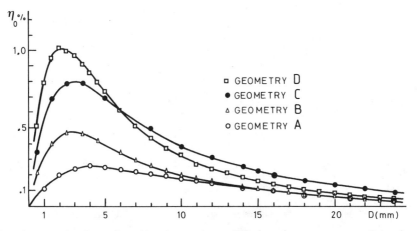

Fig. 4. Physical excitation-detection efficiencies, η_0, as a function of the distance, D, for a Fe-55 XBF-3 source exciting a Ca target and various geometries with R=12.5mm and R_0=17.5mm detectors filled with Xe at atmospheric pressure.

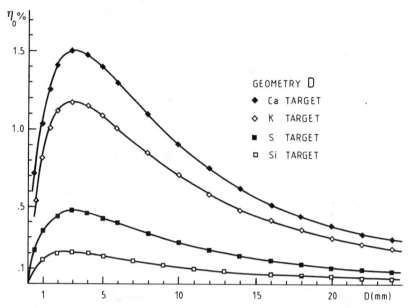

Fig. 5. Physical excitation-detection efficiencies η_0, as a function
of the distance, D, for a Fe-55 XBF-3 source exciting Ca, K,
S and Si targets with a recessed source geometry with R=25mm
and R_0=30mm detector filled with Xe at atmospheric pressure.

CONCLUSIONS

    We have shown that as far as the efficiency is concerned the receded
and the recessed source geometries are superior to the central and side (or
annular) source geometries for source excited energy-dispersive X-ray fluorescence analysis systems. The recessed source geometry is the best of them
all with peak efficiencies a factor of 3.3 better than those for the
standard side/annular source geometries.

ACKNOWLEDGMENTS

    This work was carried out with the support granted by INIC (Instituto
Nacional de Investigação Científica).

REFERENCES

1. ATX 100, available from: Aurora Tech. Inc., 620 West Center St., Bld.1,
   North Salt Lake, Utah 84054, U.S.A..
2. X-MET 840, available from: Outokumpu, Electronics Division, POB 27,
   SF-02201 Espoo, Finland.
3. 9604 Gold Assayer, available from: EG&G ORTEC, 100 Midland Road, Oak
   Ridge, Tenn. 37830, U.S.A..
4. J. M. F. dos Santos and C. A. N. Conde, to be published in Nucl. Instr.
   and Meth. - Part B (1985).
5. Isotope Products Laboratories, 1800 N. Keystone Street, Burbank,
   California 91504, U.S.A..
6. R. Tertian and F. Claisse, "Principles of Quantitative X-Ray Fluorescence Analysis", Heyden and Son Ltd., London (U.K.), 1982.
7. J. W. Robinson, "Handbook of Spectroscopy", vol. III, CRC Press,
   Boca Raton, Florida, 1981.

SAMPLE HOMOGENEITY AND REPRESENTATION IN X-RAY FLUORESCENCE

ANALYSIS OF FOODS*

K.K. Nielson and V.C. Rogers

Rogers and Associates Engineering Corporation
Salt Lake City, Utah   84110

A.W. Mahoney

Utah State University
Department of Nutrition and Food Sciences
Logan, Utah   84322-8700

INTRODUCTION

Method validation is vital for regulatory and nutrient analysis of foods, and imposes criteria for representative sampling and analysis along with more common requirements for precision and accuracy.[1,2] When solid food samples are directly analyzed by x-ray fluorescence (XRF), they must be homogeneous on a mass scale at least similar to that penetrated by the interrogating and fluorescent x-ray beams. This mass varies with x-ray energy, and hence with the elements being determined, and few data are available documenting the scale on which homogeneity can be assured. This paper examines the effective sample masses that are represented in measuring 21 different elements in pelletized dry food samples, and also the homogeneity of numerous foods and biological standard reference materials.

Two bodies of analytical data were used to evaluate the effective sample masses being analyzed, and the variations resulting from sample inhomogeneity and analytical representation. One resulted from analyzing National Bureau of Standards biological Standard Reference Materials (SRM's), and the other resulted from analyzing numerous other food materials to provide a broader range of food matrices. Analyses of variance on the resulting concentration data were used to partition inter-aliquot variations from other sources, and hence to examine the homogeneity effects.

EXPERIMENTAL

Nine SRM's were analyzed as obtained, including powdered milk (1549), oyster tissue (1566), wheat flour (1567), rice flour (1568), orchard leaves (1571), citrus leaves (1572), tomato leaves (1573), pine needles (1575), and bovine liver (1577a). Ninety-eight food samples from a previous

* This work supported in part by PHS Grants 1-R43-CA38519-01 and 5-P01-CA34243-02, National Cancer Institute, DHHS.

study,[3] were also analyzed. These had been freeze-dried and ground to a 0.1-1 mm particle size. About 0.5 g of dry sample powder was pressed into 3.2 cm diameter pellets under 2300 kg/cm^2 to yield self-supporting samples without backings or binders. Four replicate pellets were prepared from each SRM, and three replicate pellets were prepared from each food material.

To partition interaliquot variations from other analytical variations, four analyses were performed on each of the four SRM pellets, and one analysis was performed on each of the other food pellets. Each "analysis" consisted of four separate XRF spectra collected under vacuum using Gd, Ag, and Ge secondary excitation and 5 kV direct excitation (30, 20, 10 and 10 minutes, respectively, with a Kevex Model 700 spectrometer system). Spectrum analysis, compositing of intensities, and quantitation utilized the CEMAS program,[4,5] which automatically computed matrix corrections and calibrations for each sample based on its measured constituents and backscatter intensities. The resulting concentration data for Mg, Al, Si, P, S, Cl, K, Ca, Ti, Mn, Fe, Ni, Cu, Zn, As, Br, Rb, Sr, Mo, Ba, and Pb were stored directly on disk for subsequent statistical analysis. Mass absorption coefficients and sample thicknesses computed by CEMAS were also tabulated for computing the effective sample mass represented for each element in each sample matrix. The effective sample masses (M_{eff}) represented in the analysis of each element were computed from the intensity-weighted average depth as:

$$M_{eff} = \frac{2A}{\mu} \left[1 - \frac{\mu x \ \exp(-\mu x)}{1 - \exp(-\mu x)} \right]$$

where

A = intensity-weighted average beam area (1.17 cm^2)

x = total sample thickness (g/cm^2)

μ = $\sum(\mu_0 \csc\theta_0 + \mu_1 \csc\theta_1)$ = effective sample mass absorption coefficient (cm^2/g) (summed over all sample elements; subscripts 0 and 1 refer to incident and exit beams; θ are beam angles relative to sample plane).

Statistical analyses were performed for each element in each SRM material. They consisted of a one-way analysis of variance[6] to evaluate the significance of the interaliquot variations and to provide partitioned estimates of the inter-aliquot variations separate from the variations among replicate analyses of a given aliquot. Inter-aliquot variations were also estimated for the 98 other foods as the root-mean-square difference between the standard deviation among the three replicate pellet analyses and the values computed by CEMAS for statistical precision in an individual analysis. All variations were expressed as relative standard deviations (RSD = 100 S.D./Mean) for simplicity of interpretation.

RESULTS AND DISCUSSION

The effective masses of sample represented for each element in each analysis for the nine SRM's are summarized in Figure 1. The masses increase smoothly with x-ray energy due to greater x-ray penetration at higher energies (0.006 mm for Mg to 5 mm for Mo). Small discontinuities are also apparent due to changes in energy of the excitation beams for the four groups of elements. The lower line above 5 keV represents the actual masses computed for the 0.5 g pellets used in this study. The upper line gives an estimate of the maximum effective sample masses that could be obtained with thick samples (1000 mg/cm^2). The effective sample mass obtainable for elements in the Mg-Mn range is inherently limited to the 1-50 mg range, regardless of the excitation source or sample thickness. A larger beam collimator could potentially increase the sample masses by about a factor of two by viewing a larger area of the sample surface; however, the resulting effective masses are still well below the 250-500 mg aliquots recommended by NBS.[7] Replicate samples and analyses of both sides of sample pellets will also increase the total effective sample mass; however even this is impractical for elements such as Mg, Al, Si, and P, with effective masses about one percent of those recommended.[7] Matrix composition affects the effective masses by altering the sample's x-ray absorption properties; however the ranges illustrated by the bars in Figure 1 show that these variations are small compared to the dominant x-ray energy dependence.

The actual variations that resulted from inhomogeneities in analyzing such small sample masses were evaluated in terms of the RSD's among concentrations from different sample pellets (Table I). These inter-aliquot variations were partitioned from the total measured variation among the 16 replicate analyses (4 analyses x 4 pellets) using a one-way analysis of variance for each element in each of the nine SRM's. The resulting inter-aliquot variation was thus independent of the analytical variability from repeated analyses of the same sample, and should only reflect differences due to sample preparation and those resulting from element inhomogeneity. Since sample pelletizing did not utilize binders, backings, or supports that were observable by the x-ray spectrometer, inter-aliquot concentration variations due to inhomogeneity should dominate the resulting partitioned uncertainty.

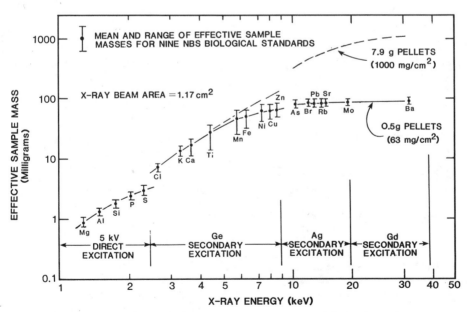

Figure 1.   Effective Sample Masses Analyzed for Nine Biological
            Standard Reference Materials Using 0.5 g Sample Pellets
            (solid line), and Computed Masses for 7.9 g Pellets
            (dashed line).

TABLE I

INTER-ALIQUOT VARIATIONS IN NINE BIOLOGICAL REFERENCE MATERIALS

| | Inter-Aliquot Relative Standard Deviations (%)* | | | | | | | | | | Mean Eff. Sample Mass (mg) |
	Powder Milk 1549	Oyster Tissue 1566	Wheat Flour 1567	Rice Flour 1568	Orchard Leaves 1571	Citrus Leaves 1572	Tomato Leaves 1573	Pine Needle 1575	Bovine Liver 1577a	Geom. Mean	
Mg	11.5	10.0	--	--	1.9	2.9	(3.2)	(1.4)	--	3.8	.9
Al	--	--	--	--	(5.7)	--	(7.4)	3.5	--	5.3	1.3
Si	--	(2.9)	(20.8)	(10.2)	(3.2)	(6.8)	(9.7)	(5.2)	--	6.8	1.8
P	(1.2)	(.5)	(.8)	(.2)	1.4	.6	.9	3.4	1.3	.9	2.5
S	1.1	(.3)	(2.3)	(2.1)	(1.6)	.9	(1.2)	(3.2)	1.3	1.3	3.1
Cl	.9	(.4)	(2.2)	(2.1)	(3.0)	(3.6)	(3.8)	(1.2)	2.0	1.8	7.4
K	1.2	.7	1.8	1.5	1.1	1.4	3.3	.8	1.5	1.3	14
Ca	1.2	7.7	5.6	5.9	1.4	1.1	1.9	2.0	8.1	2.9	17
Ti	--	(12.2)	--	--	(14.3)	--	(13.9)	(5.6)	--	10.8	26
Mn	--	.4	1.9	1.6	2.1	4.1	3.6	1.2	4.3	1.9	44
Fe	--	.7	10.4	26.9	1.4	3.0	1.8	2.8	1.7	3.0	49
Ni	--	--	--	--	4.9	--	--	(5.5)	--	5.2	56
Cu	--	.4	2.9	7.8	2.2	2.6	3.0	3.9	2.9	2.5	60
Zn	1.1	.6	1.3	1.4	2.0	1.4	1.9	(2.3)	3.0	1.5	62
As	--	3.5	--	--	4.6	--	--	--	--	4.0	74
Br	(1.0)	(.4)	(19.5)	--	(2.6)	(1.1)	(1.3)	(1.5)	(4.9)	2.0	76
Rb	(1.2)	7.6	--	(2.2)	3.2	3.9	2.5	1.3	5.6	2.9	77
Sr	(5.7)	3.8	--	--	(1.4)	.9	1.0	1.6	--	1.9	78
Mo	--	--	--	(11.4)	--	--	--	--	4.6	7.2	79
Ba	--	--	--	--	(2.9)	4.7	(2.2)	--	--	3.1	81
Pb	--	--	--	--	2.1	4.8	--	5.7	--	3.9	77
Geom. Mean	1.6	1.4	3.6	3.1	2.5	2.2	2.7	2.4	2.9	2.4	

* Values in parentheses are for elements whose concentrations are not certified by NBS.

As shown in Table I, the inter-aliquot variations were typically small, exhibiting an overall average of only 2.4 percent. Seventy-four percent of these variations were shown to be statistically insignificant ($P < 0.025$) compared to the other analytical uncertainties. The overall RSD for all elements detected above their quantitation limits was 8.6 percent. Thus, the uncertainty introduced by sample inhomogeneity and small effective sample mass is of minor importance.

Further insight into the nature and magnitudes of the homogeneity effects is gained by comparing the geometric mean variations for each element in Table I with the associated effective sample masses. If heterogeneities were relatively constant for all elements (i.e., an inert carbon-oxygen component mixed with a mineral-bearing component), an inverse correlation would be expected between these two parameters. However, their variation is nearly random ($r = -0.02$). The data in Table I instead suggest that each element may be associated with some characteristic mass fraction that is near unity for the homogeneous elements and relatively small for the inhomogeneous ones. For example, the elements P, S, Cl, K, Mn, and Zn tend to occur in major cellulose or fluid fractions of biological materials and hence their distribution is quite uniform. Conversely, the elements Si, Ti, and Fe have less important roles in major structural fractions, and tend to occur in mineral-enriched precipitates and inhomogeneous inclusions. Foreign impurities such as dust particles are also major contributors of Si, Ti, and Fe and may increase their inhomogeneity in certain biological materials. Elements such as Mg may have dual modes, occurring in the bulk cellulose matrix of materials such

TABLE II

INTER-ALIQUOT VARIATIONS IN FOOD ANALYSES

	Samples[a]	Inter-Aliquot Relative Standard Deviations (%)[b]												
		Mg	Si	P	S	Cl	K	Ca	Mn	Fe	Cu	Zn	Br	Rb	Sr
Sour Cream	5	10.8	--	5.3	5.7	3.9	4.3	3.8	--	--	--	5.4	.4	.1	2.7
Shrimp	5	5.4	--	3.9	3.3	4.2	3.7	7.6	--	14.9	4.5	2.7	1.5	--	8.1
Bread	5	3.9	--	1.2	1.1	4.2	4.7	4.8	3.4	5.3	3.3	2.2	.7	--	7.3
W.W. Bread	5	8.4	--	7.1	6.8	7.1	7.4	9.2	3.3	6.1	10.7	5.0	.5	--	3.1
Oatmeal	5	4.6	16.7	4.4	2.8	1.5	3.4	3.7	4.4	5.6	3.2	4.0	3.2	.9	7.9
Rice	4	--	11.7	6.0	6.8	9.8	5.7	8.9	5.8	11.6	.6	5.2	5.2	3.6	--
Saltines	5	8.3	20.2	2.1	2.3	5.4	3.7	1.2	2.6	7.2	4.5	1.2	.1	--	--
Waffles	4	.1	1.5	3.8	3.8	4.7	4.3	3.1	3.7	13.2	3.3	3.0	5.3	2.6	--
Beets	5	1.1	--	2.0	1.7	2.6	2.9	3.5	2.1	1.5	.2	1.0	1.3	1.6	.6
Carrots	5	.4	--	4.1	4.0	.8	1.3	.9	2.4	4.2	2.1	.5	1.8	4.1	.1
Corn	5	5.0	--	2.3	1.8	6.3	6.6	4.8	4.4	13.2	.1	4.9	.1	.7	--
Gr. Beans	4	7.5	1.8	3.6	2.4	1.9	1.7	3.8	.1	2.5	1.8	1.4	3.0	4.0	2.0
Peas	5	7.9	--	1.6	1.4	1.7	1.3	3.0	1.9	2.3	5.3	1.8	1.4	2.4	2.4
Potato	4	5.0	20.5	1.9	2.6	2.5	3.4	1.6	4.1	15.2	.1	2.5	--	4.4	2.3
Broccoli	5	2.5	--	1.6	1.9	2.4	1.8	2.2	1.1	2.2	2.1	1.8	.5	.1	.1
Spinach	4	2.1	9.6	1.2	1.3	1.5	1.7	7.0	1.4	1.9	3.1	1.0	.1	.1	4.2
Squash	4	3.0	--	1.8	1.8	1.9	2.3	3.1	4.7	3.8	1.2	.3	.1	.1	.1
Tomato	5	9.3	--	2.0	2.1	4.7	5.2	6.3	6.7	7.1	6.0	5.0	.9	2.6	2.0
Zucchini	5	1.7	2.8	1.0	.8	4.3	4.9	3.1	4.3	6.4	2.8	2.7	.2	.7	1.6
Onion	4	2.5	--	1.8	2.1	3.1	2.2	1.5	.1	3.1	2.6	2.1	6.3	6.2	.1
Cake	5	12.6	15.9	6.5	4.6	6.7	7.6	5.3	.1	18.5	2.0	.1	.1	--	--
Geom. Mean		4.1	7.7	2.6	2.5	3.2	3.3	3.5	3.2	5.6	2.5	2.1	1.4	2.3	2.9

a   Number of food samples. Three aliquots of each sample were analyzed to compute inter-aliquot
    RSDs.
b   Values of 0.1 are inserted when the estimate of individual analysis uncertainty exceeded the total
    observed standard deviation.

as the leaf standards, but in a minor or anomalous phase in milk and animal tissues. It is concluded from the uncertainties in Table I that the homogeneity of the SRM's provided acceptable inter-aliquot precisions for nearly all of the certified elements despite the small masses represented in the XRF analyses. Based on the computed uncertainties, it is estimated that 97 of the 135 analyses shown in Table I would have inter-aliquot variations less than 1 percent and 119 would have variations less than 2 percent if the recommended 500 mg aliquots were represented.

Analysis of the 98 food samples showed variations similar to those of the SRM's. As shown in Table II, their partitioned inter-aliquot variations were small and exhibited an overall geometric mean of 3.0 percent (2.3 percent if the 0.1 values are included. RSD averages for P, S, Cl, K, Mn, and Fe were generally higher than for the SRM's. Although the mean particle sizes of the foods was generally in the 0.1-1 mm range, and less precaution was taken to assure homogeneity, the effective sample masses were sufficient to provide generally acceptable inter-aliquot variations. Trends were again observed similar to those for the SRM's, including the tendency for Si and Fe to have high variability in certain materials.

In addition to the present estimates of relatively good homogeneity among small-aliquot masses, the XRF analysis geometry favors representative analysis by including a significant area (1.17 cm^2) of the pellet surface (7.9 cm^2). Thus, the number of particles represented even in the extreme case of 1 mm particles is nearly 150 (15,000 0.1 mm diameter particles), even though the entire mass of each is not completely penetrated for the lighter elements. To the extent that individual particles are homogeneous, the sampling over numerous particles greatly improves sample represen-tation. This also suggests that samples should be ground to particle sizes in the sub-millimeter range.

In summary, inter-aliquot variations in biological materials were found to be generally insignificant at effective aliquot masses much smaller than the 100-500 mg masses previously used for homogeneity assessments.[7] Although effective masses of only 0.9-80 mg were utilized, inter-aliquot variations were typically about 3 percent. Replicate aliquots should always be utilized to assess actual homogeneity, especially in unknown materials and for elements such as Ti, Si, and Fe that often exhibit erratic distributions in biological materials. The area of the sample analyzed should be maximized, and particles should be in the sub-millimeter range.

REFERENCES

1.   W. Horwitz, "Evaluation of Analytical Methods Used for Regulation of Foods and Drugs," Analytical Chemistry, 54, 67A-76A, 1982.
2.   G.R. Beecher and J.T. Vanderslice, "Determination of Nutrients in Foods: Factors That Must Be Considered," Modern Methods of Food Analysis, eds., K.K. Stewart and J.R. Whitaker, West Port, CT: AVI Publishing, p. 29-55, 1984.
3.   A.W. Mahoney, et al., "Nutrient Composition of Foods Obtained from Retail Outlets in Utah," Utah Agricultural Experiment Station Research Report 53, Logan, Utah, November 1980.
4.   K.K. Nielson and V.C. Rogers, "Comparison of X-ray Backscatter Parameters for Complete Sample Matrix Definition," Advances in X-ray Analysis, 27, 449-457, 1984.
5.   K.K. Nielson and V.C. Rogers, "Accurate X-ray Fluorescence Analysis of Environmental Materials without Standards," Trans. Am. Nucl. Soc. 49, 146-147, 1985.
6.   J.C.R. Li, "Statistical Inference," Ann Arbor: Edwards Bros., Inc., 1964.
7.   U.S. Department of Commerce, National Bureau of Standards, "Certificate of Analysis," for the Standard Reference Materials: Tomato Leaves, Rice Flour, Wheat Flour, Non-Fat Milk Powder, Oyster Tissue, Orchard Leaves, Citrus Leaves, Pine Needles, and Bovine Liver.

ANALYSIS OF REFRACTORY MATERIALS BY

ENERGY DISPERSIVE X-RAY SPECTROMETRY

Jim Parker

Manville Research and Development Center
P.O. Box 5108
Denver, Colorado 80217

Wayne Watson

Tracor Xray, Inc.
345 East Middlefield Road
Mountainview, California 94043

INTRODUCTION

Silica-alumina refractory materials are used as insulation
materials in high temperature applications. Such materials are
amenable to x-ray analysis (1). Wavelength dispersive x-ray
spectrometry is used in our laboratory for the analysis of a variety
of refractory materials. An analysis procedure was needed for a
quality control project that would provide major and minor element
determinations in silica-alumina refractory materials. The
requirements for the analysis scheme were relative accuracy and
precision to be better than one percent. The method had to be rapid,
simple to use, and inexpensive. Energy dispersive x-ray spectrometry
and a borate fusion technique were chosen as the method of choice for
this application. Described in this paper are the sample and standard
preparation procedures, data reduction methods, and analytical results.

EXPERIMENTAL

Instrumentation

The Rigaku S/Max, a wavelength dispersive spectrometer, was used
to collect intensities for testing the homogeneity of the standard
samples used for calibration. The S/Max was equipped with an
end-window rhodium x-ray tube and controlled by a DEC PDP 1103
computer. LiF200, PET, and RX-4 analyzing crystals were used.

A Tracor 4020 energy dispersive spectrometer was selected as the
instrument to be used in the quality control program. The 4020 was
introduced at the Denver X-ray Conference in 1983. The 4020 is a 30

KV system and was equipped with a rhodium x-ray tube, has vacuum
capablity, sample spinning, and a ten position sample loader.

Fusion Devices

A Herzog HAG 12 automatic fusion machine was used to produce the
standard samples used for calibration. Graphite crucibles and Pt/Au
crucibles were used in the preparation of the standards. A rocking
device was used with the Pt/Au crucibles for physical mixing of the
melt during the fusion process.

A manual fusion device described by Bowling, Ailin-Pyzik, and
Jones (2) was modified for use in the quality control program. The
modified device had three additional features not in place on the OCF
manual device. A Rotron DR 303 regenerative blower was coupled to the
fusion device. This blower is used to pull a vacuum through the
bottom of the graphite crucibles used in the quality control fusion
process. The vacuum pulls any air bubbles from the melt and also
serves to quench the melt. Additional cooling came from a 1/4 inch
air line plumbed to blow air on top of the melt. The 1/4 inch air
line is connected to a compressed air line. The air flow is
controlled by a needle valve and solenoid valve. The on/off cycles of
the solenoid valve and blower are controlled by an Eptak Eagle Signal
PC 210 Controller. When initiated, the controller was programmed to
turn the blower on for three minutes. At the end of the three minute
cycle, the blower was turned off and the solenoid valve opened
allowing the air blast to further cool the melt from the top. The air
blast is automatically stopped after four minutes. The heat source
for the fusion process is a laboratory muffle furnace capable of
maintaining a constant temperature of $1100^{\circ}C$.

Selection of Fusion Conditions

This particular fusion technique was selected because it is
possible to produce a specimen that can be analyzed with no treatment
beyond cooling of the melt to room temperature (2). That is, the fire
polished top side of the melt can be used for analysis if the
analytical surface is confined to an area covered by a 25 mm diameter
mask centered on the top of the sample. Silica and alumina containing
materials produce viscous melts and are difficult to homogenize. A
homogeneous melt would require proper attention to the selection of.
the flux, the sample to flux ratio, and the time and temperature of
the fusion. The ideal flux would have a melting point low enough to
be effective in a laboratory muffle furnace, and would produce a
stable specimen. Four fluxes were initially tested for use in this
project. They are listed with their melting points in Table 1. The
two low melting fluxes did not produce stable melts with the Herzog
fusion machine. They were eliminated from consideration after several
attempts to produce stable specimens at different sample to flux
ratios failed.

Lithium tetraborate is commonly used as a flux in fusion
applications and its properties are well known. The eutectic
compostion is used less often, but was studied for use as a flux for
refractories and minerals by Bennett and Oliver (3). The exact
compositon of the eutectic and its properties can be found in a study

Table 1   **Fluxes used in Evaluation of Fusion Procedure**

Flux	MP^{o}C
Lithium Tetraborate	920
Eutectic composition	840
Lithium Tetraborate/LiF	780
Lithium Tetraborate/Li$_2$CO$_3$	740

by Sastry and Hummel (4). The eutectic composition is approximately one part lithium tetraborate to four parts lithium metaborate. It is reputed to have the superior melting characteristics of lithium metaborate for high silica samples, but retains the ability of lithium tetraborate to fuse high alumina materials.

Stable melts were obtained with lithium tetraborate from sample to flux ratios of 10:1 to 3:1. The eutectic flux produced a stable melt at a ratio of 3:1, but became very difficult to form stable melts at a ratio of 7:1. At this point it was decided to use a sample to flux ratio of 5:1. That decision was based on several points. This ratio was in the middle of the range of ratios successfully prepared from the more limited eutectic mixture. This ratio was also a good compromise between the need to use a low ratio for optimum peak to background ratios for effective energy dispersive analysis, and the high ratios needed for reduction of interelement effects and optimum melt fluidity.

An experiment was set up to optimize the fusion conditions necessary to produce homogeneous specimens. The parameters investigated are shown in Table 2. A button was produced from each combination of conditions shown in Table 2. The bottom side of the resulting buttons were ground flat with 180, 320, and 600 grit paper, and polished with CeO$_2$. Intensities from the fire polished top side and hand polished bottom sides of the buttons were compared for melt homogeneity. The "worst case" condition for each variable was used as the standard operating condition for all of the refractory matrices. The use of one set of conditions for all matrices eliminated any confusion when switching from one analytical program to the next. Lithium tetraborate was able to homogenize RM "A" and RM "B" at 1100°C, but required a temperature of 1200°C to adequately homogenize RM "C". The eutectic compostion was effective in producing homogeneous melts for all matrices at 1100°C. The eutectic composition is more expensive than lithium tetraborate, but it was felt that it would be easier to use that flux at 1100°C than it would be to try to maintain a muffle furnace at 1200°C. The final operating conditions are shown in Table 3.

Table 2   **Fusion Parameters**

Parameter	Conditions
Time (minutes)	10, 15, 20, 25, 30
Temperature (^{o}C)	1000, 1100, 1200
Flux	Lithium Tetraborate Eutectic composition
Matrix	RM "A" RM "B" RM "C"

Table 3   **Final Operating Conditions**

Parameter	Condition
Time	30
Temperature (^{o}C)	1100
Flux	Eutectic composition

Preparation of Standard Samples

Standard samples were prepared for four different refractory matrices. The nominal compositions of these refractory materials are shown in Table 4. A separate group was set up to analyze RM "C" contaminated by minor amounts of ZrO_2. Duplicate standards were prepared by weighing known amounts of spec-pure oxides into Pt/Au crucibles. The appropriate amount of flux was added and the mixture stirred with a platinum rod. Hydrobromic acid was added as a non-wetting agent. The mixture was then fused for twenty minutes at 1100°C. Physical mixing of the melt was achieved by use of a rocking device attachment on the HAG 12. The duplicate glass buttons were then crushed, mixed, split, and refused for fifteen minutes in graphite crucibles. After quenching, the bottom sides of the standard buttons were ground with 180, 320, and 600 grit paper, and polished with CeO_2. The duplicate buttons were tested for homogeniety by comparing x-ray intensities from the fire polished top side and hand polished bottom side of the button. Sufficient counts were collected to give 0.5 % relative precision for the major elements. Non-homogeneous standards were rejected and remade from the beginning step. Calibration curves were set up for each analytical group on the wavelength dispersive system to further test the integrity of the standard samples. These calibration curves were later used to cross-check the results obtained with the energy dispersive system.

Table 4   **Nominal Composition of Refractory
Materials in Percent**

	RM "A"	RM "B"	RM "C"
SiO_2	50.0	55.0	53.0
Al_2O_3	35.0	42.0	47.0
ZrO_2	15.0		(1.0)
Cr_2O_3		3.0	

Preparation of Quality Control Samples

A representative sample of approximately six grams is pulled from
the bulk sample. The sample is ground for five minutes in a tungsten
carbide ball mill. Three grams of ground sample is weighed into a
four ounce glass jar and 15 grams of flux is added to the jar. The
weight of flux has been corrected for loss-on-ignition (LOI). The
sample/flux mixture is intimately mixed by shaking for five minutes in
the ball mill. The graphite crucibles used for the fusion process
have an outside diameter of 40 mm and are 10 mm deep. That volume is
not sufficient to hold 18 grams of the sample/flux mixture and must be
pressed into a briquette. A Herzog HAP press is used in this step.
The resulting briquette is approximately 30 mm in diameter and 25 mm
high, and fits nicely into the graphite crucible. The crucible and
sample are placed in the muffle furnace and the fusion takes place
with no further manipulation by the operator. At the completion of
the fusion, the crucible is removed from the furnace and placed on the
fusion device. The vacuum is initiated by pressing the start button.
The remainder of the quenching process takes place automatically as
previously described. Upon cooling, the sample is placed in a sample
holder designed to limit the analytical area to the center of the
fusion button. This was done by machining a Tracor sample holder to
accommodate a 40 mm diameter disk. A 25 mm diameter mask area is
machined in the center of the holder. The sample is then submitted to
the spectrometer for analysis.

The precision of the sample preparation step was tested.
Refractory matrix "C" was used as the test sample as it is the most
difficult to homogenize. Five replicate samples were prepared as
described above and analyzed on the energy dispersive spectrometer.
The results are shown in Table 5.

Table 5   **Precision of Sample Preparation Step**

	Average Value	Standard Deviation
SiO_2	53.94	0.12
Al_2O_3	45.51	0.07
ZrO_2	0.498	0.009

RESULTS AND DISCUSSION

<u>Calibration and Data Reduction</u>

Ten standard samples were used for the calibration of each refractory matrix. The duplicates were retained for use as back-ups and for quality control checks. The concentration range for the major elements in each matrix was $\pm$ 5 % from nominal. The range of Cr_2O_3 concentrations in RM "B" was from 0-4.5 %, and 0-2.0 % ZrO_2 in the zirconia contaminated RM "C" matrix. The excitation conditions for each matrix are shown in Table 6. Counts were accumulated for five hundred seconds livetime. Sufficient counts were collected to provide better than 0.5 % relative precision. Gross intensities, net intensities, and intensities from a Tracor overlap correction program, XML, were fit with linear and quadratic regressions. An intensity correction algorithm was used to account for interelement effects. Analyte peaks were also ratioed against the rhodium Compton scatter peak. The results shown in Table 7 represent the best fit for each analyte in each refractory matrix. As a general rule, the intensity correction model using net or Compton ratioed intensities provided the best precison and were used for the actual analysis. The precision of all of the fitting routines and the Compton ratio technique were similar, and none offered a significant advantage.

Table 6   **Excitation Conditions**

Matrix	Analyte Line			Excitation Conditions	
RM "A"	Si Ka	Al Ka	Zr La	8 KV	0.15 mA
RM "B"	Si Ka	Al Ka	Cr Ka	10 KV	0.10 mA
RM "C"	Si Ka	Al Ka		8 KV	0.15 mA
RM "C"	Zr Ka			24 KV	0.30 mA

Table 7   **Mean Error in Percent**

	RM "A"	RM "B"	RM "C"
SiO_2	0.09	0.08	0.07
Al_2O_3	0.06	0.08	0.08
ZrO_2	0.08		0.001
Cr_2O_3		0.006	

The Compton ratio technique is a valuable tool if there is a problem with the analytical surface of the fusion button not being flat. During the calibration of the chromium refractory matrix, one of the standards was not properly seated in the sample holder, and the standard was measured with its analytical surface not flush against the sample mask. The regression analysis showed an obvious problem with that standard. The analyte intensities of all standards in that set were ratioed against the Compton scatter peak. The improvement in the problem standard is shown in Table 8.

Table 8   **Effect of Compton Ratio**

	Chemical	Calculated	Compton Ratio
SiO_2	58.07	57.20	58.07
Al_2O_3	38.44	37.40	38.62
Cr_2O_3	3.50	3.41	3.49

Accuracy of Results

The accuracy of the quantitative energy dispersive analyses were checked by analyzing thirteen samples with known or accepted chemistry. These check samples included samples spiked with pure oxides, samples cross checked by quantitative wavelength x-ray spectrometry, and standard samples not used in calibration. The thirteen samples represented all three refractory matrices. All results showed deviations less than one percent relative. The results from these analyses are shown in Table 9.

Table 9   **Average Deviation of EDX Results From Accepted Values in Percent**

	Average Deviation
SiO_2	0.09
Al_2O_3	0.19
ZrO_2	0.06
Cr_2O_3	0.01

SUMMARY

    Energy dispersive x-ray spectrometry has been used to determine major and minor elements in a silica-alumina refractory matrix. The precision and accuracy of the analysis is better than one percent relative. A manual fusion device was used in the sample preparation procedure. A laboratory muffle furnace was used successfully as the heat source for the fusion. Graphite crucibles were used in the fusion, and the resulting melt was analyzed with no grinding and polishing steps. A eutectic compositon of one part lithium tetraborate and four parts lithium metaborate was used as the fusion flux. This flux was found to be superior to lithium tetraborate in producing homogeneous melts at a low sample to flux ratio. The total analysis time for this procedure is less than ninety minutes.

REFERENCES

1) D.G. Ashley and K.W. Andrews, "Analysis of Aluminosilicate Materials by X-ray Fluorescence Spectrometry", Analyst, 97:841 (1972).

2) G.D. Bowling, I.B. Ailin-Pyzik, and D.R. Jones IV, "A Rapid, Low Cost, Manual Fusion Sample Preparation Technique for Quantitative X-ray Fluorescence Analysis", Advances in X-ray Analysis, Vol. 27, p. 491 (1984).

3) H. Bennett and G.J. Oliver, "Development of Fluxes for the Analysis of Ceramic Materials by X-ray Fluorescence Spectrometry", Analyst, 101:803 (1976).

4) B.S.R. Sastry and F.A. Hummel, "Studies in Lithium Oxide Systems: I, $Li_2O \cdot B_2O_3 - B_2O_3$", Journal of the American Ceramic Society, 41:7 (1958).

CONTAMINATION OF SILICATE ROCK SAMPLES DUE TO CRUSHING AND GRINDING

Mark A. Tuff

New Mexico Bureau of Mines and Mineral Resources
Campus Station
Socorro, NM   87801

ABSTRACT

This study was done to learn which elements are being contributed
as contaminants in the sample preparation equipment used at the New
Mexico Institute of Mining and Technology (NMIMT). The apparatus investigated included a steel jaw crusher, a ceramic jaw crusher, two aluminum
plates as a crusher and grinder, a steel spinning plate pulverizer, an
alumina ceramic spinning plate pulverizer, an automated agate mortar and
pestle, a high speed spectromill grinder with a tungsten carbide mortar
and pestle, and the halves of a sliced quartzite cobble. Major-element
oxides and trace elements were analyzed by x-ray fluorescence. Some of
the trace elements were analyzed by instrumental neutron activation
analysis. The expected contamination of ferroalloy elements from steel
were present in varying quantities, as were tungsten, tantalum and cobalt
from the tungsten carbide. There was unexpected contamination of sulfur
and lead from the steel pulverizer.

INTRODUCTION

In any chemical analysis of geological materials the first preparation step is the reduction of the rock to a fine powder. This is accomplished by crushing and grinding the sample to successively smaller
particles. During each of these crushing and grinding steps contaminants
will be introduced that will affect the accuracy of the analysis. The
contaminants may come from improper or incomplete cleaning of the apparatus or they may come from materials in the apparatus. If the observed
contamination comes from the apparatus itself the analyst should know
which elements are being contributed and plan his preparation method and
analyses accordingly. This study was done to determine which elements
are being introduced as contaminants, in what concentrations they are
present, and which apparatus they came from.

The study was limited to the crushing and grinding equipment that is
used at NMIMT. To insure that the observed contamination tests simulated
normal sample preparation, the following steps were taken with each
apparatus: 1) it was cleaned by preparing a blank sample of silica sand;
2) an equal amount of the study rock was prepared and then discarded; 3)
the study rock was then prepared with the apparatus and retained for

565

analysis.

SELECTION OF THE STUDY MATERIAL

A quartzite cobble was selected as the study material for four
reasons. First, a quartzite cobble of more than 99.5% SiO_2 with less than
0.5% Al_2O_3 is quite hard (approximately seven on the MOHS scale); any
contamination of the sample from the crushing or grinding media would be
more likely to show up in such a hard rock. Second, quartzite is homogeneous allowing ready detection of trace contaminants. Third, quartzite
has a simple composition because most of it is SiO_2; therefore any
variation in the concentrations of the elements studied would be more
readily aparent. Finally a quartzite cobble can be sliced and used as a
crusher and grinder. Using this quartzite "grinder" will produce a
sample free from any outside source of contamination. After splitting
the samples for analysis not enough of the cobble remained for use as a
"grinder", so a quartzite of similar composition was substituted.

The quartzite cobble selected for this study was eight inches in
diameter and pale gray. The cobble was broken with a sledge hammer on a
steel plate and the rind was removed and discarded. Representative
splits were taken and visually inspected for homogeneity.

THE APPARATUS INVESTIGATED

The crushers investigated were a steel jaw crusher, an alumina
ceramic jaw crusher, two aluminum plates with the crushing force supplied
by a two pound sledge hammer, and the halves of a sliced quartzite
cobble.

The pulverizers investigated were a steel spinning plate pulverizer,
an alumina ceramic spinning plate pulverizer, the two aluminum plates,
and the two halves of the quartzite cobble.

The final grinders investigated were an agate automatic mortar and
pestle, a high speed spectromill using a tungsten carbide mortar and
pestle, and the two halves of the sliced cobble.

Contamination increases with increased exposure time to the crushing
and grinding media. During this study the crushing and pulverizing was
limited to two passes through each of the jaw crushers and pulverizers.
The crushing with the aluminum plates and the quartzite cobbles was done
by hand, and the it was stopped when the particles were crushed or ground
to a size equal to that of the sample prepared with the automated equipment. The time allowed for fine grinding was entirely dependent on the
apparatus used, and therefore varied widely. Five minutes was sufficient
time for the spectromill to reduce 10 grams of the sample to a fine
powder, while it took more than 30 minutes for the automated mortar and
pestle to achieve the same result. The fine grinding was dependent on
the particle size, and grinding was halted when the sample could pass
through a 200 mesh sieve.

PREPARATION OF THE SAMPLES

The first sample was prepared using the ceramic jaw crusher, the
ceramic pulverizer, and then ground with the automated mortar and pestle.
The second sample was prepared using the steel jaw crusher, the steel
pulverizer, and then ground in the automated mortar and pestle. The

third sample was crushed in the steel jaw crusher, pulverized with the
two aluminum plates, and then ground with the automated mortar and pestle. The fourth sample was crushed and pulverized with the aluminum
plates, and then ground with the automated mortar and pestle. The fifth
sample was crushed, pulverized, and ground with the two halves of the
quartzite cobble. The sixth sample was crushed and pulverized with the
quartzite cobble halves, and then ground with the automated mortar and
pestle. The seventh sample was crushed and pulverized with the two
cobble halves, then ground with the tungsten carbide spectromill. A
separate sample of the crushing cobble was prepared by crushing and
pulverizing it with the ceramic equipment, and then grinding it in the
automated mortar and pestle. The apparatus used to prepare the samples
are listed in Table 1.

ANALYTICAL METHODS USED

Major element oxides were determined by x-ray fluorescent spectrometry (XRF) using the method of Norrish and Hutton (1969). The samples
were fused using type 105 spectroflux and a flux-to-sample ratio of five
to one (5:1). The spectrometer was calibrated using the mean count
ratios of five separate runs of 10 international rock standards.

Major element oxides and sulfur were also determined by XRF using
the fundamental parameters program of Criss Software (Criss,1980). Two
standards were used in this determination: the international rock standard SY-3 and an ultra pure silica of Spex Industries.

Rb, Sr, Zr, Y, Nb, Th, Pb, U, Cu, Ni, Zn, and Ga were determined by
XRF using a matrix correction based on the measurement of the Rhodium
Compton radiation . Cr, V, and Ba were determined by XRF using calculated mass absorption coefficients to correct for matrix effects. Each
of these trace element programs utilizes a ten point linear regression
calibration, using 10 international rock standards. W, Ta, Cr, and Co
were determined by instrumental neutron activation analysis (INAA).

OBSERVATIONS

Results from the XRF and INAA analyses are presented in Table 1.
Significant aluminum contamination was not observed (Table 1) from either
the aluminum plates or the alumina ceramic equipment.

Although the expected contamination of ferroalloy elements from
steel was present in the sample pulverized in the steel pulverizer, it
was not present in the sample that was only crushed in the steel jaw
crusher. There was also significant contamination of lead and sulfur in
the sample crushed and pulverized in the steel equipment (Table 1).

There was a significant amount of tungsten, tantalum, and cobalt in
the sample ground in the tungsten carbide (Table 2).

There is an increase in the amount of lead, barium, zinc, and copper
in the sample ground with the quartzite cobble. However our determination
limit for copper is 20 ppm, and the copper is only 8 ppm over our determination limit in this sample.

The quartzite cobble that was used as a crusher and grinder was not
the same rock as the study rock, however, it has a very similar composition as that of the study rock (Table 3).

Table 1

OVERVIEW OF RESULTS

The first three lines of the table list the equipment used to prepare the
sample. Al is for the aluminum plates. W. C. is for tungsten carbide.
The * values are referenced in the text. Values with ND are lower than
the lower limit of determination. Fe2O3t is total iron expressed as
Fe2O3.

Preparation details

CRUSHER	Ceram	Steel	Steel	Al	Cobble	Cobble	Cobble
PULVER.	Ceram	Steel	Al	Al	Cobble	Cobble	Cobble
GRINDER	Agate	Agate	Agate	Agate	Cobble	Agate	W.C.

Oxide or element in %

SiO2	99.89	98.52	99.92	99.84	99.81	99.90	99.74
TiO2	0.053	0.047	0.083	0.076	0.085	0.049	0.054
Fe2O3t	0.17	1.60*	0.22	0.20	0.22	0.22	0.24
Al2O3	0.30	0.24	0.32	0.28	0.24	0.29	0.27
MnO	0.001	0.009*	0.003	0.001	0.002	0.002	0.001
MgO	<0.01	<0.01	<0.01	<0.01	<0.01	<0.01	<0.01
CaO	<0.01	<0.01	<0.01	<0.01	<0.01	<0.01	<0.01
Na2O	<0.01	<0.01	<0.01	<0.01	<0.01	<0.01	<0.01
K2O	<0.01	<0.01	<0.01	<0.01	<0.01	<0.01	<0.01
P2O5	0.01	0.01	0.01	0.01	0.01	0.01	0.01
S	0.008	0.024*	0.002	0.015	0.008	0.002	0.004
SUM	100.42	100.44	100.56	100.42	100.38	100.47	100.32

Element in ppm

Rb	ND	ND	ND	ND	ND	ND	ND
Sr	8	8	8	7	12	9	9
Zr	36	33	38	32	40	32	34
Y	5	4	4	4	5	5	4
Nb	ND	ND	ND	ND	ND	ND	ND
U	ND	ND	ND	ND	ND	ND	ND
Th	ND	ND	ND	ND	ND	ND	ND
Pb	9	29*	8	6	17	9	9
Cr	0.6	43.6*	1.5	0.3	3.7	1.4	2.1
V	1.8*	2.2*	2.7	1.1	1.0	1.0	1.5
Ba	10	9	10	11	28*	13	11
Ga	ND	ND	ND	ND	ND	ND	ND
Zn	10	12	6	6	19*	8	2
Cu	ND	ND	ND	ND	28*	ND	ND
Ni	5	6	2	4	6	19*	8
W	ND	ND	ND	ND	ND	ND	422*
Ta	0.1	0.1	0.1	0.1	0.1	0.1	6.0*
Co	0.1	0.9*	0.1	0.1	0.2	0.1	34.6*

Table 2

COMPARISON OF FINAL GRINDING APPARATUS

The notable difference between the final grinding apparatus are presented below. Values listed as ND are below the determination limit for this instrument. Fe203t is total iron expressed as Fe203.

Final grinder	Quartzite Cobble	Agate mortar and pestle	Tungsten crabide Spectromill
Al203 (%)	0.24	0.29	0.27
Fe203t (%)	0.22	0.22	0.24
MnO (%)	0.002	<0.001	0.001
S (%)	0.008	0.002	0.004
Sr (ppm)	12	9	9
Zr (ppm)	40	32	34
Pb (ppm)	17	9	9
Ba (ppm)	28	ND	ND
Zn (ppm)	19	8	2
Cu (ppm)	28	ND	ND
Ta (ppm)	0.1	0.1	6.0
W (ppm)	ND	ND	422
Co (ppm)	0.2	0.1	34.6

CONCLUSIONS AND RECOMMENDATIONS

The expected contamination of iron, manganese, cobalt, and chromium from the steel equipment were present. However it was present only in the sample that was crushed in the steel jaw crusher and pulverized in the steel pulverizer. This contamination was absent in the sample that was crushed in the steel and pulverized with the aluminum plates. The expected contamination of vanadium from the steel equipment was present in both of the samples that were crushed in the steel jaw crusher. The expected contamination of tungsten, tantalum, and cobalt were present in the sample ground in the tungsten carbide, but in much higher concentrations than were expected (Table 2).

Some surprising conclusions can be drawn from this study. Crushing with aluminum plates does not contribute a significant amount of aluminum contamination. The Al203 values for all of the samples are very close, and the Al203 contributed from the aluminum plates is far less than might be expected. The higher barium and zinc values in the sample prepared by crushing, pulverizing, and grinding with the cobble is surprising.

Lead and sulfur contamination persists even after scouring the steel plates with quartz. Although some contamination of sulfur from steel should be expected, the level of sulfur is much higher than is suggested from the stoichiometry. This contamination is probably due to the previous preparation of sulfide ores in this apparatus. The majority of the ores prepared at this lab contain a larger amount of pyrite than other sulfide ore minerals. Pyrite is also frequently associated in the gangue minerals. The iron contribution from the steel would dwarf the contribution from the pyrite. This would account for the the elevated sulfur value, as well as the lower than expected values of lead, copper, and zinc.

Table 3

COMPARISON OF STUDY ROCK AND CRUSHING ROCK

Both rocks were prepared with the ceramic jaw crusher, the ceramic
pulverizer, and ground in the agate mortar and pestle. Values listed
with ND are below the determination for this instrument. Fe2O3t is total
iron listed as Fe2O3.

	STUDY ROCK	CRUSHING COBBLE
Oxide or element as %		
SiO2	99.89	99.94
TiO2	0.053	0.055
Al2O3	0.30	0.10
Fe2O3t	0.17	0.10
MnO	0.001	0.001
MgO	<0.01	<0.01
CaO	<0.01	<0.01
Na2O	<0.01	<0.01
K2O	<0.01	<0.01
P2O5	0.01	0.01
S	0.008	0.007
SUM	100.43	100.21
Element in ppm		
Rb	ND	ND
Sr	8	8
Zr	36	30
Y	5	5
Nb	ND	2
U	ND	ND
Th	ND	ND
Pb	9	13
Cr	0.6	0.5
V	1.8	ND
Ba	10	9
Ga	ND	ND
Zn	10	9
Cu	ND	ND
Ni	5	5
W	ND	ND
Ta	ND	ND
Co	0.1	0.2

Because of the lack of enough sample the cobble used as a crusher and grinder was not the same cobble as that used in the study. However, with the exception of Al2O3, the agreement between the composition of the study rock and the crushing cobble is very good (Table 3).

There is a high nickel value in the sample ground in the agate mortar and pestle. The nickel value would be elevated in all of the samples crushed with the cobble if the contamination were from the cobble. The contamination is probably from the agate mortar and pestle, but does not show up in any of the other samples ground with it because of dilution effects.

It is recommended that each analyst investigate the preparation equipment that is used in his lab. He should discover which elements are being introduced as contaminants, in what quantities, and from which apparatus they come. The elements to investigate should include those that are present in the crushing or grinding surfaces, and those that are used in the production of the alloys in the equipment. Elements that may inhibit or enhance the element(s) analyzed for should also be investigated. The analyst should also investigate any elements that he analyzes routinely. He should select a study sample of a simple composition with a high degree of hardness that will reflect contaminants most readily. With this information the analyst can more effectively plan his preparation method and his analyses.

REFERENCES

Criss, J., 1980, "Fundamental-Parameters Calculations on a Laboratory Microcomputer", Advances in X-ray Analysis, Vol. 23, pp. 93-97.
Norrish K., and B. Chappell, 1977, in: Physical Methods in Determinative Mineralogy, J. Zussman, ed.: Academic Press
Norrish K., and J. T. Hutton, 1969, "An Accurate X-ray Spectrographic Method for the Analysis of a Wide Range of Geological Samples": Geochim. Cosmochim. Acta 33, pp. 431.
Thompson, G. and D.C. Bankston, 1970, "Sample Contamination From Grinding and Sieving Determined by Emission Spectroscopy": Applied Spectroscopy, Vol. 24, Number 2, pp. 210-219.

ON-STREAM WDXRF-ANALYSIS OF

LOW CONCENTRATIONS

Heikki Sipilä and Jouko Koskinen

Outokumpu Electronics
P.O. Box 27, 02201 Espoo
Finland

INTRODUCTION

Outokumpu Company has gone a long way in the last 20 years in the development and application of WDXRF-techniques in on-stream analysis of slurries in ore concentrators. The applications in which the samples were in solution form came only about 4 years ago when these analyzers were used for the first time in hydrometallurgical plants. During the past 15 years many improvements have been made in most parts of the equipment to make the performance and reliability better. The most important work to improve the analytical capability of the analyzer has been done on the spectrometers and detectors. This has resulted in better sensitivities and has given the possibility to go down to ppm levels in many determinations. The technical details of the improvements have been presented in several papers (1-4).

ON-STREAM WDXRF ANALYZERS

The results presented in this paper have been obtained using two commercially available on-stream WDXRF analyzers manufactured by Outokumpu. They have so many features in common that they are called by the same name, Courier. As an indication of the size of the equipment one is called Courier 300 and the other one Courier 30.

Courier 300, shown schematically in Figure 1, is a centralized system which means that the primary samples are brought from various parts of the plant close to the analyzer where they are split further to suitably small streams to be presented to the analyzer. This happens outside the analyzer room. The analyzer is located close to the secondary sampler. The other necessary cabinets like X-ray generator, detector electronics, computer and power distribution are in other parts of the room that is normally air conditioned. Figure 2 shows the equipment in the analyzer room in more detail. In the spectrometer cabinet there are 14 flow cells side by side and the measuring head scans them in programmable sequence. The side window X-ray tube, that is normally run at 1.5 or 3 kW power, is in the middle in upright position and the spectrometer crystal blocks are around the tube. Each block has in this case a scintillation detector on top of it.

The main parts of the Courier 30 analyzer system are shown in Figure 3. Now the analyzer unit is actually only a probe that is located in the plant in some convenient place close to the sample streams to be measured. Up to 5 analyzer units can be connected to the same system. Because the WDXRF analyzer tends to be somewhat expensive it is not feasible to put such an analyzer for each stream and therefore also in Courier 30 it is possible to measure many, in fact up to 5, streams with each analyzer unit. The secondary sampler is quite small and it acts as a sample multiplexer because the measuring head is fixed. The analyzer unit is so well shielded against the environment that it can be installed even on the floor of a concentrator plant or in a hydrometallurgical plant.

Figure 4 shows part of the inside of the analyzer unit. The X-ray tube is here of the end-window type and it is located in the middle. Normally it is run at 50 kV and 1.8 mA. Around the X-ray tube there are again the same spectrometer blocks as in Courier 300 but now equipped with proportional counters and high voltage -preamplifier units.

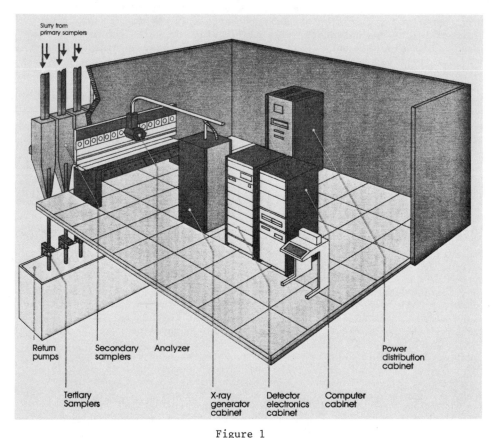

Figure 1

Analyzer architecture

Figure 2

Analyzer system

Spectrometer cabinet has 14 flow cells for 14 slurry streams and a moving measuring head with 8 simultaneous channels. Rigid mechanical construction and precise positioning of spectrometer head ensure accurate and reproductible spectrometer outputs.

The essential part that is common for both analyzers is the crystal spectrometer. In the applications where low concentrations have to be measured, the same kind of detectors, namely proportional counters, are also used. The inside of the spectrometer is shown schematically in Figure 5. The crystal is of Johansson type and is made of either quartz or LiF.

The quality of the crystal spectrometer is of primary importance in the analysis of low concentrations although the form of the exciting primary spectrum and the quality of the detecting system are certainly important, too.

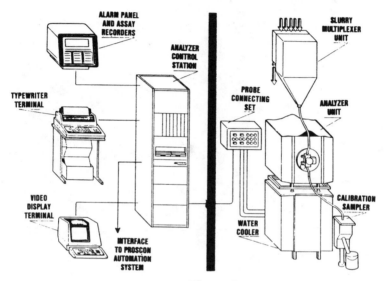

Figure 3

Inside view of a Courier-30 showing 1. the X-ray tube, 2. crystal spectrometer, 3. counter and 4. pre-amplifier.

Figure 4

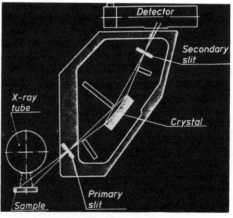

Figure 5

APPLICATION EXAMPLES AND RESULTS

All the results shown in this paper are directly related to real
practical problems and have been obtained in connection with the application work done for the customers. In almost every case the samples have
been obtained from the customer as well as their analysis, too. Only
certain solution samples have been prepared to some extent artificially.
The fact that the chemical analyses have in many cases been made by the
customer generates inevitably some additional error in the results because
the sample that has been taken to chemical analysis is not necessarily
excactly representative of that sample that has flown in the test circuit.
However, this procedure has usually been satisfactory and because the
purpose of the tests has not been the setting of new records, new chemical
analyses have been made only occasionally.

The mathematical models that transform the measured intensities into
concentrations have been kept as simple as possible.

This means that often even simple linear equations are good enough.
They are also usually less liable to give seriously erroneous results
under abnormal conditions than mathematically complex models. One should
not, however, be afraid of using even complex models when there is some
reasonable physical basis to do so.

Fortunately the processes in which the on-line analyzers are applied
are normally run under computer control. This tends to keep most process
flows fairly constant which also justifies the use of simple models.

Let us next look at the results that have been obtained. Let us
start with solution samples as they are nominally much easier to measure
than slurry samples because the sample is homogeneous and no segregation
or sedimentation is to be expected. In practice, however, care must be
taken to avoid precipitation and recrystallization. The following results
have been obtained using clear solutions at room temperatures.

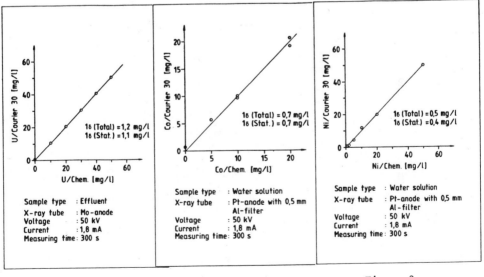

Figure 6   Figure 7   Figure 8

Figure 6 shows the result of a uranium measurement with Courier 30. The samples were made so that we got from the customer one sample with very low U content and we then added U-salt solution to cover the range of expected variations. The statistical error is almost entirely responsible for the total error but it is obvious that to control purposes the precision is sufficient.

Figures 7 and 8 give the results of Co and Ni solution measurements. The samples were made artificially and can be called water solutions. Again the statistical inaccuracy sets the limit but the limit is quite low.

Figure 9 refers now to real process samples that were received from a customer and measured as such with C-300. One obvious difference compared with the previous results is that in spite of the essentially shorter measuring time the statistical inaccuracy remains low.

Next 3 figures refer to silver determinations from solid samples. All the samples were obtained from the same customer. In 2 sets the mass of each sample was insufficient to make a circulating slurry flow and therefore solid briquettes were made without any further grinding of the sample to see what would happen. The results given in Figures 10 and 11 looked so good to the customer that he wanted to know what the results would be in actual slurry measurements. Therefore he sent more samples and thus we were able to make the tests in the ordinary way. This means that we introduce deliberately a new variable to the system, namely the varying solids content. In normal processes this variable makes the intensities vary often by 10 to 30 % and this variation has to be compensated for somehow, because the the interesting quantity is the concentration in the solids, not in the slurry. A serious drawback in slurry measurements is that background scattering becomes high due to water and it varies as solids content changes.

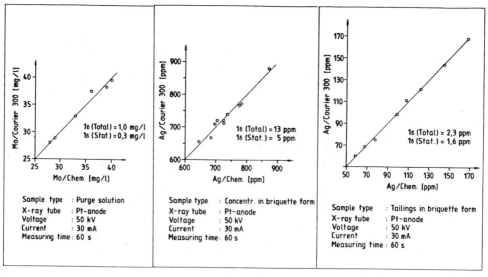

Figure 9   Figure 10   Figure 11

To make our measurements realistic we therefore use 2 or 3 different
solids contents per sample by adding proper amounts of solids into the
water in the test circuit. Figure 12 shows what we got in this case.
The result is, frankly said, unexpectedly good and may not be true from
time to time in actual practice, but it is a good indication of how good
the results can be under favorable circumstances.

The above words "unexpectedly good" do not actually say anything bad
of the equipment. The point is that often the mineralogy of silver
causes problems in the sample itself. As said earlier we normally don't
make any other sample preparation than split the sample to suitable size.
Therefore the mineralogical effects, particle size distributions and similar
nuisances are still present in the same way as in dry powder samples.

The graph in Figure 13 looks at first sight much worse than the one
in Figure 3, but now the total absolute variation of the Mo concentration
is much smaller, too. As we can see in some cases all 3 different marks
referring to 3 different solids contents coincide and in some cases the
points are widely apart. This kind of behaviour often means that there
is at least one somewhat erroneous reference analysis involved, but the
reason can be also, for instance, varying mineralogical composition.
Anyway, the error is not very high.

Figure 14 gives a result for a more common metal. In this case it
was essential to get a good accuracy and precision and therefore we used
a long measuring time to make sure that the statistical inaccuracy would
not set the limit. As can be seen the time was unnecessarily long as the
statistical error is quite insignificant.

The graph in Figure 15 was of certain economical interest to the
customer because he wanted to know how much gold was going to this particular concentrate. It is obvious that he could get useful information
concerning this slurry but it is equally obvious that the direct measurement of gold on ppm and sub-ppm levels would not be feasible with this
device without further improvements.

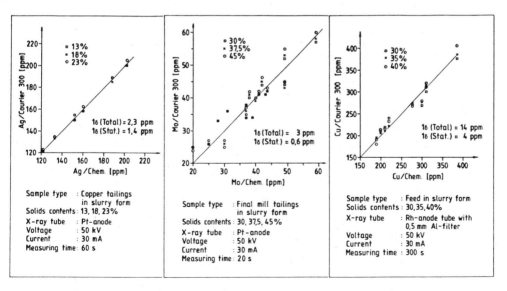

Figure 12   Figure 13   Figure 14

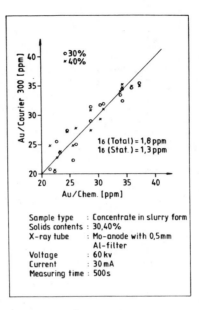

Figure 15

DISCUSSION

The minimum detection limit in the WDXRF-analysis is about one order of magnitude better than in the EDXRF-analysis (4,5). The reason for this is the superior energy resolution of the crystal spectrometer and hence better peak-to-background ratio. Also higher total count rates can be used in the WDXRF.

The good peak-to-background ratio and the high total count rate are important in the on-stream analysis because samples are very often slurries or solutions with high scattering power.

Examples of this paper show that the determination of low concentrations in solutions and even slurries using on-stream WDRXF-techniques is practical.

REFERENCES

1. H. Sipilä, Advances in X-Ray Analysis, 24, 333-335 (1981)

2. H. Sipilä and M-L Järvinen, Nucl. Instr. and Meth. 217 (1983) 298

3. M-L Järvinen and H. Sipilä, IEEE Trans. Nucl. Sci. Vol. NS-31 (1984) 356

4. M-L Järvinen and H. Sipilä, Advances in X-Ray Analysis, 27 (539-546)

5. P. Wobrauschek and H. Aidinger, X-Ray Spectrum., 12 (1983) 72

A FUSION METHOD FOR THE X-RAY FLUORESCENCE ANALYSIS OF
PORTLAND CEMENTS, CLINKER AND RAW MATERIALS UTILIZING
CERIUM (IV) OXIDE IN LITHIUM BORATE FLUXES

Gregory S. Barger

Central Process Laboratory
Southwestern Portland Cement Company
P. O. Box 937
Victorville, California 92392   U.S.A.

ABSTRACT

This method describes the addition of cerium (IV) oxide to a lithium borate flux. CeO_2 provides a non-analytic glass former to the
melt production. CeO_2 also acts as an interelemental buffer replacing
the use of lanthanum oxide for long wavelength absorption. With cerium
oxide addition, excellent results are produced, resulting in part from
the elimination of recrystallization problems encountered with lanthanum
oxide use. Analytical results easily meet the ASTM C-114 qualification
requirements for rapid method analysis of hydraulic cements.

INTRODUCTION

The accurate analysis of portland cement and related raw materials
continues to be a primary function of the cement quality control laboratory. Additional analytical responsibilities often include the evaluation of samples from geological prospecting efforts and samples of
industrial by-products. For analyses of materials having a wide range
in mineral and chemical compositions, it is impractical to establish and
maintain powder calibration curves for the many types of materials which
may be received. Fusion of samples homogeneously mixed with lithium
borate fluxes has, therefore, become a widely used analytical procedure
involving the demineralization and ionic suspension in glass of raw
materials for analysis by x-ray fluorescence [1,2,3,4]. This method,
with the addition of ceric oxide, has greatly enhanced our analytical
capabilities and reduced the need for interelemental corrections.

In 1974, Sudoh and co-workers utilized small amounts of ceric oxide
as a means of reducing bubble entrapment in glass pellets prepared for
x-ray fluorescence analysis.[5] As a result of our investigations, we
have found that ceric oxide can also (a) provide a glass forming substance to the flux mixture which allows broad dilution ratios (flux to
sample) to be used, thus reducing matrix interferences; and (b) act as
an interelemental buffer (long wavelength absorber) similar to lanthanum
oxide[6]. The recrystallization problem often encountered with
lanthanum oxide does not occur when ceric oxide is used.

EXPERIMENTAL

Samples are first oven-dried at 110°F for 2 hours to remove surface moisture. This is followed by particle size reduction to approximately 5μm (i.e., pulverizer, swing-type mill, etc.).

The sample is mixed with lithium tetraborate, lithium metaborate and ceric oxide in a 10 to 1 (flux to sample) ratio with lithium iodide added as a releasing or "wetting" agent. See Table 1. The mixture is placed in a platinum crucible (5% gold), and subsequently heated in a muffle furnace at 1100°C for 1 hour. The melt is swirled intermittently during fusion to release entrapped gas bubbles. The melted material is then cast into platinum molds (5% gold) while still in the furnace. The molds are removed from the furnace immediately after casting and annealed at about 400°C for 15 minutes on a hot plate. The annealing process relieves internal glass stresses which can develop during cooling. The glass pellets are cooled to room temperature with 100°C reductions in hot plate temperature being made every 5 minutes.

TABLE 1

Flux Composition and Proportions

Composition of Flux (% by weight)*		Mixture for Fusion (grams)	
Ceric Oxide	5.00	Flux	10.000
Lithium Tetraborate	63.27	Lithium Iodide	0.500
Lithium Metaborate	31.63	Sample	1.000
Total	100.00	Total	11.500

*Note: Flux preparation consists of grinding/homogenizing the components in a swing mill (50 grams total) for 3 minutes using Freon-TF (50 ml) as a grinding aid; flux is then air dried prior to use.

Polishing of the pellet to a plane and smooth surface is accomplished with semi-automatic polishing equipment (e.g., Buehler Minimet). The final surface is polished to a 6 μm finish using diamond polishing paste.

The samples are analyzed by wavelength dispersive x-ray fluorescence. Ratios of the resulting intensities with intensities from an external reference standard are determined. During calibration, the ratios are plotted against known concentrations for each element in each calibration standard.

SUMMARY

The method has been qualified for the analysis of portland cements in accordance with ASTM C-114 utilizing a Philips PW-1600/10 simultaeous x-ray spectrometer. No interelemental corrections were used in this qualification. The data are presented in Table 2. Note that the pellets were analyzed with (a) unpolished surfaces, (b) surfaces polished to a 15 μm finish, and (c) surfaces polished to a "final" 6 μm finish. Providing a plane specimen surface (15μm) increases analytical precision and accuracy while further polishing to a 6 μm surface will continue to enhance these parameters.

TABLE 2

Analysis of Seven NBS Standard Portland Cements
Maximum Permissible Variations in Results
(ASTM C-114; Table 1)

Component (Column 1)	(Column 2)[a]	Cerium Oxide/Lithium Borate Method		
		Unpolished Surface	15μm Surface[c]	6μm Surface
SiO_2	0.16	0.07	0.06	0.06
Al_2O_3	0.20	0.02	0.02	0.01
Fe_2O_3	0.10	0.03	0.01	0.01
CaO	0.20	0.32	0.06	0.07
MgO	0.16	0.10	0.03	0.03
SO_3	0.10	0.02	0.02	0.02
K_2O	0.03	0.007	0.004	0.003
TiO_2	0.02	0.003	0.004	0.007

Component (Column 1)	(Column 3)[b]	Cerium Oxide/Lithium Borate Method		
		Unpolished Surface	15μm Surface[c]	6μm Surface
SiO_2	$\pm$ 0.20	0.08	0.07	0.04
Al_2O_3	$\pm$ 0.20	0.04	0.04	0.04
Fe_2O_3	$\pm$ 0.10	0.03	0.02	0.02
CaO	$\pm$ 0.30	0.27	0.07	0.11
MgO	$\pm$ 0.20	0.05	0.04	0.02
SO_3	$\pm$ 0.10	0.02	0.02	0.03
K_2O	$\pm$ 0.05	0.006	0.006	0.007
TiO_2	$\pm$ 0.03	0.009	0.009	0.008

[a] Maximum difference between duplicates (precision statement)
[b] Maximum difference of the average of duplicates from SRM Certificate values (accuracy statement)
[c] ASTM C-114 Rapid Method qualifications were met (at 15 μm surface) using no interelemental corrections.

Calibration accuracy data were gathered for materials having various matrices such as cement raw mix, silicates (clay, quartz, coal ash, etc.), iron ores and limestone materials. (See Tables 3 and 4.) Data generated within tightly defined ranges, as in Table 3, showed no analytical enhancement when Dejong equation corrections (Philips software program) were applied. Broad chemical range calibration curves, while less accurate, do reduce the calibration workload required. (See Table 4.) Oxidizers can be used in the flux mixture when reduced mineral forms are present (e.g., magnetite, pyrite, etc.); lithium nitrate was found to be an acceptable oxidizing additive.

TABLE 3

Kiln Feed Matrix Calibration Data

Element	Concentration Range	Standard Deviation of Least Squares Regression	No. of Standards
SiO_2	10-25%	0.04	10
Al_2O_3	1-7	0.03	10
Fe_2O_3	1-7	0.05	10
CaO	32-47	0.08	10
MgO	0-4	0.05	10
SO_3	0-7	0.02	10
K_2O	0-1.6	0.01	10
TiO_2	0-2.5	0.02	10
BaO	0-1	0.01	10

Note: Interelemental corrections did not enhance regression results.

TABLE 4

Quartzite/Clay Calibration Data

Element	Concentration Range	Standard Deviation of Least Squares Regression	No. of Standards
SiO_2	20-100%	0.35	33
Al_2O_3	0-100	0.22	33
Fe_2O_3	0-10	0.25	33
CaO	0-20	0.21	33
MgO	0-15	0.05	33
SO_3	0-2	0.08	33
TiO_2	0-3	0.02	33
K_2O	0-4	0.03	33
BaO	0-4	0.03	33

Note: Dejong interelemental corrections were used on SiO_2 and Al_2O_3 only.

CONCLUSIONS

This analytical method provides a rapid and accurate means for analysis of cement, clinker, and raw materials. "Rapid method" qualification for cement analysis, in accordance with ASTM C-114 specifications, can be accomplished utilizing seven of the National Bureau of Standards Portland Cement Standard Reference Material Samples Nos. 633-639 and Nos. 1880-1881. The most significant advantages of the method are:

1. A combination of lithium borate fluxing materials results in a more effective dissolution of samples having a wide range of mineral forms than does lithium tetraborate alone. This combination provides a more universal flux for cement, clinker, and raw materials.

2. Ceric oxide provides a non-analyte glass former which gives broad dilution capabilities. In addition, it functions as an inter-elemental buffer (replacing lanthanum oxide) for long wavelength absorption.

3. Lithium borate/cerium oxide pellet calibration standards have an almost indefinite shelf-life requiring only occasional resurfacing before use.

4. The need for interelemental corrections due to element concentration variations (e.g., Dejong equation) is significantly reduced.

ACKNOWLEDGEMENTS

The author extends his sincere gratitude to Lawrence D. Adams and Waldemar A. Klemm (Southwestern Portland Cement Company) for their technical support. Thanks are also due to J. A. Anzelmo (Applied Research Laboratories), E. E. Larkin and T. M. Rader (Southwestern Portland Cement Company) for their helpful suggestions.

REFERENCES

1. Moore, C. W., "Spectrochemical Analysis of Portland Cement by Fusion with Lithium Tetraborate using an X-ray Spectrometer" (E-2 SM 10-26), pp. 911-17, Methods for Emission Spectrochemical Analysis, Seventh Edition, ASTM, Philadelphia, 1982.

2. Anderman, G., "Spectrochemical Analysis of Cement by Lithium Tetraborate-Lanthanum Oxide Fusion Technique Using an X-ray Spectrometer", (E-2 SM 10-20), pp. 885-92, Methods for Emission Spectrochemical Analysis, Seventh Edition, ASTM, Philadelphia, 1982.

3. Claisse, F., Instruction Manual for Claisse Fluxer VI, 1979, Corporation Scientifique Claisse, Inc., Quebec, Canada.

4. X-Ray Spectrochemical Accessories, "X-Ray Fusion Fluxes" Section, 1981 Catalog, Chemplex Industries, Inc., Eastchester, N.Y.

5. Sudoh, G.; Asahara, N.; Kitsuda, K.; and Nakayama, T., "Automatic Bead Preparation Techniques for X-ray Fluorescence Analysis", Semento Gijutsu Nempo, pp. 28, 80-83, Cement Association of Japan, Tokyo, 1974.

6. Anzelmo, J. A., "X-ray Emission Spectro-Chemical Analysis of Aluminosilicate Raw Materials and Refractories by Lithium Tetraborate-Lanthanum Oxide Fusion Technique" (E-2 SM 10-33), pp. 958-61, Methods for Emission Spectrochemical Analysis, Seventh Edition, ASTM, Philadelphia, 1982.

PARTICLE SIZE EFFECTS IN GEOLOGICAL ANALYSES BY X-RAY FLUORESCENCE

K.K. Nielson and V.C. Rogers

Rogers and Associates Engineering Corporation

Salt Lake City, Utah   84110

INTRODUCTION

Particle-size effects can cause significant errors in x-ray fluorescence (XRF) analysis of particulate materials. The effects are usually removed when samples are fused or dissolved to standardize the matrix for quantitative analysis. Recent improvements in numerical matrix corrections reduce the need to standardize the sample matrix via fusion or dissolution, particularly when the CEMAS method[1,2] is used to estimate unmeasured light-element components of undefined materials for matrix calculations. A new method to correct for particle-size effects has therefore been examined to potentially avoid the need for destructive preparation of homogeneous samples.

Particle-size effects have been described theoretically and mathematically for many practical cases.[3] Mathematical particle size corrections are usually difficult, however, because they require seldom-known sample parameters such as mean particle thickness, particle packing fraction, and the volume fractions and x-ray absorption properties of both fluorescent and non-fluorescent particles. If the sample can be treated as homogeneous, however, the resulting equation for particle-size corrections is much simpler, and only depends on the mean particle thickness, packing fraction, and absorption properties of the material. This paper presents the simplified correction equation along with a simple method for approximating the corrections without having to explicitly measure the sample particle sizes or packing fractions. The method utilizes redundant silicon measurements from different excitation x-ray energies to estimate the particle-size corrections, and utilizes the CEMAS method for bulk sample matrix definition. The feasibility of the method is demonstrated by analyses of loose and briquetted powders of four diverse rock materials.

THEORY AND EXPERIMENTS

The general intensity equation reported by Berry et al.[3] for particulate materials was simplified by assuming particle homogeneity and the absence of binders or diluents with the particles. The concentration of fluorescent particles in their equation was accordingly set equal to unity, and the ratio of non-fluorescent to fluorescent particles became zero. The resulting simplified equation for intensity with particle size effects was then expressed as a ratio of intensities with and without particle size effects. The ratio was used as a correction factor in the matrix-correction loop of the CEMAS program,[1,2] similar to the approach used earlier in incorporating particle-size corrections with other matrix corrections.[4] The intensity correction ratios had the form:

$$\frac{I_f}{I_f^o} = \frac{\eta\left[1 - e^{-(\mu_f + \mu_f')\bar{d}}\right]}{1 - (1 - \eta + \eta e^{-\mu_f \bar{d}})(1 - \eta + \eta e^{-\mu_f'\bar{d}})} \qquad (1)$$

where

I_f/I_f^o = ratio of intensities with and without Particle Size effects

η = packing fraction (bulk density/specific gravity)

μ_f, μ_f' = absorption coefficients for incident and fluorescent x-rays

$\bar{d}$ = average particle thickness (2/3 x spherical diameter)

The intensity correction equation was evaluated for a variety of different particle sizes and packing fractions to determine the nature and magnitude of particle effects expected for rock powders (SiO_2 matrix) with different excitation sources. Several representative cases are illustrated in Figure 1, and exhibit prominent changes with the different incident

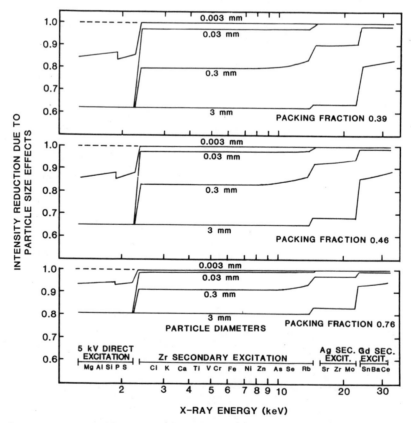

Figure 1.   Calculated Particle Size Effects For Various Particle Sizes, Packing Fractions, and Excitation Sources.

energies but small changes with fluorescent x-ray energy. This contrasts the stronger fluorescent-energy dependence observed earlier when a non-fluorescent binder was utilized in the rock matrix.[4] As noted previously, the particle size effects increase with both particle thickness and porosity $(1-\eta)$. The maximum effect (lowest intensity ratio) can be seen from Equation 1 to approach $1/(2-\eta)$, and will not extend below 0.5 (50 percent intensity loss) for the worst-case effect with low packing fraction.

An important feature of the intensity reduction curves in Figure 1 is that they are generally constant (horizontal), especially near the low-energy end, except at the discontinuity from changing incident x-ray sources. In fact, the discontinuity between 2 and 3 keV also disappears if the Zr excitation source is used to determine Mg, Al, Si, P, and S instead of the 5 kV source. This suggests that these elements potentially can be determined in small (0.003 mm) particles with no particle-size effect if the Zr source is used (dashed lines in Figure 1). Although the Zr source gives lower sensitivity and precision for these five elements, it provides a suitable second measurement of major elements such as Si in geological materials. By means of the redundant Si measurement, with and without a particle-size effect, the magnitude of the particle-size error for the 5 kV source can be estimated. The error is first estimated as a Si concentration ratio from separate CEMAS analyses with and without the 5 kV source. The Si concentration ratio can then be used to find the corresponding Si x-ray intensity ratio in order to approximate the particle size correction parameters in Equation 1. However, since the particle size corrections are nearly identical for all of the five elements, the correction also can be accomplished by simply normalizing the 5 kV spectrum by the Si intensity ratio.

Experimental analyses were conducted on four diverse rock materials to evaluate both particle packing characteristics and the redundant-silicon method for particle size corrections. The materials included jasperoid and jasperoidal breccia from northern Nevada and granite and sandstone from northern Utah. Four non-weathered chips (10-20 grams each) were selected from each of the four rock materials and used to represent sampling variation throughout the experiment. The chips were each ground and analyzed twice as loose powders and twice again as self-supporting pellets. The loose powders (~1 gram) were supported on 2.5 μm mylar films, and the pellets were pressed from 0.7 g of powder under 23 kg/mm^2. The jasperoid and breccia were hand-ground with a diamonite mortar and pestle and sieved to include only -400 mesh particles. The granite and sandstone were ground for 30 sec in a shatterbox (SPEX, Metuchen, New Jersey) grinder. Subsequent particle size analyses (ASTM-D-422) gave a median diameter of 7 μm for the granite and 2 μm for the sandstone. Each analysis consisted of four separate XRF spectra collected under vacuum using Gd, Ag, and Zr secondary excitation and 5 kV direct excitation (10, 8, 7.5, and 2 minutes respectively with a Kevex 0700 spectrometer). The direct excitation with 5 kV on the x-ray tube was dominated by its 2.7 keV Rh Lα x-ray lines. Spectrum analysis and quantitation utilized the CEMAS program,[1,2] which automatically computed matrix corrections and calibrations for each sample based on the measured constituents from all four spectra and on the backscatter intensities from the Ag-secondary spectrum. Sample mass absorption coefficients computed by CEMAS were used in Equation 1 for evaluating particle size corrections.

The silicon concentration ratios measured from separate CEMAS analyses with and without the 5 kV spectra were tabulated for each sample and used to estimate corresponding intensity ratios to use for particle size corrections. The expression that was empirically found to approximate the

TABLE I

MEASURED POWDER PACKING FRACTIONS

	Specific[a] Gravity (g/cm^3)	Powder Packing Fractions $(\eta \pm S.D.)$[b]		Pellet Packing Fraction[c] $(\eta \pm S.D.)$
		20 Taps	60 Taps	
Jasperiod	2.76	0.44+.03	0.50+.02	--
Jasperiodal Breccia	2.73	.43+.03	.49+.03	--
Granite	2.82	.37+.03	.45+.03	0.73+.02
Sandstone	2.73	.33+.02	.41+.02	.79+.02
Mean		.39+.05	.46+.04	.76+.04

a ASTM-D-854.
b Eight measurements averaged.
c Five measurements averaged.

silicon intensity ratio from the silicon concentration ratio, R_c, was $R_c(1-P)/(1-R_cP)$, where P is the sum of the fractional concentrations of elements affected by the correction (Mg, Al, Si, P, S). The intensity ratio was then used to normalize the 5 kV spectra upward in a final CEMAS analysis.

Particle packing characteristics were also measured for the four rock materials to define their nominal packing fractions and associated uncertainties. The packing fractions were measured in duplicate on each aliquot of each of the four rock materials. Each measurement consisted of loading 2.5 g of dry sample powder into a 10 cm^3 graduated cylinder, and allowing the cylinder to drop 20 times over a 2.2 cm distance. The volume was recorded, and forty additional drops were repeated before measuring the final volume. Packing fractions for each test were computed as the ratio of the measured bulk density to the specific gravity. Packing fractions for the pellets were similarly determined from pellet mass and bulk dimensions, and from the rock specific gravity.

RESULTS AND DISCUSSION

The packing fraction measurements are summarized in Table I, and illustrate that significant differences result from different preparation and handling of sample powders. Analyses of variance[5] on the eight measurements averaged for each powder indicate that 95 percent of the variance for each rock resulted from inter-aliquot variations, and only 5 percent resulted from measurement variations. Although standard handling procedures improve the packing fraction estimates, intra-sample variations, grinding variations, and ultimate support of the powder on a flexible plastic film lead to inherent variability in sample packing that is probably approximated by the 20- to 60-tap ranges in Table I. Considering the moderate packing variabilities between the diverse rock materials in Table I, it is suggested that most powdered rock samples will achieve a packing fraction in the nominal 0.39 to 0.46 range if tapped repeatedly on a solid surface, and that the uncertainty in the packing fraction can be further reduced by a factor of two to three via packing measurements with the sample powder. Pelletizing doubles the packing fractions and reduces their intra-sample variability, but maintains considerable inter-rock variations that probably depend on hardness, shape, and other mechanical characteristics.

TABLE II

GEOMETRIC MEANS OF FOUR SOURCES OF VARIATION
AS PARTITIONED BY ANALYSES OF VARIANCE

| | No. of Elements | Relative Standard Deviations[a](%) | | | | |
		Analyses	Aliquots	Preps.	Aliq. x Prep.	Total
Jasperiod	7	4.9	13.8	2.0	3.7	21.7
Jasperiodal Breccia	22	2.9	17.5	3.1	3.5	20.6
Granite	17	1.8	3.6	2.7	2.2	7.3
Sandstone	18	2.6	3.7	2.7	2.2	7.4
Mean		3.0	9.6	2.6	2.9	14.2

a   100 S.D./mean, based on two replicate analyses of each sample, four
     replicate Aliquots of each rock material, and two Preparations of
     each aliquot (powder and pellet), all averaged over all detected
     elements. Aliq. x Prep. is from convariance between Aliquots and
     Preparations.

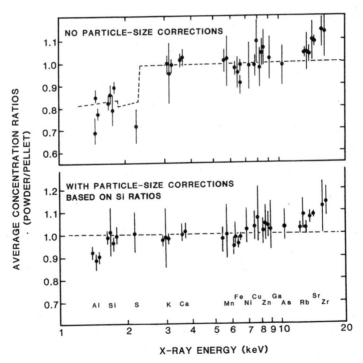

Figure 2.   Comparison of Measured and Predicted Concentration Ratios
Before and After Particle Size Corrections.

The duplicate XRF analyses of the four aliquots of each rock as loose and pelletized powders were summarized by statistical analyses, both with (Table II) and without particle size corrections. The analyses generally followed the particle size trends predicted in Figure 1. Geometric means of the variations for all detected elements are summarized with the components of variation in Table II. The jasperoid samples exhibited the most variation, mainly due to inhomogeneity (aliquot variations). Rock inhomogeneity also dominated the variations for granite and sandstone, but was much lower for these materials. Variations due to preparation averaged only 2.6 percent, or about 3.9 percent with covariance, suggesting that powder particle size errors, along with sample thickness differences and other powder/pellet differences, were not major sources of error. Variations among analyses were similar to averages from peak counting statistics. Since particle size effects mainly affected only Mg, Al, Si, P and S the average preparation variations shown in Table II were only slightly greater for the non-corrected data. Figure 2 illustrates the particle size effects more explicitly as the average ratios of measured concentrations for the powder/pellet samples as a function of x-ray energy. As shown, a bias of about 15-30 percent was observed in the powder samples for the Al, Si and S, and this bias was mostly corrected by the redundant silicon method. The residual 10 percent bias in Al is being studied further, and may reflect inaccurate correction for Si overlap or absorption in the mylar powder supports, or neglect of the changed Si absorption edge ratio in the corrections. The elevated Rb, Sr, and Zr in Figure 2 resulted from a sample mass bias unrelated to particle-size effects.

In summary, particle size corrections can be made using Equation 1 with computed (CEMAS) sample absorption coefficients when particle sizes and packing fractions are known. They can also be made from redundant silicon measurements to normalize the low-energy spectrum. The maximum error from particle size effects is 50 percent for undiluted rock powders, and the energy range over which the error applies depends strongly on the average particle size and the excitation energy. Typical errors from particle-size effects in the present rock analyses were about 15-30 percent, primarily expressed in the low-energy (5 kV) spectrum. The incident x-ray energy dominates over variation in fluorescent x-ray energies in determining particle size effects. Packing fractions are typically in the 0.39-0.46 range for loose powders, and can be estimated in the laboratory. Redundant measurement of the silicon peak using different excitation energies provides the simplest basis for more directly estimating the particle size correction without more explicitly estimating particle size and packing data. The redundant-silicon method can be used for relatively fine powders (<.03 mm) and virtually any packing fraction.

REFERENCES

1. K.K. Nielson and V.C. Rogers, "Comparison of X-Ray Backscatter Parameters for Complete Sample Matrix Definition," Advances in X-ray Analysis, 27, 449-457, 1984.
2. K.K. Nielson and V.C. Rogers, "Accurate X-Ray Fluorescence Analysis of Environmental Materials without Standards," Trans. Am. Nucl. Soc., 49, 146-147, 1985.
3. P.F. Berry, T. Furuta and J.R. Rhodes, "Particle Size Effects in Radio-Isotope X-ray Spectrometry," Advances in X-ray Analysis, 12, 612-632, 1969.
4. K.K. Nielson, "Matrix Corrections for X-Ray Fluorescence Analysis of Environmental Samples with Coherent/Incoherent Scattered X-rays," Analytical Chemistry, 49, 641-648, 1977.
5. J.C.R. Li, Statistical Inference, Ann Arbor: Edwards Bros., 1964.

AUTHOR INDEX